Study Guide

Study Guide

to accompany

HELENA CURTIS
BIOLOGY

FOURTH EDITION

David J. Fox
University of Tennessee

James J. Cockerill

WORTH PUBLISHERS, INC.

Study Guide
to accompany
Curtis: Biology, Fourth Edition
By David J. Fox and James J. Cockerill

Copyright © 1983 by Worth Publishers, Inc.

All rights reserved. No part of this
publication may be reproduced, stored in a
retrieval system, or transmitted in any form
or by any means, electronic, mechanical,
photocopying, recording, or otherwise,
without the prior written permission of the
publisher.
Printed in the United States of America

ISBN: 0-87901-187-4
First Printing: August 1983

Worth Publishers, Inc.
444 Park Avenue South
New York, New York 10016

Preface

This Study Guide, to accompany the fourth edition of BIOLOGY by Helena Curtis, is unlike any study guide currently available. The format and content of the Guide emerged as the most workable and effective after considerable classroom testing.

We make the assumption that you, the user of this Guide, have a serious interest in learning the subject matter of biology as well as in passing the examinations. It goes without saying that the best grades on exams are attained by students who really *understand* the course subject matter. To this end, we have written this Study Guide to aid you in your intellectual development.

Each chapter begins with an outline of the text chapter. This allows you to scan the contents and organization of the chapter before reading it. Next, we present the major concepts in concise form. With these in mind, you should read the chapter through rather rapidly. Do not worry about picking up every detail on the first reading. Solidify your understanding of the major concepts by reading the chapter summary.

Your second reading of the chapter will constitute your major learning experience. As you read the chapter this time, answer on a separate sheet of paper the questions in the *Focus on Chapter Details* section of this Study Guide. These questions have been carefully designed to assist you in extracting the important information from each paragraph of the text. Your written responses should be kept in a notebook for future reference. They will constitute a valuable aid for review prior to examinations.

No study guide would be complete without a section that allows you to test your understanding of the course material by answering questions such as might be asked on an exam. Such questions are particularly useful if a few of them are demanding and really put your understanding to the test. Then the following section, *Performance Analysis*, can be a very valuable learning experience. Here we discuss which answer is correct, *why it is correct, and why the other responses are incorrect*. This pedagogical technique further reinforces the correct answer in your mind and helps you to clarify information contained in incorrect responses.

You will undoubtedly find your own ways of studying from the text and this Study Guide. We would like to suggest, however, that you read the following essay, *How to Manage Your Time Efficiently and Study More Effectively* (pages vii–xiv). This Study Guide was written with these learning principles in mind, so your use of some of the essay's many helpful suggestions will undoubtedly enhance your success in this course.

We are excited about this new approach and hope that you will find our methods effective. If you have suggestions, comments, or criticisms, please do not hesitate to write us—care of Worth Publishers, Inc.

David J. Fox
July 1983

How to Manage Your Time Efficiently and Study More Effectively

Richard O. Straub
The University of Michigan at Dearborn

How effectively do you study? Good study habits make the job of being a college student much easier. Many students who *could* succeed in college fail or drop out because they have never developed an efficient way of managing their time. Even the best students can usually benefit from an in-depth evaluation of their current study habits.

There are many ways to achieve academic success, of course, but your approach may not be the most effective or efficient. Are you sacrificing your social life, or your physical or mental health, in order to get A's on your exams? Good study habits result in better grades *and* more time for other activities.

EVALUATE YOUR CURRENT STUDY HABITS

To improve your study habits, you must first have an accurate picture of your current behavior. Begin by putting together a profile of how you study. Answer the following questions by writing yes or no on each blank.

_____ 1. Do you usually set up a schedule to budget your time for studying, recreation, and other activities?

_____ 2. Do you often put off studying until time pressures force you to cram?

_____ 3. Do other students seem to study less than you do but get better grades?

_____ 4. Do you usually spend hours at a time studying one subject, rather than dividing up that time among several subjects?

_____ 5. Do you often have trouble remembering what you have just read in a textbook?

_____ 6. Before reading a chapter in a textbook, do you skim through it and read the section headings?

_____ 7. Do you try to predict exam questions from your lecture notes and reading?

_____ 8. Do you usually attempt to paraphrase or summarize what you have just finished reading?

_____ 9. Do you find it difficult to maintain your concentration when you study?

_____ 10. Do you often feel that you studied the wrong material for an exam?

Thousands of college students have participated in similar surveys. Students who are fully realizing their academic potential usually respond as follows: (1) yes,

(2) no, (3) no, (4) no, (5) no, (6) yes, (7) yes, (8) yes, (9) no, (10) no.

Compare your responses with those of successful students. The greater the discrepancy, the more you could benefit from a program to improve your study habits. The questions are designed to identify areas of weakness. Once you have identified your weaknesses, you will be able to set specific goals for improvement and implement a program for reaching them.

MANAGE YOUR TIME

Do you often feel frustrated because there isn't enough time to do all the things you must and want to do? Take heart. Even the most productive and successful people feel this way at times. But they establish priorities for their activities and learn to budget their time. There's much in the saying, "If you want something done, ask a busy person to do it." A busy person *knows* how to get things done.

If you don't now have a system for budgeting your time, develop one. Not only will your academic performance improve, but you will actually find more time in your schedule for other activities. And you won't have to feel guilty about "taking time off," because all of your obligations will be covered.

Establish a Baseline

In preparing to budget your time, it is useful to monitor exactly how you spend it. Keep a diary for a few days to establish a summary, or baseline, of the time you spend studying, socializing, working, and so on. If you are like many students, much of your "study" time is nonproductive; you may sit at your desk and leaf through a

Table 1 Sample Time-Management Diary

	Monday	
Behavior	Time Completed	Duration Hours:Minutes
Sleep	7:00	7:30
Dress	7:25	:25
Breakfast	7:45	:20
Commute	8:20	:35
Coffee	9:00	:40
French	10:00	1:00
Socialize	10:15	:15
Videogame	10:35	:20
Review Biology	11:00	:25
Biology	12:00	1:00
Lunch	12:25	:25
Study Lab	1:00	:35
Biology Lab	4:00	3:00
Work	5:30	1:30
Commute	6:10	:40
Dinner	6:45	:35
TV	7:30	:45
Study Biology	10:00	2:30
Socialize	11:30	1:30
Sleep		

Draw a similar chart for each day of the week. Note each activity as you complete it, and record the time it was completed. Then determine the duration of the activity.

book, but the time is actually wasted. Or you may procrastinate. You are always getting ready to study, but you rarely do the studying itself.

Besides revealing where you waste your time, your diary will give you a realistic picture of how much time to allot for meals, commuting, and other fixed activities. In addition, careful records should indicate the times of the day at which you are consistently most productive. A sample time-management diary is shown in Table 1.

Plan the Term

Having established and evaluated your baseline, you are ready to devise a more efficient schedule. Buy a calendar that covers the entire school term and has ample space for each day. Using the course outlines provided by your instructors, enter the dates of all exams, class presentations, term-paper deadlines, and other important academic obligations. If you have any long-range personal plans (concerts, weekend trips, etc.), enter these dates on the calendar as well. Keep your calendar up-to-date and refer to it often. I recommend carrying it with you at all times.

Develop a Weekly Calendar

Now that you have a general picture of the school term, develop a weekly schedule that includes all of your activities. Aim for a schedule that you can live with for the entire school term. A sample weekly schedule, which incorporates the following guidelines, is shown in Table 2.

1. Enter your class times, work hours, and any other fixed obligations first. *Be thorough*; using information from your time-management diary, allow plenty of time for such things as commuting, meals, laundry, and the like.

2. Set up a study schedule for each of your courses. The study-habits survey and your time-management diary will direct you. In addition, the following guidelines should be useful.

   a. Establish regular study times for each course throughout the week. Four hours of study of one subject, for example, are most profitable when divided into shorter periods spaced over several days. If you cram your studying into one four-hour block, what you attempt to learn in the third or fourth hour will interfere with what you studied in the first two hours. Newly acquired knowledge is like wet cement—it needs some time to "harden" into memory.

   b. Alternate subjects. The type of interference mentioned above is greatest between similar topics. Set up a schedule in which you spend time on several different courses during each study session. In addition to reducing the

Table 2   Sample Weekly Schedule

Time	Monday	Tuesday	Wednesday	Thursday	Friday	Saturday	Sunday
7–8	Dress, Eat	Dress, Eat	Dress, Eat	Dress, Eat	Dress, Eat		
8–9	Psychology	Study Psych.	Psychology	Study Psych.	Psychology	Dress, Eat	
9–10	English	Study English	English	Study English	English	Study English	
10–11	Study French	Free	Study French	Open Study	Study French	Study Biology	
11–12	French	Study Biology	French	Open Study	French	Study Statistics	
12–1	Lunch	Lunch	Lunch	Lunch	Lunch	Lunch	
1–2	Statistics	Biology Lab	Statistics	Study or Free	Statistics	Free or errands	Study Psych.
2–3	Biology	Biology Lab	Biology	Free	Biology	Free or errands	Study French
3–4	Free	Biology Lab	Free	Free	Study Psych.	Free or errands	Study Statistics
4–5	Job	Job	Job	Job	Job	Free	Free
5–6	Job	Job	Job	Job	Job	Free	Free
6–7	Dinner	Dinner	Dinner	Dinner	Dinner	Dinner	Dinner
7–8	Study Biology	Study Biology	Study Biology	Study Biology	Free	Free	Study Biology
8–9	Study English	Study Stats.	Study Psych.	Study Stats.	Free	Free	Study Statistics
9–10	Study Stats.	Study English	Study Stats.	Study French	Free	Free	Study English
10–11	Study Psych.	Study French	Study English	Open Study	Free	Free	Open Study

Above is a sample schedule for a student with a 16-credit load and a 10-hour-per-week part-time job. Using this chart as a sample, make up a weekly schedule, following the guidelines in the text.

potential for interference, alternating subjects will help to prevent mental fatigue with one topic.

c. Set weekly goals to determine the amount of study time you need in order to do well in each course. This will depend on, among other things, the difficulty of your courses and the effectiveness of your methods. Many professors recommend at least two to three hours studying for each hour in class. If your time diary indicates that you presently study less time than that, don't plan to jump immediately to a much higher level. Increase study time from your baseline by setting weekly goals (see item 4 below) that will gradually bring you up to the desired level. As an initial schedule, for example, you might set aside an amount of study time for each course that matches class time.

d. Schedule for maximum effectiveness. Tailor your schedule to meet the demands of each course. For the course that emphasizes lecture notes, schedule time for a daily review soon after the class. This will give you a chance to revise your notes and clean up any hard-to-decipher shorthand while the material is still fresh in your mind. If you are evaluated for class participation (for example, in a language course), allow time for a review just *before* the class meets. Schedule study time for your most difficult (or least motivating) courses during times when you are the most alert and distractions are fewest.

e. Schedule open study time. Emergencies, additional obligations, and the like could throw off your schedule. And you may simply need some extra time periodically for a project or for review in one of your courses. Schedule several hours each week for such purposes.

3. After you have budgeted time for studying, fill in slots for recreation, hobbies, relaxation, household errands, and the like.

4. Set specific goals. Before each study session, write down several specific goals. The simple note "7–8 P.M.: Study biology" is too broad to ensure the most effective use of the time. Formulate your daily goals according to what you know you must accomplish during the term. If you have course outlines with advance assignments, set systematic daily goals that will allow you, for example, to cover 15 chapters before the exam. And be realistic: Can you actually expect to cover a 78-page chapter in one session? Divide large tasks into smaller units; stop at the most logical resting points. When you complete a specific goal, take a 5- or 10-minute study break before tackling the next goal.

5. Evaluate how successful or unsuccessful your studying has been on a daily or weekly basis. Did you reach most of your goals? If so, reward yourself in some immediate manner. You might even make a list of 5 to 10 rewards to choose from. If you have trouble studying regularly, you may be able to motivate yourself by making such rewards contingent on completing specific goals.

6. Finally, until you have lived with your schedule for several weeks, don't hesitate to revise it. You may need to allow more time for chemistry and less for some other course. If you are trying to study regularly for the first time and are feeling burned-out, it is very likely that you have set your initial goals too high. Don't let failure cause you to despair and to abandon the program altogether. Accept your limitations, and revise your schedule so that you are studying only 15 to 20 minutes more each evening than you are used to. The point is to *find a regular schedule with which you can achieve some success*. Time management, like any skill, must be practiced.

TECHNIQUES FOR EFFECTIVE STUDY

Knowing how to put study time to best use is, of course, as important as finding a place for it in your schedule. Here are some suggestions that should enable you to increase your reading comprehension and improve your note taking. A few study tips are included as well.

Increasing Reading Comprehension

How do you go about studying from a textbook? If you are like many students, you simply read and reread in a *passive* manner. Studies have shown, however, that most students who simply read a textbook cannot remember more than half the material 10 minutes after they have finished. Often, what is retained is the unessential material rather than the important points upon which exam questions will be based.

This Study Guide is designed to facilitate, and allow you to assess, your comprehension of the important facts and concepts in BIOLOGY, fourth edition, by Helena Curtis. By using the steps below, however, you can improve your comprehension of any textbook. These steps make up a program known as SQ3R, which is an abbreviation for *Survey*, *Question*, *Read*, *Recite*, and *Review*. A great deal of research has shown that students using SQ3R achieve significantly greater comprehension of textbooks than students reading in the more traditional passive manner.

Survey

Before reading a chapter, look at the chapter outline or list of chapter objectives in the text or the study guide. Then read the summary at the end of the chapter. (In this Study Guide, review the Chapter Organization section that begins each chapter, and carefully read the Major Concepts for the chapter.) Next, read the textbook chapter fairly quickly. This survey will give you an idea of the chapter's content and organization. You will then be able to divide the chapter into logical sections in order to formulate specific goals for a more careful reading of the chapter.

Question

You will retain material longer when you have a use for it. If you look up a word's definition in order to solve a crossword puzzle, for example, you tend to remember it longer than if you merely come across it while reading a textbook. Surveying the chapter will allow you to generate important questions for which the chapter will provide answers. These questions correspond to "mental files" into which knowledge will be sorted for easy access.

As you survey, jot down several of your own questions for each chapter section. Alternatively, you may formulate questions based on the chapter outline or the Study Guide. The Focus on Chapter Details section of this Study Guide provides the type of questions one might formulate while surveying each chapter, but you may want to add your own questions to these. Another simple technique is to generate questions by rephrasing a section heading. For example, the paragraph heading Capillary Action could simply be turned into "What is capillary action?" Good questions will allow you to focus on the important points in the text. Examples of good questions are "List two examples of..." "What is the function of...?" and "What is the significance of...?" Such questions give you a purpose for your reading.

Read

When you have established "files" for each section of the chapter, review your first question, begin reading, and continue until you have discovered its answer. If you come to material that seems to answer an important question you don't have a file for, stop and write down the question.

Make sure to read and look at everything. Don't skip picture captions, graphs, and the like. In some cases, what may seem vague in the text will be made clear by a simple graph. In addition, it is not uncommon for test questions to be drawn from this type of supplementary material.

Recite

When you have found the answer to your question, close your eyes and mentally recite the question and its answer. Then *write* down the answer next to the question. It is important that when you recite an answer, you put it into your own words. Don't rely upon your immediate memory to repeat word for word what the author wrote.

Recitation is an extremely effective study technique recommended by many learning experts. In addition to increasing reading comprehension, it is useful for review. Trying to explain something in your own words clarifies your knowledge, often by revealing aspects of your answer that are vague or incomplete. If you repeatedly rely upon "I know" in recitation, you really *may not know*.

Recitation has the additional advantage of simulating an exam, especially an essay exam—the same skills are required of you in both cases. Too often students study without ever putting the book and notes away. That makes it very easy to develop false confidence in your knowledge. When the material is in front of you, you may be able to *recognize* an answer; but will you be able to *recall* it later, when you take an exam that does not provide these cues?

After you have recited and written your answer, continue with your next question. Read, recite, and so on.

Review

When you have finished with the final question on the material you have designated as a study goal, go back and review. Read over each question and your written answer to it. Your review might also include preparing a brief written summary that integrates all of your questions and answers. This review need not take longer than a few minutes, but it is important. It will help you retain the material longer and will greatly facilitate a final review of each chapter before the exam. (An excellent way to review your understanding of the chapters of BIOLOGY, fourth edition, is to complete the questions at the end of each chapter and those in the Testing Your Understanding section of this Study Guide. Then go through the Performance Analysis section, which explains why the correct answers are correct. You may discover that you don't know the chapter as well as you thought you did!) One final suggestion: Incorporate SQ3R into your time-management calendar. Set specific goals for completing SQ3R with each assigned chapter. Keep a record in your calendar of chapters completed, and reward yourself for being so conscientious. Initially, it takes more time and effort to "read" using SQ3R, but with practice, the steps will become automatic. More important, research has demonstrated that students reading with SQ3R comprehend significantly more material and retain their knowledge longer than passive readers.

TAKING LECTURE NOTES

Are your class notes as useful as they might be? One way to determine their worth is to compare them with those taken by other good students. Are yours as thorough? Do they provide you with a comprehensible outline of each lecture? If not, then the following suggestions might increase the effectiveness of your note taking.

1. Keep a separate notebook for each course. Use full-size pages. Consider using a ring binder that would allow you to revise and insert notes while still preserving lecture order.
2. Take notes in the format of a lecture outline. Use roman numerals for major points, letters for supporting arguments, and so on. Some instructors make this easy by delivering organized lectures and, in some cases, by actually outlining lectures on the board for you. If a lecture is disorganized, you will probably want to revise your notes soon after the class.
3. As you take notes in class, leave a wide margin on one side of the paper. After the lecture, expand or clarify any shorthand notes while the material is fresh in your mind. Use this time to write important questions in the margin next to notes that answer them. This will facilitate later review and allow you to anticipate similar exam questions.

Evaluate Your Exam Performance

How often have you received an exam grade that did not do justice to the effort you spent preparing for the exam? This is a common experience that can leave one feeling bewildered and abused. "What do I have to do to get an A?" "The test was unfair!" "I studied the wrong material!"

The chances of this happening are greatly reduced if you have an effective time-management schedule and use the study techniques described here. But it can happen even to the best-prepared student. It is most likely to occur on your first exam with a new professor.

Remember that there are two objectives for studying. One is to learn for your own general academic development. As the author of your textbook states, biology is worth studying "for its own sake, because, like art and music and literature, it is an adventure for the mind and nourishment for the spirit." Many people believe that such knowledge is the only thing that really matters. Of course, it is possible, though unlikely, to be an expert on a topic without achieving commensurate grades, just as one can, occasionally, earn an excellent grade without truly mastering the course material. During a job interview or in the work place, however, your A in Fortran won't mean much if you can't actually program a computer.

In order to keep career options open after you graduate, you must both know the material *and* maintain competitive grades. In the short run, this means performing well on exams, which is the second main objective in studying.

Probably the single best piece of advice to keep in mind when studying for exams is to *try to predict exam questions*. This means focusing on the important questions (with your instructor's emphases in mind), knowing their answers, and ignoring the trivia.

A second point is obvious. How well you do on exams is determined by your mastery of *both* lecture and textbook material. Many students (partly because of poor time management) concentrate too much on one of these sources at the expense of the other.

One useful method of evaluating how well you are learning lecture and textbook material is to analyze the questions you missed on the first exam. Many instructors review exams during class, but if yours does not, you can easily do it yourself.

Divide the questions into two categories: those drawn primarily from lectures and those drawn primarily from the textbook. Determine the percentage of questions you missed in each category.

If your errors are evenly distributed and you are satisfied with your grade, you have no problem. If you are weaker in one area than the other, you will need to set future goals for increasing and/or improving your study of that area.

Similarly, note the percentage of test questions drawn from each category. Although most courses involve exams that cover *both* lecture notes and the textbook, the relative emphasis of each may vary from instructor to instructor. And while your instructors may not be entirely consistent in making up future exams, you may be able to tailor your studying for each course by placing *additional* emphasis on the appropriate area.

Exam evaluation will also point out the types of questions your instructor prefers. Does the exam consist primarily of multiple-choice, true-false, or essay questions? You may also discover that an instructor is fond of wording questions in certain ways. For example, an instructor may rely primarily on questions that require you to draw an analogy between a theory or concept and a "real-world" example. Or questions may often be worded in the negative, such as, "Which of the following is *not* an example of...?" Evaluate both your instructor's style and your performance with each type of format. Use this information to guide your future exam preparation.

The Testing Your Understanding sections of this Study Guide will provide you with an important aid in studying for exams and determining how well prepared you are. If these tests don't include all of the types of questions your instructor typically writes, make up your own practice exam questions. Spend extra time testing yourself with the question formats that are most difficult for you. There is no better way to evaluate your preparation for an upcoming exam than by testing yourself under the conditions most likely to be in effect during the actual test.

A Few Practical Tips
Even the best intentions for studying sometimes fail. Some of these failures occur because students attempt to work under conditions that are simply not conducive to concentrated study. To help ensure the success of your study program, here are a few suggestions that should help you reduce the possibility of procrastination or distraction.

1. If you have set up a schedule for studying, make your roommate, family, and friends aware of this commitment, and ask them to honor your quiet study time. Close your door and post a Do Not Disturb sign.

2. Set up a place to study that minimizes potential distractions. Use a desk or table, not your bed or an extremely comfortable chair. Keep your desk and the walls around it free from clutter.

3. Do nothing but study in this place. It should become associated with studying, so that it "triggers" this activity just as a mouth-watering aroma elicits an appetite.

4. Never study with the television on or with other distracting noises present. If you must have music in the background (in order to mask dorm noises, for example), play soft instrumental music. Don't pick vocal selections; your mind will be drawn to the lyrics.

5. Study by yourself. Other students can be distracting or can break the pace at which *your* learning is most efficient. In addition, there is always the possibility that group studying will become nothing more than a social gathering. Reserve that for its own place in your schedule.

6. Avoid studying in too many places. If you need a place besides your room, find one that meets as many of the above requirements as possible. Find a place, for example, in the library stacks.

If you continue to have difficulty concentrating, try the following suggestions.

7. Study your most difficult or most challenging subjects first, when you are most alert.

8. Start with relatively short periods of concentrated study. If your attention starts to wander, get up immediately and take a break. It is better to study effectively for 15 minutes and then take a break than to fritter away 45 minutes out of an hour's designated study time. Gradually increase the length of study periods, using your attention span as an indicator of successful pacing.

SOME CLOSING THOUGHTS

I hope that these suggestions not only help to make you more successful academically but also enhance the quality of your college life in general. Not having the necessary skills makes any job a lot harder and more unpleasant than it has to be. Let me repeat my warning not to attempt to make too drastic a change in your life style immediately. Start by establishing a few realistic goals, and then gradually guide your performance to the desired level. Good habits require time and discipline to develop. Once established, they can last a lifetime.

Contents

Introduction	1
PART I Biology of Cells	7
SECTION 1 The Unity of Life	7
CHAPTER 1 Atoms and Molecules	8
CHAPTER 2 Water	15
CHAPTER 3 Organic Molecules	21
CHAPTER 4 Cells: An Introduction	29
CHAPTER 5 How Cells Are Organized	37
CHAPTER 6 How Things Get into and out of Cells	45
CHAPTER 7 How Cells Divide	52
SECTION 2 Energetics	59
CHAPTER 8 The Flow of Energy	60
CHAPTER 9 How Cells Make ATP: Glycolysis and Respiration	68
CHAPTER 10 Photosynthesis, Light, and Life	75
SECTION 3 Genetics	85
CHAPTER 11 From an Abbey Garden: The Beginning of Genetics	86
CHAPTER 12 Meiosis and Sexual Reproduction	93
CHAPTER 13 Genes and Chromosomes	99
CHAPTER 14 The Path to the Double Helix	106
CHAPTER 15 The Code and Its Translation	114
CHAPTER 16 Recombinant DNA	123
CHAPTER 17 The Eukaryotic Chromosome	130
CHAPTER 18 Human Genetics	138

PART II Biology of Organisms	145
SECTION 4 The Diversity of Life	145
CHAPTER 19 The Classification of Organisms	146
CHAPTER 20 The Prokaryotes	152
CHAPTER 21 The Protists	160
CHAPTER 22 The Fungi	168
CHAPTER 23 The Plants	174
CHAPTER 24 The Animal Kingdom I: Introducing the Invertebrates	182
CHAPTER 25 The Animal Kingdom II: The Protostome Coelomates	191
CHAPTER 26 The Animal Kingdom III: The Arthropods	199
CHAPTER 27 The Animal Kingdom IV: The Deuterostomes	207
SECTION 5 Biology of Plants	215
CHAPTER 28 The Plant: An Introduction	216
CHAPTER 29 Plant Reproduction, Development, and Growth	223
CHAPTER 30 Transport Systems in Plants	230
CHAPTER 31 Hormones and the Regulation of Plant Growth	238
CHAPTER 32 Plant Responses to Stimuli	243
SECTION 6 Biology of Animals	249
CHAPTER 33 The Human Animal: An Introduction	250
CHAPTER 34 Energy and Metabolism I: The Digestive System	257
CHAPTER 35 Energy and Metabolism II: Respiration	265
CHAPTER 36 Energy and Metabolism III: Circulation of the Blood	273
CHAPTER 37 Homeostasis I: Excretion and Water Balance	281
CHAPTER 38 Homeostasis II: Temperature Regulation	288
CHAPTER 39 Homeostasis III: The Immune Response and Other Defenses	295

CHAPTER 40
Integration and Control I: The Nervous System   304

CHAPTER 41
Integration and Control II: The Endocrine System   311

CHAPTER 42
Integration and Control III: Sensory Receptors and Skeletal Muscle   319

CHAPTER 43
Integration and Control IV: The Brain   329

CHAPTER 44
The Continuity of Life I: Reproduction   338

CHAPTER 45
The Continuity of Life II: Development   346

PART III **Biology of Populations**   355

SECTION 7 **Evolution**   355

CHAPTER 46
The Evidence for Evolution   356

CHAPTER 47
The Genetic Basis of Evolution   362

CHAPTER 48
Variability: Its Extent, Preservation, and Promotion   367

CHAPTER 49
Natural Selection   373

CHAPTER 50
On the Origin of Species   379

SECTION 8 **Ecology**   387

CHAPTER 51
Population Dynamics and Life-History Strategies   388

CHAPTER 52
Interactions in Communities: Competition, Predation, and Symbioses   394

CHAPTER 53
Ecosystems   402

CHAPTER 54
Biosphere   408

CHAPTER 55
The Evolution of Social Behavior   415

CHAPTER 56
Human Evolution and Ecology   422

Introduction

CHAPTER ORGANIZATION

I. **Introductory remarks**
II. **The road to evolutionary theory**
   A. Background
   B. Evolution before Darwin
   C. The age of the earth
   D. The fossil record
   E. Catastrophism
   F. The theories of Lamarck
III. **Development of Darwin's theory**
   A. The earth has a history
   B. The voyage of the *Beagle*
   C. The Darwinian theory
   D. Essay: Darwin's long delay
IV. **Challenges to evolutionary theory**
   A. Background
   B. The nature of science
   C. The Mendelian challenge
   D. Essay: Some comments on science and scientists
   E. Essay: The evolutionary paradigm
V. **Science as process**

MAJOR CONCEPTS

In the words of Ernst Mayr, "the theory of evolution is quite rightly the greatest unifying theory in biology." In 1859, Charles Darwin set forth the basic tenet of evolutionary theory in his book *The Origin of Species*. Darwin's theory was based upon two very important principles: Chance variations occur in natural populations; and natural selection acts on these variations.

For the most part, biologists prior to Darwin subscribed to the biblical account of divine creation. Before that, Aristotle believed that all of the known organisms of his day had always existed. The fixity of species was not seriously challenged until Buffon proposed that species change over time by degenerative processes, and Erasmus Darwin suggested that animals may change in response to their environment and that these changes may be hereditary.

Darwin's basic formulation of evolution maintained that differences (variations) between individuals *within* a group gradually became differences *between* groups as groups became separated in space and time. His ideas about variation stemmed largely from his own observations while aboard the *Beagle* and from his extensive reading.

His concept of evolution was stimulated by an essay by Thomas Malthus. The amount of time required for natural selection to work on existing variation was problematical for Darwin until he read the work of Charles Lyell, which indicated that the cumulative effect of natural forces had produced great change over the course of the earth's history. This geological perspective, coupled with the observations of James Hutton and William Smith, gave Darwin's theory the time it needed.

Challenges to evolutionary theory come primarily from a widespread misunderstanding of the nature of science. Science deals with observable and repeatable causes and effects with respect to natural phenomena. Any "truth" that exists outside this limitation cannot be ascertained by the methods of science.

Only briefly did Mendelian genetics pose any threat to Darwinian evolution, and that was with regard to the nature of mutations. In the 1930s, the controversy was resolved when a synthesis of both perspectives resulted in the formulation of neo-Darwinian evolution. In its various forms and modifications, this theory is still with us, although occasional challenges assure the continued evolution of the theory itself for quite some time.

HOW TO STUDY THE CHAPTER

Read the chapter, focusing on the major concepts.

Reread the chapter, using the questions that follow to help you focus on details as they relate to major concepts. Answer the questions on a separate sheet of paper. Your answers will provide a valuable study aid in preparing for examinations.

FOCUS ON CHAPTER DETAILS

I. **Introductory remarks** (page 1)
   1. What personal traits characterized the young Charles Darwin?

II. **The road to evolutionary theory** (pages 1–5)
   A. Background
      2. Describe the intellectual climate in which Charles Darwin developed his theory of evolution.
      3. What were the views of nature held by Aristotle and Linnaeus?
   B. Evolution before Darwin
      4. Differentiate between the ideas of Buffon and Erasmus Darwin with respect to changes in species over time.
   C. The age of the earth
      5. What are the basic tenets of Hutton's theory of uniformitarianism, and what are their implications?
   D. The fossil record
      6. Although William Smith did not interpret his own findings, how did the implications of his findings change the course of geology?
   E. Catastrophism
      7. Summarize the antievolutionary positions of Georges Cuvier and Louis Agassiz.
      8. In Cuvier's view, what part did catastrophism play in the extinction of species?
   F. The theories of Lamarck
      9. What was Lamarck's "bold proposal"?
      10. Summarize each of the two components of Lamarckian evolutionary theory.

III. **Development of Darwin's theory** (pages 5–9)
   A. The earth has a history
      11. How did Lyell's theory support Hutton's theory of uniformitarianism?
      12. What was the specific effect of Lyell's book on Darwin's theory of evolution?
   B. The voyage of the *Beagle*
      13. What observation impressed Darwin most during his trip along the coast of South America?
      14. How did Darwin's observations of the Galapagos tortoises and finches and of the geology of the region contribute to the development of his theory of evolution?
   C. The Darwinian theory
      15. How did Malthus's book influence Darwin's ideas about natural selection?
      16. What are the main features of natural selection as proposed by Darwin?
      17. What are the main points of Darwin's elephant and horse examples and Lamarck's giraffe example?
      18. How did Darwin's evolutionary theory differ from the theories of his predecessors?
   D. Essay: Darwin's long delay (page 8)
      19. What similarities in Darwin's and Wallace's backgrounds probably accounted for their virtually identical theories of evolution?
      20. What events led to Darwin's completing his treatise in a little more than a year?

IV. **Challenges to evolutionary theory** (pages 9–13)
   A. Background

21. What position is taken by most biologists with respect to the history of the earth and the organisms that inhabit it?
22. What seems to be the basis of the confusion regarding the controversy surrounding the theories of creation and evolution?

B. The nature of science
23. Compare the perspective of science with that of art, religion, and philosophy.
24. Describe the process of scientific investigation.
25. How does the scientific usage of the word "theory" differ from its everyday usage?

C. The Mendelian challenge
26. What controversy did formulation of the synthetic (neo-Darwinian) theory of evolution resolve?
27. Cite two of the current challenges to the synthetic theory.

D. Essay: Some comments on science and scientists (page 11)
28. Summarize in one sentence the main point of each of these statements from well-known scientists.

E. Essay: The evolutionary paradigm (page 12)
29. What is a paradigm? Evaluate the assumptions that constitute the evolutionary paradigm.

V. **Science as process** (page 13)
30. What is the chief characteristic of modern science?

TESTING YOUR UNDERSTANDING

After you have completed the following examination, compare your answers with the annotated key in the Analysis section.

1. Given the different philosophies of these men, which one might have said, "Improvement and degeneration are the same thing, for both imply an alteration of the original constitution"?
    a. Linnaeus
    b. Buffon
    c. Erasmus Darwin
    d. Charles Darwin
    e. Aristotle
2. The first scientist to work out a systematic theory of evolution was:
    a. James Hutton.
    b. Georges Cuvier.
    c. Jean Baptiste Lamarck.
    d. Charles Darwin.
    e. Erasmus Darwin.
3. Which of the following was not true of the young Charles Darwin?
    a. wealthy
    b. unmotivated as a student
    c. a naturalist
    d. an enthusiastic hunter
    e. a promising premed
4. Which of these biologists might be classified as a strict creationist?
    a. Buffon
    b. Aristotle
    c. Linnaeus
    d. Agassiz
    e. Both c and d are correct.
5. The study of fossils by the surveyor William Smith led him to propose:
    a. the theory of uniformitarianism.
    b. that the presence or absence of certain fossils is the best way to identify geological strata.
    c. that the history of the earth is inseparable from the history of organisms.
    d. the theory of catastrophism.
    e. an early form of the theory of evolution.
6. The evolutionary views of Buffon and Erasmus Darwin differed in that:
    a. Buffon believed that changes in organisms were degenerative in nature.
    b. Darwin believed that only the lesser families exhibited any change.
    c. Buffon supported a position similar to the Lamarckian view of the inheritance of acquired characteristics.
    d. Darwin's views had a major impact on his grandson Charles, whereas those of Buffon did not.
    e. Actually, the views of these two men did not differ significantly.
7. Taken together, the work of Hutton and that of

Smith has led scientists to conclude that:
a. the world is very old.
b. the earth was molded by slow, gradual processes.
c. the best way to identify strata containing oil is by examining the distribution of fossils.
d. the present surface of the earth was formed layer by layer over the course of time.

8. Cuvier and Lamarck both observed that simpler fossil forms were generally contained in older rocks. How did they differ in their interpretation of this observation?
a. Lamarck proposed that higher forms had arisen from simpler forms.
b. Cuvier proposed that a series of catastrophes led to the extinction of most species.
c. Actually, they did not differ in their interpretation of this observation.
d. Both a and b are correct.
e. The premise is false. Both men did not make this observation.

9. The inheritance of acquired characteristics means that:
a. an animal or plant species may be permanently improved by bringing it into contact with the proper environmental conditions.
b. changes in the expression of an organism's characteristics brought about by environmental influences are inheritable.
c. an animal species that does not use a certain organ over a number of generations will gradually lose use of that organ.
d. Both a and b are correct.
e. All three answers (a, b, and c) are correct.

10. Which statement best summarizes Lamarckian evolutionary theory?
a. The lesser families of organisms are "conceived by Nature and produced by Time."
b. Changes in an organ brought about by use or disuse are inheritable.
c. A universal creative principle causes all organisms to strive upward in the hierarchy of nature.
d. Both a and b are correct.
e. Both b and c are correct.

11. The geologist Charles Lyell had a profound influence on the development of Darwin's theory in that he:

a. was the founder of the scientific study of fossils.
b. supported the theory of catastrophism.
c. opposed the theory of uniformitarianism.
d. concluded that geological change was a slow process and therefore took place over a long period of time.
e. Both a and c are correct.

12. Which of the following facts about the Galapagos islands were the most important to the development of Darwin's theory?
a. Each island had its own recognizable form of tortoise.
b. The organisms of the islands differed from those of the mainland.
c. The islands were much younger than the mainland.
d. Both b and c are correct.
e. All three answers (a, b, and c) are correct.

13. How did the ideas of Thomas Malthus contribute to Darwin's intellectual development with respect to his theory of evolution?
a. Darwin saw that Malthus's conclusions about food supply and the size of human populations were true for all species.
b. Malthus's theoretical demonstration that two elephants could give rise to 19 million elephants in 750 years convinced Darwin that natural selection must operate in natural populations.
c. Malthus proposed that natural selection was really a variation on the method used by breeders of domestic animals, an idea that Darwin incorporated into his theory.
d. Darwin accepted Malthus's idea that mutations were the source of variation in natural populations of animals.

14. According to the Darwinian theory of evolution:
a. chance is what dictates which variations occur in natural populations.
b. differences between groups are gradually converted into differences among individuals within a group.
c. variations among individuals of a population have no goal or direction.
d. Both a and c are correct.
e. All three answers (a, b, and c) are correct.

15. Match each man with his theory, book, or

appropriate description.

Georges Cuvier _____   a. *species Plantarium*

_____   b. uniformitarianism

Jean Baptiste Lamarck _____   c. catastrophism

_____   d. Life has always existed.

James Hutton _____

Charles Lyell _____   e. fossil distribution

Carolus Linnaeus _____   f. acquired characteristics

Ernst Mayr _____   g. *Principles of Geology*

Charles Darwin _____   h. *The Origin of Species*

Aristotle _____   i. Evolution is the greatest unifying principle.

William Smith _____

16. True or False? Misunderstanding of the scientific process is partially responsible for the controversy between creationists and evolutionists.
17. True or False? To be useful in the field of biology, a paradigm must be completely verifiable.
18. True or False? The concept of evolution has existed since the time of Aristotle but was not widely accepted until the twentieth century.
19. True or False? There is good evidence that some of Charles Darwin's ideas came from his grandfather Erasmus.
20. True or False? Although Georges Cuvier was one of the world's experts on reconstructing extinct animals from their fossil remains, he was nevertheless an opponent of evolutionary theory.
21. True or False? To Darwin, natural selection was the only rational explanation for the fact that natural populations did not increase at geometric rates.
22. Briefly discuss the advantages of neo-Darwinism over traditional Darwinism.
23. In addition to the communication of information, scientific meetings and publications play an important role in the scientific process. What is this role?
24. Write one or two paragraphs that answer the question, What is science?
25. Discuss Einstein's premise—that the theory decides what can be observed—in terms of the evolutionary theory presented in this chapter.

PERFORMANCE ANALYSIS

1. The key word in this question is "alteration." That leaves out Linnaeus, who believed that species were created by God and have remained fixed ever since. Charles Darwin's emphasis was on adaptation by natural selection. His ideas were better developed but similar to those of his grandfather Erasmus. The emphasis of Aristotle was on the ladder or hierarchy of nature. That leaves **b**, Buffon, who emphasized not only alteration but degenerative alteration.
2. Hutton, a geologist, proposed the theory of uniformitarianism but, unlike Charles Darwin, never made the connection to biological problems. Cuvier was an antievolutionist. Erasmus Darwin hinted at evolutionary relationships but did not work out a systematic theory, as did both Charles Darwin and Jean Baptiste Lamarck. Since Lamarck preceded Darwin in time, **c** is the correct answer.
3. This is a question intended to emphasize that not every unmotivated student is a failure in life and that one need not necessarily be a premed in order to be successful. The correct answer is **e**.
4. We already know that Buffon believed in the alteration of species and that Aristotle believed that living things had always existed. Linnaeus believed that God created all species. Agassiz was an opponent of evolution who believed that for every extinction in the fossil record there was a separate creation. Therefore, **e** is correct.
5. The theory of uniformitarianism was proposed by Hutton. Smith did not interpret his findings, which leaves out c and e. Cuvier proposed the theory of catastrophism. That leaves answer **b**, which makes sense, since Smith was the first person to make a scientific study of fossils.
6. The correct answer is **a**, which is stated in so many words in the text. The second belief is attributable to Buffon, not Darwin. There is no evidence in the text for answer c. The text implies that Charles Darwin did not hold his grandfather's views in high regard.
7. Hutton was the geologist who concluded that the earth was formed by slow, gradual processes. Smith's study of the strata led others to believe that a great deal of time was required to deposit

them. Hence, answer **a** is correct. Smith was not associated with the ideas set forth in b. Hutton had nothing to do with linking fossils to particular strata (c), nor was he associated with the ideas set forth in d.

8. The premise is not false, which eliminates e. The two men did differ in the emphasis placed on these observations, which eliminates answer c. Both a and b are true statements reflecting the ideas of the men involved. Therefore, **d** is the correct answer.

9. All three answers (a, b, and c) embody the central idea that environmental factors can cause inheritable changes. Therefore, **e** is correct.

10. The first statement is attributable to Buffon. The second statement is obviously Lamarckian, since he was the major proponent of the inheritance of acquired characteristics. Consistent with his evolutionary hierarchy, Lamarck believed that one of the driving forces of evolution was an unconscious upward striving by all organisms (c). Therefore, **e** is correct.

11. Cuvier, not Lyell, was the founder of paleontology (a). Lyell actually opposed the theory of catastrophism (b) but supported uniformitarianism. Darwin's theory needed time, and that is what Lyell's viewpoint (**d** is correct) gave him.

12. The Galapagos experience helped Darwin solidify his views on natural selection. Answers a and b are consistent with the divergence of species under differing environmental conditions. Answer c indicates that the organisms on the islands were not separate creations but were derived from mainland populations. Answer **e** is therefore correct.

13. The correct answer is **a**. The calculations in b were made by Darwin, not Malthus. Malthus said nothing about natural selection (c) or mutations (d).

14. Darwin held that evolution was not directed and that the variation upon which it relies occurs in natural populations by chance. Both a and c are consistent with this view, making **d** the correct answer. A close look at b will show that such a convergent process is not consistent with evolutionary theory.

15. The answers in order are: **c, f, b, g, a, i, h, d**, and **e**.

16. **True.** Science, by reason of its self-imposed limitations, cannot investigate supernatural events of any kind. It is a faulty leap of logic to assume that a phenomenon does not exist because science cannot study it.

17. **False.** A paradigm is a way of looking at the world that is based upon one or more assumptions. For example, can we ever prove that the world we perceive with our senses is the real world? No! And yet it is a logical assumption, and we continue to behave as if it had been proved.

18. **False.** The concept of evolution does not go back to Aristotle. Although various philosophers have no doubt wondered about the relationship between living things, the first systematic statement of evolutionary theory was made by Lamarck (1744-1829).

19. **False.** The text states that Charles Darwin did not hold his grandfather's ideas in high regard.

20. **True.** Cuvier believed that new species filled the vacancies left by species that became extinct due to natural catastrophies.

21. **True.** It was Malthus's insights that led Darwin to make calculations using elephants as a model system. Natural selection seemed to be the only rational explanation of the observed facts.

22. Your answer should point out that neo-Darwinism is a synthesis of Darwinian evolution and modern genetics and therefore answers the questions raised by Mendelian genetics.

23. In a word, criticism. The proper interpretation of data accumulated through observation and scientific experimentation is virtually insured by constant subjection to every possible doubt and exception raised by other members of the scientific community.

24. See page 10 on the nature of science and page 13 on science as a process.

25. There is no single correct answer to this question. Each student will have his or her own perspective. The important thing is that the question be given some serious thought. Compare your answer to the one on page 10 under the nature of science.

PART I Biology of Cells

SECTION 1 The Unity of Life

CHAPTER 1

Atoms and Molecules

CHAPTER ORGANIZATION

I. Introduction
II. Essay: The signs of life
III. Atoms
  A. Introductory remarks
  B. Isotopes
  C. Models of atomic structure
IV. Electrons and energy
  A. Introductory remarks
  B. The arrangement of electrons
V. Bonds and molecules
  A. Introductory remarks
  B. Ionic bonds
  C. Covalent bonds
    1. Introductory remarks
    2. Polar covalent bonds
    3. Double and triple bonds
VI. Chemical reactions
  A. Introductory remarks
  B. Types of reactions
VII. Essay: Levels of organization
VIII. The biologically important elements
IX. Summary

MAJOR CONCEPTS

The biological properties of living systems depend ultimately upon the type and arrangement of their constituent atoms. The nature of the atom is determined by the number and ratio of its subatomic particles: protons, neutrons, and electrons. As our understanding of subatomic structure has increased, new models have replaced older ones.

Isotopes are atoms of a single element that have the same number of protons but different numbers of neutrons. Unpaired electrons determine the number of bonds that an atom can form when it combines with other atoms to form a molecule. Chemical bonds may be ionic, covalent, or polar covalent. Bonds between adjacent atoms may be single, double, or triple.

Chemical reactions involve exchanges of electrons among atoms. These reactions can be represented by chemical equations. Three general types of chemical reactions are (1) the combination of two or more substances to form a different substance, (2) the dissociation of a substance into two or more substances, and (3) the mutual exchange of atoms among two or more substances.

Six elements (carbon, hydrogen, nitrogen, oxygen, phosphorus, and sulfur) constitute almost 99 percent of living material. These atoms form molecules that are the basis of a series of increasing levels of organization. The cell is the level at which the properties of life first appear. Living organisms exhibit seven properties that together distinguish them from nonliving objects: (1) they are highly organized; (2) they are homeostatic; (3) they reproduce themselves; (4) they grow and develop; (5) they take energy from the environment and change it from one form into another; (6) they respond to stimuli; and (7) they are adapted.

HOW TO STUDY THE CHAPTER

Read the chapter, focusing on the major concepts.

Reread the chapter, using the questions that follow to help you focus on details as they relate to major concepts. Answer the questions on a separate sheet of paper. Your answers will provide a valuable study aid in preparing for examinations.

FOCUS ON CHAPTER DETAILS

I. Introduction (page 17)
1. Describe the circumstances under which the universe is thought to have originated.
2. How did atoms come into being? Why is it accurate to say that you and I are composed of stardust?
3. In what energetic form is the driving force of living systems stored?

II. Essay: The signs of life (pages 18–19)
4. Summarize the properties that, taken together, are uniquely common to living systems. Cite your own examples for each of these properties.

III. Atoms (pages 20-22)
A. Introductory remarks
5. What is the relationship between an atom's atomic number, atomic weight, and the number of its protons, electrons, and neutrons?

B. Isotopes
6. Using the three isotopes of hydrogen, describe how isotopes differ from one another. Do they exhibit the same chemical properties? Why?
7. What properties of radioactive isotopes make them useful in biological and medical research?

C. Models of atomic structure
8. Each successive model reflects the history of our increased understanding of atomic structure. What are the distinguishing features of each model?

IV. Electrons and energy (pages 22-24)
A. Introductory remarks
9. Why is a boulder at the top of a hill a good analogy for the potential energy of an electron?
10. How does the word "quantum" pertain to the relationship among electrons occurring at different energy levels?

B. The arrangement of electrons
11. What is an electron orbital? How many orbitals are there in the first energy level? In the second? How many electrons can each orbital hold?
12. Why are the noble gases chemically unreactive? How do their electron distribution patterns differ from those of chemically reactive atoms?

V. Bonds and molecules (pages 25-28)
A. Introductory remarks
13. When atoms interact with each other to fill their outer energy levels, what is the result?

B. Ionic bonds
14. What is the relationship between filled outer electron energy levels and ion formation?
15. How are ionic bonds formed? Cite an example.

C. Covalent bonds

Introductory remarks
16. How are covalent bonds formed? How do they differ from ionic bonds?
17. What characteristic of an atom determines the number of covalent bonds that it can form?
18. What characteristics of the carbon atom make it a good choice for the chemical basis of living organisms?

Polar covalent bonds
19. What is the source of the polarity of covalent bonds?
20. Defend the assertion that ionic, polar covalent, and covalent bonds may be con-

sidered different versions of the same type of bond.

Double and triple bonds

21. Carbon atoms can form single, double, and triple bonds between themselves and with other atoms. How does each type of bond influence the properties of a molecule?

VI. **Chemical reactions** (pages 28-30)
  A. Introductory remarks
  22. What is the relationship between the numbers of atoms in the products and the reactants of a balanced chemical equation?
  B. Types of reactions
  23. Describe the three general types of chemical reactions.

VII. **Essay: Levels of organization** (page 29)
  24. The characteristics of each new level of organization are new and different and are not just a combination of the original characteristics. At what level of organization does life suddenly appear as a new property?

VIII. **The biologically important elements** (pages 30-31)
  25. What six elements comprise almost 99 percent of all living material?
  26. What common properties do these elements have that might be responsible for their presence in living systems?

IX. **Summary** (page 31): Read the Summary. If you are familiar with the essential features of the material presented there, you are ready to take the following diagnostic examination.

TESTING YOUR UNDERSTANDING

After you have completed the following examination, compare your answers with the annotated key in the Analysis section.

1. Atoms of different elements are distinguished by:
   a. the number of neutrons in their nuclei.
   b. the number of protons in their nuclei.
   c. the number of electrons in their outer shell.
   d. their atomic numbers.
   e. Both b and d are correct.

2. Uranium has an atomic number of 92. Therefore, an atom of uranium has _____ and _____.
   a. 92 protons; 92 electrons
   b. 92 protons; 92 neutrons
   c. 92 neutrons; 92 electrons
   d. 46 neutrons; 46 protons
   e. 46 electrons; 46 protons

3. In comparing a 50-milligram sample of H_2O with an equal volume of ^2H_2O, it would be correct to say that:
   a. the ^2H_2O has a positive charge but H_2O does not.
   b. the deuterium atoms of ^2H_2O each have two neutrons.
   c. the ^2H_2O has a greater mass than does H_2O.
   d. the deuterium in ^2H_2O has a higher atomic number than does the hydrogen in H_2O.
   e. Both b and d are correct.

4. ^{14}C is an isotope of carbon (atomic number = 6) that has:
   a. 7 protons and 7 neutrons.
   b. 14 protons and 14 electrons.
   c. 6 protons and 8 neutrons.
   d. 14 protons and 14 neutrons.
   e. 6 protons, 6 neutrons, and 2 electrons.

5. Isotopes of any given element differ in all of the following *except*:
   a. weight.
   b. stability.
   c. number of neutrons.
   d. chemical properties.
   e. the amount of energy that each emits.

6. The implication of the quantum concept for a realistic model of the atom is that:
   a. electrons may be found at any distance from the nucleus.
   b. as an electron moves away from its nucleus, it releases a quantum of energy.
   c. electrons may be found only at discrete energy levels.
   d. electrons move about in orbitals at discrete distances from the nucleus.

7. A boulder being moved to the top of a slope is analogous to an electron that is:
   a. being moved away from the nucleus.
   b. being moved toward the nucleus.
   c. releasing a quantum of energy.
   d. participating in a covalent bond.
   e. Both b and c are correct.

8. According to our current understanding of atomic structure, the atom consists primarily of:
   a. empty space.
   b. a large nucleus with the electrons held closely to it.
   c. three different kinds of particles that have different charges but are equal in size.
   d. an orderly arrangement of electrons and protons surrounding a nucleus of heavier particles.
   e. Both c and d are correct.

9. The orbital model of the atom is useful because it describes:
   a. the exact location of all electrons at any given time.
   b. the exact amount of potential energy an electron possesses.
   c. a volume of space within which an electron can be found most of the time.
   d. the distance of any electron from the nucleus.

10. The sodium atom has an atomic number of 11. Therefore, the distribution of its electrons will be:
    a. 2 in the innermost level and 9 in the outermost level.
    b. 2 in the first, 8 in the second, and 1 in the outer level.
    c. 2 in each of the first two levels and 7 in the outer level.
    d. 8 in the first, 2 in the second, and 1 in the outer level.

11. When a carbon atom forms a single covalent bond with an oxygen atom:
    a. the electrons of the bond will be shared equally by the two atoms.
    b. the bond will be somewhat polar.
    c. the electrons will spend more of their time around the carbon atom.
    d. all electrons in their outer shells will be shared.
    e. Both b and c are correct.

12. An atom is most stable when:
    a. all of its electrons occupy the first energy level.
    b. its outermost energy level contains the maximum possible number of electrons for that energy level.
    c. its electrons are at their maximum potential energy.
    d. the number of its electrons equals the number of its protons.

13. $MgCl_2$ is an ionic compound formed as a direct result of bonding between:
    a. one Mg atom and two Cl atoms.
    b. one Mg^{2+} ion and two Cl^- ions.
    c. two Mg^{2+} ions and two Cl^- ions.
    d. a molecule of MgCl and a Cl^- ion.
    e. an atom of Mg and a molecule of Cl_2.

14. Which of the following generalized reactions represents the formation of salt water from NaOH and HCl?
    a. $A + B \rightarrow AB$
    b. $A + B \rightarrow C$
    c. $AB \rightarrow A + B$
    d. $AB \rightarrow BD$
    e. $AB + CD \rightarrow AD + BC$

15. Which of these molecules contains one or more single covalent bonds that are polar?
    a. O_2
    b. CO_2
    c. NaCl
    d. H_2O
    e. H_2

16. In methane (CH_4) the position of the hydrogen atoms takes the form of a:
    a. circle.
    b. straight line.
    c. tetrahedron.
    d. rectangle.
    e. cross.

17. In terms of chemical bonding, why is vegetable oil a liquid and lard a solid at room temperature?
    a. Oil has more nonpolar bonds.
    b. Some of the C—C bonds in oil are double bonds.
    c. Atoms of the oil are held together by ionic bonds.
    d. All of the atoms of the oil are free to rotate about one another.

18. When carbon is covalently bonded to three other atoms to fill its outer shell completely, the resulting molecule will have:
    a. three single bonds.
    b. one single bond and two double bonds.
    c. two single bonds and one double bond.
    d. three double bonds.
    e. one single, one double, and one triple bond.
19. Carbon has a central role in the organization of living things because it:
    a. has a completely filled outer shell.
    b. has the same number of protons and electrons.
    c. has the largest known number of isotopes.
    d. can form stable bonds with as many as four different atoms.
20. Choose the type of bond or bonds that bind the atoms of each molecule shown.

    CO_2 _____
    H_2 _____
    H_2O _____
    $CaCl_2$ _____
    O_2 _____

    a. nonpolar single bond
    b. polar covalent single bond
    c. covalent double bond
    d. ionic bond

21. CHNOPS is a handy way of designating:
    a. the first six elements in the atomic table.
    b. a molecule composed of six different elements.
    c. the six most abundant elements found in living organisms.
    d. six elements that are found together only in living things.
22. Why does the author say that although "you and I are flesh and blood, we are also stardust"?
23. A single covalent bond between two atoms results from the:
    a. transfer of electrons from the outer energy level of one atom to the outer energy level of the other atom.
    b. sharing of a single pair of electrons so that the outer energy levels of both atoms achieve stable electron distributions.
    c. sharing of inner-energy-level electrons to assure stability not found in outer energy levels.
    d. transfer of two pairs of electrons between their respective outer energy levels in order to achieve the electron distribution characteristic of the "noble" gases.
24. What properties are shared by the elements C, H, N, O, P, and S that may have determined why they were selected to be the most biologically important atoms?
25. True or False? Ions are positively or negatively charged atoms that form ionic bonds.
26. True or False? The great strength of ionic bonds prevents solid ionic substances from dissolving in water.
27. Why are the noble gases so chemically unreactive?

PERFORMANCE ANALYSIS

1. Atoms of different elements are characterized by the number of protons in their nuclei, which is equivalent to their atomic numbers; therefore, **e** is correct. The number of neutrons determines the type of isotope, not element. The number of electrons in the outer energy level determines the bonding characteristics of the atom.
2. The atomic number refers to the number of protons in the nucleus, in this case 92. But since atoms are electrically uncharged, there must also be 92 electrons, so that **a** is correct.
3. The superscript 2 applies to the two hydrogen atoms and indicates that there are two subatomic particles in the nucleus, a proton and a neutron. (This isotope is called deuterium.) Since H_2O lacks these two additional neutrons, it is lighter in mass than ^2H_2O; thus, **c** is correct.
4. Carbon's atomic number is 6, indicating that there are 6 protons in the nucleus. The isotope with an atomic weight of 14, therefore, must have 8 neutrons. Thus, **c** is correct.
5. Isotopes differ by weight, owing to differences in the number of neutrons they possess. Some of the isotopes of an element are radioactive and unstable. As these change to a more stable form, they emit various amounts of energy. Because isotopes do not differ in charge, or number of electrons, their chemical properties (**d**) are identical.
6. According to the orbital model of the atom, the orbital is a region in which an electron may be found 90 percent of the time. At any given instant, however, it is impossible to know the electron's exact location within that orbital. An electron

absorbs a discrete amount of energy (a quantum) as it moves *away* from the nucleus to occupy a higher discrete energy level. Therefore, **c** is correct.

7. Moving a boulder uphill requires an input of energy. In order to move from the nucleus, an electron must absorb energy; so **a** is correct. A boulder moving downhill releases energy, as does an electron moving closer to the nucleus.

8. Protons, neutrons, and electrons, although not equal in size, are so small in relation to the great distance between nucleus and electron orbitals that the atom is primarily empty space. Choice **a** is correct.

9. Neither an electron's exact location nor its exact distance from the nucleus (hence, its potential energy) can be known. The orbital, however, determines the region where an electron may be found 90 percent of the time (**c**).

10. According to several models of the atom, the first (innermost) energy level can hold only 2 electrons, the second can hold up to 8, and the third (in this case, the outermost) can also hold up to 8. Since sodium has an atomic number of 11, it has 11 electrons. The first two levels are completely filled, but the last has only 1 electron (**b**).

11. The only time that atoms share electrons equally (a) is when they are atoms of the same element. The covalent bond between carbon and oxygen is polar (**b**) because oxygen more strongly attracts the bonding electrons. These electrons spend more of their time around the oxygen atom, causing the carbon atom to have a slight positive charge.

12. When the outermost energy level is completely filled, an atom is stable and unreactive because it has no unpaired electrons to share with other atoms. Therefore, **b** is correct. For most biologically important elements, it takes eight electrons in the outer level to meet this condition.

13. In forming an ionic bond with chlorine (Cl), magnesium (Mg) donates the two electrons in its outer energy level to each of two Cl atoms. The bonding is a result of attraction between an Mg^{2+} ion and two Cl^- ions. Therefore, **b** is correct.

14. The overall equation for the reaction described may be written:

$$NaOH + HCl \rightarrow NaCl + H_2O$$

And if Na = A, OH = B, H = C, Cl = D, then:

$$AB + CD \rightarrow AD + BC$$

Therefore, choice **e** is correct.

15. Only H_2O (**d**) has single polar bonds, owing to the different affinities of H and O for the bonding electrons. Na and Cl (c) are joined by ionic bonding. In CO_2 (b) the covalent bonds are double bonds. Covalent bonds between the same kinds of atoms (a and e) are always nonpolar.

16. In methane, carbon is covalently bonded to four hydrogen atoms. The molecule is symmetrical (see Figure 1–10), with the hydrogen atoms positioned at the corners of a tetrahedron (**c**).

17. Vegetable oil contains double bonds, whereas lard (a solid) does not. These double bonds prevent the oil molecules from associating closely enough to form a solid; therefore, **b** is correct. Atoms of both oil and lard are held together, for the most part, by nonpolar covalent bonds.

18. Carbon is capable of forming four covalent bonds, but here it is bonded to only three other atoms. Thus, in order that its outer level contain eight electrons, one of the three bonds must be a double bond, so that **c** is correct.

19. Choice a is not a true statement; a carbon atom has only four outer electrons. All atoms have an equal number of protons and electrons (b); atoms with higher atomic numbers than carbon have as many, or more isotopes (c). Because carbon can form four stable bonds (d), it is ideally suited to be the backbone of large molecules whose integrity must be maintained despite the variable conditions encountered by living organisms.

20. The answers in order are: **c**, **a**, **b**, **d**, and **c**.

21. CHNOPS contains the first letters of each of the six most important elements in biological systems: **c**arbon, **h**ydrogen, **n**itrogen, **o**xygen, **p**hosphorus, and **s**ulfur; therefore, **c** is correct. These same elements are also found in combination in the inanimate world, making d an incorrect choice.

22. To answer this question, review the proposed origin of matter in the first several paragraphs of the chapter.

23. A covalent bond results from the sharing of electrons, so choices a and d are not correct. Ions of different charges result from the transfer of electrons. Since chemical bonding involves outer-energy-level electrons, answer c is wrong. Therefore, **b** is correct.

24. Your answer should include these points: (1) the atoms of all these elements "need to gain electrons to complete their energy levels" and therefore are capable of forming covalent bonds; (2) because the atoms are relatively small, their electrons are held rather tightly, resulting in stable molecules; and (3) with the exception of hydrogen, all of these atoms can form more than one covalent bond, "making possible the formation of large and complex molecules."

25. **True.** Recall that ions are formed when electrons are transferred from one atom to another, creating an imbalance between the numbers of protons and electrons. The net positive or negative charge is the result of this imbalance. In addition, any molecule that is charged could also be called an ion.

26. **False.** Ionic bonds tend to dissociate when a solid ionic substance (e.g., NaCl, table salt) is placed in water.

27. Your answer should point out that the noble gases have completed outer energy levels. Since only unpaired electrons can engage in covalent bond formation, the noble gases are chemically unreactive.

CHAPTER 2

Water

CHAPTER ORGANIZATION

I. Introduction
II. Water and the hydrogen bond
   A. Introductory remarks
   B. Surface tension
   C. Capillary action and imbibition
III. Water and temperature
   A. Specific heat of water
   B. Heat of vaporization
   C. Freezing
   D. Essay: The seasonal cycle of a lake
IV. Water as a solvent
V. Ionization of water: Acids and bases
   A. Introductory remarks
   B. Strong and weak acids and bases
   C. The pH scale
   D. Buffers
   E. Essay: Acid rain
VI. The water cycle
VII. Summary

MAJOR CONCEPTS

Water is the most abundant molecule in living systems. Water molecules are polar. Their association via hydrogen bonds is responsible for water's cohesive properties and high surface tension. As water gets colder, its density reaches a maximum at 4°C; below this point, its density decreases, which is the reason that ice floats.

Water has a high specific heat and is therefore able to resist dramatic changes in temperature. It has a high heat of vaporization; 540 calories are required to vaporize 1 gram of boiling water. Water has a high heat of fusion; ice absorbs almost 80 calories per gram from its surroundings as it melts. These four properties—high surface tension, high specific heat, high heat of vaporization, and high heat of fusion—have important biological implications.

The polarity of water is responsible for its ability to act as a solvent for most biological molecules, which are also polar. Many of these molecules ionize—either partially or completely—in aqueous (water) solutions. Even water ionizes to some extent.

Strong acids and bases ionize completely. Weak acids and bases ionize only partially. The concentration of hydrogen ions in solution (pH) has an effect on most biochemical reactions. Buffers resist dramatic changes in pH by combining with excess hydrogen ions or by releasing hydrogen ions into solution.

Water is continuously available to living things because it is recirculated through the water cycle.

HOW TO STUDY THE CHAPTER

Read the chapter, focusing on the major concepts.

Reread the chapter, using the questions that follow to help you focus on details as they relate to major concepts. Answer the questions on a separate sheet of paper. Your answers will provide a valuable study aid in preparing for examinations.

FOCUS ON CHAPTER DETAILS

I. Introduction (page 33)

1. Cite a few examples of the generalization that living organisms are found wherever water is present.
2. What is the proportionate contribution of water to the earth and to living systems?

II. Water and the hydrogen bond (pages 34–35)

A. Introductory remarks

3. How can water be both polar and neutral in charge?
4. Explain the four-cornered polarity exhibited by the water molecule.
5. Other than hydrogen, what two atoms are most frequently involved in hydrogen bonding?

B. Surface tension

6. What is the relationship between hydrogen bonding, cohesion of water molecules, and the surface tension of water?
7. Differentiate between adhesion and cohesion. Why is water wet?

C. Capillary action and imbibition

8. Differentiate between capillary action and imbibition, citing at least one example of each.

III. Water and temperature (pages 36–39)

A. Specific heat of water

9. Define specific heat, using water as an example.
10. What is the relationship between the specific heat of water and its hydrogen-bonding capacity?
11. Differentiate between heat and temperature. How is each measured?
12. What are some of the biological implications of the high specific heat of water?

B. Heat of vaporization

13. Why does the evaporation of water from a surface have a cooling effect? What is the role of hydrogen bonds in this process?
14. How much energy is required to vaporize 1 gram of boiling water?

C. Freezing

15. What is the relationship between water temperature and water's density? Of what significance is this relationship to living organisms?
16. How much heat energy is required to melt 1 gram of ice (i.e., what is its heat of fusion)?
17. How is it possible to vary the temperature at which water freezes? What are some practical and biological applications of the property responsible?

D. Essay: The seasonal cycle of a lake (page 39)

18. Describe the factors that produce summer stagnation, fall and spring overturns, fall and spring circulation, and winter stratification in a lake.

IV. Water as a solvent (page 40)

19. Differentiate between a solute and a solvent. What is a solution?
20. What is the relationship between water's polarity and its suitability as a solvent?
21. Why do nonpolar molecules cluster together when added to water?
22. What is the difference between hydrophobic and hydrophilic molecules?

V. Ionization of water: Acids and bases (pages 40–46)

A. Introductory remarks

23. Describe the ionization of water in terms of both the hydronium ion and the hydrogen ion.
24. What does the term "equilibrium" mean as applied to the tendency of water to ionize?
25. Define acidic and basic in terms of relative concentrations of hydrogen ions and hydroxide ions.

B. Strong and weak acids and bases

26. Differentiate between strong acids and bases and weak acids and bases.
27. Why will there be no excess of hydrogen

ions or hydroxide ions when equal amounts of a strong acid and a strong base are mixed?

28. What two functional groups give certain biological molecules the properties of weak acids and weak bases, respectively?

C. The pH scale

29. What is the relationship between the pH of a solution and its hydrogen-ion concentration?
30. What is the definition of a mole? How many particles (atoms, ions, or molecules) are present in a mole?
31. Express molar concentrations of hydrogen ions and hydroxide ions in terms of pH and vice versa (see Table 2–3).
32. What is the special significance of pH 7.0?
33. Redefine an acid and a base in terms of the pH scale.

D. Buffers

34. What is the pH range that includes almost all living processes? Cite some exceptions.
35. What is the most significant property of a buffer?
36. Describe how the carbonic acid-bicarbonate buffer system functions to maintain the pH of the blood. What do the lungs have to do with the pH of the blood?

E. Essay: Acid rain (pages 44–45)

37. What is the origin of acid rain? What acids are involved?
38. Summarize the various effects that acid rain has on organisms living in areas in which the soil is not capable of providing adequate buffering capacity.
39. Make a list of several ways in which you think that acid rain could have a social and economic impact on our nation.

VI. **The water cycle** (pages 46–47)

40. Describe the water cycle, using all of the following terms: evaporation, groundwater, zone of saturation, oceans, rivers, solar energy, moist soil, leaves of plants, polar ice, water table, percolation, and force of gravity.

VII. **Summary** (page 47): Read the Summary. If you are familiar with the essential features of the material presented there, you are ready to take the following diagnostic examination.

TESTING YOUR UNDERSTANDING

After you have completed the following examination, compare your answers with the annotated key in the Analysis section.

1. Because of the difference in the abilities of hydrogen and oxygen to attract electrons, a water molecule is:
   a. neutral in charge.
   b. positively charged.
   c. polar.
   d. covalently bonded to other water molecules.
   e. an ionic compound.
2. If X represents any atom with a strong attraction for electrons, then which of the following accurately represents hydrogen bonding?
   a. H--------X–H
   b. X--------H–X
   c. H--------H–X
   d. X--------X–H
   e. Both a and c are correct.
3. Some objects that are denser than water are able to float on water because:
   a. they adhere to water.
   b. of the cohesion of the water molecules.
   c. they are hydrophilic.
   d. capillary action between such an object and water increases the surface tension.
   e. water is denser as a liquid than as a solid.
4. The movement of water up a column with a charged inner surface is best described as:
   a. capillary action.
   b. imbibition.
   c. adhesion.
   d. surface tension.
   e. impossible.
5. Which of these statements correctly describes the difference between heat and temperature?
   a. Heat measures the number of degrees Fahrenheit or Celsius.
   b. Temperature is a measure of average kinetic energy in calories.

CHAPTER 2 *Water* 17

c. Heat is a caloric measure of the total kinetic energy.
d. Temperature is a measurement of the total kinetic energy in degrees.
e. Both b and c are correct.

6. Water is at its greatest density at:
   a. −100°C.
   b. 0°C.
   c. 4°C.
   d. 32°C.
   e. 100°C.

7. The temperature of a beaker of water heated over a flame increases more slowly than that of most other liquids because of water's high:
   a. heat of fusion.
   b. heat of vaporization.
   c. specific heat.
   d. boiling point.

8. The amount of heat needed to bring 1 gram of water at 25°C to the boiling point and then turn it to steam is approximately _____ calories.
   a. 615
   b. 540
   c. 75
   d. 187
   e. 727

9. Because of the polarity of water molecules, any substance that _____ will dissolve in water.
   a. is smaller than water
   b. is nonpolar
   c. has charged regions
   d. is hydrophobic
   e. contains hydrogen

10. Saltwater fish will not freeze:
    a. even if the ocean itself freezes.
    b. as long as the ocean does not freeze.
    c. when ocean temperatures reach 0°C.
    d. at subzero temperatures if they produce sufficient quantities of antifreeze proteins.
    e. Both b and d are correct.

11. The strength of the hydrogen bonds between water molecules in the solid state accounts for water's:
    a. surface tension.
    b. high heat of fusion.
    c. high heat of vaporization.
    d. high specific heat.
    e. All four answers are correct.

12. A solution of water and sugars will have a freezing point of:
    a. 0°C.
    b. greater than 0°C.
    c. less than 0°C.
    d. 0°F.
    e. −32°F.

13. If equal amounts of HCl and NaOH are added to a sample of pure water:
    a. there will be a large quantity of free H^+ and OH^- ions present.
    b. more of the water will dissociate.
    c. the pH will remain the same.
    d. there will be a large quantity of undissolved NaCl.
    e. HCl and NaOH will weakly dissociate.

14. Which of these substances acts as a weak base in solution?
    a. NaOH
    b. $-NH_2$
    c. NaCl
    d. CO_2
    e. H_2CO_3

15. A solution with a pH of 3 contains _____ of hydrogen ions per liter.
    a. 3 moles
    b. 10^3 moles
    c. 3 grams
    d. 0.01 mole
    e. 10^{-3} mole

16. Which of the following correctly describes a base?
    a. an OH^- acceptor
    b. an H_3O^+ donor
    c. a solution whose pH is less than 7
    d. an H^+ acceptor
    e. Both a and b are correct.

17. In a solution of pure water, one out of every _____ particles is a (an) _____.
    a. 1 million; H^+
    b. 1 million; OH^-
    c. 10 million; OH^-
    d. Both a and b are correct.
    e. There are no free H^+ or OH^- ions in pure water.

18. Consider the major buffer system in human blood described by the following chemical equation:
$$H_2CO_3 \rightleftharpoons H^+ + HCO_3^-$$
This buffer is most effective when:
a. the pH of the blood is exactly 7.0.
b. the H_2CO_3 is completely dissociated.
c. the H_2CO_3 is very weakly dissociated.
d. the concentrations of H_2CO_3 and HCO_3^- are equal.
e. Both a and c are correct.

19. What happens when an excess of OH^- ions enters the bloodstream?
a. The concentration of CO_2 increases.
b. Some additional H_2CO_3 dissociates.
c. The pH is raised dramatically.
d. The HCO_3^- ions combine with OH^- ions.

20.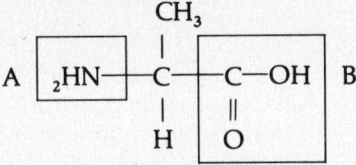

The compound shown above would be an effective buffer because:
a. group A is an H^+ donor.
b. group B is an H^+ acceptor.
c. group A is an H^+ acceptor.
d. group B is an H^+ donor.
e. Both c and d are correct.

21. Match the property of water on the left with the corresponding physical measurement on the right.

specific heat _____	a. 0°
melting point of ice _____	b. 100°C
	c. 4°C
highest density _____	d. 1 calorie per gram per degree Celsius
heat of fusion _____	
heat of vaporization _____	e. 540 calories per gram
boiling point _____	f. 79.7 calories per gram
freezing point _____	

22. Explain how bonding between hydrogen and oxygen atoms of water accounts for water's properties both as a temperature buffer and as a solvent.

23. Discuss the biological importance of each of the following properties of water.
a. specific heat
b. heat of vaporization
c. heat of fusion
d. greatest density as a liquid

PERFORMANCE ANALYSIS

1. Choice **c** is correct. See Figure 2–3a.

2. A hydrogen bond forms between an atom with a strong attraction for electrons and a hydrogen atom that is already covalently bonded to another such atom with a strong affinity for electrons. Since, by convention, hydrogen bonds are represented by a dashed line and covalent bonds by a solid line, choice **b** is correct.

3. Even objects denser than water may float because of the strong cohesive forces (**b**) provided by the hydrogen bonds that hold water molecules together.

4. Water *can* move up a column against the force of gravity. This movement by capillary action (**a**) is a result of both cohesion and adhesion, the attraction of water molecules for the particles making up the column.

5. Temperature is a measure (in degrees Fahrenheit or Celsius) of the average kinetic energy of a substance; the amount of heat is the total kinetic energy present in calories. Therefore, only **c** is correct.

6. Choice **c** is correct. As water cools, it becomes denser, reaching maximum density at 4°C. As the temperature continues to decrease, hydrogen bonding increases and water becomes less dense, eventually forming ice.

7. Temperature indicates the degree of movement of particles. Since water's hydrogen bonds are able to absorb a good deal of energy before breaking and allowing free movement of its molecules, its temperature will tend to remain stable. This property is referred to as specific heat (**c**).

8. Knowing that the specific heat of water is 1 calorie per gram per degree Celsius, you can calculate that

the amount of heat needed to raise 1 gram of water from 25°C just to the boiling point (100°C) is 75 calories. To this must be added the heat of vaporization for 1 gram, which is 540 calories, yielding a total of 615 calories (**a**). This is only an approximation, however, since specific heat is not a constant but varies slightly with temperature.

9. Any substance with charged regions (**c**) can dissolve in water because it will be attracted to the local positive or negative regions of water molecules. (See Figure 2–8.)

10. Saltwater fish have fluids with salt concentrations intermediate between pure water and ocean water. Because dissolved solutes lower water's freezing point, fish can live in the ocean until the freezing point of salt water is reached (**b**), or at lower temperatures if they are able to manufacture an antifreeze protein (**d**); therefore, choice **e** is correct.

11. Since the hydrogen bonds of water in its solid state are able to absorb a good deal of energy before the solid's latticework structure (see Figure 2–7a) gives way to the liquid phase, it takes about 80 calories to melt a gram of ice. This property is referred to as the heat of fusion (**b**).

12. The freezing point of pure water is 0°C (or 32°F), but sugars will dissolve in water and depress its freezing point; therefore, **c** is correct.

13. When added to water, both HCl and NaOH will completely dissociate, forming water and NaCl. If equal amounts are added, the pH will not change (**c**).

14. The amino group ($-NH_2$) acts as a weak base because it is a hydrogen-ion acceptor. NaOH is a strong base. NaCl will fully ionize but will not alter pH. CO_2, when dissolved in water, forms the weak acid H_2CO_3. Therefore, **b** is correct.

15. The pH of a solution may be thought of as the absolute value of the exponent in the term expressing the concentration of H^+ ions in moles per liter. Thus, a pH of 3 reflects a concentration of 10^{-3} mole per liter (**e**).

16. A base will have a pH greater than 7. When placed in a solution containing free H^+ ions, a base may act either as an H^+ acceptor or an OH^- ion donor; therefore, **d** is correct.

17. Since the pH of pure water is 7, there are 10^{-7} H^+ ions, and since these H^+ ions result from the ionization of H_2O, there must also be 10^{-7} OH^- ions. The term 10^{-7} may be rewritten as 1/10,000,000, or 1 in 10 million; therefore, **c** is correct.

18. A buffer operates best at the pH at which the H^+ donor (H_2CO_3) and the H^+ acceptor (the HCO_3^-) are equal in concentration (**d**). Choice a is incorrect because a desirable blood pH is in the vicinity of 7.3 to 7.4.

19. Hydroxide ions (OH^-) will combine with free hydrogen ions (H^+) in the solution to form water. In response, more carbonic acid (H_2CO_3) will dissociate (**b**) to reestablish the former H^+ concentration.

20. Group B, the amino ($-NH_2$) group, can accept H^+ ions (**c**) added to solution. Group A, the carboxyl group, can donate H^+ ions (**d**) to nullify the addition of OH^- ions. Therefore, choice **e** is correct.

21. The correct answers in order are: **d**, **a**, **c**, **f**, **e**, **b**, and **a**. Review the definitions in the text.

22. Your answer should explain how hydrogen bonds are formed between water molecules and how these bonds, although relatively weak, form a strong network that gives water many of its unique properties.

23. See the text, pages 36–38. The biological consequences of each physical property follow the discussion of that property in the text.

CHAPTER 3

Organic Molecules

CHAPTER ORGANIZATION

I. Introductory remarks

II. The central role of carbon
  A. Background
  B. The carbon backbone
  C. Essay: Why not silicon?
  D. Functional groups
  E. Four representative molecules
  F. The energy factor
  G. Essay: Molecular models

III. Carbohydrates: Sugars and polymers of sugars
  A. Introductory remarks
  B. Monosaccharides: Ready energy for living systems
  C. Essay: Why is sugar sweet?
  D. Disaccharides: Transport forms
  E. Storage polysaccharides
  F. Structural polysaccharides

IV. Lipids
  A. Introductory remarks
  B. Fats and oils: Energy in storage
    1. General information
    2. Sugars, fats, and calories
    3. Insulators and cushions
  C. Phospholipids
  D. Waxes
  E. Cholesterol and other steroids

V. Proteins
  A. General introduction
  B. Amino acids: The building blocks of proteins
  C. The levels of protein organization
  D. Structural uses of proteins
    1. Fibrous proteins
    2. Globular proteins
  E. Hemoglobin: An example of specificity
  F. Essay: Determining the amino acid sequence of a protein

VI. Summary

MAJOR CONCEPTS

Because it can engage in so many structurally different bonding arrangements, the carbon atom plays a central role in the shape of organic molecules. The specific properties of organic molecules are determined by the various functional groups they contain. Four main types of organic molecules predominate in living organisms: carbohydrates, lipids, amino acids, and nitrogenous bases.

Carbohydrates consist of one (mono-), two (di-), or many (poly-) sugar (saccharide) units. All simple sugars (monosaccharides) conform to the basic formula $(CH_2O)_n$ and possess an aldehyde or ketone functional group. Monosaccharides provide ready energy, disaccharides function in sugar transport, and polysaccharides serve as storage or structural molecules.

Lipids constitute a heterogeneous group of hydrophobic molecules that include the neutral fats, the steroids, and the phospholipids. Lipids serve as energy-storage molecules, as major components of cell membranes, and as hormones. Phospholipids are neutral fats that contain a phosphate group and usually another chemical group.

Proteins consist of one or more chains of amino acids linked by peptide bonds. These chains are known as polypeptides. Each amino acid contains an amino group, a carboxyl group, and some other chemical group, all attached to a central carbon atom. Proteins exhibit four levels of structural organization and are generally classified as either fibrous or globular. They function as enzymes, hormones, and structural molecules.

HOW TO STUDY THE CHAPTER

Read the chapter, focusing on the major concepts.

Reread the chapter, using the questions that follow to help you focus on details as they relate to major concepts. Answer the questions on a separate sheet of paper. Your answers will provide a valuable study aid in preparing for examinations.

FOCUS ON CHAPTER DETAILS

I. **Introductory remarks** (page 49)
   1. Relatively few elements comprise the thousands of molecules found in living systems, and relatively few kinds of molecules are represented. To what general categories do these molecules belong?

II. **The central role of carbon** (pages 49–54)
   A. Background
   2. What is the relative contribution of ions such as K^+, Na^+, and Ca^{2+} to the overall composition of living systems? How about water?
   3. What are the four different kinds of organic molecules found in large quantities in living systems? What elements does each kind of organic molecule contain?
   B. The carbon backbone
   4. In general, from what do organic molecules derive their shape and, indirectly, many of their other properties?
   5. What are hydrocarbons? Why are they of economic importance?
   C. Essay: Why not silicon? (page 50)
   6. Summarize the reasons that carbon is superior to silicon as the structural basis for biological molecules.
   D. Functional groups
   7. What important property do polar functional groups confer upon the molecules that contain them?
   8. Name at least one specialized property that is characteristic of each type of functional group. (See both the text and Table 3-1.)
   E. Four representative molecules
   9. In your own words, state how the four molecules discussed in this section differ. (See both the text and Figure 3-3.)
   10. Only about 30 molecules need to be recognized for a working knowledge of biochemistry. What are they?
   F. The energy factor
   11. Of what do covalent bonds consist?
   12. Chemical reactions always involve a change in electron configurations and therefore in bond strengths. How are bond strengths measured?
   13. Using ΔH, write an expression that represents the release of 50 kilocalories from a chemical reaction. Now change one of the components so that the equation represents the uptake of 50 kilocalories.
   14. What is meant by the word "strategy" when used in reference to the activities of living organisms?
   G. Essay: Molecular models (pages 54–55)
   15. Summarize the advantages and disadvantages of using each of the models mentioned to represent molecules.
   16. If molecules are composed mostly of empty space, how does molecular recognition take place?

III. **Carbohydrates: Sugars and polymers of sugars** (pages 54–60)
   A. Introductory remarks
   17. Define monomer and polymer as they apply to carbohydrates.
   18. What is the most common organic compound in the biosphere?
   B. Monosaccharides: Ready energy for living systems
   19. What functional groups characterize monosaccharides? What is the ratio of carbons, hydrogens, and oxygens in a monosaccharide?
   20. How can the amount of energy released from a chemical reaction be measured? (See also Figure 3-5.)

21. The oxidation of food molecules and the burning of fuels are really the same basic process. How much energy is supplied by the oxidation of a mole of glucose?

C. Essay: Why is sugar sweet? (page 57)

22. Is sugar sweet? Explain.
23. How do plants effectively use sugar to attract beneficial organisms of various sorts?
24. How is our natural taste for sugar exploited?

D. Disaccharides: Transport forms

25. What are the monosaccharide constituents of sucrose?
26. Why is it that only monosaccharides, as opposed to disaccharides and polysaccharides, follow the CH_2O ratio?
27. What are the monosaccharide constituents of lactose and trehalose?
28. Differentiate between hydrolysis and condensation.

E. Storage polysaccharides

29. What are the two principal storage forms of starch, and how do they differ from one another? (See also Figure 3-8.)
30. What is glycogen, and where is it stored in vertebrates?
31. Differentiate between glycogen and glucagon. What are their respective roles in energy mobilization in animals?

F. Structural polysaccharides

32. What is the relationship between alpha- and beta-glucose and the availability of the sugar monomers of starch, glycogen, and cellulose as energy sources? (See also Figures 3-2, 3-8, and 3-9.)
33. What is chitin, and what is it used for? How have the monomeric units been modified? (See Figure 3-10.)

IV. Lipids (pages 60-64)

A. Introductory remarks

34. Summarize the general characteristics of lipids.

B. Fats and oils: Energy in storage

General information

35. Explain the advantage to a hummingbird (or any animal, for that matter) of fat reserves as opposed to carbohydrate reserves.
36. Describe the composition of a neutral fat.
37. What property is conferred upon a fatty acid by its having one or more double bonds? What is an unsaturated fatty acid?

Sugars, fats, and calories

38. What factors determine the pattern of utilization of reserve energy stored in a body?

Insulators and cushions

39. What is the striking metabolic feature of fat deposits that are used as insulators or cushions?
40. Cite one example each of a use of fat as an insulator and as a cushion.
41. Of what advantage was excess fat to our female ancestors?

C. Phospholipids

42. Describe the structure of a phospholipid. (See also Figure 3-13.)
43. Why are the polar head and the nonpolar tail of a phospholipid important to the two-layer configuration of biological membranes? (See also Figure 3-14.)

D. Waxes

44. What is the main function of waxes in biological systems? Cite some examples.

E. Cholesterol and other steroids

45. Why are cholesterol and the other steroids grouped with the lipids?
46. Name two cellular structures that are rich in cholesterol.
47. What is the current thinking on the reduction of dietary cholesterol as a means of preventing high blood pressure, heart attacks, and strokes?

48. What are some of the functions of the various steroids?
49. Is prostaglandin a steroid? Is it a lipid?

V. Proteins (pages 64–73)

A. General introduction
50. Why do plants contain less than 50 percent protein and most animals contain more than 50 percent protein?
51. List the different functional types of proteins and give at least one example of each. (See also Table 3–2.)
52. Regardless of function, all proteins follow the same structural pattern. What is it?

B. Amino acids: The building blocks of proteins
53. What is the fundamental structure of an amino acid?
54. Describe the range of properties possessed by amino acid side groups. (See also Figure 3–17.)
55. How universal is the utilization of the 20 amino acids used to make proteins?
56. What is the name of the bond that joins amino acids into polypeptide chains?

C. The levels of protein organization
57. What constitutes the primary structure of a polypeptide? What two functional groups are involved in a peptide bond? (See Figure 3–18.)
58. What constitutes the secondary structure of a polypeptide? What kind of bond maintains this structure?
59. What effects does the nature of the R group have on the secondary structure of a polypeptide?
60. Describe the types of bonds that hold a polypeptide in a particular tertiary configuration.
61. Differentiate between fibrous and globular proteins.
62. What kinds of bonds and interactions are responsible for maintaining the quaternary structure of interacting polypeptides?

D. Structural uses of proteins

Fibrous proteins

63. Using collagen as an example, list the structural and functional features of fibrous proteins.

Globular proteins

64. Using microtubular protein as an example, list the structural and functional features of globular proteins.

E. Hemoglobin: An example of specificity
65. Summarize the structural features of the human hemoglobin molecule. (See also Figures 3–27 and 3–28.)
66. How does sickle cell hemoglobin differ structurally and functionally from normal hemoglobin?

F. Essay: Determining the amino acid sequence of a protein (pages 72–73)
67. What is the role of cleaving agents and digestive enzymes in the determination of the primary structure of a protein?

VI. Summary (pages 73–74): Read the Summary. If you are familiar with the essential features of the material presented there, you are ready to take the following diagnostic examination.

TESTING YOUR UNDERSTANDING

After you have completed the following examination, compare your answers with the annotated key in the Analysis section.

1. Which of the following organic molecules is composed of a hydrophobic chain of carbon and hydrogen atoms?
   a. alanine
   b. adenine
   c. glucose
   d. stearic acid
   e. methane
2. Molecules with polar functional groups will _____ when placed in a polar solvent.
   a. become fully ionized
   b. dissolve

c. raise the pH
   d. lower the pH
   e. not react
3. The burning of methane gas (CH_4), which yields carbon dioxide and water, has a $\Delta H°$ of -213 kilocalories per mole. This implies that:
   a. methane absorbs energy as it burns.
   b. carbon forms stronger bonds with hydrogen than with oxygen.
   c. enzymes are required in order for the reaction to proceed.
   d. methane is a carbohydrate.
4. When carbohydrates are oxidized in living cells, _____ is (are) released.
   a. both glucose and water
   b. only water
   c. both carbon dioxide and water
   d. only carbon dioxide
   e. oxygen and water
5. Which of these substances has a chemical formula that agrees with the formula $(CH_2O)_n$?
   a. lactose
   b. sucrose
   c. fructose
   d. cellulose
   e. All of the answers above are correct.
6. Which of these substances is a structural polysaccharide?
   a. keratin
   b. glycogen
   c. amylopectin
   d. cellulose
   e. collagen
7. An organism can obtain the most energy per gram by oxidizing:
   a. starch.
   b. glucose.
   c. protein.
   d. fat.
   e. glycogen.
8. Cellulose differs from storage forms of polysaccharides in that:
   a. it is a polymer made from a number of alpha-glucose molecules.
   b. its monomeric units are linear instead of rings.
   c. it consists entirely of beta-glucose molecules.
   d. it is a modified polysaccharide.
   e. it does not release energy upon hydrolysis.
9. The most striking characteristic of the general class of compounds known as lipids is:
   a. their long hydrocarbon tails.
   b. their insolubility in polar solvents.
   c. that, biologically, they are structural molecules.
   d. their phosphate functional groups.
   e. that they contain a wide variety of fatty acids.
10. Biochemically speaking, fats are primarily used _____, and phospholipids serve chiefly as _____.
    a. as transport molecules; enzymes
    b. for energy storage; structural molecules
    c. as hormones; storage forms of energy
    d. in exoskeletons; cell wall components
11. A saturated fat contains _____ than an unsaturated fat of the same size.
    a. more fatty acid molecules
    b. longer hydrocarbon chains
    c. more carbon-carbon double bonds
    d. more hydrogen
    e. fewer polar functional groups
12. A lipid that has four carbon rings and an $-OH$ functional group is known as a (an):
    a. fat.
    b. steroid.
    c. wax.
    d. phospholipid.
    e. alcoholic hydrocarbon.
13. Although phospholipids resemble fats, they differ in that they:
    a. are even more hydrophobic.
    b. lack fatty acid chains.
    c. have charged functional groups.
    d. do not contain glycerol.
    e. are not structural molecules.
14. Research into the biological importance of cholesterol has shown that it:
    a. is used as a building block for the synthesis of sex hormones.
    b. forms fatty deposits in blood vessels.
    c. plays an essential role in the conduction of nerve impulses.
    d. is a major component of cell membranes.
    e. All of the answers above are correct.

15. A peptide bond forms between the _____ of one amino acid and the _____ of a second amino acid.
    a. central carbon; central carbon
    b. amino group; carbonyl group
    c. carboxyl group; amino group
    d. side group; side group

16. The primary structure of a protein consists of:
    a. hydrophobic interactions between nonpolar R groups.
    b. hydrogen bonding between amino acids in two different parts of the polypeptide.
    c. the linear sequence of amino acids along a polypeptide.
    d. the formation of a beta-pleated sheet.
    e. the alpha-helical arrangement of amino acids in a polypeptide.

17. The collagen molecule results from a close association of:
    a. three long polypeptide chains.
    b. several globular proteins.
    c. collagen fibrils.
    d. a globular protein and a fibrous protein.
    e. two beta-pleated sheets.

18. Which of the following consists of globular protein dimers?
    a. an enzyme
    b. a microtubule
    c. hemoglobin
    d. collagen
    e. chitin

19. Hemoglobin is considered to have a quaternary protein structure because it is composed of:
    a. four different polypeptide chains.
    b. four different globular polypeptides.
    c. more than one polypeptide chain.
    d. only four different kinds of amino acids.

20. Which of the following levels of structure is fundamentally responsible for the specificity of proteins?
    a. primary structure
    b. secondary structure
    c. tertiary structure
    d. quaternary structure

21. Answer the following questions about the chemical reaction shown below.

    a. Name the reactant on the left.
       _____
    b. Name the reactant on the right.
       _____
    c. Name the product formed by this reaction.
       _____
    d. What type of compound is this product?
       _____
    e. What is the specific name for this type of chemical reaction? _____

22. Match each functional group on the left with one or more of the molecules listed at the right in which the group may be found.
    amino _____
    phosphate _____
    carboxyl _____
    aldehyde _____
    methyl _____

    a. lipid
    b. simple sugar
    c. amino acid
    d. fatty acid

23. Match each organic molecule on the left with one or more of the biological roles listed at the right.
    collagen _____
    hemoglobin _____
    cholesterol _____
    cellulose _____
    glycogen _____
    glucose _____
    insulin _____

    a. structural carbohydrate
    b. structural lipid
    c. structural protein
    d. transport protein
    e. protein hormone
    f. storage carbohydrate
    g. readily oxidizable carbohydrate

24. Answer the following questions that refer to the chemical reaction shown below.
    a. What is the general name of these two reactants? _____
    b. What is the general name of the product? _____
    c. What is the name of the bond that joins the two subunits in the product? _____
    d. If the product had many more subunits, what would it be called? _____
    e. What is the name of the functional group on the left end of the product? _____
    f. What is the name of the functional group on the right end of the product? _____

```
      R                    R
      |                    |
H—N—C—C—OH       H—N—C—C—OH
  |   |  ‖           |   |  ‖
  H   H  O           H   H  O
              ↓

           R            R
           |            |
   H—N—C—C—N—C—C—OH
       |  ‖  |   |  ‖
       H  O  H   H  O
```

25. Explain how several typical polypeptides might come to form the structure of a multimeric protein, such as hemoglobin.
26. Some people have thought that forms of life similar to our own carbon-based system could evolve under favorable conditions on planets where carbon might be scarce but silicon (atomic number = 14) is abundant. Evaluate this suggestion in terms of the atomic structure of silicon.

PERFORMANCE ANALYSIS

1. Stearic acid (d) is an example of a fatty acid. Note the long hydrocarbon chain (which is hydrophobic) shown in Figure 3-3a. Of the other choices, only methane (e) is hydrophobic, but it is not a chain.
2. Molecules with polar functional groups are hydrophilic as a result of local electrical charges and will dissolve (b) in a polar solvent, such as water. Such solutes, depending on the type of functional group, may either raise or lower the pH of an unbuffered solution.
3. The overall equation for the complete combustion of methane is:

$$CH_4 + 2O_2 \longrightarrow CO_2 + 2H_2O$$

By convention, a negative ΔH indicates that energy is released during a reaction. Consequently, there is less potential energy in the bonds of the product (CO_2) than there is in the bonds of the reactant (CH_4). Thus C—H bonds are stronger (b).

4. The overall equation for the oxidation of any carbohydrate is:

$$(CH_2O)_n + nO_2 \longrightarrow (CO_2)_n + n(H_2O)$$

Choice c is therefore correct.

5. The $(CH_2O)_n$ formula applies only to monosaccharides, such as fructose (c). Polysaccharides are formed from monosaccharides, but molecules of water are lost, which changes the relationship between the C, H, and O atoms.
6. Glycogen and amylopectin are principal forms of energy storage in animals and plants, respectively. Cellulose (d) is a structural polysaccharide found in plants. The other two substances are not polysaccharides but structural proteins used by animals.
7. Fats (d), on the average, yield about 9.3 kilocalories per gram. Proteins and carbohydrates yield little more than a third as much.
8. Glucose monomers may be either alpha (as in starches, Figure 3-8a, b) or beta in form (as in cellulose, Figure 3-9a). Choice c is correct.
9. Not all lipids have hydrocarbon tails, nor do they contain fatty acids. Only phospholipids have phosphate groups attached. Some lipids are structural, but others may be hormones (e.g., steroids) or energy-storage forms (e.g., fats). The general characteristic of insolubility in polar solvents (b), however, pertains to all lipids.
10. Fats are energy-rich compounds, though they sometimes serve as insulation from cold tempera-

tures. Phospholipids are major structural components of the cell membrane. Therefore, choice **b** is correct.

11. Unsaturated fats have C—C double bonds. When these bonds are broken and replaced with C—H and C—C single bonds, the carbon's backbone is literally saturated with hydrogen; therefore, **d** is correct.

12. Choice **b** is correct. See the structures of cholesterol (Figure 3-16a) or testosterone (Figure 3-16b).

13. Phospholipids differ in having a phosphate functional group (see Figure 3-13) that contains a negative charge at biological pH values (**c**). Choices a, b, d, and e are all false statements about phospholipids.

14. Cholesterol functions in all of the ways listed; **e** is therefore correct. However, its role in the formation of fatty deposits in blood vessels seems to be a harmful consequence of high levels rather than a normal function.

15. Choice **c** is correct. Refer to Figure 3-18.

16. Choice **c** is correct. The primary structure of a protein is a single polypeptide chain consisting of a linear sequence of amino acids (see Figure 3-19). Choices d and e apply to secondary structure only. Hydrogen bonding (b) can be responsible for all structures other than the primary, and hydrophobic interactions (a) chiefly determine the tertiary structure.

17. Choice **a** is correct. Refer to the text and Figure 3-23. The long polypeptide chains are linked by hydrogen bonding.

18. Microtubules (**b**) consist of an assembly of dimers, each of which is made up of two globular protein subunits (see Figure 3-25). Enzymes (a) are globular proteins also, but their configurations need not be dimeric. Hemoglobin (c) is a tetramer. Collagen (d) exemplifies the secondary structure of a protein. Chitin (e) is not a protein but a modified polysaccharide.

19. The quaternary structure is a result of the interaction of two or more polypeptide chains; therefore, **c** is correct. Choices a and b are both false statements because there are only two different types of subunits in hemoglobin.

20. Although higher levels of organization are important in the specific action of proteins, it is the primary structure (**a**), acted on by the local chemical environment, that ultimately determines specificity.

21. (a) **Glucose**, (b) **fructose**, (c) **sucrose**, (d) **disaccharide**, (e) **condensation reaction**.

22. The correct answers in order are **c**; **a**; **a, c, d**; **b**; and **a, c, d**. Review Table 3-1.

23. The correct answers in order are **c, d, b, a, f, g,** and **e**.

24. **Amino acids** (a) unite in a condensation reaction to form a **dipeptide** (b). The bond joining them is called a **peptide bond** (c). If there were many more amino acid subunits, we would call the product a **polypeptide** (d). On the left of the product is an **amino group** (e), and on the right is a **carboxyl group** (f).

25. Using Figure 3-22 as a rough guide, review the discussion in the text of how a linear sequence of amino acids gives rise to a tertiary structure and how several tertiary structures come together to form a quaternary protein, such as hemoglobin.

26. Silicon is similar to carbon in having four electrons that are available for bonding, but silicon cannot form long stable chains (because of its greater atomic size), nor does it form double bonds with oxygen. Because of this, a source of atmospheric silicon would be unlikely. It is therefore also unlikely that any silicon-based life form could exist. If it could, however, it would probably bear very little resemblance to carbon-based life forms.

CHAPTER 4

Cells: An Introduction

CHAPTER ORGANIZATION

I. Introductory remarks
II. The cell theory
III. The disproof of spontaneous generation
IV. The beginning of life
   A. Background information
   B. Essay: The origins of the earth
   C. The first cells
   D. Why on earth?
V. Heterotrophs and autotrophs
VI. Prokaryotes and eukaryotes
   A. General background
   B. The origins of multicellularity
VII. Viewing the cellular world
   A. Introductory remarks
   B. Types of microscopes
   C. Preparation of specimens
   D. Observation of living cells
VIII. Summary

MAJOR CONCEPTS

All living matter is composed of one or more cells; the chemical reactions of living organisms take place within cells; every cell is descended from a previous cell; and an organism's cells contain its hereditary information, which is passed from parent cell to daughter cell.

Early biologists believed that certain small organisms could arise spontaneously. Louis Pasteur put this concept to rest, except with respect to the origin of the first cells, which apparently arose by spontaneous generation on the primitive earth.

Organisms may be classified according to the manner in which they accumulate energy from their environments. Autotrophs get their energy from the sun or, in a few rare cases, from the chemical bonds of certain inorganic molecules. Heterotrophs, on the other hand, require energy in the form of energy-rich organic compounds.

All organisms are either prokaryotes or eukaryotes. Prokaryotic cells are smaller, contain no membrane-bound nuclei, and have few organelles. Most are also unicellular. Eukaryotic cells, on the other hand, contain membrane-bound nuclei, many organelles, and may be either unicellular or multicellular.

Our knowledge of cell structure and function has increased dramatically with each new technological advance in study techniques. A variety of different types of microscopes are available, each suited to a particular investigative strategy. Specimens must be prepared differently for each of the various kinds of microscopy.

HOW TO STUDY THE CHAPTER

Read the chapter, focusing on the major concepts.

Reread the chapter, using the questions that follow to help you focus on details as they relate to major concepts. Answer the questions on a separate sheet of paper. Your answers will provide a valuable study aid in preparing for examinations.

FOCUS ON CHAPTER DETAILS

I. **Introductory remarks** (page 76)
   1. Why do different atoms have different properties?
   2. What is characteristic of each level of organization?
   3. What is the basis of the differences between living and nonliving systems?

II. **The cell theory** (pages 76–77)
   4. Why were the conclusions of Matthias Schleiden and Theodor Schwann significant?
   5. How did Rudolf Virchow's generalization extend and broaden the conclusions of Schleiden and Schwann?
   6. What are the four basic parts of the modern cell theory?
   7. What interpretation does the perspective of evolution provide for the cell theory?

III. **The disproof of spontaneous generation** (pages 77–78)
   8. Summarize the thinking on the subject of spontaneous generation prior to Pasteur's experiment.
   9. Summarize the important elements of Pasteur's experiment. Why had questions about the origin of the first cells remained unasked for so long?

IV. **The beginning of life** (pages 78–82)
   A. Background information
   10. What two developments led to the reopening of investigations into the early events in the history of life?
   11. Upon what two issues concerning the composition of the primitive atmosphere is there general agreement?
   12. Describe the conditions under which chemical evolution is thought to have taken place. Why is it important that no free oxygen was present?
   13. How have the experiments of Stanley Miller confirmed Oparin's hypothesis? Do these experiments prove that organic compounds were formed spontaneously on the primitive earth?
   14. What is meant by the phrase "natural selection played a role in chemical evolution"?

   B. Essay: The origin of the earth (page 79)
   15. How are the sun, the earth, and the other planets thought to have been formed?

   C. The first cells
   16. What are proteinoid microspheres, how are they formed, and what is their supposed relationship to the first cells?
   17. Why are microfossils important to hypotheses about the origin of the first cells?

   D. Why on earth?
   18. How is the earth uniquely suited among the planets to support life?

V. **Heterotrophs and autotrophs** (page 83)
   19. What is the basic difference between heterotrophs and autotrophs?
   20. What kind of "troph" may have been the predecessor of both autotrophs and heterotrophs?
   21. Why is it thought that life on earth would soon have come to an end without the evolution of autotrophs?

VI. **Prokaryotes and eukaryotes** (pages 84–88)
   A. General background
   22. What is the one essential feature of all cells?
   23. Summarize the significant differences between the eukaryotes and the prokaryotes.

   B. The origins of multicellularity
   24. How far back in the fossil record can multicellular organisms be found?
   25. In what important way do the cells of a multicellular organism differ from single-celled organisms?
   26. How can part of an organism be phototrophic and part be heterotrophic? Cite an example.

27. What are the basic characteristics of each of the five kingdoms of organisms?

VII. **Viewing the cellular world** (pages 88–95)
   A. Introductory remarks
      28. Why has new knowledge about living systems generally come in bursts?
   B. Types of microscopes
      29. What are the advantages and disadvantages of each of the three types of microscopes?
      30. Why is resolving power more important than magnification?
   C. Preparation of specimens
      31. How is the requirement for acceptable contrast met for specimens being viewed by light and transmission electron microscopy?
      32. How do the transmission and scanning electron microscopes differ in their use of the electron beam?
      33. What is the relationship between an atom's atomic weight and its electron opacity?
      34. What is the reason for staining microscopic preparations with dyes (light microscope) or heavy metals (transmission electron microscope)?
      35. What are the advantages of fixing cells to be used for microscopic examination? Why must tissue preparations be sectioned before being viewed with all but the scanning electron microscope?
      36. What is shadowing? Why is this technique used?
      37. Why must specimens for viewing in the electron microscope first be dehydrated?
   D. Observation of living cells
      38. What is the theoretical basis for the functioning of the phase-contrast and differential-interference microscopes, and how are they used?
      39. What is the theory behind the technique of dark-field microscopy?
      40. What are two newly emerging microscopic techniques that hold great promise for advancing the field of cell biology?

VIII. **Summary** (pages 95–96): Read the Summary. If you are familiar with the essential features of the material presented there, you are ready to take the following diagnostic examination.

TESTING YOUR UNDERSTANDING

After you have completed the following examination, compare your answers with the annotated key in the Analysis section that follows the test.

1. The basis of the difference between living systems and the inanimate world is that:
   a. though the elements are the same, chemically they behave differently.
   b. different elements are involved in biochemical reactions.
   c. the subatomic particles of biologically important atoms are arranged differently.
   d. atoms and molecules in biological systems are organized differently than in nonliving systems.

2. Which of the following men performed experiments that upheld the integrity of the cell theory of Virchow?
   a. Louis Pasteur
   b. Stanley Miller
   c. A. I. Oparin
   d. Sidney Fox
   e. Robert Hooke

3. In order for Pasteur to convince proponents of spontaneous generation that their interpretations were invalid, his experimental design had to:
   a. include boiling the medium to kill microorganisms.
   b. allow oxygen to enter the experimental vessel.
   c. protect the medium from contamination.
   d. Both b and c are correct.
   e. All three answers (a, b, and c) are correct.

4. In 1860 Pasteur demonstrated to the Paris Academy of Sciences that:
   a. living things could never have evolved from nonliving things.
   b. life does not arise from the sterile environment of a laboratory flask.
   c. maggots will be found only where flies have laid eggs.
   d. microorganisms can survive at high temperatures and low oxygen levels.

5. Scientists agree that _____ was (were) initially present in the earth's primitive atmosphere.
   a. abundant free oxygen
   b. some form of carbon, hydrogen, and nitrogen
   c. few available sources of energy
   d. amino acids
   e. prokaryotes

6. Oparin's hypothesis stated that organic molecules were formed from _____ and that these molecules eventually were deposited in _____ .
   a. inorganic materials in oceans; the earth's crust
   b. the earth's crust; the ocean depths
   c. atmospheric gases; oceans and lakes
   d. interstellar gases; the atmosphere

7. We now have evidence that life on earth may have begun as long ago as _____ million years.
   a. 4600
   b. 3500
   c. 1500
   d. 750
   e. 600

8. Scientists, delving into the evolution of life from chemicals, have suggested that _____ may have given rise to the first living cells.
   a. amino acids
   b. aggregates of inorganic molecules
   c. chemoautotrophs
   d. lipid droplets
   e. protenoid microspheres

9. What two factors are apparently of fundamental importance in explaining why life was able to evolve on earth, out of all the planets?
   a. its distance from the sun; its size
   b. the abundance of water; the presence of free oxygen
   c. the presence of atoms that now comprise living tissue; the availability of water as a liquid
   d. the chemical composition of the earth's crust; the earth's intense gravitational field

10. Chemoautotrophs carry out metabolic activities using energy released from:
    a. thermonuclear reactions.
    b. inorganic reactions.
    c. organic molecules they ingest.
    d. organic molecules they are able to synthesize.

11. There is much indirect evidence to support the belief that the first cells were:
    a. eukaryotic and phototrophic.
    b. eukaryotic and heterotrophic.
    c. prokaryotic and phototrophic.
    d. prokaryotic and heterotrophic.

12. What essential feature is shared by all cells, whether prokaryotic or eukaryotic?
    a. They are bounded by a cell membrane.
    b. They have a wide variety of internal structural units called organelles.
    c. They are protected by an outer cell wall.
    d. Their genetic material is contained within a nuclear envelope.

13. You are observing a cell under a high-powered microscope with sufficient resolution and contrast to reveal a number of chromosomes and a wide variety of organelles. A cell wall is not evident. You can be certain that this cell is:
    a. a prokaryotic autotroph.
    b. from a multicellular organism.
    c. a eukaryote, but not a plant.
    d. a chemoautotrophic unicellular organism.
    e. a bacterium.

14. Which of these microscopes provides cell biologists with the highest resolution?
    a. light microscope
    b. phase-contrast microscope
    c. scanning electron microscope
    d. transmission electron microscope
    e. differential-interference microscope

15. The drawback in using the light microscope to examine the internal structure of cells is:
    a. its limited magnification.
    b. its limited resolving power.
    c. that visible wavelengths of light cannot penetrate the cell wall.
    d. that stains and dyes may not be used.

16. To overcome problems of contrast, specimens to be viewed under the light microscope should be:
    a. stained or dyed.
    b. treated with heavy metals.
    c. fixed.
    d. sectioned.
    e. dehydrated.

17. For some types of investigations, biologists prefer to use the scanning electron microscope because:
    a. heavy-metal treatments are not required.

b. its resolving power is unsurpassable.
c. it provides a three-dimensional view.
d. living material may be observed.
e. cellular features are never distorted.

18. The technique of producing a reinforced metal replica of the surface features of cells or organelles is known as:
    a. fixation.
    b. shadowing.
    c. heavy-metal depositing.
    d. scanning.
    e. deflection.

19. The optical phenomenon of interference is enhanced in _____ microscopy to observe living cells.
    a. light
    b. scanning electron
    c. transmission electron
    d. phase-contrast
    e. dark-field

20. A disadvantage of transmission electron microscopy is that:
    a. unsectioned tissues may not be viewed, since they are not electron transparent.
    b. surface features of whole organisms cannot be resolved.
    c. preparation ordinarily requires that the specimen be dead.
    d. All of the answers above are correct.

21. You are preparing a tissue for the light microscope that is too thick for the illuminating beam to pass through. Which of these is the correct sequence of specimen preparation?
    a. shadowing, embedding in resin, fixation, staining, sectioning
    b. sectioning, fixation, staining, embedding in resin, shadowing
    c. embedding in resin, sectioning, fixation, staining
    d. embedding in resin, shadowing, sectioning, staining
    e. fixation, embedding in resin, sectioning, staining

22. Name the indicated cellular components of the photosynthetic prokaryote *Anabaena azollae*.

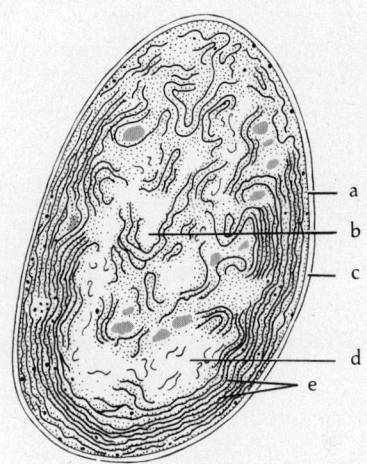

a. _____
b. _____
c. _____
d. _____
e. _____

23. Match each scientist with his achievement.
    Oparin _____
    Miller _____
    Fox _____
    Pasteur _____

    a. proposed that conditions in the primitive atmosphere could have given rise to life
    b. challenged spontaneous generation
    c. demonstrated a likely forerunner of the first living cells
    d. produced amino acids from organic gases

24. Name the indicated cellular components of this cell from a photosynthetic eukaryote (a corn plant).

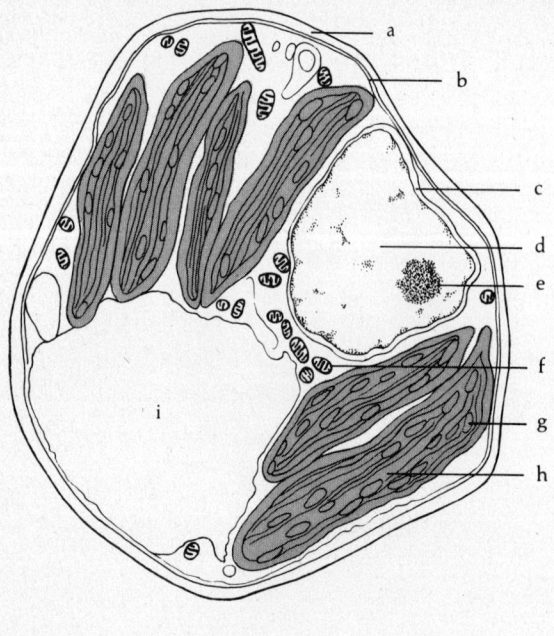

a. _____
b. _____
c. _____
d. _____
e. _____
f. _____
g. _____
h. _____
i. _____

25. Match each type of microscope with an applicable descriptive phrase.

microscope:
scanning EM _____, differential-interference microscope _____, transmission EM _____, light microscope _____

descriptive phrase:
a. uses illuminating energy in the range 0.4 to 0.7 micrometer
b. makes use of secondary emissions
c. detects differences in the diffraction of visible light waves
d. darkened regions reveal where radiation passage through the specimen is blocked

26. What are the basic similarities and differences between prokaryotic and eukaryotic organisms?
27. Compare the transmission electron microscope and the light microscope in terms of resolving power.
28. What was the mode of nutrition apparently used by the first organisms on the primitive earth? What changes took place that favored a second method of acquiring energy and thus a second group of organisms? How did the proliferation of this group produce changes that led to environmental conditions similar to those of today?
29. Match the correct letter from the "clock" with each event to indicate the time period during which the event took place.

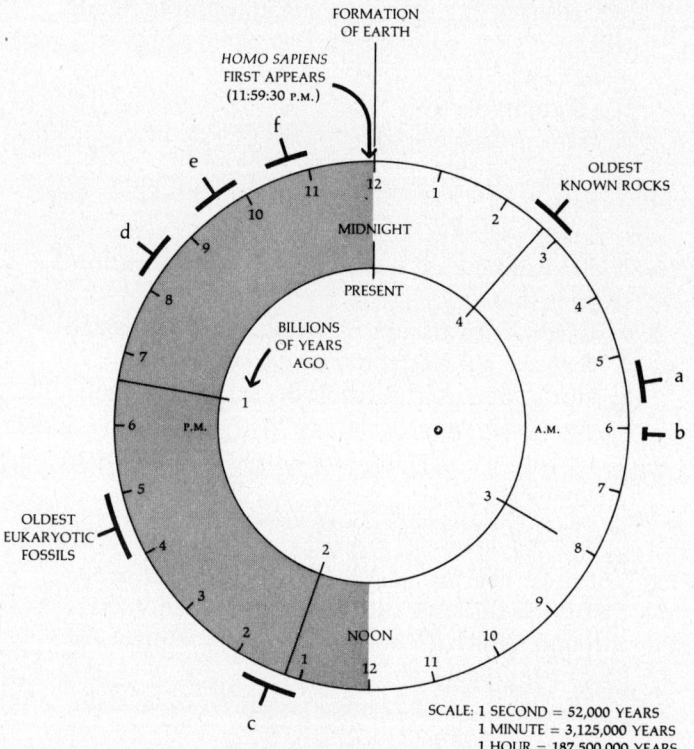

plants invade land _____
oldest fossils (prokaryotes) _____
free oxygen in atmosphere _____
oldest multicellular fossils _____
flowers _____
first photosynthetic organisms _____

34 SECTION 1 THE UNITY OF LIFE

PERFORMANCE ANALYSIS

1. The same basic elements that are found in living things are also found in nonliving things. These elements have the same subatomic organization and, hence, similar properties. It is their organization within larger molecules that constitutes the essential difference between the living and the nonliving (**d**).

2. Pasteur (**a**), in disproving the theory of spontaneous generation, upheld the tenet of the cell theory that states that all cells arise from pre-existing cells. Hooke made preliminary observations that initiated the development of the theory. All of the others tried to answer the question of how living cells might have arisen in the first place.

3. All of these conditions were necessary; therefore, **e** is correct. Choices a and c were required to counter the argument that microorganisms were initially present in the experimental vessels. Since it was believed that all life required oxygen, it was necessary to provide for its entrance into the vessels. The result of the experiment was that microorganisms were present only where condition c was not met.

4. Pasteur's work disclaimed spontaneous generation only under the specific conditions described in choice **b**. The more general contention stated in choice a was not actually addressed.

5. Although there is some controversy over the exact nature of the gases, it is generally believed that the earth's primitive atmosphere contained carbon, hydrogen, and nitrogen (**b**). It is also believed that there was no free oxygen but that there were numerous sources of energy—lightning, volcanic eruption, ultraviolet radiation, and so on. Amino acids, if formed, would have collected in the oceans. Living cells, of course, would have evolved much later.

6. Oparin hypothesized that atmospheric gases contained the elements essential to the origin of life. It has been demonstrated that with sufficient input of energy the first organic molecules would have been formed from these gases and been deposited in oceans and lakes. Choice **c** is therefore correct.

7. Microfossils have been found that date to 3500 million years ago (**b**). The other dates are important as follows:
    a. 4600 million years ago—creation of the earth
    b. 1500 million years ago—the origin of eukaryotes
    c. 750 million years ago—the first appearance of multicellular organisms
    d. 600 million years ago—formerly, the date set for the finding of the first fossil organisms

8. Protenoid microspheres (**e**) have been formed in a laboratory environment similar to the one that scientists believe existed on the primitive earth. These microspheres carry out a few chemical reactions analogous to those of living cells.

9. Because of the distance from the sun, temperatures on the earth are constant enough to ensure the stability of organic compounds. Because of the earth's size, its gravitational field is sufficient to hold a life-supporting atmosphere. Therefore, choice **a** is correct.

10. Chemoautotrophs, like phototrophs, manufacture energy-rich organic molecules using energy obtained from inorganic molecules (**b**). The organic molecules they synthesize are available as an energy source for heterotrophs.

11. The simplest explanation is that heterotrophs would have been favored in the organic "soup" environment, since they would have had little trouble obtaining energy for life processes. The first microfossils found were those of prokaryotelike organisms, making choice **d** the most likely answer.

12. Choices b and d apply only to eukaryotic cells; cell walls (c) are found only among prokaryotes, plants, and fungi. All cells, however, have cell membranes separating the cell from its external environment (**a**).

13. The variety of organelles and the organization of the genetic material provide the clues that the organism is eukaryotic, which eliminates both choices a and e. From the information given, it is impossible to determine if the cell is a whole organism or part of a multicellular organism, eliminating both choices b and d. The fact that a cell wall is absent is consistent with choice **c**, which is the only certain answer.

14. The resolution of light microscopes, including phase-contrast and differential-interference

microscopes, is severely limited in comparison to electron microscopes. The transmission electron microscope has a resolving power (about 0.5 nanometer) that is 20 times greater than that of the scanning electron microscope. Therefore, choice **d** is correct.

15. The resolving power (**b**) of the light microscope is limited by the wavelength of visible light to about 200 nanometers. Magnification can always be increased if one is industrious enough to build a more complex system of lenses. Choices c and d are not true statements.

16. Staining or dyeing (**a**) will provide the differential opacity necessary to improve contrast. Choices b and e are necessary in the use of the electron microscope. Fixation (c) is recommended before treating with stains to preserve the integrity of internal cell structures. Sectioning (d) is required when tissues are too thick for light to penetrate.

17. The only true statement among the choices is **c**. The scanning electron microscope is the preferred instrument for seeing three-dimensional qualities, such as surface features.

18. Shadowing (**b**) involves coating the specimen with metal. The organic material may be removed, leaving a metal replica that can be reinforced with carbon film and viewed under the electron microscope. Heavy-metal treatment (c) is also used with the electron microscope, but it is used to improve contrast, not to show surface features.

19. Interference refers to the differential diffraction of light rays passing through a specimen. Contrast is greatly enhanced, so that features difficult to see using ordinary light microscopy may be detected. Another advantage is that living material is not significantly disturbed. The phenomenon is employed in phase-contrast microscopes (**d**).

20. All three answers are correct (**d**). Thick tissues (a) must be sectioned so that they will be thin enough to pass the electron beam. Surface features of whole organisms (b) are visible in the scanning electron microscope, not in the transmission electron microscope. The preparations referred to (c) are fixation, heavy-metal treatments, staining, and shadowing, all of which require that the sample be dead.

21. Specimens should first be fixed to preserve the organization of cellular components. The material is then embedded in resin, sectioned, and subsequently stained; choice **e** is therefore correct.

22. These structures in order are the: **cell membrane**, **cytoplasm**, **cell wall**, **DNA region**, and the **photosynthetic membranes**. Refer to Figure 4-8.

23. The answers in order are **a**, **d**, **c**, and **b**.

24. These structures in order are the: **cell wall**, **cell membrane**, **nuclear envelope**, **nucleus**, **nucleolus**, **mitochondrion**, **starch grain**, **chloroplast**, and **vacuole**. Refer to Figure 4-11.

25. The answers in order are **b**, **c**, **d**, and **a**.

26. Your answer should include the following similarities: cell membranes, genetic material that directs the cell's activities, and the ability to reproduce. Differences include membrane-bound organelles in eukaryotes, differences in the organization of genetic material, and size (eukaryotic cells are usually larger).

27. Your answer should include the fact that the limiting factor to resolving power is the wavelength of the illuminating beam. In the case of the light microscope, the wavelength of visible light is much longer than the wavelength of the electron beam, and therefore the resolving power is much smaller.

28. It is thought that the first organisms were heterotrophs, which used the energy-rich molecules of the primordial organic soup for food. As this energy source declined, only those cells that could use an alternate source (the phototrophs) could compete. The phototrophs then became a new source of energy-rich organic compounds for the remaining heterotrophs. The phototrophs, of course, released molecular oxygen into the environment.

29. The answers in order are **e**, **a**, **c**, **d**, **f**, and **b**. Refer to Figure 4-10.

CHAPTER 5

How Cells Are Organized

CHAPTER ORGANIZATION

I. Introduction
II. **Cell size and shape**
III. **Subcellular organization**
IV. **The architecture of the cell**
   A. Introductory remarks
   B. The cell membrane
   C. The cell wall
   D. The nucleus
      1. General remarks
      2. The functions of the nucleus
   E. The cytoplasm
      1. General remarks
      2. The cytoskeleton
      3. Vacuoles
V. **Some principal organelles**
   A. Ribosomes
   B. Endoplasmic reticulum
   C. Golgi bodies
   D. Lysosomes
   E. Peroxisomes
   F. Mitochondria
   G. Plastids
VI. **How cells move**
   A. Introductory remarks
   B. Muscle protein
   C. Cilia and flagella
   D. Basal bodies and centrioles
   E. Bacterial flagella
VII. **How cells communicate**
   A. Introductory remarks
   B. Cell-cell junctions
   C. Chemical communication
VIII. **Summary**

MAJOR CONCEPTS

Cells are remarkably similar, yet just as remarkably diverse. They perform a variety of basic functions simultaneously. Cell shape and size are determined by the need for efficient intake and excretion of materials involved in metabolic activity and by the need to control this activity.

All cells are surrounded by a membrane that regulates the passage of materials. The nucleus of a eukaryotic cell is surrounded by a double membrane; the nucleus carries hereditary information and directs cellular activities. The cytoplasm contains ions, enzymes, and other molecules. The cytoskeleton maintains the shape, movements, and overall organization of the cytoplasm.

Eukaryotic cells contain many organelles, most of which are not found in prokaryotes. The organelles include ribosomes and endoplasmic reticulum, which serve as sites of protein synthesis; Golgi bodies, which package and distribute these molecules for cellular export; and lysosomes, vacuoles (or vesicles), and peroxisomes, which contain various enzymes, solutes, or toxic substances.

Membrane-bound organelles include the mitochondria, which carry out the reactions in which energy-yielding organic molecules are oxidized, and plastids, which are found only in photosynthetic organisms. Certain plastids are storage sites for plant products; chloroplasts are the sites of photosynthesis in eukaryotes.

All cells exhibit some form of movement. Internal movement at the cellular level involves microfilaments of actin and myosin, while microtubules of structures such as cilia and flagella are involved with external movement or movement of materials along cell surfaces. Cilia and flagella arise from basal bodies, and centrioles are found only in organisms with cilia and flagella.

Communication between cells is necessary for the coordination of cellular activities that contribute to the functions of tissues, organs, and the organism as a whole. Hormones, nerves, and direct cellular connections are all involved.

HOW TO STUDY THE CHAPTER

Read the chapter, focusing on the major concepts.

Reread the chapter, using the questions that follow to help you focus on details as they relate to major concepts. Answer the questions on a separate sheet of paper. Your answers will provide a valuable study aid in preparing for examinations.

FOCUS ON CHAPTER DETAILS

I. Introduction (page 97)

1. One remarkable fact about cells is their diversity. There are many different kinds of cells, but in what ways are cells similar?

II. Cell size and shape (pages 97-98)

2. How does the relationship between surface area and volume restrict the size of a cell? (See also Figure 5-1.)
3. How does a cell's metabolic requirements determine its size?
4. What factors influence the shapes of cells? Cite several examples.

III. Subcellular organization (pages 98-99)

5. How might the organelles of cells be compared to the organs of multicellular organisms?
6. What types of activities of a "simple" organism, such as *Chlamydomonas*, must be coordinated at the cellular level?

IV. The architecture of the cell (pages 100-107)

A. Introductory remarks

7. What are the two main divisions of living materials that are enclosed within the cellular membrane of eukaryotes?

B. The cell membrane

8. How does the cell membrane appear in the electron microscope?
9. What are the three major components of a cell membrane? How are they arranged? (Consult Figure 5-3.) Are they of the same type and proportions in every cell?
10. In what ways are bacterial cell membranes similar to eukaryotic cell membranes, and in what ways are they different?

C. The cell wall

11. What is the relationship between the cell wall and the cell membrane?
12. Differentiate between the middle lamella and the primary cell wall. Of what is each composed?
13. Differentiate between the primary and secondary cell walls.

D. The nucleus

General remarks

14. Describe the basic structure of the eukaryotic nucleus.
15. What is chromatin?
16. What is the role of the nucleolus?

The functions of the nucleus

17. Explain how the observations of Hertwig led him to believe that the nucleus carries hereditary information.
18. How did Hämmerling's experiment demonstrate that the nucleus exerts a controlling influence over cellular activities?

E. The cytoplasm

General remarks

19. How has our view of the cytoplasm changed over the last several decades?

The cytoskeleton

20. What protein structures make up the cytoskeleton? Describe each structure, and briefly indicate its role.

Vacuoles

21. What is a vacuole? What purposes do large central vacuoles serve for mature plant cells?
22. How do vacuoles differ from vesicles?

V. **Some principal organelles** (pages 108–113)

A. Ribosomes

23. What is the relationship between ribosomes and cellular protein synthesis?
24. How is ribosome distribution influenced by the type of protein that is made by a cell?

B. Endoplasmic reticulum

25. What is the endoplasmic reticulum? Differentiate between smooth and rough endoplasmic reticulum.
26. What governs the amount of endoplasmic reticulum made by a cell?
27. Describe the course that a newly synthesized protein follows when being exported from a cell.
28. What are some of the currently known functions of the smooth endoplasmic reticulum?

C. Golgi bodies

29. Describe the structure and function of Golgi bodies.
30. How many Golgi bodies are usually found in plant and in animal cells?
31. What is the relationship between the Golgi body and the endoplasmic reticulum? (See also Figure 5-16.)

D. Lysosomes

32. What are lysosomes? What are the functions of lysosomal enzymes?
33. What is the relationship between lysosomes and gout and rheumatoid arthritis?
34. Describe the interaction of lysosomes in a white blood cell with bacteria it has engulfed by phagocytosis.

E. Peroxisomes

35. What important detoxification function is carried out by peroxisomes?

F. Mitochondria

36. How is the number and distribution of mitochondria within a cell related to the cell's energy requirements?
37. What is the relationship between the activity level of mitochondria and the number of cristae they contain?

G. Plastids

38. What are plastids? In what types of organisms are they generally found?
39. Summarize the functions of the three kinds of plastids.

VI. **How cells move** (pages 114–117)

A. Introductory remarks

40. Cite a few examples of ways in which cells move.
41. What are the two mechanisms of cellular movement that have been discovered so far?

B. Muscle protein

42. Describe the two fibrous proteins that are associated with muscular contraction.
43. Actin is present in a great variety of cell types. What characteristic do these cells have in common? Cite a few examples of organisms that contain actin.
44. In what form and as part of what structure does actin appear to function?
45. What conclusion have cytologists come to with respect to the evolutionary position of muscle proteins?

C. Cilia and flagella

46. Differentiate between cilia and flagella.
47. Cite two examples in which cilia do not move the cell itself but move some substance or particle across the surface of the cell.

48. List a variety of cell types that contain cilia or flagella. What groups of organisms have no flagella?
49. Describe the structure of cilia and flagella. How is the organization of their various protein elements related to their overall motion?

D. Basal bodies and centrioles
50. What is the function of basal bodies? How do they differ from cilia?
51. What is the structural relationship between basal bodies and centrioles? How do they differ in distribution?
52. In what types of cells are centrioles found?
53. What is the apparent relationship between centrioles and spindles?

E. Bacterial flagella
54. What is the structure of the bacterial flagellum?
55. Differentiate between bacterial and eukaryotic flagella.
56. Describe the way in which bacterial flagella propel a cell.
57. What can you deduce about the purpose of a heterotrophic bacterium's flagella that are lost when the organism has an adequate food supply and appear when the food is used up?

VII. **How cells communicate** (pages 117-120)

A. Introductory remarks
58. If cells are to cooperate in the formation of tissues and organs, they must be able to communicate. Describe one mechanism by which cells communicate over long distances.
59. What specialization do plant cells have that allows them to communicate through the cell wall?

B. Cell–cell junctions
60. What are the functions of the three types of cell junctions in animals? Describe their structures. Where are they most commonly found?

C. Chemical communication
61. List a few examples of ways in which cells communicate with each other chemically.
62. Describe the phenomenon of contact inhibition.
63. Describe intercellular communication in the cellular slime mold *Dictyostelium discoideum*. What generalization can you make from the details of this example?

VIII. **Summary** (pages 121-123): Read the Summary. If you are familiar with the essential features of the material presented there, you are ready to take the following diagnostic examination.

TESTING YOUR UNDERSTANDING

After you have completed the following examination, compare your answers with the annotated key in the Analysis section.

1. Problems relating to diffusion and nuclear control dictate that metabolically active cells:
   a. be small in size.
   b. have a low surface-to-volume ratio.
   c. avoid mitotic division.
   d. be multinucleate.
   e. All of the answers above are correct.

2. Which of the following most adequately describes the organelles of an individual cell?
   a. They are organs similar to those found in multicellular organisms.
   b. They carry out their functions in isolation from other organelles.
   c. They function as interdependent units in a subcellular network.
   d. They are actually too small to be well described.
   e. Both b and d are correct.

3. Which of the following types of molecules have been found to be associated with the cell membrane?
   a. carbohydrates
   b. phospholipids
   c. globular proteins
   d. cholesterol
   e. All of the answers above are correct.

4. Because the growth of a plant takes place largely as a result of cell elongation, plant cells must:
   a. expand in all directions.
   b. add new material to their cell walls.
   c. constantly duplicate their organelles.
   d. construct new secondary walls.
   e. not form a cell wall until the cell matures.

5. A double membrane separates _____ and _____ from the cytoplasm of the eukaryotic cell.
   a. chromosomes; nucleoli
   b. mitochondria; chloroplasts
   c. vacuoles; vesicles
   d. lysosomes; peroxisomes
   e. Both a and b are correct.

6. Which of the following provide a network that connects all structures within the cytoplasm?
   a. intermediate fibers
   b. fibrous proteins
   c. microfilaments
   d. microtrabeculae
   e. microtubules

7. Which of the following is (are) required in the synthesis and transport of proteins that leave the cell?
   a. rough endoplasmic reticulum
   b. Golgi bodies
   c. smooth endoplasmic reticulum
   d. a sequence of hydrophobic amino acids
   e. All of the answers above are correct.

8. Ribosomes that serve as synthetic sites for materials required within a cell are:
   a. suspended in the cytoplasm.
   b. attached to the rough endoplasmic reticulum.
   c. attached to the smooth endoplasmic reticulum.
   d. anchored in the cytoskeleton.
   e. embedded in the cytoplasmic side of the cell membrane.

9. Two organelles that are found predominately in plant cells are:
   a. centrioles and basal bodies.
   b. vacuoles and plastids.
   c. chloroplasts and Golgi bodies.
   d. smooth endoplasmic reticulum and cell membranes.

10. Golgi bodies have been shown to play a major role in processing materials required:
    a. at the cell surface.
    b. in the nucleus.
    c. on the endoplasmic reticulum.
    d. by the mitochondria.
    e. All of the answers above are correct.

11. Before the contents of phagocytic vacuoles can be digested, the vacuoles must first:
    a. empty into the cytoplasm.
    b. be exported through the cell membrane.
    c. be enclosed by a second membrane.
    d. fuse with lysosomes.
    e. Both c and d are correct.

12. A cell that is extremely active metabolically will always contain:
    a. a large number of mitochondria.
    b. more than one nucleus.
    c. almost no endoplasmic reticulum.
    d. numerous cilia and flagella.
    e. All of the answers above are correct.

13. Which description of cilia and flagella applies to eukaryotic cells?
    a. long, whiplike protein filaments that attach to centrioles
    b. a 9 + 2 microtubular arrangement founded on a basal body
    c. a 9 + 0 microtubular arrangement founded on a 9 + 2 basal body
    d. groups of microtubules anchored to a baseplate in the cell membrane

14. Pairs of microtubules arranged in a circle around two central microtubules are characteristic of:
    a. bacterial flagella.
    b. centrioles.
    c. basal bodies.
    d. cilia.
    e. microfilaments.

15. Which of these represents a pair of intercellular connections allowing for the exchange of materials between two cells?
    a. tight junctions and desmotubules
    b. desmosomes and microtubules
    c. tight junctions and desmosomes
    d. nuclear pores and cisternae
    e. plasmodesmata and gap junctions

16. In which of the following tissues would tight junctions be most prevalent?

a. stems of plants
b. linings of internal organs
c. blood vessels
d. glandular tissues
e. muscles

17. Individual amoeboid cells of a slime mold respond to cyclic AMP by:
    a. looking for food.
    b. aggregating into a sluglike mass.
    c. each producing a fruiting body.
    d. dividing to form two new amoebae.

18. Normal human cells multiplying in tissue culture become _____ as a result of contact inhibition.
    a. incapable of further division
    b. spatially separated
    c. ciliated
    d. impermeable to the external environment
    e. cancerous

19. Which of these substances is (are) primarily responsible for movement of the cytoskeleton?
    a. cyclic AMP
    b. actin
    c. flagellin
    d. collagen
    e. hydrolytic enzymes

20. In what form is actin when it participates in either cell motility or internal movement of cellular contents?
    a. microtubules
    b. microfilaments
    c. intermediate fibers
    d. cilia and flagella
    e. microtrabeculae

21. Match each organelle in the list at the left with the appropriate description of its structure.
    Golgi body _____
    mitochondrion _____
    vacuole _____
    endoplasmic reticulum _____
    vesicle _____

    a. a network of interconnecting, flattened sacs, tubes, and channels
    b. a fluid-filled space bounded by a single membrane
    c. a group of flattened sacs stacked on one another
    d. contains an inner membrane with numerous cristae

22. Match each organelle on the left with a metabolic activity that takes place in it or on it.
    lysosome _____
    rough endoplasmic reticulum _____
    mitochondrion _____
    _____
    Golgi body _____
    peroxisome _____
    smooth endoplasmic reticulum _____

    a. degradation of compounds releasing H_2O_2 as a by-product
    b. oxidation of energy-yielding organic molecules
    c. sequestering of hydrolytic enzymes
    d. packaging and distribution of cell products
    e. synthesis of lipids
    f. processing of proteins for export

23. Describe the structure and function of each of the components of the cytoskeleton. Why is the cytoplasm referred to as a dynamic structure?

24. Explain the types of movement that contractile proteins are believed to bring about. What types of locomotion are brought about by microtubules?

25. Explain how Hämmerling was able to demonstrate one of the two crucial roles of the nucleus.

26. Identify the cellular components indicated in the accompanying diagram (on page 43) of a cell.
    a. _____
    b. _____
    c. _____
    d. _____
    e. _____
    f. _____
    g. _____
    h. _____
    i. What are the numerous spherical structures on the rough endoplasmic reticulum, and what is their function? _____

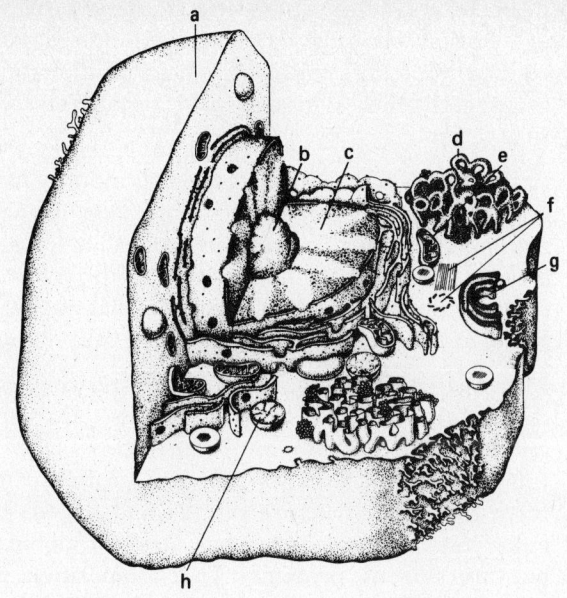

PERFORMANCE ANALYSIS

1. Because diffusion is effective only across short distances, eukaryotic cells are limited to sizes on the order of 10 to 30 micrometers. Small size also increases the amount of surface area, in relation to cell volume, through which substances can pass. Because very large cells are often multinucleate, it is believed that there is a limit on the amount of cytoplasm over which a nucleus can exert its influence. For both these reasons, choice **a** is correct.

2. Although organelles are similar to organs in having specialized structures and functions, one could point out enough differences to exclude choice a. Organelles are not only connected by the internal skeleton of the cell but have been shown to be functionally connected as well (see Figure 5-16), so choice **c** is correct.

3. According to the currently accepted model of the membrane (Figure 5-3), globular proteins (c) are present in the hydrophobic portion of the phospholipid bilayer (b). Cholesterol (d) is a constituent of many cell membranes. Carbohydrate chains (a) are attached to proteins on the outer surface of the membrane. Choice **e** is therefore correct.

4. Plant cell walls resist expansion and must continually add new materials (**b**), such as cellulose, to the walls as they elongate (see Figure 5-4b). Secondary walls are laid down as cells reach maturity and often signal the death of the cell. Organelles are produced as a cell's metabolic demand for them increases.

5. Chromosomes and nucleoli (a) are bounded by the double membrane of the nucleus, and mitochondria and chloroplasts (b) are both organelles that have outer and inner membranes. Choice **e** is therefore correct.

6. Microtrabeculae (**d**) are those parts of the cytoskeleton that connect all the organelles of the cell.

7. Proteins for export are manufactured on the ribosomes attached to rough endoplasmic reticulum (a). Choices b and c both play a role in further processing of these proteins (see Figure 5-16). Hydrophobic amino acids (d) are apparently necessary for movement of the newly synthesized protein through the lipid membrane of the rough endoplasmic reticulum. Choice **e** is therefore correct.

8. Ribosomes that make proteins for export are attached to rough endoplasmic reticulum. Advances in microscopy have shown that other ribosomes do not actually "float" free in the cytoplasm but are clustered at the interstices of the microtrabecular network, a part of the cytoskeleton. Thus, **d** is correct.

9. Plant cells have a large central vacuole. Similar structures in animals are much smaller and are generally referred to as vesicles. Plastids, such as chloroplasts, are found only among plants and their protistan relatives, the algae, so that **b** is correct. Plant cells have neither basal bodies nor centrioles. All cells have a membrane, and many eukaryotic cells have smooth endoplasmic reticulum.

10. Components of cell walls as well as glycoproteins found on the cell surface are known to be assembled and exported to the cell surface by Golgi bodies; therefore, **a** is correct.

11. Phagocytotic vacuoles formed at the cell membrane fuse with lysosomes (**d**), vesicles that contain hydrolytic enzymes.

12. Cells that are metabolically active obtain energy for biochemical processes by means of oxidation reactions carried out on the inner membranes of mitochondria. The number of mitochondria inside a cell is related to its energy requirements, so that **a** is correct. The amount of endoplasmic reticulum within a cell increases as the need for synthesis of various products increases. The cell nucleus remains constant in size during most of the life cycle of a cell. Cilia and flagella are used in locomotory activity.

13. Choice **b** is correct. Refer to Figure 5-24a. Even though centrioles and basal bodies are structurally identical, their location in the cell is different; they cannot therefore be the attachment sites for cilia and flagella (a). Choice d applies only to prokaryotic flagella. The microtubular arrangement in choice c is the reverse of what it should be.

14. Choice **d** is correct. Contrast the structure of the cilium in Figure 5-24a with that of the centriole (identical to a basal body) in Figure 5-25. A bacterial flagellum is composed of globular proteins that form a triple helix with a hollow core.

15. Desmosomes and tight junctions, generally speaking, fill spaces in between cells. Material exchange, however, can occur through cytoplasmic connections, called plasmodesmata, or across channels, called gap junctions, both of which appear in choice **e**. Nuclear pores allow for exchanges between nucleus and cytoplasm.

16. Tight junctions prevent leakage between cells. This would be required in the linings of internal organs, especially in the intestinal lining (**b**) to prevent undigested materials from leaving the digestive tract.

17. In the presence of cyclic AMP, the amoeboid cells aggregate (**b**) into a multicellular mass, and it is this mass that then sends up a single stalk tipped by a fruiting body. This sequence of events has been found to occur at times when food supplies are depleted.

18. The phenomenon of contact inhibition occurs when cells clustering together cease to multiply (**a**). Cancerous cells, however, do not exhibit contact inhibition and continue to multiply and crowd each other. (This disorganization is referred to as a tumor.)

19. Actin (**b**) has been implicated in cell motility in a wide variety of organisms, including plants. Among the other choices, only flagellin (c) is a protein involved in locomotion, but it is a component of prokaryotic flagella and has not been found in the internal structure of the cell.

20. Microfilaments (**b**), such as actin and myosin, consist of threadlike globular proteins that are capable of expansion and contraction. Microtubules are much larger globular proteins and form the basic structure of cilia and flagella of eukaryotes. Intermediate fibers are not globular but are fibrous proteins. The composition of microtrabeculae is unknown.

21. The answers in order are **c**, **d**, **b**, **a**, and **b**.

22. The answers in order are **c**; **f**; **b**; **d**, **f**; **a**; and **e**, **f**.

23. Your answer should include the sizes and functions of microtubules, microfilaments, intermediate fibers, and microtrabeculae.

24. The unifying theme in this question is movement. Actin appears to be present in all cells. Your answer should include a list of the variety of roles that actin plays in various forms of cellular movement. These roles could then be contrasted with the more restrictive roles that are played by microtubules, principally as constituents of flagella and cilia.

25. Review The Functions of the Nucleus (page 103), including Figure 5-9.

26. The indicated structures are: (a) **cell membrane**; (b) **nucleolus**; (c) **nucleus**; (d) **smooth endoplasmic reticulum**; (e) **mitochondrion**; (f) **centrioles**; (g) **Golgi body**; and (h) **lysosome**. The spherical structures on the endoplasmic reticulum are **ribosomes** (i). They function in protein synthesis.

CHAPTER 6

How Things Get into and out of Cells

CHAPTER ORGANIZATION

I. Introduction
II. The structure of the cell membrane
III. The movement of water and solutes
   A. Introduction
   B. Bulk flow
   C. Diffusion
      1. Basic principles
      2. Cells and diffusion
      3. Countercurrent exchange
   D. Osmosis
      1. Basic principles
      2. Osmosis and living organisms
      3. Osmotic potential
      4. Turgor
IV. Transport across cell membranes
   A. Basic principles
   B. The sodium-potassium pump
V. Endocytosis and exocytosis
VI. Summary

MAJOR CONCEPTS

Because of the unique structure of its membrane, a cell is able to regulate its internal environment effectively. According to the fluid-mosaic model, cell membranes are phospholipid bilayers in which globular proteins are suspended. The lipid portions of the membrane dissolve nonpolar molecules. Some of the globular proteins act as carrier molecules in transporting substances, such as glucose and important ions, either along a concentration gradient (facilitated diffusion) or against a gradient (active transport).

One of the most important active-transport systems is the sodium-potassium pump, which maintains sodium ions in low concentration and potassium ions in high concentration in the cytoplasm. Other proteins function as enzymes in the cell membrane.

Water moves by bulk flow and diffusion in response to a potential-energy difference (water potential) and crosses selectively permeable membranes by osmosis. The concentration of solutes in a solution determines its osmotic potential. In the absence of other forces, the net movement of water in osmosis is from a region of lower solute concentration (a hypotonic solution), and therefore a lower osmotic potential, to one of higher solute concentration (a hypertonic solution), and therefore of higher osmotic potential.

The cell membrane may enclose solid food particles (phagocytosis) or dissolved substances (pinocytosis) in a vacuole that then enters the cell (endocytosis). The reverse process (exocytosis) transports waste materials out of the cell. Endocytosis and carrier-assisted transport are similar in that in both processes the cell membrane is able to recognize specific substances required by the cell.

HOW TO STUDY THE CHAPTER

Read the chapter, focusing on the major concepts.

Reread the chapter, using the questions that follow to help you focus on details as they relate to major concepts. Answer the questions on a separate sheet of paper. Your answers will provide a valuable study aid in preparing for examinations.

FOCUS ON CHAPTER DETAILS

I. Introduction (pages 124–125)

1. What important capacity of cells is mediated by cell membranes?
2. What regulatory role do internal membranes play?

II. The structure of the cell membrane (pages 125–126)

3. Briefly describe the basic structure of a cell membrane according to the fluid-mosaic model. What aspect of the membrane is considered to be fluid in nature? To what components of the membrane does the term "mosaic" apply?
4. How might the membranes of different types of cells (or organelles) be different?
5. Compare the orientation of membrane proteins with respect to the outside and inside of the cell.
6. What is believed to be the function of the carbohydrate component of the cell membrane?

III. The movement of water and solutes (pages 126–131)

A. Introduction

7. What do the terms "water potential" and "hydrostatic pressure" mean?
8. Give three examples of the movement of water in response to potential energy.

B. Bulk flow

9. What is meant by the term "bulk flow"? Cite some examples of its usefulness to organisms.

C. Diffusion

Basic principles

10. In terms of the random movement of particles, explain how equal distributions of a solute and a solvent are reached.
11. How are the rate and direction of diffusion related to the strength and direction of the concentration gradient?
12. When is a solution said to be in dynamic equilibrium with its surroundings?

Cells and diffusion

13. What limitations does diffusion have in moving materials within cells? What other process helps to speed up the transport of materials?
14. Explain how steep concentration gradients for the uptake of oxygen (or the release of carbon dioxide) are maintained in a metabolically active cell.

Countercurrent exchange

15. How does the arrangement of blood vessels in the gills of a fish maintain a steep concentration gradient for maximum diffusion rates of oxygen?

D. Osmosis

Basic principles

16. What does the term "selectively permeable" mean when applied to a cell membrane? Define osmosis.
17. Discuss the meaning of the terms "isotonic," "hypertonic," and "hypotonic" with respect to the net movement of water across a membrane.

Osmosis and living organisms

18. How do organisms living in hypertonic and hypotonic environments solve the problems of osmosis?

Osmotic potential

19. Illustrate with a simple diagram what is meant by "osmotic pressure."
20. How would you measure the osmotic potential of a solution? (Consult Figure 6–8.)
21. How is the osmotic potential of a solution related to its water potential? To its solute concentration?

Turgor

22. Explain how turgor pressure is maintained at a constant level within a plant cell.
23. What are the roles of the plant cell wall and the large central vacuole in maintaining turgor?

24. What are the cellular events that lead to wilting?

IV. **Transport across cell membranes** (pages 132–135)
   A. Basic principles
      25. Which biologically important molecules can pass freely through the lipid bilayer of the cell membrane?
      26. What types of pores exist in the cell membrane? Which substances use these pores in their passage through the membrane?
      27. What is a permease? How is it like and how is it unlike an enzyme?
      28. Distinguish between the two methods of carrier-assisted transport.
      29. By what mechanism does glucose normally enter a cell? How was this mechanism deduced?
   B. The sodium-potassium pump
      30. How are Na^+ and K^+ ions distributed across the membranes of cells?
      31. What are the roles of ATP and carrier proteins in maintaining the Na^+ and K^+ gradients across the cellular membrane? Relate configurational changes that are thought to occur in integral proteins to the entry of hydrophilic molecules into the cell. (Refer to Figure 6–12.)

V. **Endocytosis and exocytosis** (pages 135–136)
   32. How do unicellular organisms (or white blood cells) digest solid food particles? What is the role of lysosomes in this process?
   33. Describe the process of pinocytosis.
   34. What role do Golgi bodies play in exocytosis?
   35. How is the membrane that is penetrated during endocytosis thought to be repaired?
   36. How is endocytosis fundamentally similar to carrier-assisted transport?

VI. **Summary** (pages 136–137): Read the Summary. If you are familiar with the essential features of the material presented there, you are ready to take the following diagnostic examination.

TESTING YOUR UNDERSTANDING

After you have completed the following examination, compare your answers with the annotated key in the Analysis section.

1. Which of the following processes requires the expenditure of energy?
   a. osmosis
   b. passive diffusion
   c. facilitated diffusion
   d. active transport

2. Which of the following most adequately describes the fluid-mosaic model of the cell membrane?
   a. a tightly packed layer of hydrophobic protein and lipid molecules
   b. a double lipid layer loosely surrounding globular proteins that extend beyond the membrane's surface
   c. two "slices" of peripheral proteins "sandwiching" a lipid bilayer
   d. a mosaic of different proteins and lipids dissolved in a semipermeable fluid

3. It has been hypothesized that the integral proteins of the cell membrane contain hydrophilic pores. This would explain how:
   a. these proteins function as carrier molecules.
   b. the bulk of glucose enters the cell.
   c. small polar molecules move through the membrane.
   d. dissolved gases are exchanged with the environment.
   e. Both a and c are correct.

4. In which of the following situations does water move from higher to lower potential?
   a. Water is moved up a mountain by means of a mechanical pump.
   b. *Paramecium* expels water using its contractile vacuole.
   c. Pond water enters through the cell membrane of a freshwater amoeba.
   d. Water moves up the stem of a plant by capillary action.

5. The diffusion of water across a selectively permeable membrane in response to a concentration gradient is called:
   a. osmotic pressure.
   b. hydrostatic pressure.

c. facilitated diffusion.
d. osmosis.
e. bulk flow.

6. The net movement of ions into a cell by diffusion occurs:
   a. against a concentration gradient.
   b. down a concentration gradient.
   c. in response to differences in water potential.
   d. as a result of all particles moving in the same direction.
   e. Both b and d are correct.

7. Solutions on either side of a membrane are said to be at dynamic equilibrium when:
   a. all movement of dissolved particles ceases.
   b. a concentration gradient no longer exists.
   c. water moves across the membrane into the solution with the lower solute concentration.
   d. bulk flow balances diffusion.

8. Fluid moves most effectively between two widely separated regions of an organism by means of:
   a. active transport.
   b. facilitated diffusion.
   c. osmosis.
   d. bulk flow.
   e. cytoplasmic streaming.

9. Which of the following substances passively diffuses across the membrane of a cell?
   a. CO_2
   b. O_2
   c. Na^+
   d. glucose
   e. Both a and b are correct.

10. Countercurrent exchange is a mechanism by which organisms:
    a. maintain a constant concentration gradient.
    b. move materials against a gradient.
    c. radically alter the concentration of substances in their environment.
    d. make use of carrier proteins to speed the ordinary rate of diffusion.

11. If a solution is isotonic, it:
    a. will neither gain fluid from nor lose fluid to its surroundings.
    b. has a higher solute concentration than its surroundings.
    c. has the same solute concentration as its surroundings.
    d. is at the same temperature as its surroundings.
    e. Both a and c are correct.

12. The addition of a solute to a fluid-filled compartment, such as a vacuole, will:
    a. increase its water potential.
    b. increase its osmotic potential.
    c. increase its hydrostatic pressure.
    d. make it more hypotonic.
    e. decrease its turgor pressure.

13. The volume gained by a saltwater fish placed into fresh water reflects the movement of water from:
    a. a region of low water potential to a region of high water potential.
    b. a hypotonic solution to a hypertonic solution.
    c. a high solute concentration to a low solute concentration.
    d. a region of high osmotic potential to a region of low osmotic potential.

14. Water is moving upward against the force of gravity through a selectively permeable membrane into a glass column containing a concentrated salt solution. The water potential of the solution is determined by:
    a. the osmotic potential of the solution.
    b. the water potential due to gravitational forces.
    c. the concentration gradient at equilibrium.
    d. Both a and b are correct.
    e. Both a and c are correct.

15. Plant cells develop turgor pressure but animal cells do not because:
    a. a cell wall can resist expansion.
    b. plants live in a hypotonic environment.
    c. plant cell membranes contain cellulose.
    d. vacuoles in plants are larger than those in animals.
    e. plant cells more readily reach equilibrium with the fluid surrounding them.

16. Which of the following statements characterizes protein molecules that transport substances through a cell membrane?
    a. They are highly specific for the substance transported.
    b. Their structures are permanently altered by the process.
    c. They are free to move between intercellular fluid and the cytoplasm.

d. They cause permanent changes in the molecules they transport.
e. Both b and d are correct.

17. The carrier molecule involved in the sodium-potassium pump operates by:
    a. maintaining equal concentration of these two ions on either side of the membrane.
    b. undergoing changes in its configuration.
    c. pumping Na^+ ions into the cell.
    d. pumping K^+ ions out of the cell.
    e. Both c and d are correct.

18. Which of the following is involved in phagocytosis?
    a. utilizing a portion of the cell membrane to form a vacuole
    b. formation of a vacuole from a cell membrane
    c. recognition of a required substance by the membrane
    d. attachment of solid particles to special sites on the membrane
    e. All of the above are correct.

19. Pinocytosis refers to the uptake of _____ molecules by a cell.
    a. membrane-bound
    b. dissolved
    c. solid food
    d. large
    e. partially digested

20. Endocytosis and active transport are similar in that in both processes:
    a. only individual molecules may be moved through the membrane.
    b. the cell membrane is temporarily broken.
    c. the membrane must recognize particular molecules.
    d. substances must move against a concentration gradient.

21. Which of these substances normally enters a cell as a result of facilitated diffusion?
    a. CO_2
    b. water
    c. glucose
    d. solid food particles
    e. Na^+ and K^+ ions

22. What seems to be the most reasonable mechanism by which protein carriers transport lipid-insoluble molecules through the cell membrane?

23. For each molecule listed at the left, find an explanation from the right-hand list for how it might enter or leave a cell.

    oxygen _____   a. lipid soluble
    water _____   b. facilitated diffusion
    Na^+ _____   c. "pumped" uphill
    large soluble   d. moves through
    proteins_____   hydrophilic pores
    glucose _____   e. pinocytotic vesicles
    fatty acids _____
    K^+ _____

24. A small plastic bag weighing 100 grams and containing a 5 percent fructose solution is placed into a water bath maintained at a constant temperature. After one hour, a lab assistant determines that the weight of the solution is 125 grams. After two hours, the solution in the bag still weighs 125 grams. Give a complete explanation for these results.

25. Complete the following diagrams by sketching the equilibrium state for each of the processes in the square provided.

a.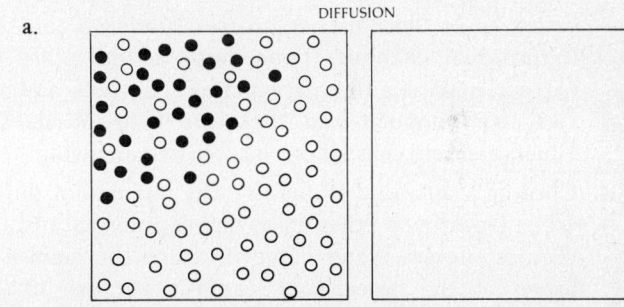

b.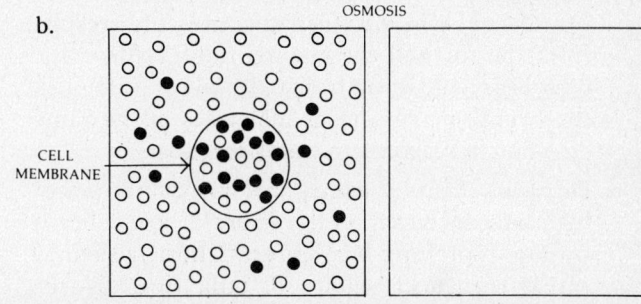

c.

ACTIVE TRANSPORT

PERFORMANCE ANALYSIS

1. Active transport (**d**) requires the energy from ATP in order to move molecules against a concentration gradient. The other three choices are processes that occur *down* a concentration gradient and therefore do not require an input of energy.

2. Choice **b** is correct. Refer to the membrane model of Figure 6–3.

3. Since polar molecules are not soluble in lipids, they would have to cross membranes through pores. It has been hypothesized that integral proteins have a hydrophilic core through which small polar molecules, such as Na^+ and K^+ ions, may be squeezed as these carrier proteins undergo configurational changes. Thus, both a and c are correct, making **e** the correct choice. Oxygen and CO_2 are nonpolar and therefore lipid-soluble. Glucose enters cells by carrier-assisted diffusion.

4. Choices a, b, and d all involve the movement of water from lower potential to higher potential and require energy. Water moves through the membrane of the amoeba by osmosis because its cytoplasm is hypertonic to the pond water (**c**).

5. Choice **d** is correct, although the diffusion of any solvent under the conditions given would be considered osmosis. Osmotic pressure results from the force of the moving fluid. Hydrostatic pressure would result from bulk flow of water, i.e., the overall movement of molecules in the same direction in response to water potential.

6. Diffusion occurs downhill, that is, from regions of high ion concentration to regions of low concentration; therefore, **b** is correct. Choice a defines active transport; choice d, bulk flow. Water potential refers specifically to the movement of water, though dissolved substances may travel along with the water.

7. When dynamic equilibrium is reached, there is no net movement of particles across the membrane, and so a concentration gradient no longer exists (**b**). Random particle movement, however, continues.

8. Differences in concentration gradients account for osmosis and facilitated diffusion only across short distances. Active transport is primarily a means of moving molecules, one at a time, between cells and the fluid surrounding them. Cytoplasmic streaming can account for movement of materials within a cell. Among the choices, only bulk flow (**d**) can move water effectively over great distances.

9. Cells that are expending energy are consuming oxygen and thus maintaining a steep concentration gradient for the diffusion of oxygen into the cell. Energy-expending cells are also producing CO_2, which diffuses down its concentration gradient across the cell membrane. Therefore, **e** is correct. Although glucose is consumed in oxidative reactions, its movement into the cell depends upon carrier-assisted transport (facilitated diffusion). The movement of sodium ions (c) is a function of active transport.

10. Countercurrent exchange is a strategy for maximizing diffusion rates and maintaining a constant concentration gradient (**a**) across membrane surfaces where substances are to be exchanged with the environment (Figure 6–6b).

11. A solution that is isotonic is surrounded by a medium having the same solute concentration (c). Because there exists no osmotic potential, the solution will neither gain nor lose fluid (a). The correct answer is therefore **e**.

12. Increasing the solute concentration of a solution will necessarily increase its osmotic potential, because there will be a greater tendency for water to enter the solution; therefore, **b** is correct.

13. In the absence of other forces, water will not move in the directions indicated in choices a, c, and d. A saltwater fish is hypertonic with respect to fresh water. The fish has a lower water potential than its environment, so water flows from the hypotonic solution to the hypertonic one (**b**).

14. Water will tend to flow up the column until the water potential of the column, because of the downward force of gravity, overcomes the effect of the osmotic potential of the solution in the column (see Figure 6-8). Thus, both these factors are involved; choice **d** is correct.

15. Plant cells exhibit turgor because their cell walls can withstand the constant osmotic pressure (**a**). This pressure is due to the fact that the large central vacuole of a turgid cell makes it so hypertonic that equilibrium is not reached.

16. Like enzymes, protein carriers are highly specific, and their structures, despite changes in configuration, are not permanently altered. Choice d, however, represents a point of departure, since enzymes do cause changes in their substrate molecules and carrier proteins do not. Carrier proteins are embedded in the membrane, and their free movement is restricted. Choice **a**, then, is the only correct answer.

17. Cells that have a sodium-potassium ion pump must maintain high internal K^+ ion concentrations and low internal Na^+ ion concentrations. The carrier protein is believed to undergo configurational changes that allow it to move these ions in opposite directions; **b** is therefore correct.

18. A look at Figure 6-13a should convince you that choices a and b are correct. In addition, active uptake of substances requires that membranes be able to recognize various molecules (c). This is provided for by special receptor sites on the membrane (d). Therefore, choice **e** is correct.

19. Choice **b** is correct, though it may be true that large or partially digested molecules in solution can be enclosed in pinocytotic vesicles. Uptake of solid food particles is called phagocytosis. The solute becomes membrane-bound only after it has contacted the cell membrane, from which the membrane of the vesicle itself is formed.

20. Endocytosis may involve the movement of large numbers of molecules and may occur down concentration gradients. It necessitates breaking the membrane, but active transport does not. One similarity, however, is that both processes depend on the capacity of the membrane to recognize particular molecules (**c**).

21. CO_2 enters a cell by passive diffusion; water enters by osmosis. Both Na^+ and K^+ ion movements depend on an active uptake mechanism. Solid food is engulfed in phagocytotic vesicles. However, glucose (**c**) relies on carrier-assisted diffusion (facilitated diffusion).

22. Configurational changes in protein molecules and the relation of these changes to the transport of substances across membranes are reviewed in Figure 6-12 and in the last paragraph of the sodium-potassium pump discussion on pages 134 and 135.

23. The answers in order are **a**; **d**; **c**, **d**; **e**; **b**; **a**; and **c**, **d**.

24. During the first hour, the solution increases in weight because water enters the bag by osmosis through its semipermeable membrane. Since the solution reaches equilibrium weight relatively quickly, we can conclude that the bag is not permeable to fructose. The weight increase does not continue during the second hour because the hydrostatic pressure inside the bag has reached equilibrium with the osmotic pressure.

25.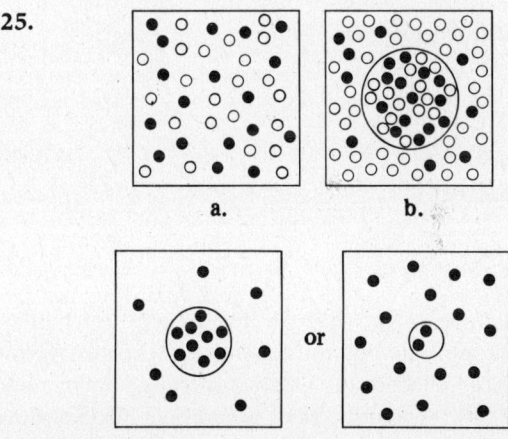

CHAPTER 7

How Cells Divide

CHAPTER ORGANIZATION

I. Introduction
II. Cell division in prokaryotes
III. Cell division in eukaryotes
IV. The cell cycle
V. Mitosis
   A. Introductory remarks
   B. The phases of mitosis
   C. The spindle
   D. The centrioles
VI. Cytokinesis
VII. Summary

MAJOR CONCEPTS

Cell division is the process by which unicellular organisms multiply and multicellular organisms grow and replace injured or worn-out cells. Each new daughter cell receives a set of hereditary information that is identical with that of the parent cell. In addition, each new daughter cell receives half of the cytoplasm of the parent cell.

The differences in prokaryotic and eukaryotic cell division are largely due to the greater amount of genetic material in eukaryotes. In dividing prokaryotes, the chromosome is replicated and each component attaches to the interior of the cell membrane at a separate site. After cell growth and elongation, the cell membrane pinches inward, forming two separate cells, each with its own chromosome.

Dividing cells have a life history, which is known as the cell cycle. During interphase, the cell grows and engages in intense biochemical activity (G_1), replicates its hereditary material (S phase), and prepares (G_2) for nuclear division (mitosis). During mitosis, the duplicated chromosomes separate in such a way that each daughter cell receives an identical copy.

The first stage of mitosis is prophase, during which the dispersed chromatin condenses and individual chromosomes (made up of paired chromatids) become visible. During metaphase, the chromatid pairs move toward the center of the cell, apparently maneuvered by spindle fibers. During anaphase, the sister chromatids separate, and each chromatid, now a separate chromosome, moves to an opposite pole of the cell. During telophase, a nuclear envelope forms around each set of chromosomes, and the chromosomes once again become diffuse chromatin. Cytokinesis, the division of the cytoplasm, is generally accompanied by equal distribution of the organelles and usually follows telophase.

HOW TO STUDY THE CHAPTER

Read the chapter, focusing on the major concepts.

Reread the chapter, using the questions that follow to help you focus on details as they relate to major concepts. Answer the questions on a separate sheet of paper. Your answers will provide a valuable study aid in preparing for examinations.

FOCUS ON CHAPTER DETAILS

I. Introduction (page 139)

1. What benefits do organisms derive from the process of cell division?
2. How must the hereditary information of the nucleus be allocated during cell division so that new cells will resemble the parent cell in structure and function? In what other ways are the daughter cells similar?

II. Cell division in prokaryotes (page 140)

3. How is the hereditary information of a dividing prokaryotic cell distributed to two new cells?

III. Cell division in eukaryotes (page 141)

4. What complicates the problem of equal distribution of hereditary material in eukaryotes?
5. Define mitosis.

IV. The cell cycle (pages 141-143)

6. What factors might account for differences in the pattern of the cell cycle for two different cell types?
7. What processes take place during interphase of the cell cycle that prepare the cell for mitosis?
8. What is the most important event in the S phase?
9. During which of the two gap (G) phases does the cell increase most in size?
10. What types of cells would be expected to repeat their cycles numerous times within their life spans? Give an example of a cell type that generally does not divide once it is mature.

V. Mitosis (pages 143-148)

A. Introductory remarks

11. In what form is the hereditary material when the cell is not dividing?
12. How does the hereditary material appear in the cell after it has condensed?
13. Can you locate the kinetochore within the centromere in the electron micrograph of Figure 7-11b?

B. The phases of mitosis

14. What events mark the beginning of prophase? How do animal and plant cells differ at this point in mitosis?
15. Locate the centrioles, spindle fibers, and aster in the drawing labeled "mid-prophase" on page 144. What separates the chromosomes from the cytoplasm at this stage?
16. What orientation do the various types of microtubules have as the cell is finishing prophase?
17. The outcome of metaphase is a precise distribution of chromosomes. Describe how this happens.
18. The word "anaphase" bears the Greek prefix that means "apart from" or "away from." Why is this an appropriate designation for the third phase of mitosis?
19. Where is the hereditary material located at the end of anaphase?
20. Distinguish a cell in early telophase from one in late telophase.
21. Note that by late telophase two membrane-bound nuclei are surrounded by a single cell membrane. How are each of these nuclei different from the original nucleus of the cell?

C. The spindle

22. How are the microtubules of the spindle oriented during mitosis?
23. In what way is the spindle believed to change its shape as mitosis proceeds? Is this change related to the overall shape of the cell at this time?

D. The centrioles

24. What is the supposed relationship between basal bodies and centrioles? How do mitotic events in *Chlamydomonas* provide evidence for this relationship?
25. What evidence is there that centrioles do not play a role in the formation of the spindle? Offer an alternative explanation for the apparent movement of the centrioles by the spindle.

VI. Cytokinesis (pages 148–149)

26. Define cytokinesis.
27. What determines the plane in which cleavage of the cell will occur?
28. What are the differences in cytokinesis as it occurs in plants and in animals?
29. What is the role of microfilaments in animal cytokinesis?
30. How are Golgi bodies involved in the formation of the cell plate during plant cell division?
31. Discuss the changes in the cell membrane that occur during animal cytokinesis, and contrast these with the membrane changes that occur in plant cells.
32. Compare the two daughter cells that result from cytokinesis. Will they truly be alike in every way if cellular operations have not gone awry?

VII. Summary (page 150): Read the Summary. If you are familiar with the essential features of the material presented there, you are ready to take the following diagnostic examination.

TESTING YOUR UNDERSTANDING

After you have completed the following examination, compare your answers with the annotated key in the Analysis section.

1. Cell division in prokaryotes differs from that in eukaryotes in that:
   a. the chromosomes of prokaryotic cells are attached to and move with the cell membrane.
   b. the nuclear membrane of prokaryotic cells is not broken down.
   c. centrioles are absolutely required for spindle formation in prokaryotes.
   d. DNA replication occurs during cell division in prokaryotes.
   e. Both b and d are correct.

2. Following mitosis, the daughter cells all have _____ of the parent cell.
   a. approximately one-half the number of organelles
   b. approximately one-half the amount of cytoplasm
   c. an exact replica of the genetic material
   d. Both a and b are correct.
   e. All three answers (a, b, and c) are correct.

3. Chromosome separation is necessarily a more complex process in eukaryotes than in prokaryotes because in eukaryotes:
   a. there is much more DNA to be distributed to daughter cells.
   b. there are many large, complex organelles.
   c. two copies of a chromosome are present at prophase.
   d. Both b and c are correct.

4. An increase in the amount of cytoplasm and in the number of organelles occurs during _____ of the cell cycle.
   a. the S phase
   b. telophase
   c. the G_1 phase
   d. the G_2 phase
   e. early prophase

5. The genetic material within the nucleus of a eukaryotic cell is replicated during:
   a. prophase.
   b. metaphase.
   c. cytokinesis.
   d. the S phase.
   e. the G_1 and G_2 phases.

6. An exact copy of every chromosome is present within the nuclear envelope during:
   a. the S phase only.
   b. the entire sequence of mitosis.
   c. the G_2 phase and part of prophase.
   d. the G_1 and S phases only.
   e. prophase and metaphase only.

7. At the beginning of prophase, the two pairs of centrioles are:
   a. at right angles to each other.
   b. to one side of the nucleus.
   c. located within the nucleus.
   d. not yet present within the cell.
   e. at opposite poles of the cell.

8. Microtubules that form the spindle are called:
   a. pole-to-pole fibers.
   b. pole-to-kinetochore fibers.
   c. astral rays.
   d. Both a and b are correct.
   e. All three answers (a, b, and c) are correct.

9. A centromere represents:
   a. the geometric center of a chromosome.
   b. the point where two chromatids are attached.
   c. the region in which spindle microtubules attach to chromatids.
   d. a chromosome during anaphase.
   e. Both b and c are correct.

10. A cell that has entered its S phase will probably:
    a. die within the next cycle.
    b. undergo mitosis.
    c. become cancerous.
    d. fail to undergo DNA replication.
    e. remain in that phase indefinitely.

11. Contact inhibition arrests cell growth in:
    a. mitosis.
    b. the G_1 phase.
    c. the S phase.
    d. the G_2 phase.
    e. tissue cultures only.

12. Which of the following accurately reflects the role of centrioles in the formation of the mitotic spindle?
    a. Centrioles are required for spindle formation in all organisms.
    b. Centrioles are required for spindle formation in all organisms except plants.
    c. Centrioles may be present at mitosis but apparently are not required for spindle formation.
    d. Centrioles direct the orientation of the microtubules forming the spindle.
    e. Both b and d are correct.

13. In examining a rapidly dividing tissue under the microscope, one would find the fewest cells in:
    a. prophase.
    b. telophase.
    c. metaphase.
    d. interphase.
    e. anaphase.

14. Anaphase of mitosis marks the:
    a. replication of DNA.
    b. formation of the mitotic spindle.
    c. division of the cytoplasm.
    d. separation of identical chromatids.
    e. breakdown of the nuclear membranes.

15. The change in size of the microtubules of the spindle during mitosis is believed to be a result of:
    a. their expansion and contraction.
    b. the pulling action of the centrioles.
    c. the removal or addition of microtubular material.
    d. cytoplasmic changes during cytokinesis.

16. The biochemical mechanism of cytokinesis in animals is thought to involve:
    a. microtubules of the mitotic spindle.
    b. the action of actin microfilaments.
    c. formation of the cell plate.
    d. synthesis of an entirely new membrane around each new cell.
    e. Both a and b are correct.

17. During cytokinesis, membrane constriction occurs:
    a. by chance, along any dimension.
    b. along an equator established by the spindle.
    c. along the longitudinal axis of the cell.
    d. only after the daughter cells have entered interphase.

18. When cytokinesis involves the formation of a cell plate, the space occupied by the plate becomes a:
    a. cell wall.
    b. cell membrane.
    c. vacuole.
    d. middle lamella.
    e. gap junction.

19. During cytoplasmic division, a cell plate appears in _____ cells after fusion of _____ .
    a. animal; membrane fragments
    b. plant; a series of vesicles
    c. prokaryotic and plant; cell walls of adjacent cells
    d. all; lysosomes with phagocytotic vacuoles
    e. all eukaryotic; the nuclear membrane with the cell membrane

(This chapter continues on page 56)

20. Rank the entries listed at the left in the order of their occurrence, and match each with the mitotic phase during which the event takes place.

The nuclear envelope breaks down. _____
The spindle is completely formed. _____
Chromatin begins to condense. _____
Pole-to-kinetochore microtubules shorten. _____
Chromosomes become diffuse. _____
Chromatid pairs become aligned. _____

a. metaphase
b. prophase
c. telophase
d. anaphase

21. Discuss the similarities and the differences exhibited by plants and animals during cell division.

22. What purpose does cell division serve in unicellular organisms? In multicellular organisms?

23. Identify the mitotic phase represented by each of the drawings below and briefly describe what is happening.

a. _____

b. _____

c. _____

d. _____

PERFORMANCE ANALYSIS

1. Compare Figure 7–3b to the plant cell in Figure 7–15 to verify for yourself that choice **a** is correct.
2. Mitotic division assures that each new cell receives an exact replica of genetic material (c). During cytokinesis, each cell receives approximately half of the parent cell's cytoplasm (b) and organelles (a); therefore, **e** is correct.

3. There are usually only two copies of a chromosome present at the start of chromosome separation in both groups, but the complexity of nuclear division in eukaryotes is necessitated by the larger quantity of genetic material present (**a**). Choice b is really not an issue, since organelles are distributed to daughter cells only after mitosis is complete.

4. Cells double in size and attain the full complement of organelles following cytokinesis, during that part of interphase known as the G_1 phase (**c**).

5. Choice **d** is correct. Refer to Figure 7-4.

6. Since replication occurs during the S phase and chromatids do not separate until anaphase, chromatids are found together inside the nuclear membrane throughout the G_2 phase and up until the time the nuclear envelope is broken down (sometime during prophase of mitosis); therefore, **c** is correct.

7. As mitosis begins, two pairs of centrioles are present (in the organisms that have them) at one side of the nucleus (**b**). Only centrioles within a pair lie at right angles to one another; the two pairs themselves need not be at right angles to each other. (See the diagram on page 143.) The pairs do not move to opposite poles until the spindle begins to form.

8. Choice **e** is correct. The spindle includes all three types of microtubules, as can be seen in Figure 7-11a and b.

9. As Figure 7-6 shows, the centromere is a constricted area of attachment between two chromatids (**b**). Note in Figure 7-11b that microtubules of the spindle become attached in this region during mitosis (**c**). The centromere need not be located at the center of the chromosome. Choice **e** is therefore correct.

10. Since the events of the S phase are preparatory to mitosis, choice **b** is correct.

11. Contact inhibition is not merely a phenomenon observed in the laboratory but occurs in living tissue as well, arresting cell growth and multiplication in the G_1 phase (**b**).

12. The exact function of the centrioles in mitosis is not known, but because they are not found in dividing plant cells and can be removed from animal cells without interrupting mitosis, they are not thought to be necessary for spindle formation (**c**).

13. Since anaphase (**e**) is the most rapid mitotic phase, there are likely to be few cells in this phase. Since cells spend most of their lives in interphase, a good number of cells, even in rapidly dividing tissue, will not be dividing.

14. Anaphase begins when the centromeres of a chromatid pair are pulled in opposite directions; therefore, **d** is correct.

15. Pole-to-kinetochore microtubules lengthen and shorten during mitosis, but since they do not get thinner or thicker, they are not believed to expand or contract. It appears more likely that material is added to them from the cytoskeleton of the cell or removed from them and returned to the cytoskeleton, so that **c** is correct. There is no evidence that centrioles influence the length of microtubules, nor would changes taking place during cytokinesis have any bearing on the movement of the spindle.

16. The mitotic spindle has already dispersed by the time of cytokinesis. Cell plates are found not in animals but in plants. The membrane furrows (see Figure 7-13) are believed to be a result of actin microfilaments (**b**) working like a purse string to draw in the membrane. The membrane breaks under the strain, and individual membranes form around the two daughter cells as ends of each membrane fragment fuse.

17. Cytokinesis may begin before telophase is complete, with constriction of the membrane always occurring in the cross section of the cell that contained the equator of the mitotic spindle (**b**).

18. The cell plate is a membrane-bound space that soon becomes impregnated with pectins to form the middle lamella (**d**). On either side of the lamella, new cell walls are constructed.

19. Cell plates are formed only in plants. (There is little reason for prokaryotes to have such a structure, since they are all single-celled.) The plate is formed as vesicles provided by nearby Golgi bodies fuse together to form a membrane-bound space; therefore **b** is correct.

20. The correct sequence of events is **2, 3, 1, 5, 6,** and **4**. The mitotic phases in which these events occur are **b, b, b, d, c,** and **a**.

21. Your answer should focus on the appearance of each kind of cell during prophase and the striking difference in the manner in which division of the cytoplasm takes place.
22. For unicellular organisms, cell division represents a means of reproduction. Multicellular organisms make use of cell division for growth and replacement of tissues.
23. Refer to the figures and text of pages 143–145 for the stages and events of mitosis.

SECTION 2 **Energetics**

CHAPTER 8

The Flow of Energy

CHAPTER ORGANIZATION

I. Introduction
II. The laws of thermodynamics
   A. Introductory remarks
   B. The first law
   C. Essay: $E = mc^2$
   D. The second law
   E. Living systems and the second law
III. Oxidation-reduction
IV. Metabolism
V. Enzymes
   A. General background
   B. Enzyme structure and function
      1. Introduction
      2. Essay: The lingering death of vitalism
      3. The induced-fit hypothesis
   C. Cofactors in enzyme action
      1. Introductory remarks
      2. Ions as cofactors
      3. Coenzymes and vitamins
   D. Effects of temperature and pH
   E. Enzymatic pathways
   F. Essay: Auxotrophs
   G. Regulation of enzyme activity
      1. Introductory remarks
      2. Allosteric interactions
      3. Essay: Some like it cold
      4. Competitive inhibition
      5. Noncompetitive inhibition
      6. Irreversible inhibition
      7. Regulation at the source
VI. The cell's energy currency: ATP
   A. General background
   B. ATP in action
VII. Summary

MAJOR CONCEPTS

Thermonuclear reactions in the sun are the ultimate source of the energy that drives living systems. In order to carry out the functions of maintenance, growth, and reproduction, living systems convert this energy into other forms. At each step, some useful energy is lost to the surroundings.

The laws of thermodynamics govern transformations of energy. The first law states that energy can be converted from one form to another but cannot be created or destroyed. The potential energy of the initial state (or reactants) is equal to the potential energy of the final state (or products) plus the energy released in the process or reaction.

The second law of thermodynamics states that the potential energy of the products of a reaction will always be less than the potential energy of the reactants. The difference between these two states is called the free energy change (ΔG), which is equal to the change in the heat (ΔH) minus the change in entropy (ΔS) multiplied by the absolute temperature (T): $\Delta G = \Delta H - T\Delta S$. Another way of stating the second law is: All natural processes tend to proceed in a direction that results in an increase in the randomness (entropy) of the system.

Metabolism refers to all of the chemical reactions that occur in a cell. Those reactions that result in the breakdown of molecules are catabolic; they supply the energy for biosynthetic (anabolic) reactions. The ATP molecule is the principal energy carrier in living systems. All chemical reactions in cells take place in steps controlled by specific protein catalysts called enzymes, which enable a cell to use molecules and energy efficiently.

Some enzymes require metal ions as cofactors. Others require coenzymes (small organic molecules that usually function as electron carriers); many vitamins are parts of coenzymes. Enzyme-catalyzed reactions are controlled primarily by allosteric interactions, which occur when a molecule other than the substrate combines with the enzyme at a site other than the active site and renders the enzyme functional or nonfunctional. Feedback inhibition occurs when a product molecule (allosteric effector) combines with an enzyme catalyzing an earlier step in the pathway and temporarily inhibits that enzyme.

Other forms of enzyme control involve the competition of the normal substrate with another, similar molecule (competitive inhibitor) and noncompetitive inhibition, in which the inhibitor molecules bind to the enzyme at a site other than the active site. Other enzyme controls include irreversible inhibitors and regulation of enzyme production. Temperature and pH also affect enzyme activity.

HOW TO STUDY THE CHAPTER

Read the chapter, focusing on the major concepts.

Reread the chapter, using the questions that follow to help you focus on details as they relate to major concepts. Answer the questions on a separate sheet of paper. Your answers will provide a valuable study aid in preparing for examinations.

FOCUS ON CHAPTER DETAILS

I. **Introduction** (page 157)
  1. Describe what happens to that portion of the sun's energy that strikes our planet.
  2. Explain the importance of the transformation of light energy by photosynthetic organisms into chemical energy.

II. **The laws of thermodynamics** (pages 157–164)
  A. Introductory remarks
    3. Define thermodynamics.
  B. The first law
    4. State the first law. Cite some examples of the types of energy conversions that are commonplace in the living world.
    5. Why can a source of potential energy never be completely converted to useful energy?
    6. State the first law in terms of its effects on the equation of a chemical reaction.
  C. Essay: $E = mc^2$ (page 159)
    7. What does Einstein's equation have to do with the very slight nonadditivity of the atomic numbers of the elements?
  D. The second law
    8. State the second law.
    9. According to the second law, what types of reactions can take place spontaneously? What types of reactions require an input of energy before they can proceed?
    10. The spontaneous decomposition of dinitrogen pentoxide is endothermic. Is this in contradiction to the second law? Explain your answer.
    11. Define entropy.
    12. In your own words, paraphrase the mathematical expression for the change in free energy, ΔG.
    13. How does the value of ΔG relate to a process being either exergonic or endergonic?
    14. Identify each term of the energy equation for the combustion of glucose.
    15. In terms of the concept of entropy, how might the second law be restated?
  E. Living systems and the second law
    16. Life is a natural process and therefore subject to the second law of thermodynamics; yet living things are continually increasing the orderliness of the molecules within them. How can these facts be resolved with the second law (as it was restated on page 163)?

III. **Oxidation-reduction** (pages 164–165)
    17. Define oxidation and reduction.

18. What changes involving electrons (or hydrogen atoms) occur during the course of a typical oxidation-reduction reaction?
19. Explain why the oxidation reactions of glucose and photosynthesis are considered to be redox reactions.
20. How are living cells able to harness the energy released in an exergonic reaction before it is all dissipated as heat?

IV. **Metabolism** (pages 165-166)
21. What is metabolism?
22. What are some principles of cellular metabolism?
23. What is surprising about the metabolism of extremely dissimilar organisms?
24. Differentiate between anabolism and catabolism. What use is made of catabolic reactions?
25. What accounts for the fact that the many different reactions taking place within a cell do not interfere with one another?
26. What factors limit the ability of molecules to engage in chemical reactions within living cells?

V. **Enzymes** (pages 167-181)

A. General background
27. Why do chemical reactions require an "energy of activation"?
28. What is the purpose of a catalyst, and how does it function?
29. What are biological molecules that serve as catalysts called?
30. What is a substrate?
31. What determines whether or not a cell can carry out a given function?

B. Enzyme structure and function

Introduction
32. Describe a typical enzyme and its active site. See Figure 8-8.
33. What is the function of the active site?

Essay: The lingering death of vitalism (page 172)
34. What was the basic difference in the opinions of the reductionists and the vitalists? How did vitalism finally die?

The induced-fit hypothesis
35. What is the relationship between an enzyme and its substrate, as proposed by the induced-fit hypothesis? (See Figure 8-11.)

C. Cofactors in enzyme action

Introductory remarks
36. What is a cofactor?

Ions as cofactors
37. Cite some examples of how ions function as cofactors.

Coenzymes and vitamins
38. Take note of the oxidized and reduced forms of NAD in Figure 8-12. How does the difference in the two forms relate to their roles as coenzymes?
39. What is a vitamin? In what way are vitamins essential to efficient enzyme activity?
40. Why does a small amount of NAD^+ go such a long way?

D. Effects of temperature and pH
41. Interpret Figure 8-13 in terms of the kinetic energy of substrates and the structural integrity of enzymes.
42. What is meant by the "denaturation" of an enzyme?
43. What is the effect of a pH change on enzyme structure?
44. On what does the optimum pH of an enzyme depend?

E. Enzymatic pathways
45. Chemical reactions take place in series or pathways. What are the advantages of this to a cell?
46. Under what conditions is equilibrium reached in a reaction?
47. What do equilibrium concentrations of reactants and products reveal about the ΔG

for the reaction? About their potential energies?

48. Explain how the pathway concept and a negative free energy change work together to ensure metabolic efficiency in a cell.

F. Essay: Auxotrophs (page 176)

49. What is an auxotroph? How are auxotrophic microorganisms useful in the elucidation of enzymatic pathways?

G. Regulation of enzyme activity

Introductory remarks

50. Name four ways in which enzyme activity may be controlled.

Allosteric interactions

51. What is the relationship between an enzyme and its allosteric effector?

52. Cite an example of an allosteric interaction that causes feedback inhibition.

Essay: Some like it cold (page 179)

53. What is the molecular basis for black and white Siamese cats and Himalayan rabbits, as well as for white Arctic foxes and baby seals?

Competitive inhibition

54. What is competitive inhibition? What factors determine the effect of an inhibitor on enzyme activity?

55. Give an example of how a drug might work as a competitive inhibitor.

Noncompetitive inhibition

56. Why is it thought that lead poisoning results from noncompetitive inhibition of an enzyme? How is lead poisoning best treated?

Irreversible inhibition

57. In what two ways do irreversible inhibitors operate? Name two such inhibitors.

Regulation at the source

58. What is characteristic of those enzymes that are produced only when needed?

VI. **The cell's energy currency: ATP** (pages 181–184)

A. General background

59. Describe the structure of the ATP molecule.

60. What is the role of coupled reactions in initiating endergonic reactions?

61. What reaction is most frequently coupled to an endergonic series?

62. What is the source of the energy provided, and how is it released?

63. Why is ADP at a lower potential energy than ATP?

B. ATP in action

64. What is the function of an ATPase?

65. What is the purpose of the phosphorylation of a reactant using a phosphate group from ATP?

66. Give an example of an anabolic reaction that involves phosphorylation.

67. Why is the ATP/ADP system referred to as an energy shuttle?

VII. **Summary** (pages 184-186): Read the Summary. If you are familiar with the essential features of the material presented there, you are ready to take the following diagnostic examination.

TESTING YOUR UNDERSTANDING

After you have completed the following examination, compare your answers with the annotated key in the Analysis section.

1. The implication of the first law of thermodynamics for the following chemical reaction is that:

   $$A + B \rightarrow C + D \qquad \Delta H = -673 \text{ kcal/mole}$$

   a. the potential energy of A and B equals the potential energy of C and D.
   b. less chemical energy is available in C and D than in A and B.
   c. C and D are at a higher potential energy level than A and B.
   d. the total energy of the system has decreased.
   e. some heat is released, as in any chemical reaction.

2. The second law of thermodynamics predicts that:

a. reactions cannot go downhill without some input of energy.
   b. if a reaction is spontaneous, then it will release energy.
   c. within a closed system, entropy will always decrease.
   d. any exergonic reaction will occur without the addition of energy.
   e. Both b and d are correct.

3. The oxidation of glucose is a biologically useful reaction because it is a(an) _____ reaction.
   a. exergonic
   b. endergonic
   c. endothermic
   d. spontaneous
   e. Both b and d are correct.

4. The melting of ice and the boiling of water take place spontaneously under appropriate conditions because these reactions:
   a. have a positive ΔG.
   b. have a positive ΔH.
   c. bring about a large increase in entropy.
   d. are exothermic.
   e. increase the orderliness of a system.

5. Which of the following best summarizes what happens when glucose is completely oxidized?
   a. Hydrogen atoms are transferred from glucose to oxygen, forming water and CO_2.
   b. Electrons are transferred from glucose to water, and CO_2 is formed.
   c. Protons are removed from glucose and transferred to oxygen to form water.
   d. Hydrogen atoms and electrons are exchanged by glucose and oxygen, forming water.

6. During the process of photosynthesis, _____ is oxidized, while _____ is reduced.
   a. glucose; oxygen
   b. carbon dioxide; water
   c. water; carbon dioxide
   d. water; glucose
   e. glucose; water

7. Chemical reactions taking place within living organisms:
   a. require highly specific enzymes.
   b. are carried out through biochemical pathways.
   c. occur in specialized areas of the cell.
   d. represent either catabolism or anabolism.
   e. All of the above are correct.

8. The energy of activation of biochemical reactions is:
   a. achieved by slight increases in temperature.
   b. lowered by the presence of a catalyst.
   c. low enough so that most reactions occur spontaneously.
   d. supplied by the potential energy of colliding substrate molecules.

9. The shape of an enzyme's active site is complementary to that of:
   a. its substrate.
   b. the product formed in the reaction.
   c. a quaternary protein structure.
   d. one of the enzyme's polypeptide chains.
   e. the enzyme's linear sequence of amino acids.

10. The induced-fit hypothesis regarding the relationship between enzyme and substrate states that:
    a. a substrate may cause a change in an enzyme's conformation.
    b. substrate molecules undergo a strain while bound to the active site.
    c. substrate and active site fit together like a lock and key.
    d. Both a and b are correct.
    e. Both a and c are correct.

11. A nonprotein organic molecule that serves as an electron acceptor in a biochemical reaction is classified as:
    a. an enzyme.
    b. a nucleotide.
    c. a coenzyme.
    d. a cofactor.
    e. a vitamin.

12. The rates of most enzyme-catalyzed reactions drop off quickly in the vicinity of 40°C because at this temperature:
    a. the enzyme becomes denatured.
    b. substrate molecules possess insufficient energy to react.
    c. active sites are saturated.
    d. cofactors become inactivated.

13. Because biochemical reactions occur as a series of steps in an enzymatic pathway:
    a. important intermediate products accumulate rapidly.

b. chemical equilibrium is quickly achieved.
   c. individual reactions may be pulled forward by those succeeding them.
   d. the reactions may go in either direction.
   e. Both a and b are correct.

14. The direction and relative length of the arrows in the following reaction indicate that:

    $$A + B \rightleftharpoons C + D$$

    a. equilibrium will not be reached.
    b. it will take only a short time for the reaction to go to completion.
    c. ΔG of the reaction is negative in value.
    d. at equilibrium the forward rate will be greater than the reverse rate.

15. Allosteric regulation of enzyme activity is said to occur when:
    a. the active site is occupied by an inhibitory substance.
    b. a second substrate competes with the usual substrate.
    c. an effector binds at a site other than the active site.
    d. the enzyme is continually being degraded.

16. Competitive inhibition is most effective when:
    a. the inhibitor does not closely resemble the normal substrate.
    b. the inhibitor is able to denature the enzyme quickly.
    c. the normal substrate is at a relatively low concentration.
    d. the effect of the inhibitor is reversible.

17. In a molecule of ATP, _____ groups are linked together and one of them is attached to a _____ .
    a. adenine; phosphate group
    b. adenine; five-carbon sugar
    c. two phosphate; nitrogenous base
    d. three phosphate; five-carbon sugar
    e. three phosphate; nitrogenous base

18. The term "coupled reactions" refers to the fact that metabolic processes often require the linking together of:
    a. individual reactions in a complex pathway.
    b. exergonic and endergonic reactions.
    c. two or more reaction steps catalyzed by the same enzyme.
    d. the phosphate groups of ATP.

19. The removal of a phosphate group from ATP is an example of a(an) _____ reaction.
    a. hydrolysis
    b. condensation
    c. phosphorylation
    d. endergonic
    e. uncatalyzed

20. Phosphorylation of a reactant molecule is often necessary in order to:
    a. decrease its potential energy.
    b. transfer energy to it from ATP.
    c. regenerate ATP from ADP.
    d. release energy for the reaction.
    e. Both a and b are correct.

21. Match each concept at the left with a descriptive phrase at the right.

    entropy _____
    ΔH _____
    free energy change _____
    oxidation _____
    anabolism _____
    energy of activation _____

    a. total of biosynthetic reactions
    b. loss of an electron or H atom
    c. input required to initiate a chemical reaction
    d. the degree of randomness of a system
    e. total increase or decrease in energy during a reaction
    f. change in heat content

22. Label the three arrows indicated in the following energy diagram for an exergonic chemical reaction.

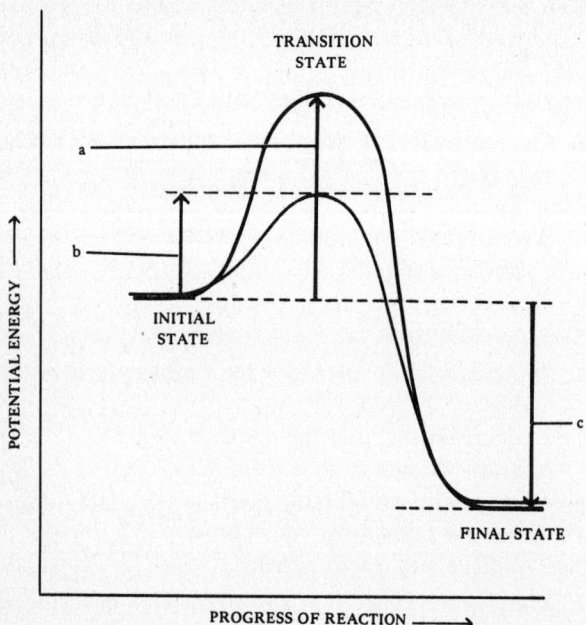

23. Does the orderliness created by living systems defy the second law of thermodynamics? Explain.
24. Explain why uncatalyzed chemical reactions would occur too slowly to be of any benefit to an organism.
25. What advantage does a cell derive from carrying out its metabolic reactions in a biochemical pathway?

PERFORMANCE ANALYSIS

1. Because the first law of thermodynamics states that energy can neither be created nor destroyed, the heat energy produced in the course of the reaction must be balanced by an equivalent loss in chemical energy. This means that there is less chemical energy in the products than in the reactants, which is what choice **b** says.
2. The second law states that if no energy is put into or leaves a system, then the potential energy of the final state in a reaction will be less than the potential energy of the initial state. Such a process must necessarily yield energy, i.e., it is exergonic. And since spontaneous reactions are those that proceed without an energy input (other than activating energy), it follows that a spontaneous reaction will release energy without any energy input. Therefore, choice **e** is correct.
3. The oxidation of glucose is exergonic, yielding 673 kilocalories per mole. Some of this can be used by an organism to perform work, so that **a** is correct.
4. Spontaneous reactions, by the second law, have a negative ΔG. According to the Gibbs equation, $\Delta G = \Delta H - T\Delta S$, this results from a large decrease in heat content (large negative ΔH) and/or a large increase in entropy (large positive ΔS). Since ice absorbs heat as it melts, the heat content of the system is increased (positive ΔH). But molecules of water in the liquid state are much more disordered than they are as ice, so that there is a large increase in the system's entropy that offsets the positive ΔH. This makes ΔG negative, so that the melting of ice, under appropriate conditions, namely, temperature = $0°C$ and pressure = 1 atmosphere, will proceed spontaneously; therefore, **c** is correct.
5. Oxidation always involves a loss of an electron or a hydrogen atom (a proton plus an electron). When glucose is oxidized, its hydrogen atoms are removed and transferred to oxygen. Thus, water is formed and the carbon of the glucose molecule remains as CO_2 (**a**).
6. Oxidation is a loss of electrons (or hydrogen atoms), and reduction is a gain. During photosynthesis (see page 164), water loses hydrogen atoms (is oxidized), and CO_2 gains hydrogen atoms (is reduced). Choice **c** is therefore correct.
7. Cells carry out chemical reactions using specific enzymes (a) that are located in a biochemical series (b) along specialized membranes or in membrane-bound spaces (c). These reactions may be either anabolic or catabolic (d). Choice **e** is therefore correct.
8. At temperatures that can be tolerated by living cells, the kinetic energy of the reactant molecules would be insufficient to provide the energy of activation. The required energy of activation is instead lowered by a catalyst (an enzyme) (**b**) that

orients the reacting molecules in such a way that their energies are sufficient to cause a reaction.

9. Enzymes are specific for certain substrate molecules because the geometry of the active site allows only certain molecules to fit; therefore, **a** is correct.

10. The lock-and-key hypothesis has been preempted by the induced-fit hypothesis. The conformation of an enzyme is believed to be altered (a) by the binding of the substrate to the active site. This conformational change causes a strain in the reacting molecules (b). Choice **d** is correct.

11. In many organic reactions, enzymes by themselves are unable to catalyze the reaction. Nonprotein organic molecules, often serving as electron acceptors, are required. These are called coenzymes (**c**). Any nonprotein essential for enzyme function is considered to be a cofactor. Coenzymes, such as NAD^+, are composed of nucleotide units and/or vitamins.

12. The conformation of enzymes necessary to bind substrate molecules is sensitive to temperatures. At 40°C, the majority of enzyme molecules will begin to lose their three-dimensional structure as a result of the breaking of hydrogen and other weak bonds by heat; therefore, **a** is correct. (See Figure 8-13.)

13. Along a reaction sequence, the product of one reaction becomes the substrate of the next. This substrate is quickly converted to product, driving the first reaction toward completion. As a result, intermediate products do not accumulate and chemical equilibrium is not attained. Therefore, only **c** is correct.

14. The direction of the longer arrow indicates that, at equilibrium, C and D will be in greater concentration than A and B. At this time, the forward and reverse rates are equal. Because more molecules of A and B have been converted to C and D than vice versa, C and D are at a lower potential energy than A and B. Thus, the reaction is exergonic (ΔG is negative); choice **c** is correct.

15. Allosteric interactions provide one way in which enzyme activity can be regulated. The allosteric effector, often the end product of a reaction sequence, binds at a special site other than the active site (**c**), causing a conformational change that inactivates the enzyme.

16. Competitive inhibition occurs when a molecule closely resembling the normal substrate of an enzyme competes with it for occupation of the active site. The inhibitory effect will be greatest when the normal substrate is in low concentration relative to the inhibitor (**c**).

17. Choice **d** is correct. Refer to Figure 8-20.

18. Coupling refers to the simultaneous occurrence of an exergonic reaction, such as the hydrolysis of ATP, and an endergonic reaction, such as photosynthesis. The energy yielded in the first reaction is used to raise the potential energy of the reactants to a level that will induce the second reaction. Choice **b** is therefore correct.

19. The removal of a phosphate group from ATP requires water to break the bonds and thus is an example of hydrolysis (**a**). The hydrolysis of ATP is exergonic. The addition of a phosphate group to ADP, known as phosphorylation, is an example of a condensation reaction, since water is removed. The hydrolysis of ATP requires an enzyme, ATPase.

20. In the coupling of an endergonic reaction to the hydrolysis of ATP, a phosphate group is transferred to the reactant molecule (**b**). This increases the molecule's potential energy.

21. The answers in order are **d, f, e, b, a**, and **c**.

22. The arrows in this figure should be labeled (a) **energy of activation of uncatalyzed reaction**; (b) **energy of activation of catalyzed reaction**; and (c) **overall energy change of reaction**.

23. The key here is that the second law applies only to closed systems, such as the universe as a whole.

24. Your answer should include a discussion of the interaction that must take place between reacting molecules. Enzymes not only lower the activation energies of chemical reactions but, in doing so, put the various reactions of a pathway into the same time frame. Enzymes also prevent the accumulation of unwanted by-products by eliminating side reactions.

25. Review the concept of equilibrium (pages 176–177) within metabolic pathways; then show why equilibrium is never achieved for any of the reactions within a pathway. See also the analysis of Question 13.

CHAPTER 9

How Cells Make ATP: Glycolysis and Respiration

CHAPTER ORGANIZATION

I. Introduction
II. An overview of glucose oxidation
III. Glycolysis
  A. Introductory remarks
  B. Steps 1 through 9
  C. Summary of glycolysis
IV. Anaerobic pathways
V. Respiration
  A. Introductory remarks
  B. A preliminary step: The oxidation of pyruvic acid
  C. Essay: Dissecting the cell
  D. The Krebs cycle
  E. Electron transport
  F. The mechanism of oxidative phosphorylation: Chemiosmosis
    1. General background
    2. How chemiosmosis works
  G. Control of oxidative phosphorylation
VI. Overall energy harvest
VII. Other catabolic pathways
VIII. Essay: Ethanol, NADH, and the liver
IX. Biosynthesis
X. Summary

MAJOR CONCEPTS

Glucose is the chief source of energy in most cells. As glucose is oxidized in a series of enzymatic steps, the energy released is repackaged in the phosphate bonds of ATP molecules.

The first phase of glucose oxidation takes place in the cytoplasm and is called glycolysis. Each six-carbon glucose molecule is split into two three-carbon molecules of pyruvic acid. The net energy yield of glycolysis is two molecules of ATP and two of NADH. An investment of two ATP molecules is required to get the process started.

In the absence of oxygen, pyruvic acid is converted into ethanol or lactic acid by a process known as anaerobic fermentation, in which NAD^+ is regenerated, allowing glycolysis to continue.

In the presence of oxygen, further breakdown of pyruvic acid occurs by the process of respiration that takes place in the mitochondria. Three-carbon pyruvic acid molecules are converted into two-carbon acetyl groups. Along with other fuel molecules, these acetyl groups enter the Krebs cycle and are completely oxidized to carbon dioxide.

Each acetyl group oxidized in the Krebs cycle results in an energy yield of three molecules of reduced NAD^+, one molecule of reduced FAD, and one molecule of ATP. The final stage of respiration is terminal electron transport, which involves a chain of electron carriers and enzymes embedded in the inner membrane of the mitochondria. The high-energy electrons of NADH and $FADH_2$ generate large quantities of free energy as they are passed along the chain of electron carriers downhill to oxygen. The free energy released powers the pumping of hydrogen ions out of the mitochondrial matrix. The electrochemical gradient thus generated is used to power the formation of ATP molecules from ADP and phosphate. The mechanism of oxidative phosphorylation is called chemiosmosis.

The complete oxidation of one molecule of glucose results in the formation of a maximum of 38 molecules of ATP. This is a free energy yield of almost 40 percent efficiency. Other food molecules, including fats, polysaccharides, and proteins, are utilized by being degraded to compounds that can slip into these central pathways at various steps.

HOW TO STUDY THE CHAPTER

Read the chapter, focusing on the major concepts.

Reread the chapter, using the questions that follow to help you focus on details as they relate to major concepts. Answer the questions on a separate sheet of paper. Your answers will provide a valuable study aid in preparing for examinations.

FOCUS ON CHAPTER DETAILS

I. **Introduction** (page 188)
   1. A cell carrying out the oxidation of carbohydrates captures the energy released and stores it in the phosphate bonds of ATP molecules. What is the role of ATP in living systems?

II. **An overview of glucose oxidation** (pages 188–189)
   2. What is the fate of electrons in the typical oxidation-reduction reaction?
   3. Write the balanced overall equation for the oxidation of glucose, and indicate the value of ΔG. What does this value mean? How much of ΔG becomes available to the cell?
   4. The overall oxidation of glucose can be divided into three stages. What are these?
   5. Describe what happens to the electrons during glycolysis and respiration.

III. **Glycolysis** (pages 189–192)
   A. Introductory remarks
      6. What is the initial reactant and the final product of glycolysis as an enzymatic pathway? What is the overall energy harvest of the process?
   B. Steps 1 through 9
      7. As glycolysis begins, the energy level of the glucose molecule must first be raised. How is this activation energy supplied?
      8. Explain how ATP acts to regulate the rate of glycolysis at Step 3 in the sequence.
      9. Examine the structures of each of the reactants in Steps 1 through 4. Can you deduce why two ATPs were added and why glucose 6-phosphate was converted to fructose 6-phosphate? (Hint: Notice the symmetry of fructose 1,6-diphosphate and the structural similarities of the products of the aldolase reaction in Step 4.)
      10. Why must Steps 5 through 9 be counted twice?
      11. The oxidation of glucose to pyruvic acid is accompanied by the reduction of what other organic compound? Identify the step during which this takes place. Of what significance is this step for the cell in terms of energy?
      12. Explain how (and during which steps) ATP is formed.
   C. Summary of glycolysis
      13. What has happened to the six-carbon glucose molecule during glycolysis?
      14. What has the sequence meant in terms of the cell's overall energy harvest?

IV. **Anaerobic pathways** (pages 192–193)
   15. What pathways can pyruvic acid follow when oxygen is not present in the cell?
   16. Which one of these pathways is referred to as fermentation, and how has it been exploited commercially?
   17. Under what conditions is pyruvic acid converted to lactic acid? How does this reaction help a hard-working athlete to continue exercising when he or she has accumulated an oxygen debt?
   18. After oxygen is replenished, what happens to the lactic acid that has built up in the bloodstream?

19. Explain how the conversion of pyruvic acid to lactic acid functions in driving the glycolytic pathway forward to meet the demands of organisms.

V. Respiration (pages 193–203)

A. Introductory remarks

20. Distinguish between ordinary respiration and cellular respiration.
21. Describe the organelle where cellular respiration takes place. What is the role of the inner membrane of the mitochondrion?

B. A preliminary step: The oxidation of pyruvic acid

22. How is acetyl CoA formed from the pyruvic acid generated in glycolysis? (See also Figure 9–8.)
23. In what form is energy released from the oxidation of pyruvic acid prior to its entry into the Krebs cycle?

C. Essay: Dissecting the cell (page 195)

24. How is the compartmentalization of biochemical pathways within organelles advantageous for the cell?
25. Differentiate between the techniques of differential centrifugation and zonal or density gradient centrifugation with respect to their roles in the preparation of isolated cell organelles.

D. The Krebs cycle

26. Using a simple chemical equation (or its verbal equivalent), describe the first reaction that takes place within the Krebs cycle.
27. As oxaloacetic acid is regenerated, what compounds (including those involved in energy exchanges) are produced within the cycle?
28. Can these reactions take place anaerobically? Explain your answer.
29. Account for the energy originally present in the carbon-hydrogen and carbon-carbon bonds of the acetyl groups. Has any more ATP been formed at this stage of respiration?

E. Electron transport

30. How are the electrons held by electron-carrier molecules disposed of during electron transport?
31. What role do the cytochromes play in electron transport?
32. What happens to the energy levels of the electrons as they are passed along the electron transport chain? How does this bring about a net gain of energy for the cell?
33. Define oxidative phosphorylation and explain why it is an appropriate term to describe the outcome of respiration.
34. How many molecules of ATP can be formed from the oxidation of NADH? Of $FADH_2$?

F. The mechanism of oxidative phosphorylation: Chemiosmosis

General background

35. How was the phosphorylation of ADP in the electron transport chain formerly thought to occur? Why did this hypothesis prove unacceptable?
36. Explain how the idea of a proton gradient that could power ATP production occurred to Peter Mitchell, the British biochemist.

How chemiosmosis works

37. On what two basic premises is the chemiosmotic theory founded?
38. In which direction is a proton gradient established across the inner mitochondrial membrane?
39. How does the impermeability of the membrane help to maintain this gradient?
40. Explain how a flow of protons down the gradient provides energy for the addition of a phosphate group to ADP. Be sure to include the role of F_0 and F_1 factors.
41. For what other biochemically important transport process does the chemiosmotic theory provide a general explanation?

G. Control of oxidative phosphorylation

42. Explain how oxidative phosphorylation is regulated by the cell's demand for ATP.

VI. Overall energy harvest (pages 203–205)

43. Take an inventory of the energy gained in each of the major steps of glucose oxidation. Account for the discrepancy in the number of ATP molecules that result from the energy released in glycolysis. Check your calculations against Table 9-1.

VII. Other catabolic pathways (page 205)

44. The Krebs cycle is a Grand Central Station for energy metabolism. Interpret this analogy with assistance from Figure 9-21.

VIII. Essay: Ethanol, NADH, and the liver (page 206)

45. Describe the complete reaction by which ethanol is metabolized by the liver.
46. What is the mechanism by which alcohol interferes with the utilization of a normal, balanced diet?

IX. Biosynthesis (page 207)

47. What is the relationship between anabolism and catabolism?
48. Describe the central role that acetyl CoA plays in anabolism and catabolism.

X. Summary (pages 207–208):
Read the Summary. If you are familiar with the essential features of the material presented there, you are ready to take the following diagnostic examination.

TESTING YOUR UNDERSTANDING

After you have completed the following examination, compare your answers with the annotated key in the Analysis section.

1. During the oxidation of glucose, 40 percent of the chemical energy of sugar molecules is _____ , while the rest _____ .
   a. used to form ATP; is lost as heat
   b. used to hydrolyze ATP; converted to mechanical energy
   c. lost as heat; is used to phosphorylate ADP
   d. released; remains in the bonds of glucose
   e. lost as heat; is stored in glycogen

2. Fill in the blanks to complete the equation for the oxidation of glucose.

   $C_6H_{12}O_6$ + _____ ⟶ _____ + $6H_2O$

   a. 38 ADP; 38 ATP
   b. O_2; CO_2
   c. $6O_2$; $6CO_2$
   d. CO_2; O_2
   e. $6CO_2$; $6O_2$

3. The cell's chief means of regulating the rate of glycolysis involves the use of _____ as an allosteric effector.
   a. pyruvic acid
   b. Na^+ ions
   c. H atoms
   d. NADH
   e. ATP

4. As a result of glycolysis:
   a. there is a small net yield of ATP.
   b. NAD^+ is reduced.
   c. two molecules of pyruvic acid are formed from one molecule of glucose.
   d. Both a and c are correct.
   e. All three answers (a, b, and c) are correct.

5. In the initial step of glycolysis, energy from the hydrolysis of ATP is:
   a. used to increase the potential energy of glucose.
   b. used to reduce NAD^+.
   c. never recovered.
   d. lost as heat.
   e. transferred to $FADH_2$.

6. Pyruvic acid formed in glycolysis will be _____ in the presence of oxygen.
   a. converted to ethyl alcohol
   b. converted to lactic acid
   c. oxidized in the Krebs cycle
   d. reduced to glycogen

7. The production of ethanol from glucose is an example of:
   a. carbohydrate reduction.
   b. hydrolysis.
   c. fermentation.
   d. respiration.
   e. phosphorylation.

8. In paying off an oxygen debt following a period of exercise, aerobic organisms convert _____ to pyruvic acid.
   a. lactic acid
   b. CO_2
   c. acetyl CoA
   d. glucose 6-phosphate
   e. NADH

9. Oxidative reactions of the Krebs cycle take place in (on):
   a. the cytoplasm.
   b. the inner matrix of the mitochondrion.
   c. the outer mitochondrial membrane.
   d. the endoplasmic reticulum.
   e. chloroplasts.

10. Krebs cycle reactions begin with the binding of _____ to oxaloacetic acid to form _____ .
    a. CO_2; a five-carbon acid
    b. hydrogen atoms; a four-carbon acid
    c. the two-carbon acetyl group; a six-carbon acid
    d. pyruvic acid; a seven-carbon acid

11. During the Krebs cycle, _____ atoms are removed from carbon compounds and transferred to _____ .
    a. carbon; CO_2
    b. hydrogen; oxygen
    c. oxygen; water
    d. hydrogen; NAD^+ and FAD
    e. phosphate-group; ADP

12. In the electron transport chain, _____ is (are) oxidized and _____ is (are) reduced.
    a. cytochromes; NADH and $FADH_2$
    b. NADH and $FADH_2$; oxygen
    c. water; NAD^+ and FAD
    d. pyruvic acid; CO_2
    e. NADH; FAD

13. The electron acceptor in the electron transport chain that is at the lowest potential energy is:
    a. NAD^+.
    b. FAD.
    c. ADP.
    d. cytochromes.
    e. oxygen.

14. Most of the carbon dioxide evolved during respiration is formed during:
    a. glycolysis.
    b. the electron transport chain.
    c. the Krebs cycle.
    d. the conversion of pyruvic acid to acetyl CoA.
    e. glycolysis and the electron transport chain.

15. The chemiosmotic hypothesis states that ATP production is driven by:
    a. the transfer of phosphate groups by intermediate compounds.
    b. a proton gradient.
    c. the ATP synthetase complex.
    d. the oxidation of NAD.

16. The F_0 and F_1 units of the inner mitochondrial membrane contain:
    a. enzymes that normally catalyze hydrolysis of ATP.
    b. enzymes that add phosphate groups to ADP.
    c. large proteins known as cytochromes.
    d. the organic acids of the Krebs cycle.

17. Oxidative phosphorylation is controlled by:
    a. the cell's demand for ATP.
    b. feedback inhibition of NADH on the cytochromes.
    c. the presence of a competing substrate.
    d. permeability changes of the outer mitochondrial membrane to protons.
    e. the concentration of CO_2.

18. The quantity of NADH formed in the Krebs cycle per molecule of glucose results in the production of _____ molecules of ATP in the electron transport chain.
    a. 38
    b. 36
    c. 24
    d. 18
    e. 8

19. Which of the following represents a biochemical crossroads for glucose oxidation, fatty acid catabolism, and the catabolism of some amino acids?
    a. glucose 6-phosphate
    b. oxaloacetic acid
    c. acetyl CoA
    d. pyruvic acid
    e. glucose

20. In order for two types of organelles to be separated from a cellular extract by centrifugation, they must differ in:
    a. size.
    b. density.
    c. electrical charge.
    d. solubility.
    e. Both a and b are correct.

21. Because the breakdown of alcohol causes the reduction of NAD^+, _____ accumulate(s) in the _____ of heavy drinkers.

a. sugars; urine
b. NADH; mitochondria
c. fats; liver cells
d. alcohol; bloodstream

22. Identify the reactant and product molecules in the following reaction sequence. What is the importance of this reaction?

$$\underset{\text{OH}}{\overset{\text{CH}_3}{\underset{\text{C=O}}{\text{C=O}}}} \xrightarrow{CO_2} \underset{\text{H}}{\overset{\text{CH}_3}{\text{C=O}}} \xrightarrow{\text{NADH} \quad \text{NAD}^+} \underset{\text{H}}{\overset{\text{CH}_3}{\text{H-C-OH}}}$$

_____ ACETALDEHYDE _____

23. What is the purpose of converting pyruvic acid to lactic acid when oxygen is not available in sufficient quantity?

24. Describe in detail the chemiosmotic theory, proposed by Peter Mitchell to explain how ATP is produced in the electron transport chain. If possible, accompany your discussion with a diagram.

25. Explain what is meant by the statement, The Krebs cycle is the Grand Central Station for energy metabolism.

PERFORMANCE ANALYSIS

1. Of the 686 kilocalories released during oxidation of a mole of glucose, about 266 kilocalories are reserved in ATP molecules (38 ATP formed × 7 kilocalories/mole). This is an efficiency of approximately 40 percent. The remaining 60 percent (approximately 420 kilocalories) is dissipated as heat. Choice **a** is therefore correct.

2. In respiration, oxygen gas, O_2, is consumed, and carbon dioxide gas, CO_2, is given off. Since glucose is a six-carbon sugar, 6 molecules of CO_2 on the right-hand side will balance the number of carbon atoms on the left-hand side. There are now 18 oxygen atoms on the right side. To balance these, 6 molecules of O_2 (6 × 2 = 12 atoms) must be added to the left side, since 6 are already present in glucose. Choice **c** is correct.

3. If a large quantity of ATP is present within the cell, then ATP itself (**e**) acts as an allosteric inhibitor of phosphofructokinase, the enzyme that catalyzes Step 3 in the glycolytic sequence. This prevents the breakdown of glucose.

4. Review Figure 9–3 and the overall equation for glycolysis on page 192 to verify that **e** is correct.

5. The hydrolysis of ATP is coupled to the first reaction step in glycolysis, the phosphorylation of glucose to form glucose 6-phosphate. Glucose 6-phosphate is at a higher potential energy level than glucose (**a**).

6. Choices **a** and **b** are pathways taken when oxygen is absent. Glycogen will be formed from glucose only if the cell has met its ATP requirements. The only aerobic pathway for pyruvic acid is its oxidation in the Krebs cycle (**c**).

7. The production of ethanol from glucose occurs anaerobically and requires both glycolysis and fermentation (**c**).

8. When oxygen is in short supply, as during periods of heavy exercise, the pyruvic acid formed during glycolysis does not enter the Krebs cycle but is reduced to lactic acid. This is accompanied by the oxidation of NADH and NAD^+ (see Figure 9–6). The availability of NAD^+ drives the glycolysis sequence forward, supplying energy for the working muscles. When demand for oxygen falls off, this lactic acid is converted back to pyruvic acid with the formation of NADH. Choice **a** is therefore correct.

9. The enzymes of the Krebs cycle lie in a dense solution (the matrix) surrounded by the highly folded inner membrane of the mitochondrion; therefore, **b** is correct.

10. Pyruvic acid is first oxidized to a two-carbon acetyl group (see Figure 9–8) before entering the Krebs cycle. Since oxaloacetic acid is a four-carbon compound, the binding of an acetyl group to it produces a six-carbon compound, citric acid (**c**).

11. Review Figure 9–9 to see that hydrogen atoms are removed from organic acids during the Krebs cycle and given to the electron acceptors NAD^+, which is reduced to NADH, and FAD, which is reduced to $FADH_2$ (**d**). These reduced compounds represent a harvest of energy from the Krebs cycle that can be used to form ATP from ADP.

12. During electron transport, NADH and FADH$_2$ give up their electrons and so become oxidized. A series of cytochromes, and finally oxygen, receive these electrons and so are reduced; therefore, **b** is correct.

13. The final electron acceptor in the chain is the one with the least potential energy. See Figure 9-13 to verify that this is oxygen (**e**). NAD$^+$ and FAD are the electron acceptors with the highest potential energy. Cytochrome molecules within the chain are at a lower potential energy than NAD$^+$ and FAD but are at a higher level than O$_2$.

14. In every turn of the Krebs cycle (**c**), two molecules of CO$_2$ are produced; therefore, for every molecule of glucose oxidized, four molecules of CO$_2$ are given off in the Krebs cycle. This is two-thirds of the total amount of CO$_2$ evolved during respiration. The other two molecules appear during the oxidation of pyruvic acid prior to the Krebs cycle (see Figure 9-9).

15. There are currently two hypotheses for the chemiosmotic movement of protons (**b**) through the inner mitochondrial membrane (see pages 199 and 200). This movement, which must occur against a concentration gradient, makes available a source of potential energy for ATP production. Intermediate compounds that were once thought to transfer phosphate groups to ADP have never been isolated, despite years of effort.

16. The F_0 and F_1 units provide a channel for the return flow of protons across the inner membrane. These units contain a large enzyme complex, ATP synthetase, that catalyzes the addition of phosphate groups to ADP (**b**).

17. Reactions of the electron transport chain go forward only if there is ADP to be phosphorylated. When most of the ADP has been converted to ATP, the reaction sequence shuts down. In effect, it is the demand for ATP (**a**) that controls oxidative phosphorylation.

18. It can be seen from Figure 9-9 that for every turn of the cycle, 3 molecules of NADH are produced. Per molecule of glucose, then, 6 molecules of NADH are produced. Each molecule of NADH carries sufficient energy to convert 3 molecules of ADP to 3 moles of ATP. So 18 molecules (6 × 3) of ATP (**d**) result from the reduction of NAD$^+$ in the Krebs cycle.

19. Both fatty acids and some amino acids can be degraded to acetyl CoA (**c**). Acetyl groups also serve as the starting point for the anabolism of these compounds. See Figure 9-21.

20. Particles of different sizes and densities (a) may be separated by differential centrifugation. If the particles are of similar size but have differing densities (b), they may be separated by zonal centrifugation. Choice **e** is correct.

21. The oxidation of alcohol by liver enzymes reduces NAD$^+$ to NADH. This accumulation of NADH inhibits glycolysis (see Step 5, page 192), and glucose instead is converted to glycogen and then to fat, which begins to build up in the liver tissues of a heavy drinker. Therefore, **c** is correct.

22. Consult Figure 9-5 to see that **pyruvic acid** is the reactant and **ethanol** is the product. The legend accompanying the figure discusses the importance of the reaction.

23. The key to answering this question is understanding the significance of the reduction of NAD$^+$ in the glycolytic sequence and the oxidation of NADH when pyruvic acid is converted to lactic acid. NAD$^+$ must continually be supplied by the oxidation of NADH by pyruvic acid.

24. For the details of the chemiosmotic theory, refer to Figures 9-14, 9-15, 9-16, 9-17, and the supporting discussion in the text.

25. The Grand Central Station referred to is the hub of the major metabolic pathways shown in Figure 9-21.

CHAPTER 10

Photosynthesis, Light, and Life

CHAPTER ORGANIZATION

I. Introduction
II. The nature of light
   A. General background
   B. Essay: Van Niel's hypothesis
   C. The fitness of light
III. Chlorophyll and other pigments
IV. Photosynthetic membranes: The thylakoid
   A. Introductory remarks
   B. The structure of the chloroplast
V. The stages of photosynthesis
VI. The energy-capturing reactions
   A. The photosystems
   B. The light-trapping reactions
   C. Cyclic electron flow
   D. Photosynthetic phosphorylation
   E. Essay: Photosynthesis without chlorophyll
VII. The carbon-fixing reactions
   A. Introductory remarks
   B. The Calvin cycle: The three-carbon pathway
   C. The four-carbon pathway
VIII. The products of photosynthesis
IX. Essay: The carbon cycle
X. Summary

MAJOR CONCEPTS

Photosynthesis is a good example of the first law of thermodynamics in action. Light energy is converted into chemical energy, first in the form of ATP and NADPH and then in the form of glyceraldehyde phosphate. Since this molecule is the one that represents net synthesis from CO_2, it has the distinction of being the precursor of all other organic molecules in living systems (remember the relationship between autotrophs and heterotrophs).

Phototrophs use light of the visible portion of the electromagnetic spectrum because that is what is most available and because its energy level is high enough to excite electrons, but not so high that it will break molecular bonds.

Light energy is captured in pigments, which, in eukaryotes, include the chlorophylls and carotenoids. Light absorbed by the pigments boosts their electrons to higher energy levels. Because of the way the pigments are packed into membranes, they are able to transfer this energy to reactive molecules, probably specially packed chlorophyll a.

Photosynthesis occurs in chloroplasts. The pigments and other molecules responsible for capturing light energy are located on flattened membranous sacs, called thylakoids, that are contained within the chloroplast.

Photosynthesis takes place in two stages—a light-capturing stage and a carbon-fixing stage. The first stage begins when light strikes antennae molecules of Photosystem II and causes electrons from reactive chlorophyll a molecules to be boosted to acceptor molecules. The electrons are passed along an electron transport chain, in a manner comparable to mitochondrial electron transport, forming ATP. Similar events occur at the next step, during which the electrons are boosted again, this time to Photosystem I. As the electrons pass along the electron transport chain, NADH is formed.

The energy-rich ATP and NADH are needed for the carbon-fixing stage, during which glyceraldehyde phosphate is synthesized from CO_2 in a series of reactions known as the Calvin cycle. When sufficient NADPH is present to meet cellular demands, Photosystem I is capable of engaging in cyclic electron flow, in which only ATP is produced.

Some plants (C_4) that are adapted to hot, dry climates conserve their water by opening their stomata less often. This strategy is made possible by the fact that CO_2 can be fixed in these plants at extremely low concentrations and then stored as four-carbon acids. Later, after being released from the four-carbon acid, CO_2 can enter the Calvin cycle for conversion into glyceraldehyde phosphate.

HOW TO STUDY THE CHAPTER

Read the chapter, focusing on the major concepts.

Reread the chapter, using the questions that follow to help you focus on details as they relate to major concepts. Answer the questions on a separate sheet of paper. Your answers will provide a valuable study aid in preparing for examinations.

FOCUS ON CHAPTER DETAILS

I. **Introduction** (pages 210–211)
  1. What profound effects did the origin of photosynthetic organisms have on the evolution of life?
  2. Summarize photosynthesis using a generalized chemical equation.
  3. What is the importance of the products for other organisms?

II. **The nature of light** (pages 212–214)
  A. General background
  4. What are the subcategories of the electromagnetic spectrum? What is the sequence of colors in the visible portion of the electromagnetic spectrum (starting at the shortest wavelength)?
  5. What two models are used to describe the phenomenon of light?
  6. Explain what changes take place in a piece of metal subjected to light of a critical wavelength. What name has been given to this phenomenon?
  7. How do the wavelength and intensity of light relate to the degree to which electrons are emitted from an illuminated metal? Which theory of light proposes to explain these effects?

  B. Essay: Van Niel's hypothesis (page 211)
  8. What was van Niel's hypothesis? What observation led him to this formulation?
  9. How did later investigators make use of a radioactive tracer to confirm the hypothesis?

  C. The fitness of light
  10. What reasons can you advance for the "selection" of visible light by living things?
  11. Elaborate on the phrase "the fitness of the environment" with respect to the use of visible light energies for life processes.

III. **Chlorophyll and other pigments** (pages 214–216)
  12. What does the absorption spectrum of a pigment reveal?
  13. Relating Figures 10–5 and 10–6 to Figure 10–4, compare the absorption properties of chlorophylls *a* and *b* and the carotenoid pigments.
  14. What explanation does the text provide for the use by plants of several different pigments in photosynthesis?
  15. What is the relationship between a pigment's absorption spectrum and its action spectrum?
  16. Is there any difference in the way isolated chlorophyll molecules and chlorophyll molecules in living tissues behave when they absorb light energy? How can this be explained in terms of the energies of the molecules' electrons?

IV. **Photosynthetic membranes: The thylakoid** (pages 217–219)

A. Introductory remarks

17. Describe the structural unit of photosynthesis. Where are these units located in prokaryotes? In eukaryotes?

B. The structure of the chloroplast

18. In the micrographs of the chloroplasts in Figure 10-9b and c, identify thylakoids, grana, the thylakoid space, the stroma, and the inner and outer chloroplast membranes.

19. How does the orientation of the thylakoids allow for optimum light reception?

V. **The stages of photosynthesis** (pages 219–220)

20. Describe Blackman's observations that led to the conclusion that photosynthesis occurs in two separate stages.

21. What designations are now preferred for the two stages of photosynthesis? Briefly describe what happens in each stage.

VI. **The energy-capturing reactions** (pages 220–224)

A. The photosystems

22. In terms of the electrons of chlorophyll, explain what happens when a photon strikes a photosystem.

23. Describe the absorption properties of the reactive centers of Photosystem I and Photosystem II.

B. The light-trapping reactions

24. Describe the process of photophosphorylation.

25. What events in Photosystem I are simultaneous with the events of Photosystem II?

26. What are the replacement sources of electrons for the reactive centers in each photosystem?

27. Chart the course of electron flow in the light-trapping reactions. (Refer to Figure 10-12 if you get stuck.) Do you see why Szent-Györgyi thought of these reactions as producing a "little electric current, kept up by the sunshine"?

28. To what two chemical forms has the energy of the sun been converted during the light-trapping reactions?

C. Cyclic electron flow

29. In terms of the events occurring in photosystems, how does cyclic electron flow differ from the noncyclic flow described above?

D. Photosynthetic phosphorylation

30. How is photophosphorylation similar to the oxidative phosphorylation that takes place inside a mitochondrion during respiration?

31. How does the direction of the proton gradient established in chloroplasts differ from that in mitochondria?

32. Which of the details of photophosphorylation are currently only speculative?

33. What *in vitro* evidence supports the chemiosmotic theory of photophosphorylation?

E. Essay: Photosynthesis without chlorophyll (page 224)

34. What was the nutritional significance of a purple pigment found on the membranes of halobacteria? What is this pigment? How does its role in photosynthesis provide support for the chemiosmotic mechanism of photophosphorylation?

VII. **The carbon-fixing reactions** (pages 225–228)

A. Introductory remarks

35. How are the energy-capturing reactions of photosynthesis related to the carbon-fixing reactions?

36. What is the source of carbon for photosynthetic organisms that live on land? That live in water?

B. The Calvin cycle: The three-carbon pathway

37. Where do the Calvin cycle reactions take place inside the cell?

38. What compounds of the Krebs cycle can be considered analogous in function to compounds in the Calvin cycle? (Compare Figures 9-11 and 10-18.)

39. What is the initial reaction of the cycle?

What compound will be continually regenerated as the cycle proceeds?

40. Why is RuDP carboxylase the most abundant protein in living systems?
41. Using Figure 10-18, trace three revolutions of the cycle. What product is formed?

C. The four-carbon pathway

42. What events precede the Calvin cycle in a C_4 plant? After what compound are the C_4 plants named? (You should, at this point, be able to answer this for C_3 plants also.)
43. What are the C_4 functions of malic and aspartic acids?
44. What purpose does C_4 metabolism serve, and under what environmental conditions is it found?
45. What is meant by photorespiration? When is it likely to occur and why? What is its effect on photosynthetic efficiency?
46. Why would a C_4 plant be less subject to photorespiration than a C_3 plant?
47. Cite several examples of common C_4 plants.

VIII. **The products of photosynthesis** (pages 228–230)

48. Why is the manufacture of glyceraldehyde phosphate considered such an important biological event?
49. Name several important organic molecules for which glyceraldehyde phosphate is a precursor. What are the functions of these molecules?

IX. **Essay: The carbon cycle** (page 229)

50. The carbon cycle diagram (page 229) illustrates that a balance between two biological processes determines the concentrations of carbon dioxide to be found in reservoirs (air, oceans, etc.). What are these two processes?
51. Under what conditions might the amount of carbon in the atmosphere increase?

X. **Summary** (pages 230–232): Read the Summary. If you are familiar with the essential features of the material presented there, you are ready to take the following diagnostic examination.

TESTING YOUR UNDERSTANDING

After you have completed the following examination, compare your answers with the annotated key in the Analysis section.

1. When light of a critical wavelength strikes a metal plate, the metal will:
   a. emit electrons.
   b. become negatively charged.
   c. become reduced.
   d. discharge photons.
2. To speed up the velocity at which electrons are ejected in a photoelectric circuit, one must:
   a. increase the brightness of the light.
   b. decrease the intensity of the light.
   c. decrease the wavelength of the light.
   d. turn the light on and off rapidly.
   e. Both a and d will speed up electron velocity.
3. In the summer, leaves are green because chlorophyll pigments:
   a. absorb green light.
   b. reflect green light.
   c. absorb all wavelengths.
   d. reflect all wavelengths.
   e. reflect red light.
4. Chlorophyll b and the carotenoid pigments are arranged within the photosystems so that they:
   a. absorb light at different wavelengths than chlorophyll a.
   b. can pass energy on to the reactive center.
   c. are the reactive centers of Photosystem I and Photosystem II, respectively.
   d. absorb visible light maximally.
   e. Both b and d are correct.
5. If isolated chlorophyll molecules are placed in a test tube containing CO_2 and H_2O, and the tube is exposed to blue-violet light:
   a. the light will serve only to raise the temperature.
   b. chlorophyll will fluoresce.
   c. sugar will be manufactured.
   d. chlorophyll will become permanently charged.
   e. Both c and d are correct.
6. All of the pigments, electron carriers, and enzymes necessary for _____ of photosynthesis are located in (on) _____ .

a. the light-trapping reactions; the thylakoids
   b. the light-trapping reactions; the stroma
   c. the Calvin cycle; the thylakoids
   d. both reactions; the thylakoids
   e. both reactions; the stroma
7. As a result of the oxidation of chlorophyll, _____ is (are) available to provide energy for _____ .
   a. ATP and NADPH; the light-trapping reactions
   b. ADP and NADP; the light-trapping reactions
   c. ATP; the reduction of NADP
   d. NADP; the phosphorylation of ADP
   e. ATP and NADPH; the fixation of carbon
8. An increase in temperature will increase the photosynthetic rate provided that:
   a. light is not available.
   b. light intensity is also increased.
   c. dim light conditions are maintained.
   d. the temperature is decreased whenever it gets dark.
   e. the CO_2 concentration is low relative to the O_2 concentration.
   f. the temperature is increased above 30°C.
9. The reactive center molecules of Photosystem I and Photosystem II are:
   a. chlorophyll *a* and chlorophyll *b*, respectively.
   b. chlorophyll *b* and chlorophyll *a*, respectively.
   c. identical chlorophyll *a* molecules.
   d. identical chlorophyll *b* molecules.
   e. two different types of chlorophyll *a*.
10. If electron flow during the light-trapping reactions is not cyclic, then the electrons removed from water molecules are ultimately passed to:
    a. $NADP^+$.
    b. ADP.
    c. electron carriers of Photosystem II.
    d. reactive centers of Photosystem I.
    e. chlorophyll *a*.
11. During the light-trapping reactions in photosynthesis, _____ energy is converted to _____ .
    a. light; chemical energy in the form of an "excited" reactive chlorophyll molecule
    b. light; electrical energy in the form of electron flow
    c. electrical; chemical energy in the form of ATP
    d. chemical; electrical energy in the form of electron flow
    e. All of the above are correct.
12. The operation of Photosystem I independent of Photosystem II produces:
    a. three-carbon sugars.
    b. four-carbon acids.
    c. only oxygen.
    d. only ATP.
    e. both ATP and NADPH.
13. The energetic basis of photophosphorylation differs from that of oxidative phosphorylation in that:
    a. the former does not involve a proton gradient.
    b. the direction of the proton gradient in the former is opposite to that in the latter.
    c. thylakoid membranes are permeable to H^+ ions.
    d. the former does not involve membrane-bound cytochromes.
    e. They are not different; both rely on an identical chemiosmotic mechanism.
14. The initial source of the carbon fixed in a C_4 plant during the Calvin cycle is:
    a. malic or aspartic acid.
    b. glyceraldehyde phosphate.
    c. oxaloacetic acid.
    d. glucose.
    e. carbon dioxide.
15. In a C_3 plant, carbon dioxide entering the Calvin cycle binds with:
    a. oxaloacetic acid
    b. RuDP.
    c. PGA.
    d. PEP.
    e. NADPH.
16. After three revolutions of the Calvin cycle, _____ molecule(s) of _____ is (are) produced.
    a. three; CO_2
    b. one; glucose
    c. three; glucose
    d. one; glyceraldehyde phosphate
    e. three; glyceraldehyde phosphate
17. In a plant with C_4 metabolism, malic acid is first synthesized in _____ , then transported to _____ .
    a. mesophyll cells; bundle-sheath cells
    b. bundle-sheath cells; thylakoids
    c. stem tissue; leaf tissue
    d. mesophyll cells; stomata
    e. the Calvin cycle; the Hatch-Slack pathway

18. Under hot, dry conditions, a C_4 plant will do better than a C_3 plant because:
    a. C_4 metabolism enhances the rate of photorespiration.
    b. C_4 plants take in CO_2 quickly and can store it.
    c. the enzymes of the C_3 cycle function poorly in these conditions.
    d. the oxidation of malic acid yields more energy than that of RuDP.
    e. All of the above are correct.
19. Generally speaking, C_4 plants have _____ than do C_3 plants.
    a. more efficient mechanisms for conserving water
    b. higher photorespiration rates
    c. lower photosynthetic rates
    d. different Calvin cycle reactions
    e. All of the above are correct.
20. Glyceraldehyde phosphate is a useful product of green plants because it:
    a. is the precursor of all organic molecules.
    b. is the ultimate source of chemical energy for all living things.
    c. closely resembles ATP.
    d. yields more energy than glucose when oxidized.
    e. Both a and b are correct.

21. Fill in the blanks in the accompanying photosynthesis diagrams.

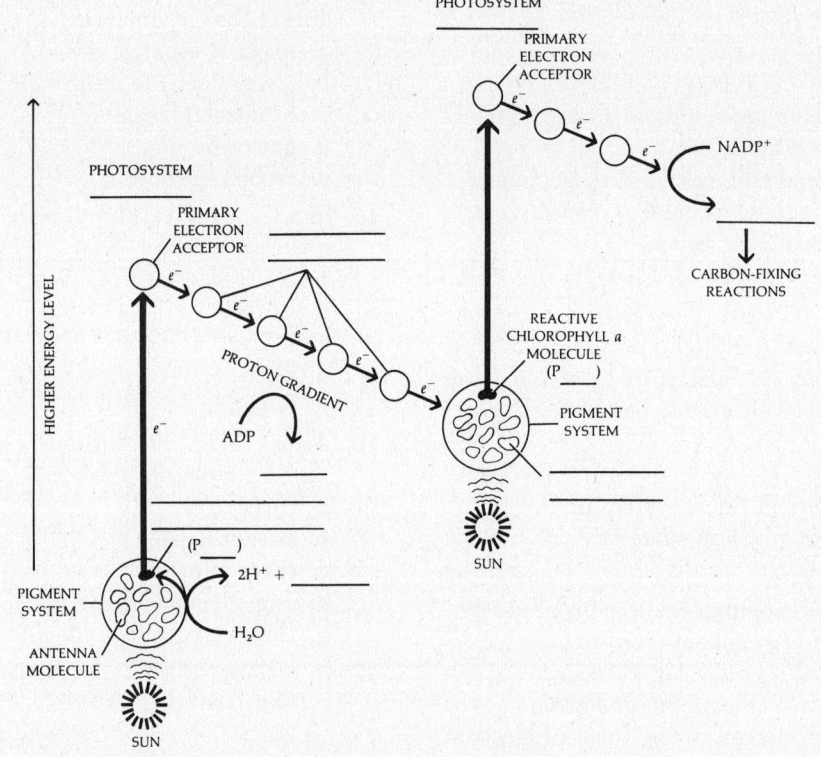

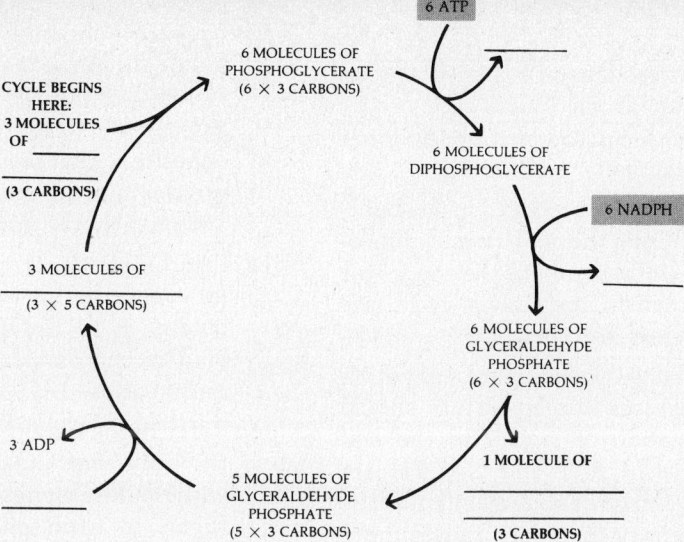

22. Summarize the light-trapping reactions and tell why they are important to the carbon-fixing reactions.
23. How does cyclic electron flow differ from the interdependent reactions of Photosystems I and II?
24. Point out the similarities and differences in the chemiosmotic mechanism as it applies to mitochondria and chloroplasts.
25. Explain how the process of photorespiration limits photosynthetic efficiency.
26. What is the connection between C_4 metabolism and the conservation of water by plants?

PERFORMANCE ANALYSIS

1. All metals exhibit the photoelectric effect: When exposed to light of a critical wavelength, they emit electrons (**a**) to become positively charged. This is not reduction but oxidation. The electrons are dislodged as a result of absorbing particles of light energy called photons.

2. In a photoelectric circuit, light striking a metal detector induces an electric current. If the light is increased in brightness, more electrons will be emitted from the metal, but in order to speed up electron velocity, shorter wavelengths of light must be used (**c**), i.e., the frequency must be increased. Turning the light on and off rapidly will only interrupt the flow of electrons.

3. Chlorophyll pigments are, by far, the most numerous leaf pigments in summer. Leaves appear green because chlorophylls reflect light of green wavelengths back to the observer (**b**). See the absorption spectra in Figure 10–5.

4. Chlorophyll *b* and the carotenoid pigments do absorb differently than chlorophyll *a*, but this is an intrinsic property of the pigments and is not due to their arrangement within the chloroplast. They are, however, arranged so that their energies may be passed on to the reactive center molecule (**b**), which for both Photosystems I and II is a form of chlorophyll *a*. The thylakoid membranes on which photosynthetic pigments are located lie parallel to one another, allowing the pigments to absorb visible light maximally (**d**). Choice **e** is therefore correct.

5. Chlorophyll, even in the presence of CO_2 and light, cannot cause sugars to be made unless it is associated with the intact thylakoid membrane. Electrons of chlorophyll will still absorb light energy and be boosted to higher energy levels. When the electrons return to the unexcited state, chlorophyll will fluoresce (**b**) or transmit light of a wavelength longer than the absorbed wavelength.

The chlorophyll therefore is not permanently changed.

6. The light-trapping reactions are carried out on thylakoid membranes (**a**); the Calvin cycle occurs in the stroma, which is surrounded by the inner chloroplast membrane.

7. Both ATP and NADPH are produced in the light-trapping reactions during the oxidation of chlorophyll (see Figure 10-13) and provide the energy for the fixation of carbon in the Calvin cycle (see Figure 10-18); therefore, **e** is correct.

8. Since the carbon-fixing reactions are enzymatically controlled, increases in temperature should speed up the formation of sugars in the leaf provided that sufficient ATP and NADPH are available to the Calvin cycle. This can be assured by increasing light intensity (**b**), which causes more chlorophyll molecules to become activated in the energy-capturing reactions that produce the ATP and NADPH.

9. As can be seen from Figure 10-12, the reactive centers of Photosystems I and II are P_{700} and P_{680}, respectively. These represent two different forms of chlorophyll *a* (**e**).

10. In noncyclic electron flow (Figure 10-12), electrons from water replace those lost at the reactive centers of Photosystem II. Electrons from these reactive centers are boosted to electron acceptors that transfer them to the reactive centers of Photosystem I. Electrons from these reactive centers are in turn boosted to electron acceptors, which subsequently transfer them to $NADP^+$ (**a**).

11. Energy from the sun is first converted to a flow of electrons in the photosystems. The energy of these electrons is trapped in the chemical bonds of ATP and NADPH. Overall, then, the light reactions represent the conversion of light energy to chemical energy. But the chemical energy stored in the chemical bonds of chlorophyll is converted to electrical energy as well. All four choices (**e**), then, are correct.

12. When Photosystem I operates alone, electron flow is cyclic (see Figure 10-14) and only ATP may be produced as electrons return downhill to the reactive center (**d**).

13. Photophosphorylation is also explained by a chemiosmotic hypothesis. Its electron transport chain uses cytochromes and establishes a proton gradient across a membrane that is impermeable to H^+ ions, but the direction of the gradient (**b**) is from the stroma into the thylakoid space. This is opposite to that seen in the mitochondrion, where protons move from the matrix across the inner, and possibly the outer, membrane.

14. Like C_3 plants, C_4 plants take in carbon dioxide (**e**) from the surrounding air. Instead of first binding it to RuDP, as in C_3 plants, C_4 plants bind carbon dioxide to PEP and store it in four-carbon acids.

15. Refer to the Calvin cycle diagram of Figure 10-18 to verify that CO_2 entering the cycle combines with ribulose diphosphate, RuDP (**b**), a five-carbon sugar, to form phosphoglycerate, PGA. In C_4 plants, CO_2 first combines with phosphoenolpyruvate, PEP, to form oxaloacetic acid.

16. Refer to Figure 10-18, which summarizes three turns of the Calvin cycle. The three molecules of CO_2 entering the cycle result in the production of one molecule of the three-carbon compound glyceraldehyde phosphate (**d**).

17. Refer to Figure 10-21, which shows that CO_2 is stored in malic acid in the mesophyll cells and released to the Calvin cycle in the bundle-sheath cells (**a**).

18. C_4 plants need not keep their stomata open long in order to obtain a healthy supply of CO_2. Thus there is little opportunity for water to escape. The CO_2 taken in is stored in organic acids, such as malic acid, in the bundle-sheath cells. It is released to the same Calvin cycle reactions that take place in C_3 plants. Choice **b** is therefore correct.

19. Choice **a** is correct. Refer to the analysis of Question 18.

20. Glyceraldehyde phosphate, being the end product of the Calvin cycle, is the precursor of all other organic molecules (a) and is the source of energy for all life (b). It is the precursor of glucose, starches, and fats, which plants use for their energy, as do the animals that eat them. Therefore, **e** is correct.

21. Consult Figures 10-12 and 10-18 for the appropriate labels.

22. See Figure 10-12 for a visual summary of the light-trapping reactions. The important products of these reactions are ATP and NADPH, which are used in the Calvin cycle for the reductive synthesis of glyceraldehyde phosphate from CO_2.

23. Compare Figures 10-12 and 10-14. The main difference lies in the fact that in cyclic electron flow, electrons from Photosystem I are cycled back into the electron transport chain that leads from Photosystem II, thereby generating ATP instead of being passed to $NADP^+$ to form NADPH.

24. Chemiosmosis in the two is virtually identical except that protons are pumped out of the mitochondrial inner membrane but into the chloroplast through its inner membrane.

25. In a C_3 plant when CO_2 concentrations are low, RuDP carboxylase will react with oxygen instead of carbon dioxide, forming glycolic acid. This is then oxidized, yielding CO_2. This photorespiration does not produce ATP or NADPH. Since CO_2 is released where none has been taken in, the overall efficiency of photosynthesis is reduced, particularly since no ATP or NADPH is produced.

26. Since C_4 plants are more efficient at fixing CO_2 than are C_3 plants, the stomata do not have to stay open nearly as long, and therefore the loss of water is reduced.

SECTION 3   Genetics

CHAPTER 11

From an Abbey Garden: The Beginning of Genetics

CHAPTER ORGANIZATION

I. Introduction
II. Early ideas about heredity
III. The first experiments
IV. Blending inheritance
V. The contributions of Mendel
   A. Introductory remarks
   B. The principle of segregation
      1. Introduction
      2. Consequences of segregation
   C. The principle of independent assortment
      1. General background
      2. Essay: Mendel and the laws of chance
      3. A testcross
   D. The influence of Mendel
VI. Summary

MAJOR CONCEPTS

Ideas concerning the nature of inheritance have very early origins. Many of these were grounded in mythology, folklore, and legends because the true nature of sexual reproduction and inheritance had not yet been discovered. It was not until the 1670s that sperm and egg were discovered. Even then, their respective roles in the reproductive processes were misunderstood. The conceptual breakthrough that established modern genetics as a science was made less than 150 years ago by the Austrian monk Gregor Mendel.

Mendel's principle of segregation states that hereditary factors (genes) appear in pairs—alternative forms of the same gene are known as alleles—and that members of the pair segregate into different sex cells (gametes). An individual organism receives one allele of each of its pairs from each parent. Some alleles are dominant, others are recessive. A dominant allele prevents the expression of its recessive counterpart when both exist together in an organism.

The genetic makeup of an organism is known as its genotype; the outward appearance of the organism is its phenotype. When both alleles for a characteristic are identical, the genotype is said to be homozygous. When they are different, the genotype is heterozygous. A testcross is used to determine whether an individual with the dominant phenotype is homozygous or heterozygous.

Mendel's principle of independent assortment applies to two or more genes. It states that the alleles of each gene segregate independently of one another when gametes are formed. (We know now, thanks to more recent work, that this is true only when the gene pairs are on different pairs of chromosomes.)

The ratio of dominant to recessive phenotypes in the progeny of a cross of two heterozygotes is 3:1. When two gene pairs are heterozygous (and on separate pairs of chromosomes) in both parents, the phenotypic ratio of the progeny is 9:3:3:1.

HOW TO STUDY THE CHAPTER

Read the chapter, focusing on the major concepts.

Reread the chapter, using the questions that follow to help you focus on details as they relate to major concepts. Answer the questions on a separate sheet of paper. Your answers will provide a valuable study aid in preparing for examinations.

FOCUS ON CHAPTER DETAILS

I. **Introduction** (page 237)
  1. The idea of inheritance has been acknowledged throughout history, but when did man finally begin to study the way in which hereditary processes operate?

II. **Early ideas about heredity** (pages 237–238)
  2. What can we deduce about man's early knowledge of inheritance based on his breeding of plants and animals and on the many legends that evolved as a result?
  3. What notion, first put forth by Aristotle regarding fertilization, persisted until the seventeenth century?
  4. What other notions did man have about the way in which new life forms might originate?

III. **The first experiments** (pages 238–239)
  5. What reproductive structures first came into focus under the microscopes of van Leeuwenhoek and de Graaf?
  6. Summarize the ideas of the seventeenth-century spermists and ovists about heredity.

IV. **Blending inheritance** (page 239)
  7. How did the work of nineteenth-century horticulturists lead to the formulation of a blending hypothesis of inheritance?
  8. What did the blending hypothesis fail to explain?
  9. What objections might natural selectionists, such as Darwin, raise to the blending hypothesis?

V. **The contributions of Mendel** (pages 239–247)

A. Introductory remarks
  10. What was Mendel's great contribution to the fledgling science of genetics?
  11. List the features of the garden pea plant that made it an ideal organism for Mendel's experiments.
  12. In what four ways did Mendel's approach differ from that of his predecessors, enabling him to formulate the basic principles of heredity?
  13. What advantages did Mendel's methods have for researchers who might try to verify his work?

B. The principle of segregation

General background
  14. Explain how Mendel was able to carry out interbreeding of plant varieties that were normally self-pollinating.
  15. On what basis did Mendel refer to a trait as being dominant or recessive?
  16. In a cross involving yellow-seeded and green-seeded pea plants, in what ratios do the dominant and recessive traits appear in the F_1? In the F_2?
  17. What hypothesis did Mendel advance to explain the appearance and disappearance of hereditary characteristics in his experimental crosses?

Consequences of segregation
  18. Define true-breeding, homozygous, and heterozygous in terms of the word "allele."
  19. Give working definitions for genotype and phenotype.
  20. What were the two additional experiments that Mendel performed to test his segregation hypothesis? What genetic ratios in each were necessary for confirmation of the hypothesis?
  21. Based on the definition of a testcross given in this section, would you classify either of the two experimental crosses in Question 20 as a testcross? If so, which one? See also Figures 11-9, 11-10, and 11-11.

C. The principle of independent assortment

General background

22. Explain how the outcome of the self-pollination of F_1 plants with the genotype $RrYy$ led Mendel to propose the principle of independent assortment. Clearly state this principle.

23. Match each element of the 9:3:3:1 ratio with the phenotypic combination it represents. (Pick any two of the pea plant traits you like.) What "P and F_1" genotypes could have produced this ratio?

Essay: Mendel and the laws of chance (page 244)

24. How do you calculate the probability of any number of independent events occurring together?

25. What is ensured by a large sample size?

A testcross

26. This subsection provides step-by-step instructions for the analysis of a testcross involving two gene pairs. Do several problems using the example of Figure 11–11 as a guide. Note that the phenotypic ratio of such a testcross, called a dihybrid testcross, is always 1:1:1:1.

D. The influence of Mendel

27. What was the fate of Mendel's work? When and how was it finally rediscovered?

VI. **Summary** (pages 247–248): Read the Summary. If you are familiar with the essential features of the material presented there, you are ready to take the following diagnostic examination.

TESTING YOUR UNDERSTANDING

After you have completed the following examination, compare your answers with the annotated key in the Analysis section.

1. The seventeenth-century spermists and ovists contended respectively that sperm and that egg cells:
   a. fused together at conception.
   b. contained miniature human beings.
   c. united in such a way as to produce blended characteristics.
   d. carried discrete hereditary factors.
   e. existed only in animals, not in plants.

2. The hypothesis of blending inheritance was not acceptable to followers of Darwin because it could not explain how:
   a. hybrid individuals could be formed.
   b. both male and female contributions to the offspring could come about.
   c. variation was maintained in a population.
   d. the homunculus became an embryo.

3. Mendel was successful in testing his own hypothesis regarding inheritance because he:
   a. studied only clear-cut differences between varieties.
   b. analyzed second-generation results as well as first.
   c. employed mathematics in evaluating experimental outcomes.
   d. Only b and c are correct.
   e. All three answers (a, b, and c) are correct.

4. The flower of the garden pea plant was particularly well suited to Mendel's purposes because:
   a. accidental crossbreeding is unlikely.
   b. there is but a single flower color.
   c. it attracts a wide variety of beneficial insect pollinators.
   d. male and female sex organs are located in separate flowers.

5. Mendel considered those traits that are hidden in the F_1 generation to be:
   a. heterozygous.
   b. recessive.
   c. incompletely dominant.
   d. incapable of breeding true.
   e. incapable of being inherited.

6. Alternative forms of the same gene are known as:
   a. gametes.
   b. heterozygotes.
   c. genotypes.
   d. alleles.

7. Mendel's principle of segregation states that factors separate during:
   a. fertilization.
   b. gamete formation.

c. cross-pollination.
d. seed dispersal.

8. Expressions of a trait, such as smooth versus wrinkled seeds, are referred to as:
   a. genotypes.
   b. varieties.
   c. phenotypes.
   d. factors.
   e. alleles.

9. The genetic makeup of an organism is called:
   a. a gene pool.
   b. its phenotype.
   c. a trait.
   d. an allele.
   e. its genotype.

10. An organism with two identical alleles for a given trait is:
    a. homozygous.
    b. dominant.
    c. independently assorting.
    d. self-pollinating.
    e. heterozygous.

11. When a pea plant that has yellow seeds is pollinated by a plant that has green seeds, all of the F_1 plants have yellow seeds. This means that the allele for yellow is:
    a. homozygous.
    b. true-breeding.
    c. dominant.
    d. present in both parents.
    e. assorting independently.

12. The manager of a greenhouse is trying to fill an order for garden pea plants that have white flowers by allowing F_1 plants to self-fertilize. These F_1 plants have been bred from true-breeding purple-flowered and true-breeding white-flowered parents. What portion of the stock from this cross can the manager use to fill the order?
    a. all of it
    b. three-fourths of it
    c. one-half of it
    d. one-fourth of it
    e. none of it

13. Which of the following is an example of a testcross?
    a. $TT \times Tt$
    b. $Aa \times Aa$
    c. $pp \times pp$
    d. $Dd \times dd$
    e. Both a and d are testcrosses.

14. Self-pollination in a plant that is heterozygous with respect to two gene pairs (each with one dominant allele) will cause _____ phenotypes to appear among the offspring.
    a. eight
    b. four
    c. three
    d. two
    e. no

15. The cross in the preceding problem can be solved using a Punnett square containing _____ squares.
    a. 4
    b. 8
    c. 9
    d. 16
    e. 36

16. An organism that is heterozygous for three pairs of genes can form _____ different kinds of gametes with regard to these genes.
    a. two
    b. three
    c. six
    d. eight
    e. nine

17. In using a Punnett square, the _____ appear at the head of each row and column.
    a. genotypes of the two parents
    b. phenotypes of the two parents
    c. allelic combinations of possible gametes
    d. genotypes of the offspring
    e. expected probabilities in the offspring

18. A cross involving two gene pairs results in a ratio of 9:3:3:1. This means that _____ of the offspring will exhibit _____ .
    a. 3/8; the two dominant traits
    b. 3/16; the two recessive traits
    c. 3/16; one dominant and one recessive trait
    d. 6/16; one dominant and one recessive trait
    e. 3/8; just one recessive trait

19. An organism of unknown genotype is crossed with a homozygous recessive having a genotype of *bb*.

Fifty percent of the progeny are similar in appearance to the unknown. Its genotype therefore is _____ .
   a. *BB*
   b. *Bb*
   c. *bb*
   d. still undetermined
20. Which of the following statements is historically accurate with respect to the status of Mendel's genetic principles?
   a. They were applied by plant and animal breeders of his time.
   b. They were not made known to his colleagues.
   c. They were published but were ignored until after his death.
   d. They were not officially published in a widely circulated journal.
   e. Both b and d are correct.
21. Parents heterozygous for two gene pairs can produce progeny that are different from themselves. This illustrates:
   a. a testcross for two traits.
   b. the effects of crossbreeding.
   c. that dominance is a reason for increased variability.
   d. the principle of independent assortment.
   e. that recessive alleles can be expressed in the heterozygous state.
22. This matching problem applies to the following cross:

   Ww Dd × *ww Dd*

   where
   W is a dominant allele producing smooth seeds
   w is a recessive allele producing wrinkled seeds
   D is a dominant allele producing tall stems
   d is a recessive allele producing short stems

   Match the phenotypic classes at the left with the frequency of their occurrence in the progeny of the cross, as listed at the right.

tall _____	a. 3/8
wrinkled _____	b. 1/4
smooth and tall _____	c. 9/16
wrinkled and short _____	d. 1/2
carry an unexpressed allele for short stems _____	e. 3/4
heterozygous for both traits _____	f. 1/8
	g. 3/16

23. What was Mendel's single most important contribution to the fledgling science of genetics?
24. State in your own words the two principles that Mendel set forth. What observations led him to these formulations?
25. Outline a single experimental cross (or perhaps a set of crosses) that could clearly distinguish between Mendelian inheritance and blending inheritance.
26. What ideas about heredity that seem almost laughable to us today had been seriously considered prior to the time of Mendel? Were these conceptions merely folklore, or were some justifiable in view of the lack of fundamental knowledge of the biology of reproduction? Explain.

PERFORMANCE ANALYSIS

1. The spermists and the ovists were followers, respectively, of van Leeuwenhoek and de Graaf. Spermists believed that a tiny creature, the homunculus, was contained within human sperm; ovists, that it was contained within human egg cells. The homunculus was thought to be the future human being in miniature; therefore, **b** is correct.
2. Followers of Darwin reasoned that after one generation of blending inheritance, the hereditary particles would have been completely mixed in the offspring, resulting in uniformity. This was clearly not the case. Darwin's idea of natural selection depended on the maintenance of genetic variation within populations; therefore, **c** is correct.
3. Choice **e** is correct. Mendel studied differences (a) such as flower color, length of stems, texture of the seed, etc., which in the garden pea are clear-cut and easily measurable. He allowed the F_1 to self-pollinate so that he could analyze results of a second generation (b). And Mendel counted his results and subjected them to quantitative analysis (c).
4. Choices b and d (refer to Figure 11-5) are false. Because the reproductive organs remain enclosed by petals until after fertilization, Mendel was able

to obtain crosses that would not be biased by an accidental entry of pollen (**a**) into his experimental flowers. When desired, he could cross-pollinate the plants by hand.

5. Traits that are hidden in one generation but reappear in a succeeding generation are determined by recessive (**b**) alleles. When a recessive allele is present with a dominant allele, as in a heterozygous individual, its effect is masked.

6. Alleles (**d**) are alternative forms of the same gene.

7. As a result of his experimental crosses, Mendel hypothesized (the principle of segregation) that hereditary factors separated at the time gametes were formed (**b**) and reunited when the gametes fused (fertilization).

8. The outward appearance of a trait is a phenotype, so **c** is correct. Genotype refers to the genetic makeup as determined by what alleles (or, as Mendel would have called them, factors) are present. Varieties or breeds refer to groups of organisms with a common genetic heritage, such as the pure-breeding dwarf plants that Mendel used.

9. Choice **e** is correct. Refer to the above explanation.

10. An individual whose genotype can be represented by alleles of the same type, such as *AA*, is termed homozygous (choice **a**). Depending on the relative effects of the alleles, such an individual may be homozygous dominant or homozygous recessive.

11. In crossing true-breeding yellow-seeded plants with true-breeding green-seeded ones, the offspring will each contain two different alleles. Since, however, all offspring in this cross are of the yellow phenotype, the allele for yellow must be dominant over the allele for green (**c**).

12. The cross the manager is carrying out to fill the order is $Ww \times Ww$ (purple flowers), which in the next generation gives one-fourth *WW* (purple flowers), one-half *Ww* (purple flowers), and one-fourth *ww* (white flowers). Therefore, choice **d** is correct; the manager can use the one-fourth that are white to fill the customer's order.

13. In a testcross, an individual of dominant phenotype but unknown genotype is crossed with the homozygous recessive. Only answers c and d involve homozygous recessives, but c cannot be correct because the other mating partner does not contain a dominant allele. The correct answer is therefore **d**. The heterozygous mating partner in d would have the dominant phenotype, but an observer would not know whether it was heterozygous or homozygous dominant without test-crossing it.

14. Review Figure 11-10 to see that there will be four phenotypes in the progeny (**b**).

15. Again, see Figure 11-10. The Punnett square contains 16 squares (**d**), since four gametic possibilities from each parent must be combined.

16. The probability of one factor of any pair entering a gamete is 1/2. Since, by Mendel's law of independent assortment, the segregation of different pairs of factors is independent, the probability of a given gametic combination, where three gene pairs are involved, is $1/2 \times 1/2 \times 1/2$, or 1/8; and so there will be eight (**d**) different combinations.

17. Refer to the simple four-square Punnett squares of Figure 11-8. The genotypes of all possible gametes for each parent (in this case two) are shown outside the rows and columns, respectively; therefore, **c** is correct. Genotypes of the offspring are presented inside the squares, and the probability of a given genotype in the offspring is easily computed by counting the number of squares containing that genotype and dividing it by the total number of squares.

18. Refer again to the dihybrid cross of Figure 11-10. As you can count, 3/16 of the offspring are both round (a dominant trait) and green (a recessive trait), and 3/16 are both yellow (a dominant trait) and wrinkled (a recessive trait). Therefore, $3/16 + 3/16 = 6/16$ (**d**) is correct.

19. If 50 percent of the offspring resemble the unknown, then it follows that 50 percent are homozygous recessive. The only possible explanation for these results is the cross $Bb \times bb$, indicating that the unknown genotype is *Bb* (choice **b**).

20. To set straight the historical record regarding Mendel: Mendel made his results public at an 1865 meeting of the Brünn Natural History Society, which published his paper in the society proceedings the following year (both choices b and d are

incorrect). However, Mendel's findings were not immediately applied because their significance was not understood, and they were subsequently ignored until after his death (c).

21. Choice a is incorrect by definition of a testcross, while choice e is incorrect because of the definition of recessive. Not all crossbreedings (b) produce nonparental phenotypes. Dominance (c) actually reduces the variability of a trait in the offspring, because it masks the effect of other alleles. Only d is correct; because of the law of independent assortment of alleles, new combinations will be produced in the offspring.

22. The correct answers in order are e, d, a, f, d, and b.

23. As stated on page 240, Mendel's most important contribution was that he interpreted his results as indicating that inherited characteristics are determined by discrete factors.

24. Check your own words against the formal statements of the two principles on pages 241 and 244.

25. Self-pollination of an F_1 (or crossing two F_1 animals of opposite sex) should do the trick. Try it and see. Blending inheritance would not allow the recovery of the recessive phenotype in the F_2.

26. The bizarre hybrids of ancient folklore stretch even the most fertile imagination, but ideas such as Aristotle's, which incidentally appeared in seventeenth-century medical texts, would have been quickly set aside had man had more insight into human reproductive biology.

CHAPTER 12

Meiosis and Sexual Reproduction

CHAPTER ORGANIZATION
I. Introduction
II. Haploid and diploid
III. Meiosis and the life cycle
IV. Meiosis vs. mitosis
V. The phases of meiosis
   A. General description
   B. Meiosis in the human species
   C. Essay: The consequences of sexual reproduction
VI. Cytology and genetics meet: Sutton's hypothesis
VII. Summary

MAJOR CONCEPTS

Meiosis is a unique type of nuclear division that reduces the number of chromosomal sets of a cell from two to one. A diploid cell (two sets of chromosomes) divides to produce four haploid (one set of chromosomes) cells. In the first meiotic division, homologous chromosomes are separated. In the second division, sister chromatids separate as in mitosis.

Fertilization results when two haploid gametes (of opposite sexes or mating types) unite to form a diploid zygote. Meiosis and fertilization occur at different times during the life cycles of different organisms. In primitive unicellular organisms, the haploid stage is predominant. In plants there is an alternation of haploid and diploid generations, and in animals, the diploid stage is predominant.

In humans (and other vertebrates), meiosis occurs as part of gamete formation. In males, one diploid spermatogonium gives rise to four haploid sperm. In females, one diploid oogonium gives rise to one haploid ovum and three haploid polar bodies.

As a result of his observations on grasshopper meiosis, William Sutton hypothesized that chromosomes are the carriers of genes and provide the physical mechanism for their separation according to Mendel's principle of segregation.

HOW TO STUDY THE CHAPTER

Read the chapter, focusing on the major concepts.

Reread the chapter, using the questions that follow to help you focus on details as they relate to major concepts. Answer the questions on a separate sheet of paper. Your answers will provide a valuable study aid in preparing for examinations.

FOCUS ON CHAPTER DETAILS

I. **Introduction** (page 249)
   1. What two events are necessary in the life cycle of a sexually reproducing eukaryote? What is the role of each event?

II. **Haploid and diploid** (pages 249–250)
   2. All species have a characteristic chromosome number. What is that number in humans?

3. How does the chromosome number of the gametes differ from that of the somatic cells? What terms denote these differences?

4. Using the symbols *n* and *2n*, demonstrate what happens in fertilization. What is a zygote?

5. What is the relationship between homologous chromosomes?

6. What is the origin of each of the members of a homologous pair?

7. What is the primary role of meiosis? What other contribution does it make?

III. Meiosis and the life cycle (pages 250–252)

8. When does meiosis occur in such organisms as *Chlamydomonas* and *Neurospora*? In this type of life cycle, what does meiosis accomplish?

9. Paying particular attention to the timing of fertilization and the role of meiosis, describe the life cycle of a fern.

10. What is alternation of generations, and in what organisms is it typically found?

11. In the typical animal life cycle, when does meiosis occur? What is the result of fertilization?

IV. Meiosis vs. mitosis (page 252)

12. Explain the differences between meiosis and mitosis in terms of the number of cells produced and the genetic constitution of these cells.

V. The phases of meiosis (pages 253–257)

A. General description

13. What are the main features of meiosis I and meiosis II?

14. What events in prophase I lead to the formation of tetrads?

15. Explain the process of crossing over. What is its significance?

16. Follow the diagrams of the stages of meiosis on pages 254 and 255 while reading the accompanying explanation in the text. List the significant features of each stage.

17. What feature of anaphase I distinguishes meiotic division from mitotic division?

18. In what genetically important ways does a cell finishing the first meiotic division differ from the original cell?

19. What events may or may not take place between meiosis I and II, depending upon the species?

20. What is the chromosomal configuration of the cell at the beginning of meiosis II?

21. The events that take place during the phases of meiosis II (see diagrams on page 255) resemble those of mitosis, except, of course, that the end results differ. How do they differ?

22. Use Figure 12–10 to review meiosis and to compare it with mitosis.

B. Meiosis in the human species

23. In what human tissues (and in what specific cells) does meiosis occur?

24. How do the meiotic products formed in the female differ from those of the male? (Compare Figure 12–12 to Figure 12–11.) What is the advantage of this difference to the egg?

C. Essay: The consequences of sexual reproduction (page 258)

25. List several methods that organisms use to reproduce asexually.

26. What are two disadvantages of sexual reproduction and the one advantage that, apparently, strongly outweighs them? (To convince yourself of this advantage, review the diagrams accompanying the essay.)

27. State a general formula for determining the number of genetic combinations that can arise in gametes following meiosis, given the haploid number of chromosomes.

VI. Cytology and genetics meet: Sutton's hypothesis (page 259)

28. What important observation did William Sutton make about grasshopper chromosomes engaged in meiosis?

29. What parallel did Sutton draw between the events of meiosis and the work of Mendel?
30. What did Sutton propose as a result of his observations?
31. In order to help integrate what you have learned in Chapters 11 and 12, study the Punnett square of Figure 12-14 (reconstructed from Figure 11-10 on page 245 to include the idea that Mendelian factors are actually located on genes). Explain how Mendel's two laws are obeyed if we accept Sutton's hypothesis.

VII. Summary (pages 259-261): Read the Summary. If you are familiar with the essential features of the material presented there, you are ready to take the following diagnostic examination.

TESTING YOUR UNDERSTANDING

After you have completed the following examination, compare your answers with the annotated key in the Analysis section.

1. In a species that is diploid during most of its life cycle, fertilization of gametes will:
   a. restore the diploid number.
   b. reduce the diploid number by one-half.
   c. produce a tetraploid zygote.
   d. result in a zygote with two separate nuclei.
2. An organism that has 12 homologous pairs of chromosomes has a diploid number of _____ .
   a. 2
   b. 6
   c. 12
   d. 24
   e. 48
3. By the end of telophase II of meiosis:
   a. there are two nuclei in each of four cells.
   b. each nucleus has n chromosomes.
   c. all four cells each have $2n$ chromosomes.
   d. each cell has $2n$ chromatids.
   e. there are four cells, each having $2n$ pairs of homologous chromosomes.
4. In a sexually reproducing species that is haploid during most of its life cycle, meiosis:
   a. restores the diploid condition.
   b. does not occur.
   c. I precedes fertilization; meiosis II follows it.
   d. follows fertilization.
   e. does not reduce the chromosome number.
5. In ferns, meiosis is directly followed by the production of:
   a. a haploid sporophyte.
   b. a diploid sporophyte.
   c. haploid spores.
   d. haploid gametes.
   e. a gametophyte.
6. During _____ of meiosis, _____ separate.
   a. anaphase II; homologues
   b. anaphase II; chromatids
   c. prophase I; homologues
   d. prophase II; chromatids
   e. telophase I and II; chromatids
7. During the tetrad stage of meiosis, interchange of genetic material takes place between:
   a. nonhomologous pairs of chromosomes.
   b. chromatids of homologous chromosomes.
   c. male and female pronuclei.
   d. identical alleles.
   e. male and female gametes.
8. In metaphase I of a meiotic division, chromosomes are:
   a. distributed randomly within the nucleus.
   b. dispersed in the form of chromatin.
   c. beginning to migrate to opposite poles of the cell.
   d. engaged in crossing over.
   e. grouped by homologous pairs along the equator.
9. The phenomenon of crossing over can be observed during:
   a. prophase I.
   b. prophase II.
   c. both anaphase I and anaphase II.
   d. anaphase I only.
   e. the period between meiosis I and II.
10. Which of the following events is *never* observed between telophase I and prophase II?
    a. cytokinesis
    b. reconstruction of nuclear membranes
    c. replication of DNA
    d. dispersing of chromosomes
    e. Both a and c are correct.

11. In the males of vertebrate species, meoisis occurs in:
    a. all somatic cells.
    b. primary spermatocytes.
    c. secondary oocytes.
    d. spermatids.
    e. sperm cells.

12. Explain how the contents of a primary oocyte are distributed immediately after the cytokinesis that follows meiosis in the human female.
    a. There are four haploid cells of approximately equal size.
    b. There are four haploid cells, one containing most of the original cytoplasm.
    c. There is but one large cell, the ovum, which contains all of the original genetic material.
    d. There is one haploid cell, the ovum, and three diploid polar bodies.

13. In a cell undergoing meiosis, chromatid pairs are held together at their centromeres through:
    a. metaphase I.
    b. anaphase I.
    c. telophase I.
    d. metaphase II.
    e. telophase II.

14. The notion that Mendelian factors (genes) might _____ occurred to William Sutton as he observed meiosis in grasshopper testes.
    a. be carried on chromosomes
    b. lie within the cytoplasm
    c. cross over during prophase I of meiosis
    d. assort independently during anaphase
    e. be present in gametes

15. William Sutton observed that chromosome behavior during _____ was strikingly similar to Mendel's principle of segregation.
    a. mitosis
    b. cytokinesis
    c. meiosis
    d. crossing over
    e. differentiation of the gametes

16. Had Mendel been able to witness "the dance of the chromosomes," he would undoubtedly have realized that his principle of independent assortment would be obeyed:
    a. only if each chromosome contained but a single allele.
    b. when the alleles in question were on different pairs of homologous chromosomes.
    c. if the alleles in question were on the same pair of homologous chromosomes.
    d. only if there was no crossing over during meiosis.
    e. Both b and d are correct.

17. Asexual reproduction is considered less costly in terms of energy than sexual reproduction because reproduction in asexual organisms never involves:
    a. cytoplasmic division.
    b. mitosis.
    c. replication of genetic material.
    d. meiosis.

18. Which of the following represents the correct chronological order of cellular events that permits a single spermatogonium to give rise to four sperm cells?
    a. meiosis, DNA replication, differentiation of spermatids
    b. DNA replication, mitosis, meiosis
    c. DNA replication, mitosis, differentiation of spermatids
    d. DNA replication, meiosis, differentiation of spermatids

19. A human female is capable of producing _____ genetically different gametes strictly as a result of independent assortment.
    a. 46^2 b. 23^2 c. 2^{23} d. 2^{46} e. 46

20. Match each event outlined below with a term from the list that follows.

    diploid stage ⟷ haploid stage _____
    diploid cell ⟶ haploid cells _____
    haploid gamete + haploid gamete ⟶ diploid zygote _____
    products of meiosis ⟶ haploid spores _____
    one diploid cell ⟶ two diploid cells _____
    one two-nucleate cell ⟶ two diploid cells

    a. fertilization
    b. differentiation
    c. alternation of generations
    d. cytokinesis
    e. meiotic division
    f. mitotic division

21. Match each label on the left with the appropriate description of the human cell from the list at the right.

    gamete _____
    tetrad _____
    somatic cell _____
    zygote _____
    polyploid _____

    a. grouping of 4 chromatids
    b. cell with 92 chromosomes
    c. cell with 23 chromosomes
    d. cell with 23 homologous pairs
    e. the first diploid cell of the next generation

22. Diagram the life cycle of a single-celled organism that alternates between sexual reproduction and asexual reproduction. Label all major events, such as mitosis, meiosis, fertilization, differentiation, etc. Label all stages as either haploid or diploid.

23. Mitosis and meiosis use practically the same cellular machinery, yet there are some important differences between the two. List these differences grouped according to (1) movements of the chromosomes and action of the spindle apparatus; and (2) the final consequence of the process for number of cells and genetic makeup of the cells.

PERFORMANCE ANALYSIS

1. In species that are diploid, meiosis results in haploid gametes. When gametes fuse at fertilization, the zygote that results has the diploid number (a).

2. Choice d is correct; the diploid number of chromosomes is the total number of chromosomes in a somatic cell. Twelve pairs times 2 chromosomes per homologous pair equals 24 chromosomes.

3. See the diagram on page 255 labeled telophase II. Note that there are four nuclei, each of them haploid (n chromosomes); choice b is therefore correct.

4. Choice d is correct. *Chlamydomonas* is an example of an organism that is haploid during most of its life cycle. Refer to Figure 12-3 to verify for yourself that in such cases meiosis must follow fertilization in order to restore the original haploid number.

5. If you consult the fern life cycle in Figure 12-4, you will see that meiosis occurs in the sporangia, the spore-producing bodies of the fern; therefore, c is correct.

6. A look at the diagrams on pages 254 and 255 should convince you that homologues separate at anaphase I and chromatids separate at anaphase II (b).

7. Crossing over, the interchange of genetic material between chromatids of homologous chromosomes (b), takes place during the tetrad stage (prophase I) of meiosis.

8. The diagram of metaphase I on page 254 shows that pairs of homologous chromosomes are grouped along the cell's equator (e).

9. Choice a is correct. Refer to the explanation for Question 7 above.

10. Depending on the species involved, cytokinesis (a), reconstruction of nuclear membranes (b), and dispersing of chromosomes (d) may take place. However, the purpose of meiosis—the reduction of the diploid number to the haploid number—would be defeated if the genetic material were replicated before the completion of meiosis; therefore, c is correct.

11. Choice b is correct. Refer to Figure 12-11.

12. Figure 12-12 shows that one haploid cell, the ovum, contains the bulk of the cytoplasm of the oogonium. As many as three smaller haploid cells, the polar bodies, may be present. Choice b is therefore correct.

13. Although homologues separate at anaphase I, individual chromatids do not separate from one another until anaphase II; therefore, choice d is correct.

14. Since Sutton observed that chromosomes resembling one another in size and shape paired during meiosis and then became separated, he hypothesized that chromosomes carried Mendelian factors (a).

15. Since Sutton observed the separation of homologous chromosomes during anaphase I of meiosis (c), he suspected that it was at this point in gamete formation that Mendel's principle of segregation was obeyed.

16. Mendel considered genetic factors (genes) to be discrete entities. However, had he been able to look over Sutton's shoulder to witness the meiotic dance of the chromosomes, he would have realized that if two genes lay on the same chromosome, their alleles would not assort independently as would alleles on different chromosomes (**b**). If each chromosome contained but a single allele (a), the principle would be obeyed for every pair of alleles; but choice a is qualified by *only if*, which means that no other gene arrangement would work and makes this choice incorrect. Obviously, independent assortment takes place in situations where there are many different genes on each chromosome. Choice d is incorrect because crossing over deals with the separation of different genes on the same chromosome, while independent assortment deals with the separation of alleles of different genes on nonhomologous chromosomes.

17. Asexual reproduction does require cytoplasmic division (a), mitosis (b), and DNA replication (c). Only sexually reproducing organisms require meiosis (**d**), which is considered a complex and expensive process. But if you look around at all the sexually reproducing species that exist, you could infer that meiosis must be worth the costs.

18. Follow along using Figure 12-11. After replicating its DNA, a spermatogonium first divides mitotically to form primary spermatocytes. Following mitosis, the newly formed diploid spermatocytes undergo replication. Meiosis I results in secondary spermatocytes, and meiosis II produces haploid spermatids. These differentiate into haploid sperm cells. Choice **d** reflects this sequence.

19. Since the haploid number in the human female is 23 and 1 of 2 chromosomes for each of the 23 pairs is alloted per gamete, there will be 2^{23} (**c**) possible gametic combinations for a female egg.

20. The correct answers in order are **c**, **e**, **a**, **b**, **f**, and **d**.

21. The correct answers in order are **c**, **a**, **d**, **e**, and **b**.

22. Compare your diagram to Figure 12-3.

23. Consult Figure 12-10 for a comparison of mitosis and meiosis.

CHAPTER 13

Genes and Chromosomes

CHAPTER ORGANIZATION

I. Introduction

II. **Broadening the concept of the gene**
   A. Introductory remarks
   B. Gene interactions
      1. Incomplete dominance
      2. Epistasis
      3. Genes and the environment
   C. Mutations
      1. General introduction
      2. Mutations and genetic research
      3. Mutations and evolutionary theory
   D. Polygenic inheritance and pleiotropy
      1. Continuous variation
      2. Pleiotropy
   E. Multiple alleles

III. **Chromosomes and genes**
   A. Introductory remarks
   B. Sex determination
   C. Essay: Calico cats, Barr bodies, and the Lyon hypothesis

IV. **The golden age of** *Drosophila*
   A. Introductory remarks
   B. Sex-linked characteristics
   C. Linkage
   D. Recombination

V. **Mapping the chromosome**

VI. **Giant chromosomes**

VII. **Summary**

MAJOR CONCEPTS

The phenotype of any organism is the result of the interaction between that organism's genes and its environment. But genes themselves have many different levels of interaction. Mendel's work indicated that the relationship between alleles was one of dominance and recessiveness. Later work by others indicated that some alleles exhibit incomplete dominance, which results in a phenotype in heterozygotes that is intermediate between the phenotypes of the two homozygotes, reminding one of the older theory of blending inheritance.

Polygenic inheritance is a situation in which many different genes contribute to the expression of a given phenotype. It should be distinguished from the situation in which a given gene is represented by many different alleles within a given population (multiple alleles). The situation that exists when a given gene affects many different traits or characteristics is called pleiotropy. Finally, epistasis is an interaction of two or more different genes such that an allele of one or the other of them (or both) is prevented from being expressed. Care should be taken to distinguish epistasis from dominance, in which the gene being masked is an allele (recessive) of the gene doing the masking (dominant).

Mutations are the ultimate and continual source of the genetic variations that are the basis of evolution. Certain factors (mutagens), including x-rays, ultraviolet light, and some chemicals, can increase the rate of genetic mutations.

A great deal of progress was made in genetics before scientists discovered the location of genes. Sutton proposed that genes are located on chromosomes. The correlation between sex determination and chromosomes, by Sutton, and the discovery of a sex-linked trait in *Drosophila*, by Morgan, convinced most geneticists that genes are, in fact, located on chromosomes.

Two genes located on the same chromosome (in the same linkage group) do not segregate independently. However, they may separate and recombine during crossing over. As originally hypothesized by Sturtevant, the frequency with which a pair of genes recombines can be used to determine the relative locations of genes on a chromosome. This process is called chromosome mapping.

HOW TO STUDY THE CHAPTER

Read the chapter, focusing on the major concepts.

Reread the chapter, using the questions that follow to help you focus on details as they relate to major concepts. Answer the questions on a separate sheet of paper. Your answers will provide a valuable study aid in preparing for examinations.

FOCUS ON CHAPTER DETAILS

I. Introduction (page 262)

1. How was the "golden age" of genetics launched?

II. Broadening the concept of the gene (pages 262–267)

A. Introductory remarks

2. Most genetically determined traits are more complex than the simple inheritance patterns demonstrated by Mendel. Explain.

B. Gene interactions

Incomplete dominance

3. What is incomplete dominance? Cite an example.

Epistasis

4. Define epistasis.
5. How can a 9:7 ratio in the F_2 generation of different pure-breeding varieties be explained by epistasis? See Figure 13-2.
6. Using a Punnett square like that of Figure 13-2, explain the results given in the text (page 263) for flower color in the F_2 generation of *Salvia horminum*.

Genes and the environment

7. Are phenotypes determined solely by genotypes? Explain.
8. Cite several examples of how environmental conditions influence genetically determined traits.

C. Mutations

General introduction

9. Explain how the idea of mutants first came to be recognized.
10. How did Hugo de Vries define mutation? (Remember this definition, for it will later be qualified.)

Mutations and genetic research

11. What is a mutagen? How did knowledge of the action of mutagens enhance genetic research?

Mutations and evolutionary theory

12. How did Mendel's principles support the theory of the early evolutionists? What major question did these principles fail to answer?
13. What role do mutations play in evolution?

D. Polygenic inheritance and pleiotropy

Continuous variation

14. What is polygenic inheritance? How do individuals of a population differ with regard to traits determined polygenetically? See Table 13-1.

Pleiotropy

15. What is pleiotropy? In what ways can a single mutation alter an organism's phenotype?

E. Multiple alleles

16. How can there be more than two alleles of a gene if organisms are diploid?

III. Chromosomes and genes (pages 267–268)

A. Introductory remarks

17. What attitudes toward the gene as a physical entity persisted in spite of Sutton's work?

B. Sex determination
18. What difference did Sutton notice between the chromosomes of males and females of diploid species? What terms are used to denote this difference?
19. Describe the chromosomal makeup of the human male and female. Is this same pattern present in other species?
20. Why is it said that the sperm cell determines the sex of human offspring?
21. How does the fact that male and female chromosomes have a different appearance support the hypothesis that genes are located on chromosomes? (Look for even stronger evidence of this sort in the remainder of the chapter.)

C. Essay: Calico cats, Barr bodies, and the Lyon hypothesis (page 271)
22. According to Mary Lyon, what are Barr bodies?
23. In what organisms have Barr bodies been found? How many Barr bodies are there in the somatic cells of these organisms? In the gametes?
24. How is the Lyon hypothesis able to explain coat color in calico cats?

IV. **The golden age of** *Drosophila* (pages 268–272)

A. Introductory remarks
25. Why did *Drosophila* turn out to be such an excellent tool for genetic research in Morgan's Columbia lab?
26. Describe the chromosome composition of *Drosophila*.

B. Sex-linked characteristics
27. What genetic ratio did Morgan obtain when he crossed a white-eyed male with a normal, red-eyed female? What ratio did he obtain by crossbreeding the F_1 offspring? Why did he consider this surprising? (Do a Punnett square for the cross based on simple Mendelian inheritance, and you should see why.)
28. What additional cross did Morgan make, and what results did he obtain?
29. How did the Columbia team explain the results of these two crosses? Diagram each of these crosses, following the assumptions of Morgan. (Write the allele for eye color carried on the X chromosomes as X^W or X^w. Indicate the absence of the allele on the Y chromosome as Y^o.)
30. Diagram a cross between a white-eyed female and a red-eyed male. (Check your results against Figure 13–9.) Would you expect to obtain males and females in equal proportions? Explain.
31. Traits determined by genes on the sex chromosomes are said to be sex-linked. How does the idea of sex-linked traits provide support for the gene-chromosome hypothesis?

C. Linkage
32. In what case is Mendel's law of independent assortment not upheld? Why was Mendel fortunate in being able to demonstrate this law using pea plants?
33. What is a linkage group?

D. Recombination
34. In crossing normal flies with black-bodied, short-winged mutant ones, what F_2 ratios are expected if (a) the gene pairs are on different chromosomes or (b) the gene pairs are linked?
35. What deviation from this expectation did Morgan observe?
36. What ratios did Morgan expect in testcrossing F_1 (heterozygous for both gene pairs) according to situations (a) and (b) above?
37. Based on the testcross results, did Morgan accept hypothesis (a) or (b)? What qualification did he make in order to explain the differences between expected ratios and those he actually tallied?
38. How does this explanation fit in with what you learned about meiosis in Chapter 12?

V. **Mapping the chromosome** (pages 272–274)
39. How are genes organized on a pair of homologous chromosomes?

40. What did Morgan discover about the degree of recombination between linked genes?
41. What did Sturtevant think was the relationship between recombination frequency and physical distance between genes?
42. How did Sturtevant propose to map a chromosome? (See Figure 13-13.)
43. Why is a chromosome map not necessarily an accurate representation of the chromosome?

VI. **Giant chromosomes** (pages 274-275)

44. What accounts for the appearance of giant chromosomes in the salivary glands of *Drosophila* larvae?
45. How does the staining of these chromosomes allow geneticists to detect changes in chromosome structure?
46. What chromosomal changes (aberrations) have been observed? How did the study of these aberrations make it possible to assign genes to specific physical locations on a chromosome?
47. In what ways did *Drosophila* make significant contributions toward solving the mystery of chromatin's role in heredity?

VII. **Summary** (pages 275-276): Read the Summary. If you are familiar with the essential features of the material presented there, you are ready to take the following diagnostic examination.

TESTING YOUR UNDERSTANDING

After you have completed the following examination, compare your answers with the annotated key in the Analysis section.

1. The ultimate source of genetic variation, upon which evolutionary change depends, is:
    a. independent assortment.
    b. mutation.
    c. sexual reproduction.
    d. natural selection.
    e. crossing over.
2. The expression of a genetically determined trait may depend upon:
    a. genotype.
    b. gene interaction.
    c. the influence of the environment.
    d. Both a and b are correct.
    e. All three answers (a, b, and c) are correct.
3. A trait that is continuously variable is most likely to be determined by:
    a. more than one gene.
    b. a dominant allele.
    c. pleiotropic inheritance.
    d. linked genes.
    e. gene interaction.
4. A cross between a plant with red flowers and one with white flowers produces an F_1 generation that is all pink. The simplest genetic explanation for this is:
    a. pleiotropic inheritance.
    b. polygenic inheritance.
    c. incomplete dominance.
    d. epistasis.
    e. sex-linked inheritance.
5. Which of these statements accurately applies to sex determination in humans?
    a. Females are heterogametic.
    b. X chromosomes have no influence on sexual characteristics, since they are inactivated during development.
    c. Sex of the offspring is determined by the male.
    d. Males receive their X chromosomes from their fathers.
6. Sutton and a number of other cytological geneticists observed that in many organisms, the sexes differ in:
    a. the diploid number of chromosomes.
    b. a single pair of chromosomes.
    c. the total number of autosomes.
    d. the number of genes present on the X chromosome.
    e. outward appearance, even though there are no distinguishing features in their chromosomal makeup.
7. All genes on a single chromosome are referred to as:
    a. homologous.
    b. a linkage group.
    c. autosomal.

d. epistatic.
   e. allelic.

8. In order to explain the distribution of white eyes among male and female offspring of a white-eyed male fruit fly, Morgan postulated that the allele for white eye color is:
   a. dominant and X-linked.
   b. dominant and autosomal.
   c. recessive and X-linked.
   d. recessive and present on both sex chromosomes.
   e. recessive and autosomal.

9. If a white-eyed *Drosophila* male ($X^w Y^o$) is crossed with a red-eyed female carrier ($X^W X^w$), then _____ percent of the offspring will be white-eyed males.
   a. 50
   b. 25
   c. 12.5
   d. 100
   e. There will be no white-eyed males.

10. Genes that seldom exhibit independent assortment are:
    a. located on different homologous pairs.
    b. never found in eukaryotes.
    c. located on the same chromosome but are widely separated.
    d. located close to one another on the same chromosome.

11. The diploid number in garden peas is 14. The number of linkage groups, therefore, is _____.
    a. 56
    b. 28
    c. 14
    d. 7
    e. The number of linkage groups cannot be determined from the diploid number.

12. Two dominant alleles, *A* and *B*, are closely linked. A cross between organisms that are heterozygous at both loci (*AaBb*) should result in _____ of the progeny being recessive for both trait *A* and trait *B*.
    a. none
    b. 1/16
    c. 1/4
    d. 3/8
    e. 1/2

13. The appearance of _____ is considered to be cytological evidence that crossing over is taking place during meiosis.
    a. translocations
    b. inversions
    c. chiasmata
    d. linkage
    e. mapping

14. If locus *X* is determined to be at a map distance of 10 units from both loci *Y* and *Z*, and if the map distance between *Y* and *Z* is approximately 20, then:
    a. locus *X* is closer to the centromere than is *Y* or *Z*.
    b. the physical distance from *X* to *Y* is exactly equal to that from *X* to *Z*.
    c. the crossover frequency between *X* and *Y* is equal to that for *X* and *Z*.
    d. alleles *X* and *Z* will assort independently more often than *Y* and *Z*.
    e. Both a and b are correct.

15. The giant salivary gland chromosomes are a result of:
    a. numerous DNA replications without a mitotic division.
    b. a single DNA replication followed by continual mitotic division.
    c. mitosis without cytokinesis.
    d. a number of translocations involving the same homologous pair.
    e. an unlimited amount of crossing over during meiosis.

16. Chromosomal aberrations may be detected by:
    a. examining chromosomal banding patterns.
    b. analyzing phenotypic ratios in a genetic cross.
    c. observing the gross appearance of condensed chromosomes.
    d. Both a and c are correct.
    e. All three answers (a, b, and c) are correct.

17. During early prophase I of meiosis in a human spermatocyte, an observer detects 22 homologous chromosome pairs, 1 unmatched normal chromosome, and 2 chromosomal fragments. The observer should suspect that a(n) _____ has occurred.
    a. crossover
    b. replication
    c. translocation

    d. inversion
    e. deletion
18. It was difficult for many cell biologists at the time of Sutton to believe that chromatin held the key to the mystery of hereditary variation because:
    a. it was difficult to show that chromatin and chromosomes were, in reality, the same material.
    b. under the microscope, the chromatin of one organism looked much like that of any other.
    c. chromatin was known not to contain DNA.
    d. during nuclear division, the chromatin is not equally apportioned.
19. The existence of Barr bodies in mammalian females means that:
    a. for some genes, females are functionally haploid.
    b. only one-half of female gametes will contain an active X chromosome.
    c. somatic cells of the female are not all identical.
    d. Both a and c are correct.
    e. All three answers (a, b, and c) are correct.
20. Match the descriptive phrase on the left with the appropriate term on the right.

an abrupt change in hereditary material _____	a. mutation
having two X chromosomes _____	b. mutagen
	c. autosomal
	d. sex-linked
an organism with a pair of recessive alleles _____	e. locus
	f. map distance
the exact position of a gene _____	g. homogametic
	h. homozygous
reflects the probability of a crossover event _____	
alleles that assort independently of those on X and Y chromosomes _____	

21. Cite one example of polygenic inheritance and one of pleiotropic inheritance.
22. Demonstrate with an experimental example why it might be difficult to determine whether a pair of genes is linked or unlinked.
23. How did Morgan's experimental crosses demonstrating the pattern of eye color inheritance in *Drosophila* support the Sutton hypothesis?

PERFORMANCE ANALYSIS

1. Mutations (**b**) provide the genetic variation upon which natural selection acts. Recombinations that occur by independent assortment and crossing over in sexually reproducing organisms are, by themselves, insufficient to bring about major evolutionary changes.
2. Genetic expression depends upon the organism's genotype (**a**) at one or more loci, interactions (**b**) among these loci, and the way the genotype will be expressed in a given environment (**c**); therefore, choice **e** is correct.
3. A trait determined by more than one gene (**a**) will be distributed in a population according to the shape of the curve in Figure 13–5. Extend the two-locus example of Table 13–1 to three or more loci, and you will readily see why.
4. Pleiotropic inheritance can be ruled out, since only one trait is discussed in the problem. Since the distribution of the trait among sexes is not provided, sex-linked inheritance need not be considered. Polygenic inheritance and epistasis both involve more than a single gene. You can, on the other hand, account for the results given with a single gene, if it is assumed that the effects of both alleles are demonstrated. Therefore, the simplest explanation is incomplete dominance (**c**).
5. In humans, it is the male, not the female, who is heterogametic. Just one of the X chromosomes, represented by a Barr body, becomes inactive. (Also, if an extra X chromosome is present in either a male or a female, sexual characteristics are likely to be affected.) Males receive their X chromosomes (**d**) from their mothers and their Y chromosomes from their fathers. By passing on the Y chromosome, it is the male who determines the sex of the offspring; therefore, choice **c** is correct.
6. Sutton observed (correctly) that the sex of an individual is associated with the appearance of a single pair of chromosomes (**b**), the sex chromosomes.

7. Genes on the same chromosome are said to be linked, or to constitute a linkage group (**b**), because ordinarily they do not assort independently during gamete formation. It is members of a gene pair that are referred to as homologous.

8. Because a cross between the mutant male and red-eyed females produced no white-eyed offspring of either sex, Morgan postulated that the allele was recessive. Because only males in the F_2 were white-eyed, Morgan hypothesized that the allele was carried only on the X chromosome, where its effect on the male would not be masked if it was recessive. Choice **c** is therefore correct.

9. The cross between a red-eyed female heterozygote and a white-eyed male can be outlined as $X^W X^w \times X^w Y^o$. Using a four-square Punnett square, you should be able to determine that one-fourth of the offspring will be red-eyed females, one-fourth will be white-eyed females, one-fourth will be red-eyed males, and one-fourth will be white-eyed males; therefore, **b** is correct.

10. Genes that do not assort independently as predicted by Mendel's principle are linked. But because of crossing over during meiosis, linked genes sometimes assort independently. Those that do least often are positioned closest together on the same chromosome (**d**).

11. The number of linkage groups is equal to the number of homologous pairs of chromosomes—in this case, 7 (**d**).

12. One-fourth of the progeny (**c**) will be recessive for both traits. (The answer would be the same if only one locus were involved.)

13. Chiasmata (**c**) (see Figure 13–11) are actual points of physical attachment of homologous chromosomes engaged in crossing over.

14. The three loci may be mapped as follows:

    $$Y \longleftarrow 10 \longrightarrow X \longleftarrow 10 \longrightarrow Z$$

    Their location with respect to the centromere is not known. Because map distance does not relate exactly to physical distance on the chromosome, the loci may not be equally spaced, even though they are 10 units apart. What the equality of map distances *does* mean is that the crossover frequency of X and Y will be equal to that of X and Z (**c**).

15. Because mitosis does not follow replication of DNA (**a**) in the salivary gland cells of larval fruit flies, daughter chromosomes are not separated and give the appearance of single large chromosomes.

16. Chromosomal aberrations involving a large enough portion of the chromosome will be detectable under a microscope as gross structural changes (**c**). Banding patterns in *Drosophila* (**a**) were used to detect the aberrations that were suspected after examining phenotypic ratios (**b**) in mapping studies. Therefore, **e** is correct.

17. A deletion (**e**) in a single chromosome would produce two fragments totaling together the size of the original chromosome and leave one full-sized chromosome (its homologue) unmatched.

18. Choices **a**, **c**, and **d**, even at the time of Sutton, would have been regarded as false. Choice **b**, then, is correct; chromatin was considered to be a homogeneous substance.

19. Barr bodies are X chromosomes that become inactivated early in embryonic life (except in those cells that will become gametes) and are not expressed genetically. This means that for the genes on the X chromosome, the female mammal is functionally haploid (**a**). Since either of the two X chromosomes may become inactivated, somatic cells may not be identical (**c**). Choice **d** is therefore correct.

20. The correct answers in order are **a**, **g**, **h**, **e**, **f**, and **c**.

21. Examples of polygenic inheritance are height in humans, kernel color in wheat, and skin color in humans. An example of pleiotropy is the frizzle trait in fowl.

22. You can solve this problem by remembering that even unlinked genes assort independently only 50 percent of the time. A linkage situation involving 50 percent recombination would be impossible to distinguish from a nonlinkage situation.

23. Sutton had suggested that Mendelian factors, or genes as we now call them, were located on chromosomes. Morgan later correlated the occurrence of different chromosomes with a hereditary characteristic, namely, the sex of an organism.

CHAPTER 14

The Path to the Double Helix

CHAPTER ORGANIZATION

I. Introduction
II. The chemistry of heredity
  A. Introductory remarks
  B. The language of life
III. The DNA trail
  A. The transforming factor
  B. The nature of DNA
  C. The bacteriophage experiments
  D. Further evidence for DNA
    1. General information
    2. Chargaff's results
  E. The hypothesis is confirmed
IV. The Watson-Crick model
  A. Introductory remarks
  B. The known data
  C. Building the model
  D. Essay: Who might have discovered it?
V. DNA replication
  A. General information
  B. A confirmation
  C. The mechanics of DNA replication
  D. The energetics of DNA replication
VI. DNA as a carrier of information
VII. Summary

MAJOR CONCEPTS

Early studies in genetics prompted scientists to search for the means by which genetic information is passed from one generation to the next. At first, many scientists suspected that proteins (and their constituent amino acids) comprised the "language of life." This proposal was later disproved when overwhelming evidence showed that deoxyribonucleic acid (DNA) carries the genetic information.

Work in several different areas performed by different scientists contributed to the evidence that DNA carries the hereditary information. These studies included work with a "transforming factor" that was passed from encapsulated to nonencapsulated pneumococci (Griffith); identification of Griffith's transforming factor as DNA (Avery and colleagues); the discovery that all somatic cells of the same species contain the same quantity of DNA (Mirsky); and the revelation that the proportion of the nitrogenous bases in DNA is constant within a species but varies between species (Chargaff).

In 1953, James Watson and Francis Crick proposed that the structure for DNA consists of a double helix composed of two chains of nucleotides that are joined together by hydrogen bonds. The helical, ladderlike DNA molecule consists of subunits called nucleotides. Each nucleotide consists of a phosphate group, a sugar group (deoxyribose), and a nitrogenous base. The rails of the ladder are composed of alternating sugars and phosphate groups. The rungs of the ladder are composed of two nitrogenous bases, one protruding from each sugar-phosphate rail. There are four bases in all: adenine and cytosine, which are known as purines; and thymine and guanine, which are pyrimidines. Adenine always pairs with thymine, and guanine always pairs with cytosine.

In DNA replication, the two strands of the double helix separate and a new strand forms along each old one. Therefore, each new helix has one old strand and one new strand. Specific enzymes are needed at each step to break and/or form bonds. The energy for DNA replication is supplied by triphosphate compounds (such as ATP and GTP). The large amount of energy released during DNA replication makes the reaction highly exergonic, thereby precluding the reverse reaction (breakdown of DNA as it is formed).

HOW TO STUDY THE CHAPTER

Read the chapter, focusing on the major concepts.

Reread the chapter, using the questions that follow to help you focus on details as they relate to major concepts. Answer the questions on a separate sheet of paper. Your answers will provide a valuable study aid in preparing for examinations.

FOCUS ON CHAPTER DETAILS

I. Introduction (page 278)
1. What marked a turning point for genetics in the 1940s?

II. The chemistry of heredity (page 278)
A. Introductory remarks
2. Vitalists believed that the complexities of life processes (heredity in particular) could not be understood in terms of "lifeless" chemicals. When was this view finally put to rest?

B. The language of life
3. Why were proteins thought to be the genetic material?
4. By what mechanism were chromosomal proteins thought to guide the many activities of the cell?

III. The DNA trail (pages 279-284)
A. The transforming factor
5. What was the goal of Frederick Griffith's research?
6. Why did these experiments attract the attention of geneticists?
7. How was the "transforming factor" of Griffith's experiments shown to be DNA? Who performed this research?

B. The nature of DNA
8. How was DNA discovered and by whom?
9. How could DNA be traced to specific locations within an organism?
10. Become familiar with the basic structure of a nucleotide (see Figure 14-4a). What variations in this basic structure occur? (See Figure 14-4b.)
11. How did P. A. Levene contribute to the notion that DNA was *not* the genetic material?

C. The bacteriophage experiments
12. What are bacteriophages? What group of bacteriophages are associated with *E. coli*?
13. What advantages do these bacteriophages have as a genetic research tool?
14. Describe the infection cycle set up by these bacteriophages in their *E. coli* hosts.
15. How does the chemical nature of viruses make them ideally suited for solving the problem of the nature of the gene?
16. Describe the method Hershey and Chase used to determine what component of the bacteriophages had entered the host cells.
17. What did the results of the Hershey-Chase experiment signify in terms of the chemical nature of phage T4 genes?

D. Further evidence for DNA

General information
18. What findings of Alfred Mirsky gave additional support to the hereditary role of DNA?

Chargaff's results
19. What did Erwin Chargaff discover about the relative quantities of the four nitrogenous bases? How do these findings compare to Levene's findings?
20. What did Chargaff's results indicate about the potential of DNA for carrying hereditary information?
21. What striking features regarding base proportions in DNA are evident in Table 14-1?

E. The hypothesis is confirmed
22. What requirements must be met by the genetic material?
23. Which two of these requirements did Watson and Crick demonstrate that sub-

stantiated the hypothesis that DNA is the genetic material?

IV. The Watson-Crick model (pages 285-288)

A. Introductory remarks

24. What general approach did Watson and Crick use in determining the structure of DNA?

B. The known data

25. What data on DNA were already known from the work of Levene, Pauling, Wilkins and Franklin, and Chargaff? What pieces of data were conflicting?

C. Building the model

26. Describe the Watson-Crick model of DNA structure.

27. How did the final structure indicate that DNA was heterogeneous enough to carry a large amount of coded information?

28. What is the role of hydrogen bonding in the DNA structure?

29. Why are the two ends of a strand referred to as 3' and 5'? (See Figure 14-11.) Define the term "antiparallel."

D. Essay: Who might have discovered it? (page 289)

30. What scientists other than Watson and Crick might have made the discovery of DNA?

31. How does Crick feel about his personal achievements in comparison to the actual object of the discovery, the DNA helix?

V. DNA replication (pages 290-292)

A. General information

32. Following the logic proposed by Watson and Crick, use a hypothetical sequence of base pairs to show how two identical molecules of DNA might be formed from one.

B. A confirmation

33. Compare the semiconservative hypothesis of Watson and Crick (Figure 14-13b) with two other possibilities (Figure 14-13a and c) tested by Meselson and Stahl. How would heavy isotopes distinguish between these three methods of replication?

34. Explain how Meselson and Stahl were able to obtain samples of DNA that represented a certain number of generations of *E. coli*.

C. The mechanics of DNA replication

35. What enzymes must be present before DNA replication can occur in a test tube solution?

36. What other enzymes are associated with the ability of DNA to replicate?

D. The energetics of DNA replication

37. In what form are nucleotides required for DNA synthesis?

38. What bonds are broken as these nucleotides are added to the newly forming strand? What further reactions occur?

39. Is this really a waste of energy? Explain.

VI. DNA as a carrier of information (page 292)

40. If connected in a single thread, what is the length of DNA from a single human cell? What is its information equivalent?

VII. Summary (page 293): Read the Summary. If you are familiar with the essential features of the material presented there, you are ready to take the following diagnostic examination.

TESTING YOUR UNDERSTANDING

After you have completed the following examination, compare your answers with the annotated key in the Analysis section.

1. The assembly of DNA from nucleotide triphosphates instead of from nucleotide monophosphates ensures that:
   a. there will be a threefold increase in the normal amount of available nucleotides.
   b. DNA replication will proceed with a large positive ΔH.
   c. the production of ATP will be a beneficial by-product of replication.
   d. nucleotides will not be removed from DNA by DNA polymerase.
   e. Both b and d are correct.

2. If DNA is placed in a solution (at ideal pH and temperature) that is generously supplied with nucleotides, replication will:
   a. proceed according to the semiconservative hypothesis.
   b. not take place until DNA polymerase is added.
   c. cease after one generation.
   d. take place endergonically.
   e. None of the above is correct; DNA replication occurs only in living tissues.

3. If *E. coli* is removed from an ^{15}N growth medium and allowed to replicate in an ^{14}N medium, the second generation grown in the ^{14}N medium will contain:
   a. all heavy DNA.
   b. all light DNA.
   c. all half-heavy DNA.
   d. one-half half-heavy and one-half light DNA.
   e. one-fourth heavy and three-fourths light DNA.

4. A strand of DNA may be considered a _____, since it can generate a new and _____.
   a. replicate; identical strand
   b. template; complementary strand
   c. master; completely different strand
   d. double; identical double helix
   e. variable; completely different helix

5. Which of the following details emerged from the Watson-Crick model of the double helix?
   a. DNA is wider than had been expected based on x-ray diffraction pictures.
   b. The two strands run in the same direction.
   c. The width of the DNA helix is 10 angstroms.
   d. Phosphate groups are joined to either 3'- or 5'-carbon atoms.

6. The Watson-Crick hypothesis for a model of DNA could account for Chargaff's data by assuming that:
   a. any of the four bases could pair with any other base.
   b. purines pair only with purines and pyrimidines pair only with pyrimidines.
   c. adenine pairs only with thymine and guanine pairs only with cytosine.
   d. any base sequence within the molecule is possible.

7. Hydrogen bonding plays an important role in the DNA structure by joining together:
   a. the two strands of the helix.
   b. adjacent purine bases.
   c. phosphate groups and five-carbon sugars.
   d. 3'- and 5'-carbon atoms.
   e. Both a and b are correct.

8. The Watson-Crick model suggested that _____ could carry coded information.
   a. the position of the sugars
   b. the sequence of the bases
   c. the linear array of amino acids
   d. high-energy phosphate bonds
   e. the turns of the double helix

9. The text compares DNA to a "twisted ladder" whose rails are _____ and whose rungs are _____.
   a. sugar-phosphate backbones; nitrogenous base pairs
   b. pyrimidines; purines
   c. helical; antiparallel
   d. nucleotides; phosphate groups
   e. phosphate groups; deoxyribose sugars

10. As they began their studies on the structure of DNA, James Watson and Francis Crick were aware that DNA:
    a. is a double helix.
    b. contains base pairs.
    c. is composed of equal amounts of the four bases.
    d. is made up of nucleotides.
    e. is variable enough to carry a genetic code.

11. In order to carry a code for vast amounts of hereditary information, a DNA molecule must be:
    a. large.
    b. complex in its structure.
    c. capable of self-replication.
    d. heterogeneous.
    e. All of the above are correct.

12. In order for molecular geneticists to explain how mutant traits could be passed from one generation to the next, they had to show that:
    a. errors in DNA could be faithfully copied.
    b. similar changes could arise in the chromosomes of both parent and offspring.
    c. DNA is capable of repairing itself.
    d. Both b and c are correct.
    e. All three answers (a, b, and c) are correct.

13. Alfred Mirsky was able to add support to the hypothesis that genes were made up of DNA by demonstrating that:
    a. cells of all organisms, regardless of species, contain equal amounts of DNA.
    b. both somatic cells and gametes contain equal amounts of DNA.
    c. gametes contain one-half the amount of DNA of somatic cells.
    d. all chromosomes of a given species contain the same amount of DNA.
    e. the phenomenon of transformation occurs in all bacteria.

14. Which of the following details concerning an *E. coli* bacteriophage infection cycle is correct?
    a. The entire virus particle enters the host cell.
    b. Viral DNA is replicated within the host cell many times.
    c. Viruses released from the host cell absorb protein coats left attached to the cell wall.
    d. Viral infection cannot take place in a radioactive medium.

15. If the protein coat and DNA core of a bacteriophage are radioactively labeled (so that they may be identified in a cellular extract), what substances will be found inside the host cell?
    a. ^{32}P
    b. ^{35}S
    c. both ^{32}P and ^{35}S
    d. ^{15}N
    e. ^{14}C

16. Which of Levene's findings about DNA was later proved to be incorrect?
    a. DNA contains four different nitrogenous bases.
    b. The basic unit of DNA is composed of a sugar, a phosphate group, and a nitrogenous base.
    c. DNA has a helical structure analogous to that of many proteins.
    d. DNA has the structure of a tetranucleotide.

17. O. T. Avery's isolation of DNA proved to be of great significance because:
    a. DNA was shown to be a hereditary transforming factor.
    b. it was the first time that DNA had been isolated from living tissue.
    c. it showed conclusively that genes are composed of DNA.
    d. it allowed Frederick Griffith to find a vaccine for pneumonia.

18. If heat-killed virulent pneumococci are injected into a host containing living nonencapsulated pneumococci, then:
    a. the host will not manifest disease.
    b. living encapsulated bacteria may appear.
    c. the nonencapsulated bacteria will be destroyed.
    d. the heat-killed bacteria may, in some unknown way, regain function and cause the disease.

19. Originally, DNA was *not* strongly suspected to be the carrier of hereditary information because:
    a. it had not been detected on chromosomes.
    b. it was believed to have a simple structure.
    c. chromosomal proteins were thought to be master copies of cellular proteins.
    d. All of the above are correct.

20. Match the researcher (or team) with their contribution to genetics.

    Miescher _____
    Watson and Crick _____
    Mirsky _____
    Luria and Delbruck _____
    Levene _____
    Hershey and Chase _____
    Griffith and Avery _____
    Meselson and Stahl _____

    a. viral replication is due to injection of DNA into the host
    b. first to extract a nucleic acid
    c. prepared a density gradient that separated DNA of different weights
    d. DNA is a transforming factor
    e. showed that gametes contained only half the quantity of DNA of other cells
    f. suggested a mechanism by which DNA copies itself
    g. studied the T phages of *E. coli*
    h. steered attention away from DNA as a hereditary substance
    i. DNA is a double helix

21. Explain how the technique of density-gradient

centrifugation was used to confirm the semiconservative hypothesis of DNA replication.

22. What advantages did bacteriophages offer to geneticists investigating the nature of the genetic material?

23. Watson and Crick hoped that the structure they proposed for DNA would reveal how the molecule could fulfill its biological roles. Discuss these roles, and explain how it became apparent to Watson and Crick that their structure could meet some of these requirements.

24. Below is a drawing of the chemical structure of a portion of a double-stranded DNA molecule. Name the components of the molecule that are labeled a through n.

PERFORMANCE ANALYSIS

1. Nucleotide triphosphates such as ATP or GTP are used in DNA replication. The energy of the bonds is released and ensures that the replication process, otherwise only slightly exergonic, will go strongly in the forward direction, resulting in the addition of nucleotides to the growing strand; therefore, **d** is correct.

2. DNA replication can take place exergonically outside of living tissue as long as DNA polymerase is present (**b**). Replication will continue until the supply of available nucleotides is exhausted.

3. Because DNA replication occurs semiconservatively (see Figure 14-13b), the second (F_2) generation of *E. coli* replicating in an ^{14}N medium will contain one-half half-heavy DNA and one-half light DNA. This was demonstrated by the Meselson-Stahl experiment (see Figure 14-14e). Choice **d** is therefore correct.

4. During replication, each strand of the double helix acts as a template guiding the formation of a complementary strand (**b**). For example, where adenine is present on the parent strand, the new strand will contain thymine, and so forth. In this way two exact replicas of the parent helix will be produced.

5. Figure 14-10 shows that phosphate groups are joined to the 3'- and the 5'-carbon atoms of the ribose sugar molecules (**d**). Figure 14-11 shows that the two backbones are antiparallel (run in opposite directions). Consult Figure 14-9 to verify that there are 10 base pairs per turn and that the width of the helix (20 angstroms) agrees with Wilkins's measurements based on Rosalind Franklin's x-ray diffraction photography.

6. Erwin Chargaff had shown that the amount of adenine in any sample of DNA is approximately equal to the amount of thymine and that the same relationship was also true for cytosine and guanine. Watson and Crick realized that if in their model a purine base could pair only with a pyrimidine—for example, A with T or G with C (**c**)—the model could explain Chargaff's rule.

7. Figure 14-11 shows the hydrogen bonds that form between the complementary base pairs and hold together the two strands of the helix (**a**).

8. Watson and Crick were quick to realize that because there was no biochemical restraint on the order of nitrogenous bases in a strand of DNA, the base sequence (**b**) of the molecule could carry the vast amount of hereditary information required by the cell.

9. This twisted ladder is diagrammed for you in Figure 14-9. Verify for yourself that choice **a** is correct.

10. Watson and Crick knew from Levene's chemical analysis of DNA that the basic structural unit of the molecule was a nucleotide unit (**d**). Levene's proposal that DNA contained equal amounts of the four bases was in contradiction to Chargaff's findings. It was not until the final model was confirmed that a, b, and e were demonstrated.

11. In order to carry a code for making the vast amount of highly complex proteins found in an organism, DNA must be both large and complex itself. Both of these attributes imply that DNA is heterogeneous. Since each daughter cell inherits an exact copy of the genetic material when a cell divides, DNA has to be replicated prior to mitosis. Therefore, choice **e** is correct.

12. Since DNA is passed from one generation to the next via the gametes, changes in the genetic material responsible for mutant phenotypes must be faithfully reproduced (**a**) prior to gamete formation if the offspring is to exhibit the mutant phenotypes.

13. Choice **c** is correct. This finding of Mirsky's agreed with Mendel's assumption that for each trait, each gamete receives one of the two factors present in the diploid cells that give rise to the gametes.

14. Hershey and Chase demonstrated (using a radioactive medium) that when a bacteriophage infects a population of *E. coli*, only the DNA of the virus enters the cell and is replicated (**b**) by the host cell machinery. The viral DNA can be transcribed and translated into the proteins necessary for a new viral protein coat.

15. This is precisely what was done by the Hershey and Chase team using *E. coli* as a host. If the virus inserted only its DNA into the host cell, then radioactive phosphorus (^{32}P) (**a**) would be traced to the intracellular extract. If radioactive sulfur (^{35}S)

appeared inside the cell, then Hershey and Chase would have concluded that only protein had been inserted by the phage. (This is because proteins contain sulfur but not phosphorus and DNA contains phosphorus and not sulfur.)

16. Levene was eventually proved to be correct in all of his findings regarding the chemical nature of DNA except for his suggestion that the concentrations of all nitrogenous bases were equal. It was because of this erroneous finding that Levene was led to suggest that the structure of DNA was that of homogeneous tetranucleotide (**d**). Levene did not report a helical structure for DNA.

17. Avery repeated Griffith's experiments and showed that the transforming factor that had converted nonvirulent bacteria into virulent bacteria was DNA; therefore, **a** is correct.

18. In Griffith's experiment, the heat-killed virulent pneumococci supplied a transforming factor (DNA) to the living nonvirulent pneumococci. The transformed bacteria could then produce capsules that protected them from the host cell's defenses, enabling them to cause the disease; therefore, **b** is correct.

19. As explained in the answer to Question 16, Levene contributed to the assumption in choice b. Since chromosomes were known to contain protein but not known to contain DNA, choices a and c become acceptable; therefore, **d** is correct.

20. The correct answers in order are **b**; **f**, **i**; **e**; **g**; **h**; **a**; **d**; and **c**.

21. Your explanation should include a thorough discussion of the work of Meselson and Stahl, who evaluated the Watson-Crick hypothesis of DNA replication. The experiment is outlined in the figures on page 291.

22. The T-even bacteriophages introduced by Delbruck and Luria could be easily cultured, were readily identified microscopically, and had phenomenal reproductive rates. The Hershey-Chase experiment could be used to point out several other minor advantages.

23. The biological roles DNA had to fulfill if it was to be accepted as the genetic material are listed on page 284. Watson and Crick were able to demonstrate, as a consequence of their model, how the first two roles could be performed.

24. The correct answers are:
    a. **phosphate group**
    b. **deoxyribose**
    c. **hydrogen bond**
    d. **adenine**
    e. **deoxyribose**
    f. **thymine**
    g. **adenine**
    h. **phosphate group**
    i. **guanine**
    j. **adenine**
    k. **thymine**
    l. **guanine**
    m. **cytosine**
    n. **nucleotide**

CHAPTER 15

The Code and Its Translation

CHAPTER ORGANIZATION

I. Introduction
II. Genes and proteins
  A. One gene-one enzyme
  B. The structure of hemoglobin
  C. The virus coat
III. The genetic code
  A. Introductory remarks
  B. The code is a triplet
IV. The RNAs
  A. Introduction
  B. RNA synthesis
    1. General information
    2. Messenger RNA
    3. Transfer RNA
    4. Ribosomal RNA
V. Protein synthesis
VI. Breaking the code
  A. General information
  B. Hemoglobin reexamined
  C. Essay: Bacteriophage φX174 breaks the rules
VII. Regulation of protein synthesis
  A. Introduction
  B. The *lac* operon
VIII. Summary

MAJOR CONCEPTS

Once DNA was accepted as the carrier of hereditary information, work turned toward discovering the specific mechanism by which the information is transferred. The essence of the process is that genetic information is coded in the sequence of nucleotides in DNA molecules, and these, in turn, determine the sequence of amino acids in protein molecules.

Three types of RNA are transcribed from DNA: messenger RNA (mRNA), which is complementary to the DNA strand coding for a particular series of amino acids and serves as the template for protein synthesis; transfer RNA (tRNA), which carries amino acids to mRNA; and ribosomal RNA (rRNA), which is contained within the ribosomes. The ribosome provides binding sites for both mRNA and tRNA. Protein synthesis (translation) occurs on the ribosome in three stages: initiation, elongation, and termination.

Two groups of scientists (Nirenberg and Matthaei, and Khorana) were pioneers in determining which nucleotide triplet codes for which amino acid. Of the 64 possible triplet combinations of the four-nucleotide code, 61 have been identified with one of the 20 amino acids; the other 3 serve as "stop signals."

In bacteria, one method of regulating protein synthesis involves an operon (a cluster of functionally related genes that are transcribed to the same mRNA). The operon may be inhibited by a repressor protein. An inducer can inactivate the repressor protein, inducing protein synthesis by allowing RNA polymerase to attach to the operon's promoter. This action initiates the formation of mRNA.

HOW TO STUDY THE CHAPTER

Read the chapter, focusing on the major concepts.

Reread the chapter, using the questions that follow to help you focus on details as they relate to major concepts. Answer the questions on a separate sheet of paper. Your answers will provide a valuable study aid in preparing for examinations.

FOCUS ON CHAPTER DETAILS

I. Introduction (page 295)

1. Watson and Crick demonstrated how DNA could carry hereditary information and how this information could be passed from one generation to the next. What was the next question to occupy geneticists?

II. Genes and proteins (pages 295-296)

A. One gene-one enzyme

2. What is meant by "one gene-one enzyme"? How did Beadle and Tatum's Nobel Prize work provide support for this hypothesis?

3. Why is the Beadle and Tatum hypothesis more precisely referred to as "one gene-one polypeptide chain"?

B. The structure of hemoglobin

4. How was Linus Pauling able to detect differences in the hemoglobin of sickle cells and normal blood cells?

5. How did Pauling's results (see Figure 15-2) relate to the Beadle and Tatum hypothesis? (Hint: Remember that sickle cell anemia is a hereditary disease.)

C. The virus coat

6. How do infection cycles of bacteriophages demonstrate the relationship between DNA and proteins?

III. The genetic code (pages 296-297)

A. Introductory remarks

7. What concept was introduced to explain how DNA could direct protein synthesis?

B. The code is a triplet

8. How can the 20 amino acids found in proteins be coded for by only four different nucleotides? What is a codon?

9. What question needed to be answered before the triplet codon hypothesis could be proved?

IV. The RNAs (pages 297-299)

A. Introduction

10. What five lines of evidence suggested that ribonucleic acid (RNA) might play a genetic role?

11. Which evidence indicated that RNA might itself carry a hereditary code?

12. Why were cell-free extracts used in the experiments to uncover the role of RNAs in protein synthesis?

B. RNA synthesis

General information

13. What is transcription? What enzyme is required for this process?

14. How does the copying of RNA from a strand of DNA differ from DNA replication?

15. What are the three types of RNA that may be transcribed from DNA?

16. What is the role of each type?

Messenger RNA

17. Describe the structure of an mRNA molecule.

18. How do mRNAs convey genetic information?

Transfer RNA

19. Describe the structure of a tRNA molecule. Describe the specialized site at which the amino acid attaches.

20. What is the anticodon?

Ribosomal RNA

21. What is the composition of ribosomes?

22. Describe the two subunits that make up prokaryotic ribosomes.

23. What binding sites are characteristic of each ribosomal subunit?

V. Protein synthesis (pages 300-301)

24. What is translation?

25. What are the three steps of protein synthesis? Outline what happens in each step (see Figure 15-9). Pay close attention to the bonds that form and break at the A and P sites of the ribosome.

26. When you feel that you understand the

process of protein synthesis, make up a hypothetical mRNA sequence from 5′ to 3′. Match each triplet codon of the sequence with the appropriate tRNA anticodon and its associated amino acid (refer to Figure 15-13).

27. What is a polysome?

VI. Breaking the code (pages 302–303)

A. General information

28. A cell-free extract of *E. coli* contains all the ingredients necessary to conduct protein synthesis. What happens when a "foreign" mRNA is added to such an extract?
29. How did Nirenberg and Matthaei use a synthetic message to crack the genetic code?
30. What general information about the genetic code did Khorana formulate as a result of his experiments?
31. What is characteristic of a group of codons that specify the same amino acid? (See Figure 15-13.)

B. Hemoglobin reexamined

32. How can the difference between normal and sickle cell hemoglobin be accounted for at the nucleotide level of DNA?
33. How can we now modify de Vries's definition of mutations?
34. How do additions or deletions of nucleotides within a gene affect protein synthesis? Consult Figure 15-14.

C. Essay: Bacteriophage φX174 breaks the rules (page 304)

35. What was startling about the genome of phage φX174 in light of the number of different proteins the virus could manufacture? Explain how the problem was resolved.

VII. Regulation of protein synthesis (pages 304–306)

A. Introduction

36. What are three ways a bacterial cell can regulate its production of proteins?
37. What is an operon?
38. What is the relationship between structural and regulatory genes?

B. The *lac* operon

39. What information is contained in the *lac* operon of *E. coli*? How is the expression of this information influenced by the presence or absence of lactose?
40. What is the disadvantage to an *E. coli* cell of a mutation in the repressor site of the *lac* operon?
41. How are the regulatory and structural genes organized in the *lac* operon of *E. coli*? (See Figure 15-18.)
42. How does cyclic AMP help *E. coli* regulate its energy use more efficiently? (See Figure 15-18.)

VIII. Summary (pages 307–308): Read the Summary. If you are familiar with the essential features of the material presented there, you are ready to take the following diagnostic examination.

TESTING YOUR UNDERSTANDING

After you have completed the following examination, compare your answers with the annotated key in the Analysis section.

1. In studying the hereditary disease known as sickle cell anemia, Linus Pauling speculated that people with sickle-shaped blood cells differ from normal individuals in having:
   a. a single recessive allele at the locus for hemoglobin production.
   b. a completely different amino acid sequence in their hemoglobin.
   c. differently charged hemoglobins.
   d. alpha units of hemoglobin present in their blood cells but not beta units.

2. Beadle and Tatum demonstrated that in *Neurospora* a change in a single gene is correlated with:
   a. a loss of function in an enzyme.
   b. a change in the mRNA sequence.
   c. a single amino acid change in a protein.
   d. the failure of *Neurospora* to carry out protein synthesis.

3. The process of transcription involves the synthesis of _____ from _____ .
   a. RNA; DNA
   b. RNA; RNA polymerase
   c. DNA; RNA
   d. proteins; DNA
   e. anticodons; codons

4. Nucleotides containing uracil are transcribed from nucleotides that *must* contain a(n):
   a. thymine.
   b. adenine.
   c. deoxyribose.
   d. ribose.
   e. Both b and c are correct.

5. Analysis of the structure of a tRNA molecule reveals that it is specific for a certain type of:
   a. mRNA molecule.
   b. rRNA molecule.
   c. amino acid.
   d. protein.
   e. nucleotide.

6. A tRNA molecule attaches to the mRNA molecule:
   a. via hydrogen bonding.
   b. at the codon of the mRNA.
   c. via complementary base pairing.
   d. on the ribosome.
   e. All of the above are correct.

7. Prokaryotic ribosomes are composed of:
   a. tRNA and protein.
   b. two subunits of different size.
   c. polysomes.
   d. a promoter and an operon.
   e. DNA and RNA.

8. The following question applies to this mRNA sequence:

   3'   5'
   AUGGAGACCGUA

   If a tRNA carrying the amino acid methionine (codon AUG) is attached to the mRNA at the P site, then the anticodon of the tRNA at the A site will be:
   a. AAG.
   b. UAC.
   c. CUC.
   d. UGG.
   e. GGU.

9. Proteins can be thought of as a sort of "language of life" whose words (amino acids) are encoded by the DNA molecule in the form of:
   a. a single nitrogenous base.
   b. pairs of nitrogenous bases.
   c. a triplet of nitrogenous bases.
   d. the operon.
   e. a single gene.

10. When both the A and P sites of the ribosome are occupied by tRNA molecules, a _____ bond forms between adjacent _____ .
    a. hydrogen; anticodons
    b. peptide; amino acids
    c. high-energy phosphate; nucleotides
    d. disulfide; polypeptides
    e. covalent; ribosome subunits

11. The termination of protein synthesis occurs:
    a. upon recognition of an mRNA stop codon.
    b. as soon as mRNA transcription stops.
    c. when the first codon of an mRNA sequence is released from the P site.
    d. at the 5' end of the mRNA strand.

12. The "poly-U" synthesized by Nirenberg and Matthaei was a type of:
    a. polypeptide.
    b. mRNA.
    c. bacterial gene.
    d. polysaccharide.
    e. viral DNA.

13. A long sequence of mRNA that contains only two of the four possible bases in an alternating sequence will produce a polypeptide composed of _____ different amino acids.
    a. two
    b. three
    c. four
    d. six
    e. The alternation of only two bases would be insufficient to produce a polypeptide.

14. The number of codons that can be translated into amino acids is:
    a. 4.
    b. 16.
    c. 20.
    d. 61.
    e. 64.

15. Analysis of hemoglobin from patients with sickle cell anemia reveals that the disease is related to:
    a. a substitution for a single amino acid.
    b. a single nucleotide substitution.
    c. deletion of part of a DNA sequence.
    d. addition of a nucleotide sequence to DNA.
    e. Both a and b are correct.

16. In light of current knowledge of the genetic code and mechanisms of protein synthesis, a mutation may be regarded as *always* producing a change in:
    a. phenotype.
    b. genotype.
    c. protein structure.
    d. a nucleotide sequence.
    e. enzyme function.

17. A single mRNA molecule is translated into four different enzymes involved in the same sequence of reactions. The sequence of DNA from which this molecule is transcribed is referred to as a(n):
    a. gene.
    b. polypeptide.
    c. operon.
    d. regulator.
    e. biochemical pathway.

18. Because the amount of protein in a prokaryotic cell must be adjusted to the cell's metabolic demands, _____ is regulated.
    a. protein translation
    b. DNA replication
    c. mRNA transcription
    d. ribosome synthesis
    e. Both a and c are correct.

19. Match each substance with its description from the list that follows.
    RNA polymerase _____
    transfer RNA _____
    inducer _____
    regulatory gene _____
    artificial mRNA _____
    ribosomal RNA _____
    a. a cloverleaf-shaped transport molecule
    b. a structural component in protein synthesis
    c. attaches to the promoter sequence when an operator is repressed
    d. facilitated the breaking of the genetic code
    e. binds to a repressor when a metabolic pathway must be opened
    f. controls transcription of an operon

20. Match each genetically important sequence with the type of molecule, from the list that follows, in which it generally appears.
    ATCGCTCAT _____
    $3'$
    ......CCA- _____
    UCUCUCUCUC _____
    val-arg-met-pro _____
    UAGCGAGUA _____
    a. messenger RNA
    b. DNA
    c. transfer RNA
    d. synthetic messenger RNA
    e. a polypeptide

21. Why was ribonucleic acid believed to play an important role in protein synthesis?

22. Using your knowledge of protein structure (as discussed in Chapter 3), explain why Beadle and Tatum's "one gene-one enzyme" hypothesis could not be considered technically correct.

23. The diagram on the facing page summarizes protein synthesis in a bacterial cell. Identify the indicated components.

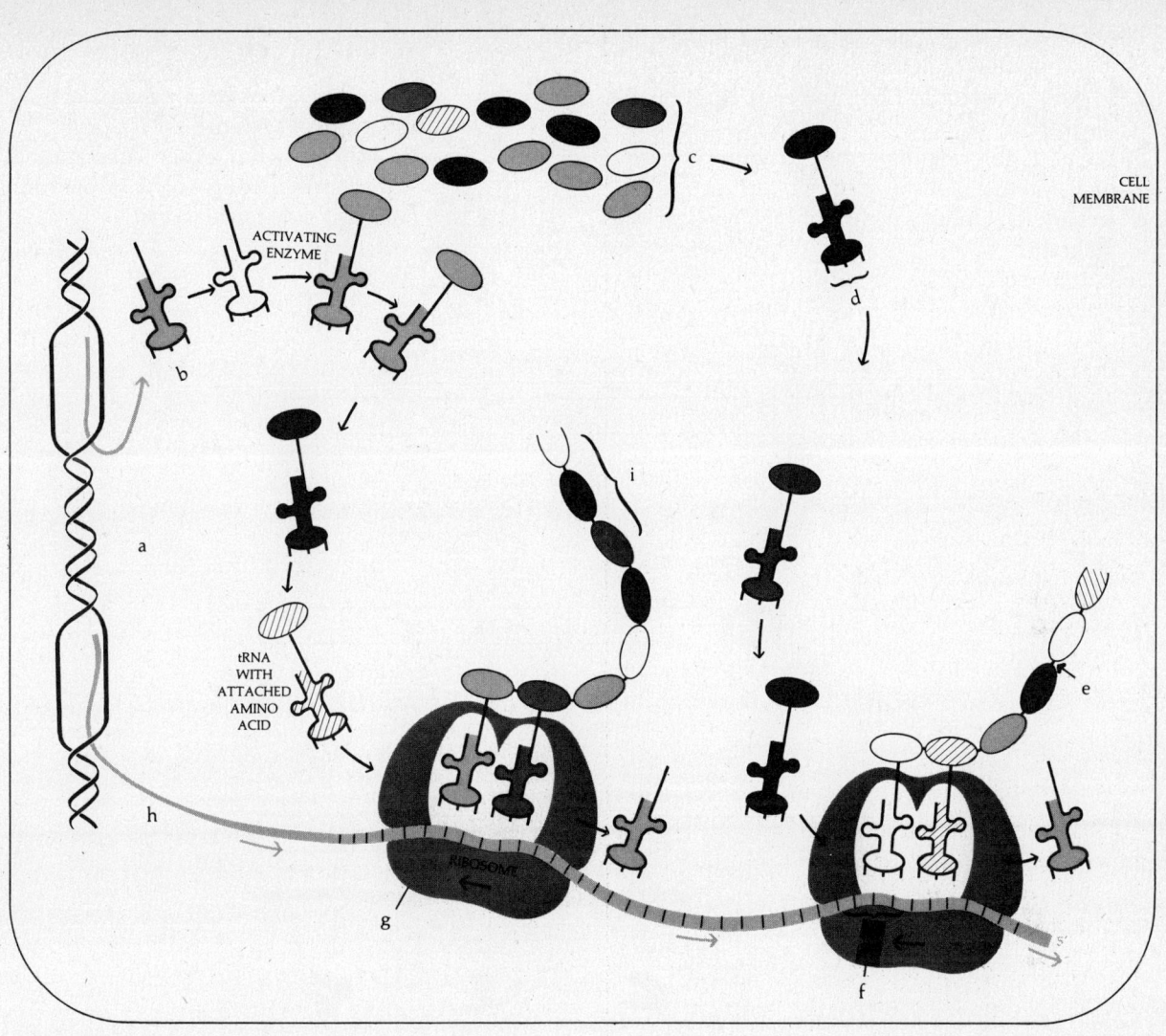

CHAPTER 15 *The Code and Its Translation*

24. The following diagram summarizes the operation of the *lac* operon of *E. coli*.
    a. In part (a), identify the numbered parts of the diagram.
    b. In part (b), identify the structure at the question mark. Is the operator "on" or "off"?
    c. In part (c), identify the numbered parts of the diagram.
    d. In the case of the *lac* operon, what substance inactivates the repressor?
    e. How does the *lac* operon exemplify a classic feedback control system?
    f. Besides lactose, what other substance is involved in the regulation of the *lac* operon, and how does this substance work?

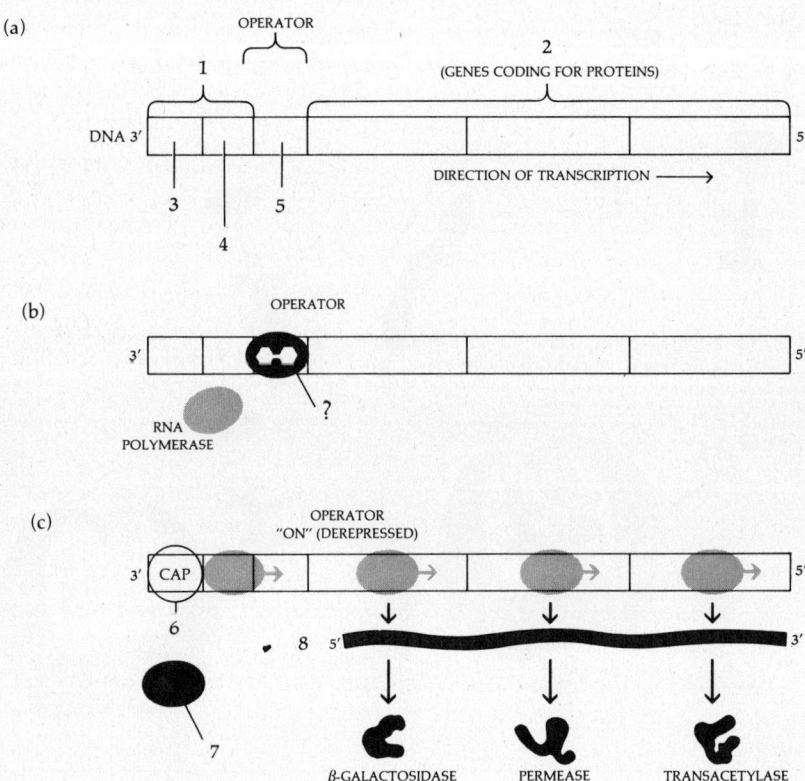

PERFORMANCE ANALYSIS

1. Speculating that differences in hemoglobin structure would reflect differences in the electrical properties of hemoglobin, Pauling tried to separate the proteins of three different sickle cell phenotypes (normal, carrier, and affected) by electrophoresis. The proteins fell into three groups with respect to charge (see Figure 15-2); therefore, **c** is correct.

2. At the time that Beadle and Tatum carried out their work (the early 1940s), very little was known about the details of protein synthesis or of the hereditary role of mRNA. Had protein sequencing techniques been available at the time, Beadle and Tatum might have been able to equate the mutations they induced with single amino acid changes. However, what they were able to show was important enough: A mutation in one gene is correlated with the loss of function of an enzyme (**a**).

3. Transcription is the copying of RNA from DNA (**a**). It is catalyzed by the enzyme RNA polymerase.

4. RNAs contain the nitrogenous base uracil in place of thymine. During transcription of DNA, which contains nucleotides made up of deoxyribose (c) sugars, uracil-containing nucleotides pair with nucleotides containing adenine (b); therefore, **e** is correct.

5. Note in Figure 15-6 that a tRNA molecule, depending on the anticodon it contains, is specific for a certain type of amino acid (**c**), which binds at its 3' end.

6. After the mRNA becomes attached on the ribosome (d), the anticodon of the tRNA molecule pairs with a sequence, the codon (b), on the mRNA. This pairing is a result of hydrogen bonding (a) between complementary base pairs (c); therefore, choice **e** is correct.

7. Choice **b** is correct. See Figure 15-7.

8. Translation of the mRNA sequence takes place from 5' to 3'. The first codon, AUG, coding for the amino acid methionine, is at the P site on the ribosome. The second codon, at the A site, is CCA, which attracts the anticodon GGU (**e**).

9. Amino acids of a protein—more accurately, a polypeptide—are coded for by triplet codons (three-base sequences); therefore, **c** is correct.

10. Choice **b** is correct (review Figure 15-9). A peptide bond forms between the amino acids that are attached at the 3' end of the two tRNA molecules occupying the P and A sites on the ribosome.

11. There are three triplet codons (see Figure 15-13) that specify "stop," the termination of the polypeptide chain (**a**). Choice b is incorrect, for as long as mRNA molecules are present in the cytoplasm, protein synthesis can continue. Choice c is wrong because this step occurs at the beginning of protein synthesis. Choice d is incorrect because translation occurs from the 5' end of the mRNA strand to the 3' end.

12. "Poly-U," or poly-uracil, was a synthetic mRNA (**b**) used to determine the coding significance of genetic sequences containing only the nitrogenous base uracil. This was the initial successful step in the breaking of the genetic code. Poly-U translated into a polypeptide containing only phenylalanine residues.

13. An mRNA made up of repetitions of just two bases, such as AGAGAGAG, contains two possible codons, AGA and GAG, which can code for no more than two (**a**) different amino acids. Can you think of two different repeating bases that would produce a polypeptide containing but a single type of amino acid?

14. Using a triplet codon and four possible nitrogenous bases as a code, a DNA sequence could encode $4^3 =$ 64 different amino acids. However, as pointed out in Figure 15-13, three of the codons are used to "stop" protein synthesis, so that 61 (**d**) are left to code for the 20 amino acids.

15. The gene that produces sickle cell hemoglobin contains a valine residue in place of glutamic acid (a). Since GAG specifies glutamic acid and GUG specifies valine, the disease can result from a single nucleotide substitution (b); therefore **e** is correct.

16. Mutations were formerly considered to be changes in the phenotype, but now any inheritable change in a nucleotide sequence (**d**) constitutes a mutation, whether or not that change is translatable into a mutant phenotype. Choice c looks tempting, but since a specific amino acid may be associated with several similar codons, a change in the nucleotide sequence does not necessarily change the amino acid sequence. Choice b is incorrect for the same reason.

17. According to the theory of Jacob and Monod, enzymes involved in the same biochemical sequence are often produced by genes that lie adjacent to one another on a chromosome and are transcribed together when the biochemical reaction sequence must occur (see Figure 15-18). These genes are known collectively as an operon (**c**).

18. A cell meets its demand for the proper quantities of proteins by closely regulating transcription (c). Since protein translation will continue only as long as new mRNA is transcribed, protein translation (a) is also regulated. Therefore, **e** is correct.

19. The correct answers in order are **c, a, e, f, d,** and **b**.

20. The correct answers in order are **b, c, d, e,** and **a**.

21. RNAs were implicated in protein synthesis because they were found both in the nucleus, where DNA is also found, and on the ribosomes, where proteins are synthesized.

22. Enzymes are very often composed of more than one type of polypeptide. Such a molecule would then require more than one gene to produce it. Another reason that Beadle and Tatum's hypothesis needed to be refined was that not all proteins are enzymes: a better hypothesis—as it was in fact later stated, and then further revised—would have been "one gene-one polypeptide chain."

23. The correct answers are:
    a. DNA
    b. tRNA
    c. amino acids
    d. anticodon
    e. peptide bond
    f. codon
    g. ribosome
    h. mRNA
    i. growing peptide chain

24. a. (1) promoter
       (2) operon
       (3) CAP site
       (4) RNA polymerase binding site
       (5) repressor site
    b. repressor; off
    c. (6) CAP-cAMP complex
       (7) inactive repressor-inducer complex
       (8) mRNA
    d. lactose
    e. The *lac* operon codes for the enzymes that an *E. coli* cell needs to utilize lactose as an energy source. In the presence of lactose, the repressor is inactivated and cannot bind to the operator, thus allowing mRNA to be transcribed. If lactose is not present, the repressor binds to the operator, which prevents the transcription of mRNA.
    f. As an energy source, glucose is more efficient than lactose. As the supply of glucose in a cell decreases, the level of cyclic AMP (cAMP) increases. As this occurs, **cyclic AMP binding protein (CAP)** binds to cAMP, and the complex binds to the promoter site of the *lac* operon. This enables the binding of RNA polymerase to the promoter. If lactose is present in sufficient concentrations to inactivate the repressor, the operator will not be occupied by the repressor and the operon will be able to transcribe mRNA.

CHAPTER 16

Recombinant DNA

CHAPTER ORGANIZATION

  I. Introduction
 II. The bacterial chromosome
III. Plasmids and conjugation
     A. Introductory remarks
     B. The F plasmid
        1. General information
        2. Chromosome mapping
     C. R plasmids
 IV. Viruses and transduction
     A. General information
     B. Temperance and lysogeny
     C. Transduction
     D. Introducing lambda
  V. Movable genetic elements
     A. Introductory remarks
     B. Insertion sequences
     C. Transposons
 VI. Restriction enzymes
VII. Some tricks of the trade
     A. Introductory remarks
     B. DNA cloning
     C. DNA sequencing
     D. Hybridization techniques
VIII. Summary

MAJOR CONCEPTS

In bacteria, DNA is usually located in two types of structures: the bacterial chromosome and plasmids. Bacterial chromosomes are circular and self-replicating, and the genes are arranged in a linear sequence around the chromosome. Plasmids are also circular and self-replicating, and are smaller than bacterial chromosomes. They are transferred during cell division and conjugation to the other cell involved. Plasmids can be spliced into and removed from the bacterial chromosome. Resistance genes on plasmids can be transferred to viruses and to bacteria of other species.

Viruses contain either DNA or RNA as their nucleic acid and are often involved in recombinant processes. In some RNA viruses, the RNA functions as mRNA and is also the template for the production of more RNA. In retroviruses, the RNA codes for the enzyme reverse transcriptase, which transcribes DNA from the viral RNA. In some DNA viruses, known as temperate bacteriophages, the DNA is incorporated into the host-cell chromosome. In a process known as transduction, viruses transfer genetic material from one cell to another.

In nature, genetic material may either be exchanged between homologous segments of DNA or be inserted into or removed from segments of DNA. Insertions can cause mutations, and the insertion sequence often makes numerous copies of itself that can then be inserted into the same molecule or into another molecule. Transposons are movable segments of DNA that function as genes (are expressed phenotypically). There is an insertion sequence at each end of the transposon segment.

Several techniques are used in recombinant DNA research. Restriction enzymes, which cleave DNA at specific sites, are used in splicing operations. DNA cloning by plasmids and viruses produces large quantities of identical nucleotide sequences. DNA-sequencing techniques are used to determine the sequence of isolated genes. Various hybridization techniques are used to detect specific nucleic acid sequences.

HOW TO STUDY THE CHAPTER

Read the chapter, focusing on the major concepts.

Reread the chapter, using the questions that follow to help you focus on details as they relate to major concepts. Answer the questions on a separate sheet of paper. Your answers will provide a valuable study aid in preparing for examinations.

FOCUS ON CHAPTER DETAILS

I. **Introduction** (page 310)
1. Describe the revolution that genetics underwent about a decade ago.

II. **The bacterial chromosome** (pages 310–311)
2. Describe the organization of genetic material in *E. coli*.
3. Why do microbiologists assume that the packaging of DNA in *E. coli* is so precisely determined?
4. What is responsible for the characteristic appearance of the *E. coli* chromosome before and during replication?

III. **Plasmids and conjugation** (pages 311–314)

A. Introductory remarks
5. What are the characteristic properties of plasmids?
6. How many genes do plasmids contain as compared to bacterial chromosomes?

B. The F plasmid

General information
7. What trait is associated with the F plasmid of *E. coli*?
8. What are F^+ and F^- cells?
9. Name and explain the process by which F^- calls can become F^+ cells (see Figure 16–4).
10. Describe the process by which an F^- cell can receive chromosomal material from an Hfr cell (see Figure 16–5).

Chromosome mapping
11. How is selective disruption of the conjugation process used to map the *E. coli* chromosome?
12. How do these mapping studies show that the bacterial chromosome is circular?

C. R plasmids
13. How is drug resistance spread within a population of bacterial cells?
14. Why does the ability of bacteria to transfer genes for drug resistance pose a serious health problem?
15. How are R plasmids that lack genes for production of pili transferred to other cells?

IV. **Viruses and transduction** (pages 314–318)

A. General information
16. Explain how the genetic material of DNA and RNA viruses is utilized once it is within the host cell.
17. On what host-cell materials is the virus dependent?
18. Why does this dependency make viruses obligate parasites?
19. How does the amount of genetic information carried by the T-even and ϕX174 bacteriophages compare to that carried by *E. coli*?
20. What unique genetic ability do retroviruses possess?

B. Temperance and lysogeny
21. What is a lysogenic bacterium? What accounts for its characteristic property?
22. What is a prophage? What might cause it to leave the host chromosome and begin an infectious cycle?
23. How does a lytic cycle differ from a lysogenic cycle? (See Figure 16–8.)

C. Transduction
24. What is transduction? How does it resemble conjugation?
25. What is the difference between general and restricted transduction?

D. Introducing lambda
26. What are the different forms taken by the DNA of phage lambda inside and outside the host cell?

27. What is the role of the viral enzyme lambda integrase?
28. By what method does lambda become a prophage?
29. What are the similarities between temperate phages and bacterial plasmids? What are the differences?

V. **Movable genetic elements** (pages 318–319)

A. Introductory remarks

30. Describe two basic ways that genetic recombination takes place naturally. List four examples for *each* of these ways.
31. How are homologous exchanges thought to occur? (See Figure 16-11.)

B. Insertion sequences

32. What are insertion sequences? Where are they found, and what unique property do they have?
33. How is the existence of an insertion sequence more limited than that of a virus or a plasmid?
34. In terms of site of insertion, how do these sequences differ from lambda?

C. Transposons

35. What are transposons? Why are they referred to as "jumping genes"?
36. How do transposons differ from insertion sequences?
37. How does the bacterium *Salmonella* make use of a transposon in regulating its production of certain flagellar proteins? What name has been applied to this mechanism?

VI. **Restriction enzymes** (pages 319–321)

38. What is the function of restriction enzymes produced by bacteria?
39. How do organisms protect their own DNA from the restriction enzymes they produce?
40. How is it that some strains of *E. coli* harbor lambda bacteriophages, while others do not?
41. What two research techniques have been made possible by the discovery of restriction enzymes?
42. How is Eco RI employed to splice together DNA segments from different sources?

VII. **Some tricks of the trade** (pages 321–324)

A. Introductory remarks

43. What organisms gave molecular geneticists the techniques to understand the molecular genetics of eukaryotes?

B. DNA cloning

44. What is DNA cloning? How is it useful to genetic researchers?
45. How can genes be spliced into the bacterial plasmid pSC101?
46. What steps must be taken before copies of the inserted gene can become available for use in research?
47. What other genetic reservoirs can be used in DNA cloning?
48. What potential does a foreign gene have inside its unwary host?

C. DNA sequencing

49. What were the first three entire genomes to be sequenced?
50. How is a segment of DNA that is to be sequenced treated before electrophoresis is used?
51. How does separation by electrophoresis create a "sequence ladder"? (Refer to Figure 16-15.)

D. Hybridization techniques

52. When is a molecule of DNA said to be denatured? How is this achieved? How can denaturation be reversed?
53. What is a hybrid DNA molecule?
54. How does DNA hybridization provide an index of similarity between two strands of DNA?
55. How is hybridization used to detect specific nucleic acids?

VIII. **Summary** (pages 324–325): Read the Summary. If you are familiar with the essential features of the material presented there, you are ready to take the following diagnostic examination.

TESTING YOUR UNDERSTANDING

After you have completed the following examination, compare your answers with the annotated key in the Analysis section.

1. Plasmids found in prokaryotic cells represent:
   a. replicates of the bacterial chromosome.
   b. a circular DNA molecule containing a small complement of genes.
   c. extranuclear genetic material.
   d. a DNA virus that has infected the cell.

2. Which of the following best describes the normal DNA replication process of the E. coli chromosome?
   a. It begins near the 5' end and moves toward the 3' end.
   b. It proceeds according to the rolling-circle method.
   c. It starts at a single point and proceeds in both directions.
   d. It begins at a number of points and may proceed in either direction.

3. An E. coli referred to as an Hfr cell:
   a. has a high frequency of resistance to antibiotic drugs.
   b. has an F plasmid incorporated into its chromosome.
   c. lacks the ability to be a donor during conjugation.
   d. has a plasmid called Hfr.

4. If conjugation occurs between an E. coli of F^+ strain and one of F^- strain, the F^- host may:
   a. acquire the capacity to produce pili.
   b. become an Hfr cell.
   c. itself become an F-plasmid donor.
   d. receive the F factor.
   e. All of the above are correct.

5. Geneticists have demonstrated that bacteria carry genes for _____ on R plasmids.
   a. reverse transcription
   b. drug resistance
   c. high recombination potential
   d. restriction-enzyme synthesis
   e. All of the above are correct.

6. Aside from having _____, a virus is dependent upon its host cell in order to reproduce.
   a. the ability to make its protein coat
   b. a copy of its own genetic material
   c. several metabolically essential enzymes
   d. its own source of ATP

7. Retroviruses are unique as genetic entities because they:
   a. can transcribe RNA to DNA.
   b. can incorporate their genetic material into a host cell's chromosome.
   c. have DNA that is not a double helix.
   d. can bring about transduction in their host cells.
   e. Both c and d are correct.

8. Lysogeny is triggered in a cell that contains a(n):
   a. temperate bacteriophage.
   b. T-even bacteriophage.
   c. RNA virus.
   d. lysogenic bacterium.
   e. transposon.

9. Transduction is a method of genetic recombination in which:
   a. a donor cell and host cell exchange genes during mating.
   b. DNA fragments pass freely between cells of different strains.
   c. viruses carry genetic material from one host to another.
   d. homologous chromosomes exchange short segments of DNA.

10. When bacteriophage lambda is in its prophage state, it carries out transcription of the gene for:
    a. a repressor protein.
    b. lambda integrase.
    c. coat protein.
    d. the triggering of lysis of the host membrane.

11. Which of the following is a method of genetic recombination in eukaryotic cells that is *not* exhibited by prokaryotic cells?
    a. transduction
    b. transformation
    c. crossing over
    d. conjugation

12. DNA sequences that are flanked by insertion sequences are known to:
    a. carry no genetic information in themselves.
    b. be movable genetic elements.
    c. code for restriction enzymes.

d. be methylated as a protection against cleavage.

13. DNA segments that can enter a genome at many different points but that remain fixed once they do enter are known as:
    a. transposons.
    b. insertion sequences.
    c. incorporated plasmids.
    d. prophages.
    e. restriction genes.

14. Restriction enzymes have proved helpful in genetic research because they:
    a. cleave DNA molecules at specific sites.
    b. can be used to create "sticky ends."
    c. make bacteria more sensitive to transduction.
    d. Both a and b are correct.
    e. All three answers (a, b, and c) are correct.

15. Which of the following terms describes a DNA molecule that contains genes from different sources?
    a. transposon
    b. chimera
    c. replicate
    d. double-stranded
    e. clone

16. Insertion of a foreign DNA sequence into an *E. coli* plasmid is done in order to _____ after several generations.
    a. produce a cloned gene that can be harvested
    b. remove the ability to transcribe the foreign gene
    c. bring about bacterial transformation
    d. form a mutant, disabled strain of *E. coli*

17. Which of the following techniques has facilitated the sequencing of DNA segments?
    a. gene cloning
    b. radioactive labeling
    c. gel electrophoresis
    d. random cleavage by restriction enzymes
    e. All of the above are correct.

18. Techniques developed to form hybrid nucleic acids rely upon the fact that:
    a. DNA will associate with DNA but not with RNA.
    b. only identical strands of nucleic acid will associate.
    c. strands having a high degree of complementarity will associate.
    d. denaturation of nucleic acids is always irreversible.
    e. hydrogen bonding between strands is strong enough to withstand the technique.

19. A restriction enzyme acts upon the following DNA segment by cleaving both strands between adjacent thymine and cytosine nucleotides.

    ...TCGCGA...
    ...AGCGCT...

    Which of the following pairs of sequences indicates the sticky ends that are formed?
    a. ...AGCGC   CGCGA...
    b. ...TCGC   TCGG...
    c. ...T   T...
    d. ...GA   AG...

20. Match each term with its definition. Answers may be used more than once.

    pSC101 _____   a. a temperate virus
    φX174 _____   b. a completely sequenced genome
    SV40 _____   c. found to have overlapping genes
    Eco RI _____   d. a restriction enzyme
    phage lambda _____ e. an *E. coli* plasmid

21. Match each term listed at the left with its description. Answers may be used more than once.

    rolling circle _____   a. proposed mechanism for gene exchange
    theta replication _____ b. transposon
    jumping gene _____   c. method of bacteriophage infection
    lytic cycle _____   d. precedes cell division in *E. coli*
    single-strand switch _____ e. serves to regulate protein synthesis
    flip-flop switch _____   f. form of replication of an *E. coli* donor chromosome during conjugation

22. What resemblance does a temperate phage have to

a plasmid? What additional abilities does the phage have?

23. What is the difference between general and restricted transduction? Under what conditions does each occur?
24. Explain the techniques involved in the sequencing of DNA molecules. How has the development of recombinant DNA research paved the way for this type of analysis? How is a sequence ladder employed?
25. Explain how conjugation in *E. coli* may be manipulated to provide a complete map of this bacterium's single chromosome.

PERFORMANCE ANALYSIS

1. Virtually all bacteria carry a part of their genetic complement on small circular DNA molecules (**b**) called plasmids.
2. DNA replication in *E. coli* starts at a single point and proceeds in both directions (**c**), giving the duplicating chromosome the appearance of the Greek letter theta; hence its name, theta replication. Rolling-circle replication (**b**), which moves from the 5' end to the 3' end (**a**), occurs as a prelude to conjugation.
3. The chromosome of an Hfr (high frequency of recombination) cell has an incorporated F plasmid (**b**), so that during conjugation it is able to donate portions of its own chromosome along with some (or all) of the F factor. Recombination of donor and recipient bacterial DNAs results in new gene combinations in the host.
4. The F⁻ host may receive the F factor (**d**) during conjugation. As a result, it becomes able to produce pili (**a**) and therefore able to donate the genes on the F plasmid (**c**). If the F plasmid becomes incorporated into its chromosome, the bacterial cell becomes an Hfr cell (**b**). For these reasons, choice **e** is correct.
5. R plasmids have been shown to carry genes conveying resistance to antibiotic drugs (**b**). Genes responsible for frequency of recombination are carried on the F plasmid.
6. When a virus encounters a suitable host cell, it inserts a copy of its DNA (**b**) and uses the cell's machinery and materials (nucleotides, amino acids, enzymes, ATP, etc.) to reproduce and rebuild its protein coat.
7. Choice **c** is false because retroviruses contain not DNA but RNA, which they can transcribe into DNA (**a** is correct) after they have entered a host cell. (This ability is due to the fact that retroviruses can make the enzyme reverse transcriptase.) Both **b** and **d** are true not only of retroviruses but of other viruses as well.
8. Temperate bacteriophages (**a**) are DNA viruses that become integrated into the host-cell chromosome (which is then known as a lysogenic bacterium) and are later released to infect other host cells. The best studied of the temperate phages is phage lambda. The T-even phages are lytic phages.
9. When a prophage breaks loose from its host chromosome, it sometimes takes with it adjoining host genes. Additionally, many viruses pick up host chromosome fragments that result from the infectious cycle. When viruses transfer genetic material from a previous host to a new one, transduction is said to occur. Transduction is a source of genetic recombination for the host cell. Choice **c** is therefore correct.
10. When phage lambda becomes a prophage in its host cell (an event made possible through the activity of the enzyme lambda integrase), a repressor protein (**a**) is transcribed. This repressor protein shuts down the reading of the operon whose genes control the expression of a lytic cycle.
11. Choices a, b, and d are methods of recombination for prokaryotes. Crossing over (**c**) occurs during meiosis in eukaryotic cells.
12. Genetic elements such as transposons require insertion sequences for their mobility; therefore, **b** is correct. Transposons, commonly referred to as "jumping genes," *do* carry genetic information.
13. Insertion sequences (**b**) may not leave the genome once they have entered it; but once they have replicated, the replicated sequence may insert itself anywhere in the genome.
14. Molecular geneticists use restriction enzymes to cleave strands of DNA at specific base sequences (**a**), thus creating protruding ends that are

complementary "sticky ends" (b); therefore, **d** is correct. See the diagram on page 320 of the text.

15. DNA molecules that are hybrids—that is, that contain sequences from different organisms or contain artificial sequences (such as in genetic engineering with *E. coli*)—are called chimeras (**b**).

16. If a foreign gene is inserted into a bacterial plasmid and the bacterium is stimulated to reproduce, multiple copies of the gene can be harvested (**a**) when the replicated plasmids are isolated from the cell.

17. All of the techniques listed (**e**) are involved in DNA sequencing. Cloning (a) provides a sufficient quantity of the sequence to work with. Fragments are created by randomly cleaving the sequence with restriction enzymes (d). Radioactive labeling (b) can be used to identify fragments according to their length; gel electrophoresis (c) separates them by length. Using the information gained from all these techniques, a sequence ladder (see Figure 16-15) may be prepared from which the order of the nucleotides can be read directly.

18. Nucleic acid hybrids may be used not only to detect the presence of specific nucleic acids (by providing exact complementary sequences) but may also indicate the extent to which two nucleic acid sequences are similar (**c**). Before hybridization can be achieved, however, it is necessary to break the double helix into its complementary strands, that is, to denature it. This is done by gentle heating and can be reversed to re-form the original helix.

19. Cleavage of each of the two strands between the T and C would leave the pairs of sticky ends indicated in choice **a**. As complementary bases of the protruding, sticky ends pair together, parts of the complementary sequences could be joined.

20. The correct answers in order are **e**; **b**, **c**; **b**; **d**; and **a**.

21. The correct answers in order are **f**; **d**; **b**, **e**; **c**; **a**; and **b**, **e**.

22. Temperate phages and plasmids both self-replicate and become incorporated into their host chromosome. Temperate phages, though, assemble a protein coat, which gives them the ability to lyse the host-cell membrane and to take up a temporary existence outside the host cell.

23. The difference between general and restricted transduction lies in which portions of the host-cell chromosome are taken up by a virus before it lyses the host-cell membrane. Review the section on transduction (pages 316–317).

24. Your discussion should develop the analysis of Question 17 in greater detail.

25. The genes on the *E. coli* chromosome are ordered in a linear array, and replication always starts at the point where the F plasmid is incorporated. Because of this, conjugation can be interrupted at various times to see at what points in the conjugation sequence mutant phenotypes appear in the recipient cells as a result of recombination. The chronological appearance of the mutants is directly related to the order of the genes responsible.

CHAPTER 17

The Eukaryotic Chromosome

CHAPTER ORGANIZATION

I. Introduction
II. Cytological observations
   A. Introductory remarks
   B. The nucleolus
III. The chromosomal proteins
   A. Background information
   B. The DNA helix and proteins
   C. The nucleosome
   D. Chromosomal proteins and gene regulation
IV. Essay: The totipotent cell
V. DNA of the eukaryotic chromosome
   A. General information
   B. DNA replication
   C. Classes of DNA
   D. Multiple genes
   E. Gene amplification
   F. The hemoglobin family
      1. General background
      2. Evolution of the hemoglobins
   G. Introns and exons
      1. General information
      2. The function of introns
      3. Other mRNA editing
   H. Jumping genes
   I. Sex-determining cassettes in yeasts
   J. Antibody formation
   K. Genes that cause cancer
      1. General information
      2. Viruses in gene transfer
      3. Crown gall disease
VI. The DNA of mitochondria
VII. Summary

MAJOR CONCEPTS

In eukaryotes, the genetic material is packaged into chromosomes, which are visible when condensed at mitosis or meiosis. The chromatin in these condensed chromosomes occurs in the form of loosely packed euchromatin or densely packed heterochromatin. The nucleolus, which is composed of a cluster of chromatin loops from several chromosomes, is the site of ribosome production.

Eukaryotic chromosomes differ greatly from prokaryotic chromosomes. Eukaryotic DNA is always associated with proteins, most of which are in the form of histones. DNA wrapped around histone cores forms nucleosomes, the basic packaging units of eukaryotic DNA. Several lines of evidence suggest that DNA transcription is related to the degree of condensation of the chromatin.

Eukaryotes have much more DNA than prokaryotes. DNA replication is bidirectional, as it is in prokaryotes, but there are thousands of initiation sites on each eukaryotic chromosome.

There are three classes of DNA in eukaryotes: short repetitive sequences arranged in tandem; moderate repetitive sequences arranged in tandem or dispersed throughout the DNA; and long sequences occurring in only one or a few copies. Most structural genes in eukaryotes occur singly, but some are known to be repeated many times. Of these multiply occurring genes, some are uniform, such as those coding for the rRNAs, tRNAs, and histones. Others are diverse, such as those of the myoglobin-hemoglobin family, which code for proteins with slightly different properties. Genes that occur in multiples facilitate the appearance and survival of genetic variations.

Most eukaryotic structural genes are composed of two types of segments: those protein-coding sequences that are expressed (exons) and those intervening sequences (introns) that are not translated into proteins.

In the eukaryotic chromosome, as in that of prokaryotes, there are many different mechanisms for transfer of genetic material. DNA segments can change positions on a chromosome. Cancers may result from the transposition of a normal gene from one site to another, resulting in changes in cell regulation. Viruses may be involved in these transpositions, carrying genes from one organism to another. A plant tumor has been shown to be the result of the incorporation of a bacterial plasmid into the host chromosome.

Mitochondrial DNA differs from both prokaryotic and eukaryotic DNA. This is an exception to the universality of the genetic code.

HOW TO STUDY THE CHAPTER

Read the chapter, focusing on the major concepts.

Reread the chapter, using the questions that follow to help you focus on details as they relate to major concepts. Answer the questions on a separate sheet of paper. Your answers will provide a valuable study aid in preparing for examinations.

FOCUS ON CHAPTER DETAILS

I. **Introduction** (page 327)
   1. Once the universality of the genetic code was established, molecular biologists speculated that the eukaryotic chromosome would prove to be a larger version of the prokaryotic chromosome. Has this turned out to be the case?

II. **Cytological observations** (pages 327–328)
   A. Introductory remarks

   2. What differences in the organization of DNA have been observed in the eukaryotic nucleus? What terms are used to denote these differences? What are the staining properties of the different categories?

   B. The nucleolus
   3. What is the composition of the nucleolus?
   4. How are eukaryotic and prokaryotic ribosomes similar, and how do they differ?
   5. What is the connection between the role of the nucleolus and materials in the cytoplasm?

III. **The chromosomal proteins** (pages 329–332)
   A. Background information
   6. Describe the dominant type of protein found in association with chromatin. When in the cell cycle are these proteins synthesized?
   7. What are the roles of other proteins in the nucleus?

   B. The DNA helix and proteins
   8. Describe the four forms the DNA helix can assume.

   C. The nucleosome
   9. Why has the sheer amount of DNA found within the eukaryotic nucleus generated such interest?
   10. How are DNA and associated histones arranged within a nucleosome?
   11. How much DNA is packaged in a nucleosome? How does the packaged DNA compare to its unfolded length?
   12. How many nucleosomes are there in a gene?

   D. Chromosomal proteins and gene regulation
   13. What is the presumed regulatory role of chromosomal proteins?
   14. What embryological evidence is there that transcription of DNA is associated with euchromatin rather than heterochromatin?
   15. What two observations indicate that RNA is *not* transcribed from the DNA of tightly condensed chromosomes?

16. What is the significance of "puffs" on the chromosomes of larval insects? What does the effect of ecdysone on puffing imply about the type of activity that occurs in these puffed regions?
17. What are "lampbrush chromosomes"? What do they demonstrate about the degree of condensation of DNA and about gene expression?
18. What do transpositions induced in *Drosophila* indicate about the genetic activity of euchromatin?

IV. **Essay: The totipotent cell** (page 333)

19. What experiments did Briggs and King perform? What conclusion did they draw about gene expression in the developing embryo?
20. Explain how the work of Gurdon and Steward demonstrated a general relationship between cellular differentiation and gene regulation.

V. **DNA of the eukaryotic chromosome** (pages 332–345)

A. General information
21. What is the relationship between a molecule of DNA and a chromosome?
22. How does the amount of DNA differ from cell to cell within an organism? Within a species? From one species to another?
23. How does prokaryotic DNA differ from eukaryotic DNA with regard to the proportion of the DNA that is used in gene expression?

B. DNA replication
24. At what point in the cell cycle does replication occur?
25. How does replication in eukaryotes differ from bacterial replication?

C. Classes of DNA
26. What is satellite DNA? Where is it found in the chromatin, and what is known about its function?
27. Describe the other kind of repeating sequences found in the eukaryotic genome.
28. In what form does the rest of the eukaryotic DNA occur?

D. Multiple genes
29. Describe the genes that code for the four separate types of rRNA (see Figure 17–11).
30. How are the tRNAs transcribed?
31. What class of proteins is among the few proteins coded for by multiple genes?

E. Gene amplification
32. What is gene amplification? With what metabolic activities is it often associated?

F. The hemoglobin family

General background
33. How does diploidy preserve variability?
34. How does gene duplication increase the opportunities for genetic variation?

Evolution of the hemoglobins
35. How is the structure of myoglobin related to that of hemoglobin?
36. Why is the hemoglobin molecule considered more evolutionarily advanced than myoglobin?
37. What is the beta family of hemoglobin genes? How are these genes arranged in the genome of adult primates?
38. Why are some of the genes of the alpha and beta families not expressed? What are these types of genes called?

G. Introns and exons

General information
39. What are introns? How frequently do they occur in the eukaryotic genome? What are exons?
40. How did Pierre Chambon attempt to isolate the gene for ovalbumin? Why did he fail?
41. Describe the hnRNAs. What is their relationship to intervening sequences (introns) and exons?

42. What happens to introns between the occurrence of transcription and the beginning of translation?
43. What are snRNPs? What is thought to be their role in ensuring that only exons are expressed and that they are properly expressed?

The function of introns

44. What speculations have been made concerning the function of introns in promoting recombination?
45. What does the position of introns in the globin genes imply about their function?

Other mRNA editing

46. What two additions are made to eukaryotic mRNA molecules before they leave the nucleus? What is the function of each addition?

H. Jumping genes
47. How was Barbara McClintock able to infer the existence of jumping genes in eukaryotes? What does her work indicate about the function of movable genetic elements?

I. Sex-determining cassettes in yeasts
48. How do yeasts reproduce sexually? What determines the particular mating type of a yeast? How does a yeast change its mating type?
49. How do experiments with yeast demonstrate that gene expression may be under the influence of a "cassette" mechanism?

J. Antibody formation
50. What are antibodies? Why is their production at odds with the "one gene-one polypeptide" hypothesis?
51. What are the important structural characteristics of a typical antibody molecule? (See Figure 17-22.)
52. What is the significance of the constant and variable regions of antibodies?
53. How does the idea of movable genetic elements account for the extreme diversity of antibodies manufactured by a single organism?

K. Genes that cause cancer

General information

54. What is the mutation theory of cancer? What evidence supports this theory?
55. How do DNA and RNA viruses affect the host-cell genome in a way that might cause cancer?
56. What are oncogenes? What is their origin?
57. How is a general theory of cancer related to the concept of movable genetic elements?

Viruses in gene transfer

58. What evidence is there that retroviruses have been involved in gene transfer?
59. What does the plant protein leghemoglobin indicate about the possible role of viruses in evolution?
60. How has the evolutionary role of viruses been characterized by Luria and by Davis?

Crown gall disease

61. What evidence is there that cancer can be induced in a eukaryote by a prokaryote?
62. How does the action of the bacterium resemble that of a cancer-causing DNA virus?
63. What benefits do the bacteria receive as a result of the infection?

VI. **The DNA of mitochondria** (page 345)
64. What evidence is there that the genetic code is *not* universal?
65. From which parent is the genetic material of the mitochondrion inherited? How might this help to explain certain inherited diseases?

VII. **Summary** (pages 345–346): Read the Summary. If you are familiar with the essential features of the material presented there, you are ready to take the following diagnostic examination.

TESTING YOUR UNDERSTANDING

After you have completed the following examination, compare your answers with the annotated key in the Analysis section.

1. RNA that supplies part of the structural foundation of newly forming eukaryotic ribosomes is:
   a. transcribed directly from DNA in the nucleolus.
   b. copied from mRNA in the cytoplasm.
   c. copied from older rRNA along the endoplasmic reticulum.
   d. translated from mRNA on old ribosomes.

2. Histones are genetically important because they constitute the:
   a. protein complex of ribosomes.
   b. majority of chromosomal protein.
   c. intervening sequences found in DNA.
   d. regulating elements of gene transcription.

3. Several lines of cytological evidence point to _____ as the region where active transcription of DNA takes place.
   a. heterochromatin
   b. euchromatin
   c. tightly condensed chromosomes
   d. satellite DNA

4. Tightly condensed regions of the DNA helix that resemble beads on a string are called:
   a. loops.
   b. puffs.
   c. Barr bodies.
   d. nucleosomes.

5. Which of these statements about the amount of DNA present in eukaryotic cells is *not* true?
   a. The amount of DNA is the same in every diploid cell of a given species.
   b. From one species to the next, the total amount of DNA may vary greatly.
   c. Most of a cell's DNA is used for gene expression.
   d. DNA is often redundant and highly repetitive.

6. In eukaryotes, DNA replication begins at _____ and proceeds in _____ .
   a. the centromere; both directions
   b. the 5' end; the direction of the 3' end
   c. many points; both directions
   d. a single point; either one direction or the other

7. Ribosomal RNA has been found to be coded for by:
   a. short, highly repetitive sequences.
   b. multiple-copy genes.
   c. single-copy genes.
   d. All of the above are correct, but for different types of rRNA.

8. The concept of gene amplification is derived from the observation that:
   a. transcription of a gene may occur repeatedly.
   b. a nucleotide sequence may be added to a gene after transcription.
   c. the number of introns within a genetic sequence continually increases.
   d. additional copies of a gene may be produced as needed.

9. DNA sequencing has revealed that the genes coding for the nonalpha polypeptide chain(s) of hemoglobin are actually:
   a. only pseudogenes.
   b. a family of genes in a linear array.
   c. dispersed throughout the genome.
   d. all transcribed as a single unit.

10. Segments of DNA that are transcribed onto RNA but are not translated into protein are called:
    a. introns.
    b. exons.
    c. transposons.
    d. pseudogenes.

11. The existence of introns is evidence that:
    a. mRNA is edited before it attaches to ribosomes.
    b. most of the chromosome is composed of structural genes.
    c. gene amplification occurs quite frequently.
    d. mechanisms exist to limit the frequency of crossing over.

12. A 5' cap and a poly-A tail represent modifications of:
    a. DNA molecules prior to transcription.
    b. DNA molecules following replication.
    c. mRNA molecules before they leave the nucleus.
    d. mRNA molecules after they attach to the ribosomes.
    e. protein molecules that are functionally important.

13. The "cassette" mechanism proposed to explain switching of mating types by yeast cells is founded on the existence of:

a. intervening sequences.   d. a transducing virus.
   b. jumping genes.   e. pseudogenes.
   c. diversity genes.
14. The ability of lymphocytes to produce more than a million different kinds of antibodies can now be explained on the basis of:
   a. shuffling of gene sequences.
   b. V, D, and J genes.
   c. variable and constant regions of polypeptides.
   d. the one gene-one polypeptide hypothesis.
   e. Only three answers (a, b, and c) can account for this.
   f. All four answers (a, b, c, and d) can account for this.
15. Which of the following *best* describes the role of viruses in causing cancers?
   a. Carcinogenic viruses insert RNA into the host-cell DNA.
   b. Viruses contain cancer genes that they insert into oncogenes.
   c. Viruses cause transpositions of normal genes that adversely affect cell regulation.
   d. Viruses apparently have no role in causing cancer.
16. Closely related pseudogenes separated from their gene family and having the features of processed mRNA can be accounted for by the action of:
   a. retroviruses.
   b. jumping genes.
   c. lengthy intervening sequences.
   d. plasmid transfer.
17. Which of the following examples of movable genetic elements appears at this time to be an isolated phenomenon?
   a. flip-flopping of adjacent genes
   b. transposition of elements between distant points in the genome
   c. insertion of bacterial plasmids into eukaryotic chromosomes
   d. viral transfer of genes between different species
   e. reshuffling of genes that code for a single polypeptide chain
18. In which of the following ways does the DNA of a mitochondrion differ from that of the rest of the genome?
   a. It is not a double helix.
   b. It encodes for mitochondrial protein using different nitrogenous bases.
   c. It is maternal DNA only.
   d. A different genetic code is used.
   e. Both c and d are correct.

19. Match each concept on the left with its description.

   gene amplification _____   a. enhances accumulation of genetic variation
   multiply occurring genes _____   b. provides antibody diversity
   chromosomal "puffing" _____   c. regions of active mRNA synthesis
   gene shuffling _____   d. responsible for crown galls in plants
   bacteria-mediated cancer _____   e. supports periods of rapid development

20. Match the genetic jargon at the left with the entity it describes.

   "an exquisitely thin filament" _____   a. ribonucleoprotein particles
   "lampbrush" chromosomes _____   b. viruses
   "bits of heredity looking for a chromosome" _____   c. the DNA molecule of a eukaryote
   "jumping genes" _____   d. a right-handed alpha helix or beta form
   master "cassettes" _____   e. movable genetic elements
   "snurps" _____   f. sites of RNA transcription
   "naked" DNA _____   g. store genes that move to or from particular slots

21. Match each term listed at the left with an appropriate description.

   exons _____   a. tightly packed DNA
   hnRNAs _____   b. disabled members of a gene family
   pseudogenes _____

heterochromatin _____
nucleoli _____

c. nuclear form of transcribed gene
d. clusters of chromosomal loops
e. protein-coding sequences

PERFORMANCE ANALYSIS

1. Eukaryotic ribosomes are composed of proteins (synthesized from mRNA in the cytoplasm) and rRNA, which is transcribed on loops of the nucleolus (i.e., from DNA); therefore, **a** is correct.

2. Histone proteins make up the bulk of chromosomal protein (**b**) by weight. They are found associated with DNA in units called nucleosomes.

3. Euchromatin (**b**) is an unfolded form of DNA that is abundant in egg cells. It makes up the chromosomal loops or puffs, regions that have been shown to be associated with active RNA transcription.

4. Choice **d** is correct. Refer to Figure 17-4.

5. DNA contains a number of highly repetitive sequences, as well as sequences that interrupt structural genes but are themselves not transcribed. Because of this, the amount of DNA used in gene expression is limited to a small proportion (about 1 percent in humans) of the total DNA of the cell; therefore, **c** is correct.

6. As you can see from Figure 17-9, replication of DNA in a eukaryote may begin at a number of initiation sites and proceeds in both directions; therefore, **c** is correct.

7. Ribosomal RNA is transcribed from multiply occurring genes (b) and from much-repeated single genes (a and c). Therefore, **d** is correct.

8. Observations of DNA transcription in egg-forming cells have revealed that additional copies (**d**) of the genes coding for rRNA are produced prior to increased protein synthesis.

9. Mammals contain genes for a number of hemoglobin polypeptides that substitute for the beta chain at various times in the life of an organism. These families of genes (**b**) occupy long nucleotide sequences and are assumed to be the results of a number of gene duplications. Pseudogenes (a) are genes within the family that have been disabled by mutations.

10. Review the work of Chambon (pages 337–338), who was able to show that structural genes are often interrupted or enclosed by DNA sequences that are never expressed. He referred to these as intervening sequences, or introns (**a**) for short.

11. Since sequences are known to exist (see the analysis for Question 10) that are transcribed from DNA but are not present in cytoplasmic mRNA, geneticists believe that mRNA is edited (**a**) prior to its translation.

12. The poly-A tail (a long sequence of adenine nucleotides at the 3' end of mRNAs) and a 5' cap (a single molecule of 7-methylguanosine) are added to mRNAs after transcription but before the mRNAs leave the nucleus (**c**).

13. Certain strains of yeast are able to switch mating types by the insertion of various genes into a particular slot, or locus. Copies of these so-called jumping genes (**b**) are made from master genes stored in "cassettes" at sites adjacent to the locus determining mating type.

14. The variable regions of antibody molecules determine the molecules' specificity for certain antigens. These variable regions are produced from DNA segments composed of V, D, and J genes (b) that have been shuffled (a) from other parts of the genome. The genes coding for the constant regions (c) of antibody molecules become joined to this sequence of V, D, and J genes (see Figure 17-23). Thus, single polypeptide chains are produced by more than a single gene, defying the one gene-one polypeptide hypothesis (and making choice **e** correct).

15. Most cancer viruses are RNA viruses, but these transcribe their RNA into DNA, which may then take up a spot on the host-cell chromosome. Viral DNA sequences associated with cancer in animal hosts have been traced to a eukaryotic origin. The proposed role of viruses in cancer is that they cause changes in cell regulation by transposing genes from one site to another in the genome (**c**).

16. Edited pseudogenes that appear in unusual places in the genome, or that turn up in unrelated

organisms, are believed to be the result of reverse transcription of edited mRNA, a process carried out only by retroviruses (**a**). There is evidence for this from both the hemoglobin gene family and immunoglobulin gene sequences.

17. Choices a, b, d, and e have been documented as generally occurring phenomena. The insertion of bacterial plasmids into the genome of eukaryotes (**c**) has been observed only in the case of crown gall disease in plants infected by bacteria of the genus *Agrobacterium*.

18. Because male gametes, the sperm, contribute only genetic information to the zygote, the DNA of mitochondria must clearly be of maternal origin (c). Differences in the genetic code used by mitochondria in translating mitochondrial DNA into protein (d) have also been found. Choice **e** is therefore correct.

19. The correct answers in order are **e**, **a**, **c**, **b**, and **d**.

20. The correct answers in order are **c**, **f**, **b**, **e**, **g**, **a**, and **d**.

21. The correct answers in order are **e**, **c**, **b**, **a**, and **d**.

CHAPTER 18

Human Genetics

CHAPTER ORGANIZATION

  I. **Introduction**
 II. **The human karyotype**
     A. General information
     B. Essay: Preparation of a karyotype
III. **Chromosome abnormalities**
     A. Background information
     B. Down's syndrome
     C. Abnormalities in sex chromosomes
     D. Chromosome deletions
     E. Amniocentesis
 IV. **Sex-linked hereditary traits**
     A. Color blindness
     B. Hemophilia
  V. **Mapping human chromosomes**
     A. Introductory remarks
     B. Mouse-human hybrid cells
     C. Radioactive probes
 VI. **Inborn errors of metabolism**
     A. Background information
     B. Phenylketonuria
     C. Tay-Sachs disease
VII. **The hemoglobins**
VIII. **Medical implications of recombinant DNA techniques**
     A. Introduction
     B. Prenatal diagnosis of genetic disease
     C. Designer genes
     D. Gene therapy
        1. Putting genes into mammalian cells
        2. Putting genes into mammalian organisms
 IX. **Summary**

MAJOR CONCEPTS

A karyotype is the graphic representation of an organism's chromosomes. In humans, karyotypes are used as part of genetic counseling for couples with a high risk of bearing children with genetic abnormalities. Chromosomal abnormalities that can be visibly detected include deletions, translocations, and extra chromosomes (usually resulting from nondisjunction). Down's syndrome is among the human disorders linked to an extra chromosome.

Sex-linked genetic defects are caused by recessive mutations carried on the X chromosome. These characteristics are most often expressed in males, since a female would need two affected X chromosomes to express the trait. Color blindness and hemophilia are among the sex-linked disorders in humans.

Genetic diseases that are caused by malfunctions of essential proteins (including enzymes) result from mutations in the alleles that code for the proteins. These diseases are usually only apparent in the homozygote. Examples of such diseases in humans include phenylketonuria, Tay-Sachs disease, and sickle cell anemia.

Recombinant DNA techniques are of potential importance in (1) the early diagnosis of genetic diseases, (2) mass production of important proteins, and (3) the treatment of human hereditary diseases.

HOW TO STUDY THE CHAPTER

Read the chapter, focusing on the major concepts.

Reread the chapter, using the questions that follow to help you focus on details as they relate to major concepts. Answer the questions on a separate sheet of paper. Your answers will provide a valuable study aid in preparing for examinations.

FOCUS ON CHAPTER DETAILS

I. Introduction (pages 348–349)

1. How do the methods for gathering information about human genetics differ from the methods used for other species?

II. The human karyotype (pages 349–351)

A. General information

2. What is a karyotype? How is it prepared? (See also the essay on page 350.)
3. How are karyotypes used in genetic counseling?

B. Essay: Preparation of a karyotype (page 350)

4. Give the reason for each step shown in the preparation of a karyotype.

III. Chromosome abnormalities (pages 351–354)

A. Background information

5. What is nondisjunction? What are the possible consequences for the developing human embryo if fertilization involves a gamete in which nondisjunction has occurred?

B. Down's syndrome

6. What are the abnormalities exhibited by a person afflicted with Down's syndrome?
7. Describe the two chromosomal abnormalities that can give rise to Down's syndrome.
8. Determine the possible phenotypes of a child one of whose parents is a translocation carrier. (Check your work against Figure 18-5.)
9. How can the possibility that a couple will produce an affected child be assessed?
10. What may explain why the risk of Down's syndrome increases with the age of the mother?

C. Abnormalities in sex chromosomes

11. What combinations of sex chromosomes may result from nondisjunction? Draw diagrams of the final stage of meiosis that account for these combinations.
12. What defects are usually associated with sex chromosome abnormalities?

D. Chromosome deletions

13. What abnormalities are correlated with a deletion in chromosome 11?
14. How are children with a high risk of developing Wilm's tumor identified?

E. Amniocentesis

15. Describe how amniocentesis is used to detect genetic defects in an unborn child.

IV. Sex-linked hereditary traits (pages 354–356)

A. Color blindness

16. What conditions result in color blindness in human males and females?
17. When color blindness is transmitted to sons, why is the trait received only from their mothers?
18. What are the characteristics of sex-linked inheritance evident in the pedigree of Figure 18-9?

B. Hemophilia

19. What is hemophilia? How is it inherited?

V. Mapping human chromosomes (page 356)

A. Introductory remarks

20. How is it possible to assign certain genes to human sex chromosomes?
21. How must the human autosomes be mapped?

B. Mouse-human hybrid cells

22. What happens to human chromosomes in hybrid cells made up of human and mouse chromosomes?
23. How can a particular human gene be detected on the remaining chromosome of the hybrid culture?
24. How have these techniques been refined?

C. Radioactive probes

25. What are nucleic acid probes?
26. How do they indicate that a given gene (or group of genes) is located on a given chromosome?

VI. Inborn errors of metabolism (pages 357–358)

A. Background information
27. What did Garrod mean by "inborn errors of metabolism"?
28. What did Garrod postulate was the hereditary error or deficiency, at least in the case of alcaptonuria?
29. Why aren't heterozygotes adversely affected by the recessive alleles they carry?

B. Phenylketonuria
30. What "inborn error" is inherited by a person with phenylketonuria? How does this error affect the afflicted child?
31. How is PKU inherited? What is its incidence in the United States?
32. How can PKU be effectively treated?

C. Tay-Sachs disease
33. What "inborn error" is inherited by a child with Tay-Sachs disease? How does this error affect the afflicted child?
34. What is the frequency of carriers for Tay-Sachs disease in the United States?

VII. **The hemoglobins** (page 359)
35. How common is sickle cell anemia in the United States?
36. How does this beta-chain mutation of hemoglobin result in circulatory problems?
37. How can heterozygotes for sickle cell anemia be detected before they decide to have children?
38. What is the probability of two heterozygotes having an afflicted child?

VIII. **Medical implications of recombinant DNA techniques** (pages 360–363)
A. Introduction
39. What general precautions are being taken to ensure that recombinant DNA research does not adversely affect human health?
40. How have recent research achievements alleviated fears of widespread disaster produced by runaway recombinant bacteria?
41. In what three areas are recombinant techniques expected to be useful?

B. Prenatal diagnosis of genetic disease
42. What recombinant DNA tools are employed in the early detection of sickle cell anemia? How are these tools used to distinguish between normal and mutated globin genes?
43. What results does the prenatal test yield for carriers? For fetuses that will contract the disease?

C. Designer genes
44. Why are human proteins that are manufactured by bacteria preferable to those isolated from animals?
45. Why did Itakura choose somatostatin as the first protein to be synthesized by bacteria?
46. What must be done to create an artificial gene?
47. How did the Itakura team insert the somatostatin gene into the *E. coli* genome?
48. Where was this *E. coli* genome spliced so that useful amounts of somatostatin could be isolated?
49. What other successes have made human proteins more available for medical applications?

D. Gene therapy

Putting genes into mammalian cells
50. What is the main goal of gene therapy?
51. What are some of the difficult manipulations that must be carried out if gene therapy is to be successful? Which of these has already been realized?
52. Familiarize yourself with the text's examples of recombination and the role of viruses in each of the cases.

Putting genes into mammalian organisms
53. How did Rudolph Jaenisch make use of a virus to establish a foreign gene in the germ line of experimental mice?
54. What recombination involving mammalian genes was carried out by Gordon and Ruddle? What additional success did they achieve?

55. What is the attitude of geneticists toward the future of gene therapy in the treatment of human disease?

IX. **Summary** (page 364): Read the Summary. If you are familiar with the essential features of the material presented there, you are ready to take the following diagnostic examination.

TESTING YOUR UNDERSTANDING

After you have completed the following examination, compare your answers with the annotated key in the Analysis section.

1. Nondisjunction is said to occur when:
   a. nonhomologous chromosomes become attached and fail to separate before meiosis is complete.
   b. homologous chromosomes fail to separate during anaphase of meiosis.
   c. a prophage fails to release itself from the host genome during DNA replication.
   d. the nuclei of male and female gametes fail to join during fertilization.

2. A karyotype may be used to detect:
   a. the number of chromosomes present.
   b. abnormalities in chromosomes.
   c. carriers of Mendelian recessives.
   d. frameshift mutations.
   e. Both a and b are correct.

3. The karyotype of a person with translocation Down's syndrome has:
   a. one less than the normal complement of chromosomes.
   b. three distinct copies of chromosome 21.
   c. a segment added to chromosome 15.
   d. one extra chromosome 15.
   e. a perfectly normal appearance.

4. A person having one Y and two X chromosomes will develop into a:
   a. sterile male.
   b. sterile female.
   c. fertile male.
   d. fertile female.
   e. hermaphrodite.

5. X-linked recessive alleles:
   a. may not be passed from father to son.
   b. may be passed from father to daughter.
   c. may not be passed from mother to son.
   d. may be passed only from mother to daughter.
   e. Both a and b are correct.

6. During amniocentesis, in preparation for a karyotype, _____ are removed from the amniotic fluid of the embryo.
   a. live maternal cells
   b. dead maternal cells
   c. live fetal cells
   d. dead fetal cells
   e. fetal enzymes

7. Human chromosomes may be mapped by a technique involving mouse-human hybrid cells because of the fact that:
   a. extensive crossing over takes place between the two species' chromosomes.
   b. most human chromosomes are lost from the culture.
   c. human genes, but not mouse genes, are expressed.
   d. most gene products of man and mouse are identical.

8. The technique of radioactive probing to determine genotype requires that:
   a. the DNA of the patient be radioactively labeled.
   b. the exact location of the suspected gene within the genome be known.
   c. a radioactive copy of the suspected gene be synthesized.
   d. Both a and b are correct.
   e. All three answers (a, b, and c) are correct.

9. Phenylketonuria and Tay-Sachs disease are similar in that:
   a. both are determined by a single recessive allele in the homozygous state.
   b. in both cases, biochemical products accumulate until toxic levels are reached.
   c. both can lead to symptoms of mental retardation.
   d. in both cases, affected individuals lack an enzyme.
   e. All of the above are correct.

10. If two carriers of the sickle cell allele marry and have children, their chance of producing diseased offspring is _____ percent.

a. 100
   b. 75
   c. 50
   d. 25
   e. 0

11. Which of the following represents an informed attitude regarding the hazards of recombinant DNA experiments using altered *E. coli*?
    a. Man is not in danger unless the bacteria escape from genetic laboratories.
    b. If these bacteria were to invade healthy intestinal tracts, they would produce lethal amounts of foreign protein.
    c. These experiments pose a constant health hazard and should be more closely regulated.
    d. The chances of widespread disaster are probably less than those for another smallpox epidemic.
    e. Both b and c are correct.

12. The first human gene product to be synthesized by a bacterium was:
    a. insulin.
    b. somatostatin.
    c. interferon.
    d. beta-galactosidase.
    e. ovalbumin.

13. Through recombinant DNA research, biologists have realized that *E. coli* can produce _____ when _____ is introduced at the proper sequence in the bacterial genome.
    a. bacterial DNA; human DNA
    b. bacterial proteins; human RNA
    c. human proteins; human DNA
    d. human DNA; bacterial DNA
    e. nearly any kind of substance; human DNA

14. Which of the following is true regarding the pioneering efforts in the field of molecular genetics known as gene therapy?
    a. Foreign genes have been taken up by mammalian cells only in tissue cultures.
    b. Viruses have been used as vectors in the transfer of genes between different species.
    c. Proviruses have been successfully introduced into a mammalian genome.
    d. Genes have been successfully transferred only between organisms of the same species.
    e. Both b and c are correct.

15. Itakura demonstrated that a human growth hormone could be produced in quantity by *E. coli* when the gene for the hormone was:
    a. inserted immediately following the operator of the *lac* operon.
    b. spliced into a structural gene of the *lac* operon.
    c. inserted immediately following the last structural gene of the *lac* operon.
    d. introduced at any of a number of points on an R plasmid.

16. Which of the following diseases is currently being detected prenatally with the use of a radioactive nucleic acid probe?
    a. Wilm's tumor
    b. sickle cell anemia
    c. phenylketonuria
    d. Tay-Sachs disease
    e. hemophilia

17. Carriers of _____ are readily detectable through karyotyping.
    a. Down's syndrome
    b. sickle cell anemia
    c. phenylketonuria
    d. hemophilia
    e. color blindness

18. All of the following diseases have been related to an enzyme deficiency *except*:
    a. phenylketonuria.
    b. alcaptonuria.
    c. Tay-Sachs disease.
    d. sickle cell anemia.
    e. The premise is false; all of these diseases are caused by enzyme deficiencies.

19. Which of the following genetic diseases can be effectively treated if detected early?
    a. Down's syndrome
    b. sickle cell anemia
    c. Tay-Sachs disease
    d. phenylketonuria
    e. hemophilia

20. Which of the following is *generally* true of a person who is a carrier of a deleterious Mendelian recessive allele?
    a. He or she often contracts the disease, though the symptoms are not severe.
    b. The effects of the "bad allele" are masked by the "good allele."

c. One-half of his or her children will be affected with the disease if he or she marries another carrier.

d. He or she is phenotypically identical to a noncarrier even at the gene-product level.

21. Match each term with its description(s).

metaphase chromosomes _____

nondisjunction chromosomes _____

stained chromosomes _____

X chromosomes _____

bacterial chromosomes _____

chromosomal deletions _____

a. contain recessive alleles for color blindness

b. used to prepare karyotypes

c. exhibit characteristic banding patterns

d. responsible for XO and XXY genotypes

e. responsible for a form of kidney cancer

f. may accept foreign genes

PERFORMANCE ANALYSIS

1. When homologous chromosomes fail to separate during anaphase I of meiosis (**b**), or when sister chromatids fail to separate during anaphase II, the gametes produced will have either one too many or one too few chromosomes.

2. Since choices c and d usually involve only a few base pairs of the chromosome, they will not be detectable in a karyotype. Both the number of chromosomes (**a**) and any structural abnormalities (**b**) will be visible, though staining may be required to see such abnormalities (**e** is correct).

3. A person with Down's syndrome caused by nondisjunction will have three distinct copies of chromosome 21 (see Figure 18-4b); that is, he or she has one more than the normal complement of chromosomes. But a person with translocation Down's syndrome has a portion of an extra chromosome 21 attached to chromosome 15 (**c**). The person also has the equivalent of one more chromosome than the normal complement of 46.

4. A person with the genotype XXY is a male and is usually sexually underdeveloped and sterile; therefore, **a** is correct.

5. Traits determined by recessive alleles linked to the X chromosome can be passed from mother to either son or daughter but cannot be passed from father to son (a). The reason for this is that if the father is color-blind, then he carries the allele for color blindness on his X chromosome. If he donates an X chromosome at fertilization, then he will have a daughter who may or may not be color-blind (b). Choice **e** is therefore correct.

6. The amniotic fluid of the fetus contains fetal cells sloughed off by the embryo. Live cells must be used to prepare the karyotype because the cells must be grown in tissue culture and stimulated to enter mitosis; therefore, **c** is correct.

7. When human chromosomes are inserted into cells of another species, the chromosomes are gradually lost (**b**). When the number of human chromosomes has been reduced to one, any human proteins being made by the hybrid cell are identified as being on the remaining chromosome.

8. Gene probes must be radioactively labeled (**c**) so that the segment of DNA with which they hybridize can be detected. In this way, the exact location of the gene within the genome can be determined.

9. Both phenylketonuria and Tay-Sachs disease are caused by enzyme deficiencies (d) that are due to the presence of a single recessive allele in the homozygous state (a). When biochemical intermediates accumulate (b), they affect brain development, resulting in mental retardation (c); therefore, **e** is correct.

10. This problem can be solved using a four-square Punnett square. Since sickle cell anemia results from the recessive allele in the homozygous state, only 25 percent (**d**) of the offspring of two carriers (heterozygotes) will be affected.

11. Recombinant *E. coli* are contained in the same way as other harmful pathogens, such as smallpox (**d**). The mutant bacteria are also disabled so that they can reproduce only in specialized cultures. Even if they were to escape from the laboratory and compete with normal *E. coli* in the human intestinal

tract, they would probably not produce consequential amounts of foreign proteins.

12. Somatostatin (**b**) was the first human protein to be synthesized by *E. coli*. The Itakura team accomplished this by inserting an artificial human gene for somatostatin into an *E. coli* plasmid.
13. Choice **c** is correct. See analysis for Question 12.
14. Choice a is false; Rudolf Jaenisch introduced a leukemia virus into the nuclei of living mouse embryos. Choice d is also false; a rabbit globin gene has been inserted into monkey cells. Choices b and c, however, are both true, for viruses have proved to be an essential element in both of these gene transfers. In the case of Jaenisch, they were incorporated into the host genome as a provirus; therefore, **e** is correct.
15. In order for Itakura to harvest useful quantities of somatostatin, it was necessary to splice the gene into an *E. coli* structural gene of the *lac* operon (**b**)—namely, the gene for beta-galactosidase. This was necessary so that the somatostatin synthesized by *E. coli* would not be recognized as a foreign substance and thus would not be enzymatically degraded.
16. Because the mutated beta globin gene associated with sickle cell anemia (**b**) hybridizes with a longer radioactive sequence than that of the normal beta globin gene, sickle cell anemia may be diagnosed prenatally from cells taken by amniocentesis.
17. Both nondisjunction and translocation Down's syndrome (**a**) are caused by chromosomal abnormalities that can be detected by karyotyping. (See Figure 18-4.)
18. Sickle cell anemia (**d**) is due *not* to an enzyme deficiency but to an alteration in the structure of an oxygen-transporting protein, hemoglobin.
19. Infants showing early symptoms of phenylketonuria (**d**) can be placed on a diet low in phenylalanine and will develop normally.
20. People carrying a single recessive allele may make less than normal of a certain gene product but still enough to maintain normal phenotypes. This is often referred to as the masking of the recessive allele by the dominant allele (**b**). If two carriers of Mendelian recessives marry, one-fourth of their offspring, on the average, will exhibit the disease.
21. The correct answers in order are **b**; **d**; **b**, **c**; **a**; **f**; and **e**.

PART II Biology of Organisms

SECTION 4 The Diversity of Life

CHAPTER 19

The Classification of Organisms

CHAPTER ORGANIZATION

I. Introduction
II. What is a species?
   A. General background
   B. The naming of species
III. Taxonomy
IV. Systematics
   A. Introductory remarks
   B. Homology and phylogeny
   C. The monophyletic ideal
V. Taxonomic methods
   A. Introductory remarks
   B. New methodologies
      1. General background
      2. Numerical phenetics
      3. Cladistics
   C. Essay: How to construct a cladogram
   D. Some criteria
      1. General background
      2. Amino acid sequences
      3. Molecular clocks
VI. A question of kingdoms
VII. Summary

MAJOR CONCEPTS

Taxonomy is the classification of organisms into formal categories. Systematics is the study of the relationships among groups of organisms. Phylogeny is the evolutionary history of organisms, inferred from the systematics of living and fossil organisms.

A species can be defined as a group of actually or potentially interbreeding organisms that is reproductively isolated from other groups of organisms. Separate but related species are grouped into genera (singular, genus).

Organisms are grouped into taxa (categories). These categories (from the most inclusive to the least inclusive) are kingdoms, phyla (or divisions), classes, orders, families, genera, and species. In the classification system used in this text, there are five kingdoms: Monera, Protista, Fungi, Plantae, and Animalia.

Three methodologies that are currently used to classify organisms are (1) traditional taxonomy, which reflects similarities and differences based on evolutionary history; (2) numerical phenetics, which depends upon a numerical scoring of *observable* similarities and differences; and (3) cladistics, which is based solely upon evolutionary branching sequences (genealogy) and disregards overall similarities.

Biochemical data, including differences in protein structure, are being used in addition to anatomy as criteria for classifying organisms.

HOW TO STUDY THE CHAPTER

Read the chapter, focusing on the major concepts.

Reread the chapter, using the questions that follow to help you focus on details as they relate to major concepts. Answer the questions on a separate sheet of paper. Your answers will provide a valuable study aid in preparing for examinations.

FOCUS ON CHAPTER DETAILS

I. Introduction (page 371)

1. Define taxonomy and systematics. What is the relationship between the two disciplines?

II. What is a species? (pages 371–374)

A. General background

2. Contrast the simpler Latin definition of species with the more rigorous definition of Ernst Mayr.
3. What is the importance of the terms "groups" and "populations" to Ernst Mayr's definition of species?
4. Summarize the instances in which it is difficult to use Mayr's definition of a species.
5. Clarify the two usages of the word "species"—one emphasizing constancy and the other recognizing change.
6. Why is the species concept retained in spite of the difficulties it presents?

B. The naming of species

7. What is a genus? When may the genus name be used alone?
8. Why is a specific epithet meaningless when used alone? Cite an example.
9. When may a genus name be abbreviated?
10. Summarize the rules (if there are any) by which organisms may be named. Cite an example for each point you make.
11. Why is the binomial system of nomenclature a necessary tool of science?

III. Taxonomy (pages 374–376)

12. Differentiate between the terms "taxon" and "category" as used in taxonomy.
13. List (from the most inclusive to the least inclusive) the categories in current usage for the classification of organisms.
14. What are the rules that govern how these categories are written?
15. How is hierarchical classification useful in storing and retrieving information about an organism? (Consult Table 19-1 before you answer this question.)
16. What is the evidence against the proposition that all taxonomists place organisms in the same series of categories? (Hint: How do the lumpers differ from the splitters?)

IV. Systematics (pages 376–378)

A. Introductory remarks

17. What is systematics? How have the goals of classification changed from Linnaeus's time to ours?
18. What two functions does classification of organisms attempt to serve? Are these functions clearly compatible?

B. Homology and phylogeny

19. What danger is inherent in using superficial similarities between organisms as a basis for grouping them together? Cite an example.
20. How has the concept of homology made it easier for classification schemes to reflect evolutionary relationships?
21. Differentiate between homology and analogy.
22. What generalization can be made about the relationship between the number of separate parts involved in a feature shared by several species and the likelihood that the feature arose independently in each species?

C. The monophyletic ideal

23. Differentiate between monophyletic and polyphyletic taxa.

V. Taxonomic methods (pages 378–385)

A. Introductory remarks

24. Summarize the steps taken by a taxonomist in classifying a newly discovered organism.

B. New methodologies

General background

25. What criticisms are voiced of the traditional methodology of classification by both pheneticists and cladists?

Numerical phenetics

26. Summarize the classification methods of numerical phenetics.
27. Upon what basic assumption does the value of this classification scheme rest? (Hint: Why are so many characteristics considered?)

Cladistics

28. What is the goal of the cladists? What argument do cladists use to support their methodology?
29. Why does cladistics produce a more revolutionary scheme of classification than does numerical phenetics?
30. What problem do the traditionalists see with the classification scheme produced by cladistics?

C. Essay: How to construct a cladogram (page 381)

31. Discuss the two essential features of classification by cladistics: branching points and the generation of testable hypotheses.

D. Some criteria

General background

32. Although all methods of systematics are based largely on anatomy, what new techniques are being used?

Amino acid sequences

33. What are two advantages of comparing the amino acid sequences of homologous proteins from different organisms?
34. What presumption is made when the evolutionary relationship between organisms is determined according to the number of amino acid differences in a homologous protein?
35. To what extent does the comparison of protein structures confirm traditionally determined evolutionary relationships?

Molecular clocks

36. What are the two schools of thought on the meaning of differences in homologous protein structure among organisms?
37. Using cytochrome *c* as an example, give the details of the molecular-clock hypothesis.
38. Why is nucleic acid sequencing preferable to amino acid sequencing in determining the relatedness of different organisms?

VI. **A question of kingdoms** (pages 385–388)

39. Summarize the current problems of classifying organisms according to kingdom.
40. On what are most contemporary proposals concerning kingdoms based?
41. Cite an example that substantiates the statement that "no system of kingdoms is really satisfactory."

VII. **Summary** (pages 388–389): Read the Summary. If you are familiar with the essential features of the material presented there, you are ready to take the following diagnostic examination.

TESTING YOUR UNDERSTANDING

After you have completed the following examination, compare your answers with the annotated key in the Analysis section.

1. Ernst Mayr's rigorous definition of a species may be strictly applied only in the case of:
   a. eukaryotes.
   b. prokaryotes.
   c. multicellular organisms.
   d. plants.
   e. animals.

2. Which of the following *always* represents an unambiguous designation for the species of fruit fly popular in genetic research?
   a. *Drosophila*
   b. *Drosophila melanogaster*
   c. *D. melanogaster*
   d. *melanogaster*
   e. *Drosophila m.*

3. Organisms are considered members of the same species *only if* they:
   a. inhabit the same geographical location.
   b. actually interbreed and produce fertile offspring in nature.
   c. are theoretically members of an interbreeding natural population.

d. are members of the same population.
e. Both b and d are correct.

4. The taxonomic category that includes the greatest diversity of organisms within a given kingdom is the:
   a. phylum (or division).
   b. class.
   c. order.
   d. family.
   e. superfamily.

5. The straightforward classification of organisms into meaningful groups falls within the domain of:
   a. systematics.
   b. taxonomy.
   c. population biology.
   d. phylogeny.

6. The underlying foundation of all taxonomic and systematic work is:
   a. the species concept.
   b. the binomial classification system.
   c. the five-kingdom system.
   d. the phylogenetic tree.

7. Post-Darwinian systematists prefer a classification scheme that is based upon:
   a. superficial similarities.
   b. external similarities.
   c. homologous structures.
   d. analogous structures.
   e. adaptations to specific environments.

8. Two terms that demonstrate the difference of opinion among taxonomists on the importance of similarities between organisms are:
   a. taxon and category.
   b. phylum and division.
   c. lumpers and splitters.
   d. binomials and common names.

9. In an ideal phylogenetic classification scheme, all species descended from their most recent common ancestor would be placed into:
   a. a monophyletic family.
   b. a monophyletic genus.
   c. a polyphyletic genus.
   d. a single subgenus.
   e. separate genera.

10. The method of numerical phenetics attempts to show degrees of relatedness between taxa based on analysis of:
    a. homologous structures.
    b. analogous structures.
    c. unit characters.
    d. genealogy.
    e. Both a and b are correct.

11. The cladistic taxonomist places major emphasis on the accurate documentation of:
    a. phylogenetic trees.
    b. homologies versus analogies.
    c. character similarity.
    d. biochemical differences.

12. The analysis of protein sequences is most useful in determining the evolutionary relationship between groups of organisms that:
    a. have vastly different metabolic pathways.
    b. differ greatly in morphology.
    c. are genotypically very similar.
    d. use similar proteins for different functions.

13. According to the "random-tick" hypothesis, two species whose cytochrome c molecules differ by a relatively great number of amino acids:
    a. must require functionally different kinds of cytochrome c.
    b. would probably not be classified in the same family.
    c. have a relatively distant common ancestor.
    d. have diverged only within the recent past.
    e. Both a and b are correct.

14. The recent five-kingdom proposal is an attempt to classify all organisms based upon:
    a. a monophyletic-kingdom approach.
    b. recent knowledge of evolutionary history.
    c. cellular organization.
    d. the nutritional mode of organisms.
    e. Both c and d are correct.

15. Amino acid sequences are conveniently viewed as _____ by biologists who place little importance on the selective value of homologous protein structures.
    a. molecular clocks
    b. poor evolutionary criteria
    c. unchanging
    d. genetically unimportant
    e. more revealing than nucleotide sequences

16. The basis of the former two-kingdom system was classification of organisms as to whether they were:
    a. prokaryotes or eukaryotes.
    b. plants or animals.
    c. unicellular or multicellular.
    d. autotrophs or heterotrophs.
    e. Both a and c are correct.

17. Which of the following is true of the five-kingdom system of classification?
    a. All one-celled organisms belong to a separate kingdom.
    b. All heterotrophic organisms belong to a single kingdom.
    c. All autotrophic organisms belong to a single kingdom.
    d. All prokaryotes belong to a single kingdom.
    e. Organisms are separated into kingdoms based on their ecological roles.

PERFORMANCE ANALYSIS

1. Ernst Mayr rigorously defined species as groups of "actually or potentially interbreeding natural populations that are reproductively isolated from other such groups." This definition would not strictly apply to those groups of organisms that reproduce only by asexual means, such as many prokaryotes and some plants. Therefore, **e** is correct.

2. According to the binomial system of nomenclature devised by Linnaeus, any species is identified by two names; the first is the name of its genus, and the second is its species name. *Drosophila melanogaster* (**b**) is therefore correct. Only where there is no chance of ambiguity (such as in literature where an organism is referred to by its full name when first mentioned and the genus name is abbreviated in subsequent references) may a species be referred to as shown in choice c.

3. A group of organisms is considered to be a species only if it constitutes an actually or theoretically interbreeding natural population (**c**). A species may be represented by many populations in a number of geographical locations. Don't be swayed by choice b; some populations of the same species, such as those living on different continents, may never have the chance to mate and reproduce.

4. A handy mnemonic for learning the hierarchy of classifying organisms is *K*ing *P*hillip *C*ame *O*ver *F*riday and *G*ave a *S*peech, where the italicized letters stand for kingdom, phylum (though protists and plants are frequently classified into divisions instead of phyla, in which case the mnemonic begins with *K*ing *D*avid), class, order, family, genus, and species. Choice **a** is therefore correct.

5. Taxonomy (**b**) is the classification of organisms, systematics is the study of their relationships, and phylogeny reflects their evolutionary history.

6. The species concept (**a**) is a foundation for taxonomy, systematics, and the study of evolution. Classification systems other than the binomial system would also work, so b is incorrect. Taxonomy and systematics are independent of the particular system of classification, making c and d incorrect also.

7. Post-Darwinian systematists are concerned with the origin of similarities or differences among groups of organisms. They base their systems of classification on the occurrence of homologous structures (those structures that have a common origin, though not necessarily a common function); therefore, **c** is correct.

8. In classifying organisms into categories, splitters attribute more importance to differences than do lumpers; therefore, **c** is correct.

9. If a classification scheme is to be an accurate reflection of phylogeny—the evolutionary history of a group of organisms—then all organisms within a given taxon should be descended from a common ancestor in the next higher level of classification; therefore, **b** is correct.

10. Numerical phenetics concerns itself with sets of observable characteristics present in a group of organisms. Thus, numerical pheneticists analyze differences in sets of unit characters (**c**), which are traits that may be assigned discrete values (see Table 19–2) and that cannot be further subdivided.

11. Cladistic taxonomists attempt to classify organisms according to their evolutionary history and so would try to construct highly accurate phylogenetic trees (**a**).

12. Organisms differing greatly in their morphology (**b**) (e.g., a fish and a fungus) would be difficult to compare except by such biochemical means as the analysis of the amino acid sequences of proteins.

13. Those biologists who propose a random-tick hypothesis to account for differences in protein structure in groups of organisms assume that a large number of amino acid differences would simply be the result of random changes over long periods of time and would indicate that the groups in question share a relatively distant ancestor (**c**).

14. Two important characteristics used in classifying organisms within one of five kingdoms (Monera, Protista, Fungi, Plantae, and Animalia) are cellular organization (c) and mode of nutrition (d). Choice **e** is therefore correct.

15. Biologists who believe that differences in protein structure are accounted for largely by random processes over time (see analysis to Question 13) think of amino acid sequences as molecular clocks (**a**) that chart the history of changes in molecular biochemistry.

16. In the original system of classifying organisms, fungi, algae, and microorganisms were classified along with plants, and protozoans were grouped with animals; therefore, **b** is correct.

17. Although unicellular organisms may be either monerans or protists, those organisms that lack a nucleus (prokaryotes) are classified within the kingdom Monera; therefore, **d** is correct.

CHAPTER 20

The Prokaryotes

CHAPTER ORGANIZATION

I. Introduction
II. The classification of prokaryotes
III. The prokaryotic cell
  A. Introductory remarks
  B. The cytoplasm
  C. The cell membrane
  D. The cell wall
    1. Introductory remarks
    2. Chemical structure of the cell wall
  E. Flagella and pili
IV. Diversity of form
V. Types of motility
VI. Essay: Sensory responses in bacteria
VII. Reproduction and resting forms
VIII. Prokaryotic nutrition
  A. Heterotrophs
  B. Chemoautotrophs
  C. Photosynthetic prokaryotes
    1. Eubacteria
    2. Cyanobacteria
IX. Viruses: A case apart
  A. Introductory remarks
  B. Structure of viruses
  C. Origin of viruses
X. Viroids: The ultimate in simplicity
XI. Microorganisms and human ecology
  A. Symbiosis
  B. How microbes cause disease
  C. Prevention and control of infectious disease
XII. Summary

MAJOR CONCEPTS

Prokaryotes are currently classified as cyanobacteria (division Cyanophyta, formerly blue-green algae) and bacteria (division Schizophyta). The cells of prokaryotes are characterized by DNA that is not associated with histones and by the absence of a membrane-bound nucleus and membrane-bound organelles.

Prokaryotes may be nonmotile, or they may move by gliding or by using flagella. Reproduction in prokaryotes is usually asexual (by binary fission, budding, or fragmentation), but genetic exchanges and recombinations can occur. Prokaryotes include both heterotrophs and autotrophs (chemoautotrophic and photosynthetic prokaryotes).

Viruses, which are composed of a protein coat and DNA or RNA, are difficult to classify within any of the five kingdoms, since it is not even clear whether they should be classified as living or nonliving.

Viroids are composed of small RNA molecules; they lack the protein coat characteristic of viruses.

Symbiotic relationships involve a close living association between two unrelated organisms. The three kinds of symbiosis are mutualism (both partners benefit); commensalism (one partner benefits); and parasitism (one partner is harmed).

HOW TO STUDY THE CHAPTER

Read the chapter, focusing on the major concepts.

Reread the chapter, using the questions that follow to help you focus on details as they relate to major concepts. Answer the questions on a separate sheet of paper. Your answers will provide a valuable study aid in preparing for examinations.

FOCUS ON CHAPTER DETAILS

I. **Introduction** (pages 390–391)
   1. Describe the three kingdoms that comprise the microorganisms. What other entities are also studied by microbiologists?
   2. What organisms are classified in the kingdom Monera? What factors have accounted for their great success?
   3. What are obligate anaerobes? What are facultative anaerobes?
   4. What is the adaptive value of the spores formed by some prokaryotes?
   5. Describe two important ecological roles performed by prokaryotes.

II. **The classification of prokaryotes** (pages 391–393)
   6. What characteristics have taxonomists depended upon to classify the prokaryotes? How does this type of classification compare to that employed for eukaryotes? Why has it been difficult to determine phylogenetic relationships among prokaryotes?
   7. Into what two divisions have prokaryotes been grouped?
   8. What methods are proving fruitful in establishing evolutionary relationships among prokaryotes?
   9. Describe those prokaryotic organisms that are so different from other monerans that a separate kingdom has been proposed for them. What is their phylogenetic relationship to other prokaryotes?
   10. What is anticipated about the future of the classification of prokaryotes?

III. **The prokaryotic cell** (pages 394–396)
   A. Introductory remarks
   11. Which group (and which organism in particular) is the most thoroughly studied of the prokaryotes?
   B. The cytoplasm
   12. What are the striking features of the cytoplasm of prokaryotes?
   13. Describe the prokaryotic chromosome.
   C. The cell membrane
   14. How does the prokaryotic membrane differ from that of eukaryotes?
   15. What cellular functions are carried out on the prokaryotic membrane?
   D. The cell wall
   Introductory remarks
   16. What types of cell walls are found among the prokaryotes?
   17. What functions do cell walls serve? Why do the mycoplasmas *not* require a cell wall?
   Chemical structure of the cell wall
   18. What is the constituent of a bacterial cell wall that combines with gentian violet stain? What is the composition of gram-negative cell walls?
   19. What other characteristics does the particular architecture of the cell wall determine?
   20. What structure secreted outside the bacterial cell wall is associated with pathogenicity?
   E. Flagella and pili
   21. Describe the structure of the bacterial flagellum (see also page 117).
   22. How is the bacterial flagellum attached to the cell wall?
   23. How would you distinguish a bacterium's flagellum from one of its pili?
   24. What functions do the pili serve?

IV. **Diversity of form** (pages 397–398)
   25. What are the three basic shapes into which all bacteria can be classified?
   26. What accounts for the filamentous nature of certain rod-shaped bacteria?

27. How are the spirochetes unique?
28. What serious disease is caused by rickettsiae?

V. Types of motility (pages 398–399)
29. Describe the different kinds of movement exhibited by flagellated prokaryotes, and relate each kind to rotation of the flagella.
30. What means of locomotion are exhibited by other prokaryotes?

VI. Essay: Sensory responses in bacteria (page 400)
31. Define phototaxis, chemotaxis, and magnetotaxis.
32. With what prokaryotic types has each of these responses been associated?
33. Where are the sensory receptors for each of these responses located in the cell?
34. Of what benefit to the cell is each of these responses?

VII. Reproduction and resting forms (pages 399–401)
35. What types of prokaryotic reproduction result in a clone?
36. What factors account for the extraordinary adaptability found among prokaryotes?
37. When do bacteria form spores? Describe the process of spore formation.
38. How is a spore protected once it is released? What events are associated with the onset of spore germination?

VIII. Prokaryotic nutrition (pages 401–405)
A. Heterotrophs
39. What ecological functions are attributed to heterotrophic prokaryotes?
40. What is a parasite? Are all parasites pathogenic? What benefits might an organism derive from an association with a heterotrophic bacterium?

B. Chemoautotrophs
41. How do chemoautotrophs obtain their food energy?
42. Describe the role of methanogens in the decomposition process. Summarize this description with a chemical equation.
43. What compounds are utilized by prokaryotes that participate in the nitrogen and sulfur cycles? What is the ecological significance of the compounds they oxidize or produce during oxidation?

C. Photosynthetic prokaryotes

Eubacteria
44. On what basis might eubacteria be separated into two photosynthetic types?
45. Identify the pigments and electron donors that are found within the two groups.
46. Make a superficial comparison of photosynthesis as carried out by eubacteria and as carried out by green plants.

Cyanobacteria
47. How is photosynthesis in the cyanobacteria similar to that in photosynthetic eukaryotes?
48. List the pigments used by cyanobacteria in photosynthesis. How are they organized within the cell?
49. Why are nitrogen-fixing cyanobacteria considered the most nutritionally simple of all organisms? For what types of environments does their nutritional independence make them ideally suited?

IX. Viruses: A case apart (pages 405–407)
A. Introductory remarks
50. Why is it difficult to categorize viruses?

B. Structure of viruses
51. What is the size range of viruses?
52. Describe the structure of a virus.
53. How is its specificity determined?
54. Cite a few examples of viruses that are known to cause infections in humans.
55. What are the methods by which viruses reproduce?
56. How does a viral infection spread from cell to cell?

C. Origin of viruses

57. What is the current hypothesis concerning the origin of viruses?

X. **Viroids: The ultimate in simplicity** (pages 407–408)

58. What is a viroid? How are viroids similar to and different from viruses?
59. Describe the structures of PSTV. What disease has it been shown to cause? What may have been the origin of PSTV?
60. Where are viroids found? Why are viroids even more dependent on their hosts than are viruses?
61. By what mechanism are viroids thought to affect their host cells?

XI. **Microorganisms and human ecology** (pages 408–410)

A. Symbiosis

62. What is symbiosis? What are the three types of symbiotic relationships?
63. What is one criterion for the evolutionary success of a parasite? What might happen if this criterion is not met?

B. How microbes cause disease

64. What are three general ways in which bacteria may produce pathogenic effects? Which method results in diphtheria? Which results in streptococcal pneumonia?
65. Describe several diseases caused by *Streptococcus pyogenes*.

C. Prevention and control of infectious disease

66. What types of control measures have been demonstrated to be effective against disease-causing microorganisms?
67. What is an antibiotic? How are they produced? To what two problems has their widespread application given rise?

XII. **Summary** (pages 410–411): Read the Summary. If you are familiar with the essential features of the material presented there, you are ready to take the following diagnostic examination.

TESTING YOUR UNDERSTANDING

After you have completed the following examination, compare your answers with the annotated key in the Analysis section.

1. In what way can prokaryotes be considered diverse?
   a. They carry out a wide variety of metabolic functions.
   b. There are more species of prokaryotes than of any other group of organisms.
   c. They perform many different ecological roles.
   d. They have numerous anatomical structures composed of interconnected parts.
   e. Both a and c are correct.

2. Which of the following groups of bacteria are often classified in the separate kingdom Archaebacteria?
   a. the photosynthetic eubacteria
   b. bacteria that lack cell walls
   c. bacteria that reduce carbon dioxide to methane
   d. bacteria that can live without oxygen
   e. the nitrogen-fixing bacteria

3. Which of the following characteristics would be useful in classifying a prokaryote according to the traditional system?
   a. nutritional mode
   b. reproductive pattern
   c. presence or absence of a cell wall
   d. similarity of nucleotide sequences
   e. Both a and c are correct.

4. Which of these statements *best* describes the organization of the Monera?
   a. The kingdom consists of two divisions, the Cyanophyta and the Eubacteria.
   b. The two major subdivisions are the cyanobacteria and the bacteria.
   c. This kingdom is organized according to a hierarchical scheme.
   d. The two major subdivisions are the autotrophs and the heterotrophs.
   e. This kingdom is organized according to evolutionary relationships.

5. A _____ is an attachment device used in bacterial conjugation.
   a. mesosome
   b. flagellum

c. pilus
d. capsule
e. spore

6. In comparing the features of gram-positive bacteria with gram-negative bacteria, it would be correct to say that gram-positives:
   a. are not as sensitive to antibiotic treatment.
   b. are not susceptible to the enzyme lysozyme.
   c. lack lipoprotein-lipopolysaccharide cell wall layers.
   d. have cell membranes that contain cholesterol, whereas gram-negatives do not.
   e. Both a and b are correct.

7. Which of the following would a microbiologist be surprised to find within a bacterial cell?
   a. plasmids
   b. internal membranes
   c. a circular chromosome
   d. ribosomes

8. Which of the following is the *best* description of a bacterial flagellum?
   a. chains of protein monomers wound around a hollow core
   b. pairs of microtubules founded on a basal body
   c. a rigid, cylindrical rod
   d. a filament composed of peptidoglycans
   e. a protein filament anchored in the cytoplasm but penetrating the cell wall

9. Bacilli, cocci, and spirilla are terms given to different types of bacteria based on:
   a. their methods of asexual reproduction.
   b. the organization of their hereditary material.
   c. the shape of their cells.
   d. the number and positioning of their flagella.
   e. their modes of motility.

10. Bacteria whose cell membranes are encircled by fibrils are called:
    a. spirochetes.
    b. mycobacteria.
    c. rickettsiae.
    d. vibrios.
    e. schizophyta.

11. When a flagellated bacterium "running" in a certain direction reverses the action of its flagella, it will begin to:
    a. move in the opposite direction.
    b. tumble.
    c. glide.
    d. slow down and come to a complete stop.
    e. rotate while maintaining a fixed position.

12. A spore that has completely formed inside a typical bacterial cell is separated from the external environment (in order, from the spore to the external environment) by:
    a. a cell wall, a spore coat, and a cell membrane.
    b. a spore membrane, a spore coat, a cell membrane, and a cell wall.
    c. a spore coat, a cell membrane, a spore coat, and a cell wall.
    d. a spore coat, a spore membrane, and a cell wall.

13. Which of the following are methods of asexual reproduction found among members of Monera?
    a. binary fission
    b. budding
    c. fragmentation
    d. release of spores
    e. All of the above are correct.

14. Members of which of these groups are responsible for the recycling of nutrients to plants?
    a. parasitic bacteria
    b. heterotrophic bacteria
    c. chemoautotrophs
    d. photoautotrophs

15. The cyanobacteria resemble photosynthetic eukaryotes in all of the following ways *except* that cyanobacteria do not:
    a. use water as an electron donor.
    b. contain chlorophyll pigments.
    c. possess chloroplasts.
    d. use carotenoids as accessory pigments.
    e. produce oxygen during the light-trapping reactions.

16. The equation
    $$CO_2 + 2H_2S \xrightarrow{light} (CH_2O) + H_2O + 2S$$
    describes photosynthesis in:
    a. the purple sulfur bacteria.
    b. the green sulfur bacteria.
    c. the cyanobacteria.
    d. single-celled eukaryotes.
    e. Both a and b are correct.

17. The primary photosynthetic pigment in the green sulfur bacteria is _____, whereas the primary

photosynthetic pigment of the purple bacteria is _____ in color.
   a. chlorophyll *a*; purple
   b. chlorobium chlorophyll; blue-gray
   c. chlorobium chlorophyll; purple
   d. bacteriochlorophyll; purple
   e. bacteriochlorophyll; red or yellow

18. Which of these statements is true regarding viruses and their connection with simple life forms?
   a. Viruses were most probably the precursors of prokaryotic cells.
   b. Viruses represent self-sustaining organisms that have lost all but a few cellular functions.
   c. Viruses are cellular fragments that are partially independent.
   d. Viruses most probably evolved from more simple forms called viroids.

19. Inside a host cell, a DNA virus directs protein synthesis using its own:
   a. nucleotides.
   b. amino acids.
   c. ribosomes.
   d. mRNA and tRNA.
   e. nucleic acid code.

20. A disease of potatoes known as spindle-tuber disease is a result of infection of healthy root-cell nuclei by:
   a. small fragments of RNA.
   b. a DNA-containing virus.
   c. an RNA-containing virus.
   d. a parasitic bacterium.
   e. insects carrying microorganisms.

21. *Escherichia coli* depends on the environment of the human intestine for its nutrition and supplies its host with vitamin K. Humans and *E. coli*, therefore, have entered into a _____ relationship.
   a. parasitic
   b. mutualistic
   c. commensal
   d. pathogenic
   e. chemotrophic

22. A successful parasite is one that:
   a. allows its host to grow and reproduce.
   b. kills its host.
   c. weakens its host but does not kill it.
   d. causes a lengthy illness.
   e. Both c and d are correct.

23. Match each life form with the appropriate description.

   commensal _____
   cyanobacteria _____
   heterotrophic bacteria _____
   nitrogen fixers _____
   obligate anaerobes _____
   facultative anaerobes _____

   a. can live with or without oxygen
   b. are not found where there is oxygen
   c. reduce atmospheric nitrogen to ammonia
   d. recycle nutrients from organic debris
   e. have no detrimental impact on their host
   f. are often the first life form on a new lava bed

24. Match each group of organisms with its distinguishing characteristic. Answers may be used more than once.

   halophiles _____
   viruses _____
   thermoacidophiles _____
   methanogens _____
   gram-positive bacteria _____
   Mycobacterium _____
   sulfur bacteria _____

   a. cell walls combine firmly with gentian violet stain
   b. do not use the Calvin cycle
   c. use H_2S as an electron donor in photosynthesis
   d. some can transcribe DNA from RNA
   e. have best chance of survival in a boiling flask of HCl
   f. filamentous prokaryotes
   g. reduce CO_2 to a simple organic compound

25. Provide a brief review of the means by which microbes cause disease in their host organisms.

26. Discuss the ecological importance of at least four major prokaryotic groups.

27. Compare prokaryotes with eukaryotes on each of the following points.

a. organization of the cytoplasm
b. cell membranes
c. cell walls
d. flagella

28. Give several reasons why you think that the cyanobacteria might be more closely related to photosynthetic eukaryotes than to eubacteria.

29. Offer several reasons for the extraordinary adaptability of prokaryotes.

PERFORMANCE ANALYSIS

1. As a group, the prokaryotes carry out many different metabolic functions, several of which are ecologically important; therefore, **e** is correct. So far only about 2,700 species of prokaryotes have been discovered, and although there are undoubtedly many more, the prokaryotes, in terms of species, cannot be considered diverse.

2. Three genera of bacteria-like organisms—the halobacteria, the thermoacidophiles, and the methanogens (**c**), which reduce carbon dioxide to methane—are sometimes classified as Archaebacteria to indicate their primitive evolutionary position.

3. Prokaryotes are traditionally classified according to such features as nutritional mode (a) and presence or absence of a cell wall (c). Choice **e** is therefore correct. They do not have complex reproductive patterns as suggested by b. The existing classification of prokaryotes is currently being revised to reflect evolutionary relationships inferred from the similarity of nucleotide sequences.

4. Choice **b** is correct. The Monera are grouped into two divisions: Cyanophyta, the cyanobacteria (formerly known as the blue-green algae), and Schizophyta, the bacteria, of which the eubacteria are a subdivision. The classification does not reflect a hierarchical or evolutionary scheme.

5. The pilus (**c**), a thin, cylindrical extension of a bacterial cell, is used both for attachment to a food source or substrate and for attachment to another cell during conjugation (see Figure 16-3 on page 312).

6. Gram-positive bacteria are characterized by cell walls that combine firmly with gentian violet; such walls lack a lipoprotein-lipopolysaccharide outer layer (**c**). Gram-positive bacteria are more susceptible to both antibiotics and lysozyme. Only the mycoplasmas use cholesterol as a component of the cell wall.

7. Choice **b** is correct; one characteristic common to all prokaryotes is that they lack an internal membrane system.

8. Choice **a** is correct. Bacterial flagella are anchored in the cell membrane (and cell wall of gram-negative bacteria), as shown in Figure 20-6.

9. Choice **c** is correct. Refer to Figure 20-8.

10. Between the cell membrane and the cell wall of spirochete bacteria (**a**) are two sets of modified flagella, the fibrils of the axial filament.

11. Bacteria sustain a directed swimming motion as long as their flagella maintain a counterclockwise direction. When the flagella reverse their direction (rotate clockwise), the cell begins to tumble (**b**).

12. Follow the sequence of activities in bacterial spore formation in Figure 20-13. You should easily be able to see why choice **b** is correct.

13. Monerans can multiply asexually in all of the ways listed (**e** is correct). Binary fission (a) is simple cell division; budding (b) refers to an outgrowth from the parent cell that develops into a new individual; fragmentation (c) refers to the breaking off of cells from filaments; spores (d), as indicated in Question 12, may be formed and can grow into a new organism when released.

14. Heterotrophic bacteria (and the fungi) are the chief decomposers of organic material. The minerals released during decomposition become available to plants; therefore, **b** is correct.

15. The cyanobacteria have a number of photosynthetic features in common with photosynthetic eukaryotes. The cyanobacteria lyse water and evolve oxygen in the light-trapping reactions, and they employ both chlorophyll *a* and carotenoid pigments. The photosynthetic reactions, however, do not take place in chloroplasts (c) but on specialized membranes distributed in the peripheral portion of the cell (see Figure 20-17).

16. Both the purple sulfur bacteria (a) and the green

sulfur bacteria (b) use sulfur-containing compounds, such as H_2S, as electron donors in the light-trapping reactions of photosynthesis. Instead of evolving oxygen as do the cyanobacteria and photosynthetic eukaryotes, the purple and green sulfur bacteria produce elemental sulfur as a waste product. Choice **e** is therefore correct.

17. The green sulfur bacteria use the green photosynthetic pigment chlorobium chlorophyll, which is similar to chlorophyll *a*. The purple sulfur bacteria use a blue-gray pigment known as bacteriochlorophyll; therefore, **b** is correct.

18. Viruses do not appear to have been chemical ancestors of living cells (a), nor are they thought to be degenerate forms of living cells, as is implied by choice b. A relation between viroids and viruses has not been established, although both are now thought to be cellular genetic fragments that, at least in the case of viruses, have taken up a partially independent existence (**c**).

19. Any DNA or RNA virus uses the host-cell machinery (ribosomes, enzymes, etc.) and the host cell's materials (tRNA, amino acids, and nucleotides) to assemble mRNA and to direct protein synthesis. The virus does, however, insert and utilize its own genetic code (**e**).

20. Spindle-tuber disease is associated with the presence of a small fragment of RNA (**a**) within the nuclei of root cells. This fragment, only 359 nucleotides long, is the viroid PSTV.

21. When both organisms benefit in some way from a symbiotic relationship, as indicated here, the association is referred to as mutualistic (**b**).

22. A parasite is only successful, evolutionarily speaking, if it can meet its nutritional requirements without interfering with the host's ability to grow and reproduce (**a**). Disease or death of the host is usually a result of a sudden change in the host-parasite relationship and is not adaptive for the parasite.

23. The correct answers in order are **e**, **f**, **d**, **c**, **b**, and **a**.

24. The correct answers in order are **b**; **d**; **e**; **b**, **g**; **a**; **f**; and **c**.

25. Your discussion should center on three types of activities: cell lysis, secretion of toxins, and the generation of an immune response.

26. Photosynthetic bacteria fix carbon; heterotrophic bacteria are important in decomposition; chemotrophs recycle nutrients; cyanobacteria, in addition to fixing carbon, also fix nitrogen. Because of their simple nutritional requirements, cyanobacteria are often the first colonizers in barren areas.

27. a. The cytoplasm of prokaryotes is not subdivided by membranes, has no membrane-bound organelles, and contains the single circular loop of DNA (and any plasmids that may be present). The cytoplasm of eukaryotes is filled with a network of microtrabeculae and contains many complex membrane-bound organelles, including a membrane-bound nucleus.
    b. Eukaryotic cell membranes are constructed of phospholipids and cholesterol and are embedded with proteins. Except for the mycoplasmas, prokaryotic cell membranes do not contain cholesterol or other steroids. The electron transport system is located on the cell membrane of prokaryotes and on the mitochondrial membranes of eukaryotes.
    c. The cell walls of eukaryotes are composed of cellulose and/or other complex polysaccharides and may have proteins embedded in them. The cell walls of prokaryotes (except for the methanogens and their close relatives) do not contain cellulose but contain many complex molecules, such as peptidoglycans and lipoproteins, that are not found in eukaryotic cell walls.
    d. Eukaryotic flagella exhibit a 9 + 2 arrangement of microtubules. Prokaryotic flagella are constructed of monomers of a protein (flagellin) assembled in chains and wound around a hollow central core.

28. See the analysis of Question 15.

29. Prokaryotes are highly adaptable because they have relatively short generation times, making them more vulnerable to mutation. Also, there are many ways they can undergo genetic recombination, such as transduction, transformation, conjugation, and plasmid transfer.

CHAPTER 21

The Protists

CHAPTER ORGANIZATION

I. Introduction
II. The evolution of the protists
   A. The endosymbiont hypothesis
   B. Classification of protists
III. The algae
   A. Introduction
   B. Characteristics of the algae
   C. Division Euglenophyta: Euglenoids
   D. Division Chrysophyta: Diatoms and golden-brown algae
   E. Division Pyrrophyta: "Fire" algae
   F. Division Chlorophyta: Green algae
     1. Introduction
     2. Paths to multicellularity
     3. Life cycles in the green algae
   G. Division Phaeophyta: Brown algae
   H. Division Rhodophyta: Red algae
IV. The slime molds
V. The protozoa
   A. Introduction
   B. Phylum Mastigophora
   C. Phylum Sarcodina
   D. Phylum Ciliophora
   E. Phylum Opalinida
   F. Phylum Sporozoa
VI. Essay: The evolution of mitosis
VII. Patterns of behavior in protists
   A. General introduction
   B. Avoidance in *Paramecium*
VIII. Summary

MAJOR CONCEPTS

Organisms in the kingdom Protista (and in the kingdoms Fungi, Plantae, and Animalia) have eukaryotic cells. Protists are mostly unicellular, with a few simple multicellular forms. Protists exhibit several responses to stimuli, including phototaxis, chemotaxis, avoidance, and habituation.

Eukaryotic cells may have evolved from symbiotic relationships between prokaryotic cells. It is hypothesized that cells that were internalized by host cells became specialized as mitochondria and chloroplasts.

Three major groups of protists are the algae, the slime molds, and the protozoa. The algae are aquatic organisms that include unicellular and simple multicellular organisms. The green algae (division Chlorophyta) are thought to be the evolutionary ancestors of plants because the following characteristics are common to both groups: Their photosynthetic pigments are beta-carotene and chlorophylls *a* and *b*; their food reserves are stored as starch; and their cell walls contain cellulose.

The slime molds are heterotrophic organisms that are either coenocytic (division Myxomycota) or multicellular (division Acrasiomycota).

The protozoans include both free-living and parasitic forms. The cells are typically complex, and members of three phyla (Mastigophora, Sarcodina, and Ciliophora) can be identified by their locomotive structures. The phyla Opalinida and Sporozoa include only parasitic protozoans.

HOW TO STUDY THE CHAPTER

Read the chapter, focusing on the major concepts.

Reread the chapter, using the questions that follow to help you focus on details as they relate to major concepts. Answer the questions on a separate sheet of paper. Your answers will provide a valuable study aid in preparing for examinations.

FOCUS ON CHAPTER DETAILS

I. Introduction (page 412)

1. Describe the general characteristics of the protists.

II. The evolution of the protists (pages 412–415)

A. The endosymbiont hypothesis

2. What is the endosymbiont hypothesis?
3. How strong is the evidence that mitochondria may be descendants of specialized bacteria?
4. What advantages accrued to the cells that absorbed mitochondria?
5. What question about eukaryotic origins remains unexplained by the endosymbiont theory?
6. What evidence do modern symbiotic relationships yield in favor of the endosymbiont hypothesis?

B. Classification of protists

7. On what do biologists generally agree regarding protistan lineages and relationships to other eukaryotes?
8. What three subdivisions of the protist kingdom are currently in use? Familiarize yourself with the major phyla or divisions within each of these three groups (see Table 21-1).

III. The algae (pages 415–427)

A. Introduction

9. What characteristics do members of the six divisions of algae have in common?
10. Speculate upon the important ecological role of the unicellular algae.
11. To what environmental conditions have the multicellular algae adapted?

B. Characteristics of the algae

12. What criteria are used in classifying the algae into their six divisions?
13. Study Table 21-2. If you were given several characteristics of a species of algae, could you correctly identify its division?
14. To what do each of the six divisions of algae owe their names?
15. Algal chloroplasts contain a wide variety of pigments. What does this reveal about the probable evolutionary history of photosynthetic organisms?
16. What characteristics of either plants or animals are seen during reproduction in the various algal divisions?

C. Division Euglenophyta: Euglenoids

17. Review the major characteristics of euglenoids in Table 21-2.
18. Describe a typical euglenoid cell.
19. To what types of stimuli do euglenoids respond? How do they move about?
20. By what method do euglenoids reproduce? How do heterotrophic euglenoids arise?

D. Division Chrysophyta: Diatoms and golden-brown algae

21. Review the major characteristics of chrysophytes in Table 21-2.
22. What is an important ecological role of the Chrysophyta?
23. How is the cytoplasm of a diatom separated from the external environment?
24. Explain the meaning of "diatomaceous earth."
25. In what ways are some chrysophytes related to protozoans?
26. Describe sexual reproduction in the Chrysophyta.

E. Division Pyrrophyta: "Fire" algae

27. Review the major characteristics of pyrrophytes in Table 21-2.
28. Describe the structures for which the dinoflagellates have been named.
29. What unique types of cell walls are found in this group?
30. What causes a red tide? What is its effect on other organisms?

F. Division Chlorophyta: Green algae

Introduction

31. Review the major characteristics of chlorophytes in Table 21–2.
32. Why are the green algae considered to be the most diverse of all the algae?
33. Why are plants thought to be descendants of the green algae?

Paths to multicellularity

34. Why are the chlorophytes of interest, evolutionarily speaking?
35. Outline the pathway to multicellularity along the volvocine line. Which forms exhibit specialization of function?
36. Why is the volvocine line called an evolutionary dead end?
37. What pathway to multicellularity is represented by coenocytic organisms? What groups of organisms are coenocytic?
38. What pathway to multicellularity is demonstrated by *Ulva* and *Spirogyra*?

Life cycles in the green algae

39. Review the life cycle of *Chlamydomonas* (refer to Figure 12–3 on page 251).
40. Describe the three different gametic conditions found in *Chlamydomonas*.
41. Describe in detail the complex life cycle found in multicellular algae.
42. To what do the terms "isomorphic" and "heteromorphic" refer?

G. Division Phaeophyta: Brown algae

43. Review the major characteristics of phaeophytes in Table 21–2.
44. What "specializations" are exhibited by the brown algae?
45. What types of life cycles are found within this group?

H. Division Rhodophyta: Red algae

46. Review the major characteristics of rhodophytes in Table 21–2.
47. What major ecological role is filled by the red algae?
48. What types of life cycles are found within this group?

IV. **The slime molds** (pages 427–428)

49. Into what two divisions are the slime molds classified? What is the major difference between the two groups?
50. In what habitats may slime molds be found?
51. How do the plasmodial slime molds obtain their food?
52. Describe the life cycle of a typical plasmodial slime mold. Compare this to the life cycle of the cellular slime mold shown on page 120.

V. **The protozoa** (pages 428–433)

A. Introduction

53. Briefly describe the protozoa.
54. How are the protozoans classified into their representative phyla?
55. What types of reproduction are known among the protozoans?

B. Phylum Mastigophora

56. Review the major characteristics of mastigophores in Table 21–3.
57. How do mastigophores obtain their food?
58. List a few important parasitic forms in this phylum.

C. Phylum Sarcodina

59. Review the major characteristics of sarcodines in Table 21–3.
60. What functions are served by the pseudopodia found in this group?
61. What types of structures surround the cytoplasm of sarcodines? What purposes might these structures serve?
62. What is the geologic significance of the Foraminifera?

D. Phylum Ciliophora

63. Review the major characteristics of ciliophores (ciliates) in Table 21–3.
64. Describe a typical ciliate.
65. What is the role of the macronuclei in conjugation?

66. What does conjugation in the ciliates indicate about the evolutionary significance of the process?

E. Phylum Opalinida

67. Review the major characteristics of opalinids in Table 21-3.

68. What features do the opalinids have in common with ciliates? With mastigophores?

F. Phylum Sporozoa

69. Review the major characteristics of sporozoans in Table 21-3.

70. Briefly describe the life cycle of *Plasmodium*.

71. What efforts have been concentrated on controlling a major disease caused by a sporozoan?

VI. Essay: The evolution of mitosis (page 432)

72. Explain how mitosis in the protists reflects an intermediate stage between prokaryotic cell division and the nuclear divisions of higher eukaryotes.

73. What trends are apparent in the evolutionary sequence described here?

VII. Patterns of behavior in protists (pages 434-436)

A. General introduction

74. What behaviors seen in the amoeba demonstrate that protist behavior is somewhat more complex than that of prokaryotes?

B. Avoidance in *Paramecium*

75. How does a *Paramecium* respond to changes in its environment?

76. Using some examples from your text, explain how this type of response is adaptive for the organism. What is the evolutionary significance of these behavioral responses in *Paramecium*?

VIII. Summary (pages 436-438): Read the Summary. If you are familiar with the essential features of the material presented there, you are ready to take the following diagnostic examination.

TESTING YOUR UNDERSTANDING

After you have completed the following examination, compare your answers with the annotated key in the Analysis section.

1. Which of these observations seems to contradict the endosymbiont hypothesis of eukaryotic origin?
    a. Cyanobacteria lacking cell walls are modern symbionts of eukaryotic algae.
    b. The ribosomes of mitochondria resemble bacterial ribosomes.
    c. Mitochondria have an insufficient quantity of DNA to code for all mitochondrial proteins.
    d. Enzymes found on chloroplast membranes are identical to those in prokaryotic membranes.

2. The eukaryotes:
    a. gave rise to the methanogens.
    b. are assumed to be anaerobic cells that served as hosts for endosymbionts.
    c. are assumed to be photosynthetic prokaryotes that were ingested by nonphotosynthetic cells.
    d. Both a and c are correct.

3. In comparison to higher plants, the algae as a group exhibit greater diversity. This is evident from the number of different _____ they utilize.
    a. photosynthetic pigments
    b. storage forms of carbohydrate
    c. cell wall components
    d. locomotory arrangements
    e. All of the above are correct.

4. Which of the following best describes the evolutionary position of the protists?
    a. All modern protists descended from a single protist ancestor.
    b. Primitive protists gave rise to all other eukaryotic organisms.
    c. Modern protists are probably not very different from their eukaryotic ancestors.
    d. The algae and the protozoans represent a single distinct phylogenetic line.
    e. Both c and d are correct.

5. Plants are believed to be most closely related to:
    a. Euglenophyta.
    b. Cyanophyta.
    c. Chlorophyta.

d. Chrysophyta.
   e. Fungi.
6. In a species of *Chlamydomonas* that is isogamous, gametes of different strains are:
   a. both motile and of the same size.
   b. both immotile and of the same size.
   c. of different sizes and may or may not be motile.
   d. genetically identical but of different size.
   e. genetically identical and of the same size.
7. A diploid cell that functions as a dormant form of a green alga is called a:
   a. zygote.
   b. gamete.
   c. sporophyte.
   d. spore.
   e. zygospore.
8. Which of the following represents genetically identical structures in the life cycle of an alga characterized by alternation of generations?
   a. gametophyte and sporophyte
   b. gametes of a single gametophyte
   c. spores of a single sporophyte
   d. spores and gametes of consecutive generations
9. If the gametophyte and sporophyte of a multicellular alga closely resemble one another, they may be referred to as:
   a. isogamous.
   b. anisogamous.
   c. isomorphic.
   d. coenocytic.
   e. genetically identical.
10. Which of these life forms is coenocytic?
    a. the gametes of a multicellular alga
    b. the plasmodium of a slime mold
    c. a *Volvox* colony
    d. *Paramecium*
11. Unicellular algae are important ecologically because they:
    a. constitute the majority of oceanic plankton.
    b. are nitrogen fixers.
    c. decompose organic compounds.
    d. are primarily parasites.
12. Which of the following pairs of protists *never* develop flagella?
    a. Sarcodina and Ciliophora
    b. Chlorophyta and Euglenophyta
    c. Sporozoa and Rhodophyta
    d. Mastigophora and Pyrrophyta
    e. Myxomycota and Ciliophora
13. Protozoans are classified as Ciliophora, Sarcodina, or _____ based on _____.
    a. Sporozoa; complexity of the life cycle
    b. Mastigophora; locomotory structures
    c. Opalinida; method of reproduction
    d. Mastigophora; mode of nutrition
    e. Sporozoa; complexity of behavior patterns
14. An amoeba that cannot escape a bright light exhibits a behavior pattern known as habituation. At first it _____, then later it _____.
    a. does not respond; contracts
    b. tries to escape; contracts for an indefinite time
    c. contracts; resumes normal activities
    d. contracts; contracts only periodically
15. Behavioral experiments with *Paramecium* have demonstrated that it reacts to changes in its environment by:
    a. turning toward an unfavorable stimulus.
    b. responding randomly to any type of stimulus.
    c. responding positively to favorable stimuli.
    d. responding negatively to unfavorable stimuli.
16. Although protozoans are a diverse group of protists, one common feature is:
    a. their heterotrophic mode of nutrition.
    b. the presence of flagella.
    c. that they are free-living.
    d. the absence of sexual reproduction.
    e. the organization of their genetic material.
17. When the plasmodium of a slime mold no longer has an adequate food supply, it:
    a. divides into a large number of single amoeboid cells.
    b. forms a peripheral layer of flagellated cells.
    c. produces spores.
    d. incorporates more amoeboid cells.
18. Match each group of algae with one or more descriptions.

    Chlamydomonas _____   a. store food reserves as starch
    red algae _____   b. have an encircling flagellum
    Euglena _____
    diatoms _____   c. have no flagellated forms

dinoflagellates
_____

d. are enclosed in double silicon shells
e. include multicellular forms
f. are autotrophs with contractile vacuoles

19. Match each of the animal-like creatures listed at the left with one or more statements from the list at the right.

mastigophores
_____

Acrasiomycota
_____

ciliophores _____

sarcodines
_____

sporozoans
_____

opalinids _____

a. protect themselves by secreting a shell
b. exist as aggregations of amoeboid cells
c. exchange genetic material through conjugation
d. include parasites with complex life cycles and no locomotory structures
e. probably gave rise to amoebas and ciliates
f. contain two or more nuclei

20. Why are plants thought to be descended from the green algae?

21. Using several examples, demonstrate why protist behavior is considered more complex than that of prokaryotes.

22. What evidence is there that eukaryotes originated as a result of a symbiotic relationship between prokaryotic cells?

23. Which protist phylum or division is most difficult to characterize as either plant-like or animal-like? Provide support for your selection.

24. Describe the three pathways to multicellularity that are found among the green algae.

Extra-Credit Question

	A	B	C	D	E
Chlorophyta	__	__	__	__	__
Euglenophyta	__	__	__	__	__
Phaeophyta	__	__	__	__	__
Mastigophora	__	__	__	__	__
Ciliophora	__	__	__	__	__
Sarcodina	__	__	__	__	__
Cellular slime molds	__	__	__	__	__

In the matrix above, the columns A through E represent different clusters of characteristics. Each individual characteristic is assigned a number in the list below. Assign the appropriate number to the correct column for each division or phylum of organisms.

A. photosynthetic (1), heterotrophic (2), or both (3)
B. cell wall present (1), cell wall absent (2)
C. method of locomotion is primarily amoeboid (1), or primarily by flagella (2), or primarily by cilia (3)
D. sexual reproduction is known (1), or unknown (2)
E. unicellular forms only (1), or multicellular forms only (2), or both unicellular and multicellular forms (3)

PERFORMANCE ANALYSIS

1. Choices b and d state evidence that both mitochondria and chloroplasts could have originated from prokaryotes that took up residence inside other prokaryotic cells. Choice a shows that even today a similar type of symbiosis is possible; this also supports the endosymbiont hypothesis. The amount of DNA found in mitochondria, however, is insufficient to code for the quantity of mitochondrial protein, which would seem to contradict the hypothesis (c is correct).

2. The urkaryotes gave rise to the eukaryotes, making choice a incorrect. The urkaryotes are assumed to be cells that depended on anaerobic fermentation for energy and that served as hosts for oxygen-utilizing prokaryotes that were the forerunners of mitochondria. Choice **b** is therefore correct.

3. Table 21-2 reflects that all of these (**e**) are reasons for the diversity of the algae as a group. Higher plants utilize only chlorophylls *a* and *b*, have cell

walls constructed of cellulose (and usually impregnated with lignin), store food only in the form of starches, and do not possess flagella.

4. Biologists are in agreement on two points with regard to the evolutionary history of the protists: (1) they represent a number of distinct phylogenetic lineages (eliminating choices a and d); and (2) primitive protists gave rise to all other eukaryotic organisms (correctly stated in choice **b**).

5. Because chlorophytes (**c**) contain chlorophylls *a* and *b*, have cell walls made of cellulose fibers, and store carbohydrates in the form of starch, they are thought to be the ancestors of modern plants.

6. Choice **a** is correct. Refer to Figure 21-16 for descriptions of isogamy, anisogamy, and oogamy.

7. In sexually reproducing green algae, gametes unite to form a zygote around which a thick protective wall forms. This structure is the zygospore (**e**), which remains dormant until environmental conditions are favorable.

8. Gametophyte and sporophyte are haploid and diploid, respectively, and so could not be genetically identical (a). Because spores are the result of meiosis, genetic recombinations are possible (therefore both c and d are eliminated). But since gametes are produced by mitosis from the haploid gametophyte, all such gametes would be genetically identical; choice **b** is correct.

9. The sea lettuce *Ulva* (Figure 21-17) is an example of a multicellular alga in which the gametophyte and sporophyte resemble each other. This condition is referred to as isomorphic (**c**).

10. The plasmodium of a slime mold (**b**) is a streaming mass of protoplasm containing many nuclei. This means that the plasmodium is coenocytic.

11. Unicellular algae of the phyla Chrysophyta and Pyrrophyta comprise the bulk of oceanic plankton (**a**), or microscopic vegetation.

12. Choice **c** is correct; the Sporozoa have no means of locomotion, and the Rhodophyta are one of the few groups of organisms that have no flagellated cells. Why not a or e? Some sarcodines develop flagella in some stages of their life cycle, though their primary means of locomotion are pseudopodia. In the Myxomycota, the spores produced by the plasmodium are flagellated.

13. Actually, any of the groups listed would qualify as protozoans. But the traditional method of classifying protozoans according to their means of locomotion readily distinguishes the Ciliophora (ciliated protozoans), the Sarcodina (amoeboid protozoans), and the Mastigophora (flagellated protozoans). The Opalinida move by either flagella or cilia. Choice **b**, then, is correct.

14. An amoeba, when first exposed to a bright light, will contract suddenly, but after a while it resumes normal activities (**c**). This common behavior pattern is referred to as habituation.

15. The locomotory behavior of *Paramecium*, generally speaking, is determined by its reactions to unfavorable stimuli, such as bright light or toxic chemicals. In the presence of such stimuli, *Paramecium* makes avoidance responses; therefore, **d** is correct.

16. Not all protozoans bear flagella (Ciliophora and Sporozoa, for example), nor are all free-living (sporozoans are all parasites). Sexual reproduction is known in all phyla. Ciliophores (ciliates) have a different nuclear organization than that found in other groups. All protozoans, however, are heterotrophic (**a**).

17. When a plasmodial slime mold runs out of food, the plasmodium divides into a number of protoplasmic masses. Each mass sends up a stalk capped by a fruiting body (the sporangium) from which spores are eventually released (**c**).

18. The correct answers in order are **a**; **c**; **e**; **f**; **d**; and **a**, **b**.

19. The correct answers in order are **e**; **b**; **c**, **f**; **a**; **d**; and **f**.

20. Refer to the analysis for Question 5. Also note that the green algae are likely relatives of plants because some forms are multicellular.

21. Types of protist behavior considered more complex than prokaryotic behavior include phototaxis, habituation, and, in amoebas, the changing of the size of pseudopodia in response to different sizes of prey encountered.

22. Most of the evidence for the so-called endosymbiont hypothesis of eukaryotic origin comes from a comparison of mitochondria (or chloroplasts) with modern-day prokaryotes. Evidence is

also provided by the fact that many modern prokaryotes have taken up residence inside eukaryotic cells.

23. If you said Euglenophyta, you made a good choice for several reasons. Euglena are plant-like in that they are autotrophs, possessing chloroplasts containing the pigments chlorophyll *a* and *b*. But they also have a number of animal-like characteristics. They can develop into heterotrophic cells, they have flagella, and they are surrounded by a flexible pellicle more similar to the cell membrane of animals than to the cell wall of plants.

24. You should identify each of the three pathways with a representative group from the green algae. *Volvox* is an example of colony formation by the association of individual cells. A second pathway is exemplified by the coenocytic alga *Valonia*. True multicellularity, in which cells remain attached to one another following cytokinesis, is demonstrated by *Ulva*.

Extra-Credit Question. The completed matrix should look like this:

	A	B	C	D	E
Chlorophyta	1	1	2	1	3
Euglenophyta	3	2	2	2	1
Phaeophyta	1	1	2	1	2
Mastigophora	2	2	2	1	1
Ciliophora	2	2	3	1	1
Sarcodina	2	2	1	1	1
Cellular slime molds	2	1	1	1	3

CHAPTER 22

The Fungi

CHAPTER ORGANIZATION

I. Introduction
II. Reproduction in the fungi
III. Classification of the fungi
IV. Characteristics of the fungi
   A. Division Chytridiomycota
   B. Division Oomycota
   C. Division Zygomycota
   D. Division Ascomycota
   E. Division Basidiomycota
   F. Division Deuteromycota
V. Symbiotic relationships of fungi
   A. Introductory remarks
   B. The lichens
   C. Mycorrhizae
VI. Essay: Predaceous fungi
VII. Summary

MAJOR CONCEPTS

Except for the unicellular yeasts, fungi are either coenocytic or multicellular heterotrophs composed of masses of filaments (hyphae). They absorb organic compounds digested extracellularly by secreted enzymes.

In most groups, the principal component of the hyphal cell walls is chitin. Cellulose is the main component of the cell walls of oomycetes.

Fungi reproduce by fragmentation or by the production of asexual and/or sexual spores. The fungi are classified on the basis of structure and patterns of reproduction into six divisions: Chytridiomycota, Oomycota, Zygomycota, Ascomycota, Basidiomycota, and Deuteromycota.

Fungi are important ecologically in the decomposition of organic matter, as components of lichens, and in association with plant roots (mycorrhizae).

HOW TO STUDY THE CHAPTER

Read the chapter, focusing on the major concepts.

Reread the chapter, using the questions that follow to help you focus on details as they relate to major concepts. Answer the questions on a separate sheet of paper. Your answers will provide a valuable study aid in preparing for examinations.

FOCUS ON CHAPTER DETAILS

I. **Introduction** (pages 439–440)

   1. Most fungi are coenocytic or multicellular. Describe their general organization.

   2. The fungi are heterotrophic but nonmotile. How are they able to seek out new sources of food?

   3. What means of acquiring food are found among the fungi?

   4. What is the primary ecological role of the fungi?

II. **Reproduction in the fungi** (page 441)

   5. Describe asexual methods of reproduction employed by the fungi. What are sporangia and sporangiophores used for?

   6. What are gametangia? How are they involved in the three methods of sexual reproduction found in fungi?

7. What is a dikaryon? How does it arise? In what fungal groups is it found?

III. Classification of the fungi (pages 441–422)

8. What basic criteria are used to classify the fungi into their five principal divisions?
9. The Deuteromycota may be considered as a sixth division. Why is this division referred to as a "waste basket"?
10. Which divisions are thought to represent distinct evolutionary lineages from the first fungi? What features do they have in common? How do they differ from the higher fungi?
11. Draw a simple evolutionary tree representing the relationships among the five principal divisions of fungi.

IV. Characteristics of the fungi (pages 442–448)

A. Division Chytridiomycota

12. Review the major characteristics of the Chytridiomycota in Table 22-1.
13. How do chytrids differ structurally from other fungi?
14. In terms of gametes, what is unique about the chytrids in general, and *Allomyces* in particular? What else is unique about this genus?
15. What are the habitats of the chytrids?

B. Division Oomycota

16. Review the major characteristics of the Oomycota in Table 22-1.
17. Why are the oomycetes appropriately named?
18. Give a brief description of asexual and sexual reproduction among the oomycetes.
19. What feature do these fungi share with the plants?
20. What nutritional modes are employed by the oomycetes?
21. What two members of this division have had a profound economic impact?

C. Division Zygomycota

22. Review the major characteristics of the Zygomycota in Table 22-1.
23. What modes of nutrition are found within this group of terrestrial fungi?
24. Why are the zygomycetes appropriately named?
25. Briefly describe the life cycle of the black bread mold, *Rhizopus stolonifer*, which can reproduce both asexually and sexually.
26. What is the adaptive value of a zygospore?

D. Division Ascomycota

27. Review the major characteristics of the Ascomycota in Table 22-1.
28. How have some of the ascomycetes been of economic importance?
29. Describe the basic structural organization of these higher fungi (see also Figure 22–5).
30. What are conidia? Where are they produced?
31. How are ascospores produced as a result of sexual reproduction? (See also Figure 22–11b.) Where does this process of sexual spore formation take place? How do you think the ascomycetes got their name?
32. How do yeasts (single-celled ascomycetes) reproduce?
33. Why has a plant parasite in this division attracted medical attention?

E. Division Basidiomycota

34. Review the major characteristics of the Basidiomycota in Table 22-1.
35. Describe a mushroom in fungal terminology. Why are mushrooms often found growing in a circle?
36. Describe the process of sexual reproduction in the basidiomycetes. Do you see why they are appropriately named?
37. What is a gill fungus? Give several common examples. What is the significance of the gill? (See Figure 22–12.)
38. Name some other types of basidiomycetes.

F. Division Deuteromycota

39. Review the major characteristics of the Deuteromycota in Table 22-1.

40. What types of fungi are placed in this division?
41. Cite several examples that are of medical or economic importance.

V. **Symbiotic relationships of fungi** (pages 449-452)

A. Introductory remarks

42. What are two symbiotic relationships involving fungi that are important in the colonization of barren areas by photosynthetic organisms?

B. The lichens

43. What organisms may be the symbiotic partners in a lichen?
44. Cite several examples of habitats in which lichens thrive. How do their activities in these environments affect other organisms?
45. What are the requirements of a typical lichen, and how are they satisfied?
46. How do lichens provide an indication of the level of air pollution?
47. How are lichens classified? In the majority of cases, what division of fungus is involved?
48. What is the mechanism of lichen reproduction?

C. Mycorrhizae

49. What kind of an association is a mycorrhizae?
50. What observations implied the existence of mycorrhizal associations?
51. Describe the two kinds of mycorrhizal associations that have been discovered?
52. What benefits are derived from mycorrhizae by the participants?
53. How widespread are such associations among modern plants?
54. What has fossil evidence suggested about the evolutionary significance of mycorrhizae?

VI. **Essay: Predaceous fungi** (page 451)

55. How do a number of highly specialized species of fungi capture nematodes?

56. What cellular mechanism has been proposed to account for this predatory behavior?

VII. **Summary** (pages 452-453): Read the Summary. If you are familiar with the essential features of the material presented there, you are ready to take the following diagnostic examination.

TESTING YOUR UNDERSTANDING

After you have completed the following examination, compare your answers with the annotated key in the Analysis section.

1. Choose the one correct statement about members of the kingdom Fungi.
   a. All are multicellular.
   b. All are either free-living or parasitic.
   c. All are known to reproduce sexually.
   d. All have either spores or gametes that are flagellated.
   e. All are heterotrophic.

2. Which of the following nutritional strategies is *not* apparent among the members of the kingdom Fungi?
   a. uptake of food particles by endocytosis
   b. absorption of small dissolved molecules
   c. extracellular secretion of digestive enzymes
   d. symbiotic relationships with other organisms
   e. predation

3. Chitin is a modified polysaccharide found in the cell walls of _____ and in the exoskeletons of _____.
   a. all fungi; sarcodines
   b. most fungi; arthropods
   c. nonmotile eukaryotes; arthropods
   d. most fungi and all plants; dinoflagellates
   e. higher fungi only; diatoms

4. It is thought that the first fungi were probably:
   a. yeasts.
   b. prokaryotes.
   c. one-celled eukaryotes.
   d. colonial algae.
   e. bacteria-like.

5. There is some doubt as to whether members of the division _____ should be classified as fungi.
   a. Zygomycota

b. Basidiomycota
   c. Ascomycota
   d. Chytridiomycota
   e. Both b and c are correct.
6. During its growth period, the bulk of the cytoplasm of a fungus is contained in:
   a. individual cells of the hyphae.
   b. the mycelium.
   c. spore-producing bodies.
   d. specialized hyphae.
   e. intercellular junctions.
7. Which of the following represents a specialized hypha that functions in asexual reproduction?
   a. haustorium
   b. gametangium
   c. sporangiophore
   d. rhizoid
   e. ascus
8. Although most fungi lack flagella, a fungus may nevertheless occupy new areas by:
   a. extending its mycelium.
   b. releasing spores.
   c. forming pseudopodia.
   d. dispersing its fruiting bodies.
   e. Both a and b are correct.
9. A fungus for which a sexual life cycle has not been observed is grouped with the:
   a. gill fungi.
   b. protists.
   c. chytrids.
   d. imperfect fungi.
   e. acrasiomycetes.
10. The higher divisions of fungi differ from the lower ones in that they:
    a. reproduce sexually.
    b. have septate hyphae.
    c. can enter a dormant stage.
    d. have cell walls made of cellulose.
    e. are multicellular.
11. A fungal hypha is referred to as _____ if it contains two distinct, unfused nuclei of complementary mating types.
    a. a dikaryon
    b. septate
    c. coenocytic
    d. haploid
    e. a zygospore
12. A fungus having flagellated gametes or flagellated spores is:
    a. strictly terrestrial.
    b. probably parasitic.
    c. dependent on water for reproduction.
    d. anisogamous.
    e. classified as an imperfect fungus.
13. Chytrids resemble green algae in having:
    a. green pigments.
    b. flagellated gametes.
    c. sporangia.
    d. a parasitic life style.
    e. Both a and c are correct.
14. Different mating forms of an oomycete are most accurately described as:
    a. positive and negative strains.
    b. male and female.
    c. complementary.
    d. isogamous.
    e. incompatible.
15. The zygospore of the black bread mold *Rhizopus stolonifer*, contains _____ just before its germination.
    a. a single haploid nucleus
    b. a single diploid nucleus
    c. unfused nuclei of different strains
    d. many diploid nuclei
    e. haploid spores
16. Inside the ascus of an ascomycete, _____ nuclei are produced as a result of _____.
    a. four diploid; mitosis
    b. four diploid; meiosis
    c. eight haploid; mitosis
    d. eight haploid; meiosis followed by mitosis
    e. eight haploid; mitosis followed by meiosis
17. Although yeasts are classified with the Ascomycota, they are atypical because they:
    a. do not reproduce sexually.
    b. do not form an ascus.
    c. are single-celled.
    d. can reproduce both sexually and asexually.
    e. Both a and b are correct.
18. The spores of the common field mushroom, *Agaricus*, are produced on a specialized hypha called a(n):
    a. basidium.

b. gill.
   c. basidiocarp.
   d. ascus.
   e. sporangiophore.
19. No matter what species are involved, a lichen is always a symbiotic relationship between a(n) _____ and a _____.
   a. ascomycete; cyanobacterium
   b. ascomycete; green alga
   c. autotroph; heterotroph
   d. fungus; protist
   e. Both b and d are correct.
20. Which of the following is true about mycorrhizae?
   a. In order for the association to be beneficial to the plant, the fungus must penetrate the root cells.
   b. Mycorrhizae are apparently commensal relationships between a plant and a fungus.
   c. Most plants that utilize mycorrhizae can grow as well without them.
   d. Mycorrhizae are commonly associated with structural changes in root tissues.
   e. Mycorrhizal associations have evolved only recently.
21. Match each division of fungi with one or two of its unique characteristics.

    Basidiomycota _____
    Zygomycota _____
    Deuteromycota _____
    Oomycota _____
    Ascomycota _____
    Chytridiomycota _____

    a. build cell walls of cellulose
    b. produce flagellated gametes
    c. form a resistant zygospore
    d. bear asexual spores called conidia
    e. have differentiated gametes
    f. produce sexual spores inside a conspicuous fruiting body
    g. do not reproduce sexually

22. Match each fungus with its reproductive structure.

    gill fungi _____
    yeasts _____
    Neurospora _____
    Rhizopus _____
    potato blight fungus _____

    a. asci
    b. basidia
    c. zygospores
    d. oogametes

23. Give several reasons why a taxonomist might argue that some of the lower forms of fungi should be classified in a division or phylum of the kingdom Protista.
24. Briefly discuss the various modes of existence (free-living, symbiotic, and parasitic) found among the fungi, and comment on the ecological importance of each mode.

PERFORMANCE ANALYSIS

1. The exceptions to choices a through d are yeasts, which are unicellular; lichens, which contain a fungus that is a mutualistic symbiont; the Fungi Imperfecti, for whom a sexual mode of reproduction has not been demonstrated; and the higher fungi, which have neither flagellated spores nor flagellated gametes. Only **e** is correct: all members of the kingdom Fungi are heterotrophs.
2. The majority of fungi obtain their nutrition by digesting molecules extracellularly or by absorbing small molecules. Some fungi benefit nutritionally by associations with algae (lichens); others (see the essay on page 451) are predaceous. However, fungi are not known to employ endocytosis (**a**).
3. Chitin is a cell wall component in most fungi, though not in the Oomycota, which contain cellulose as do the plants and green algae. Chitin is also found in the exoskeleton of arthropods; therefore, **b** is correct. The external skeletons or shells of many protozoans are built of calcium or silicates.
4. The first fungi are thought to have been unicellular eukaryotes (**c**) that have no living counterparts.
5. Since the Chytridiomycota (**d**) have flagellated gametes (an animal-like characteristic) and exhibit alternation of generations (as do plants and green algae) they do not closely resemble other members of the kingdom.
6. The cytoplasm of the fungus is contained within filamentous structures known as hyphae. All of

the hyphae taken together are a mycelium; therefore, **b** is correct. Choice **a** is incorrect because the cytoplasm of hyphae is allowed to flow freely between individual cells.

7. A sporangiophore, a stalklike structure, is a hypha specialized for supporting the sporangium, a caplike structure that contains asexual spores; therefore, **c** is correct.

8. Fungi may inhabit new areas either by growth of the mycelium (**a**) or by releasing spores (**b**) that are borne on the wind. The fruiting bodies (i.e., the sporangia) rupture to release spores but are not themselves dispersed. Fungi do not possess pseudopodia. Choice **e**, then, is correct.

9. The imperfect fungi (**d**) are all those for whom a sexual life cycle has not been demonstrated.

10. All divisions of fungi have both multicellular forms and forms that reproduce sexually (except for the Fungi Imperfecti, in which sexual reproduction is unknown). Even "lower" forms such as the Zygomycota can produce a dormant stage, the zygospore. The higher fungi, the Ascomycota and the Basidiomycota, have cell walls made of chitin and septate hyphae (**b**).

11. A dikaryon (**a**), a cell containing two unfused nuclei as a result of the mating of two different strains, is present in the sexual life cycles of Ascomycota and Basidiomycota (see Figure 22-12).

12. Those fungi having flagellated reproductive cells (gametes or spores), such as the Chytridiomycota and Oomycota, are dependent on water for reproduction (**c**).

13. Like the green algae, chytrids produce flagellated gametes. Like primitive land plants, they bear attachment devices called rhizoids. Choice **b** is therefore correct.

14. Oomycetes are the only members of the kingdom Fungi that produce gametes that are truly male and female (**b**).

15. The zygospore, a diploid cell of *Rhizopus* and other Zygomycota, results from the fusion of gametangia of different mating strains (see Figure 22-9). These nuclei eventually fuse to produce a multinucleate zygospore, in which all but a single diploid nucleus degenerate before germination (**b**).

16. Nuclei within the ascus of an ascomycete fuse, then undergo meiosis. This is followed by a single mitotic division that results in eight haploid nuclei; therefore, **d** is correct.

17. Yeasts, like other Ascomycetes, reproduce both sexually (by means of ascospores formed within an ascus) and asexually (by budding). They are atypical of higher fungi, however, because they are unicellular (**c**).

18. The common field mushroom, *Agaricus*, is a member of Basidiomycota and so produces sexual spores in a club-shaped hypha called a basidium (**a**).

19. Protists (**d**) are involved in only some lichen associations. Lichens contain either a cyanobacterium (kingdom Monera) or a green alga (kingdom Protista), but not all lichens contain an ascomycete (**a** and **b**). However, all lichens do contain an autotroph (a photosynthetic cyanobacterium or a green alga) and a heterotroph (fungus). Therefore, **c** is correct.

20. Mycorrhizae represent mutualistic (not commensal, as in **b**) relationships between fungi and plants that evolved, according to fossil evidence, early in the transition of plants to land. Plants have become so dependent on mycorrhizal associations that if transplanted to soil where their fungal symbionts are not present, they are unable to grow. The fungus of a mycorrhizal association may (the endomycorrhizae) or may not (the exomycorrhizae) penetrate the root cells of the plant. In either case, a mycorrhizal association is detectable as a change in the structure of the root tissue, such as a swelling, coiling, or branching; therefore, **d** is correct.

21. The correct answers in order are **f**; **c**; **g**; **a**, **e**; **d**; and **b**.

22. The correct answers in order are **b**, **a**, **a**, **c**, and **d**.

23. The oomycetes resemble a number of protists in having flagellated spores. Their cellulose cell wall relates them to the green algae. The chytridiomycetes have both flagellated spores and flagellated gametes and an internal organization that is characteristic of plantlike protists.

24. Free-living fungi that inhabit the soil are the most important organisms in the process of decomposition. Symbiotic fungi such as those present in mycorrhizae and lichens support colonization of an area by autotrophic organisms. Many groups of fungi also have important ecological roles as parasites.

CHAPTER 23

The Plants

CHAPTER ORGANIZATION

I. Introduction
II. The transition to land
   A. Introduction
   B. The ancestral alga
   C. The ancestral plant
III. Classification of the plants
IV. Division Bryophyta: Mosses, hornworts, and liverworts
   A. General information
   B. Bryophyte reproduction
V. The vascular plants: An introduction
   A. Evolutionary developments in the vascular plants
   B. Essay: Coal Age plants
VI. The seedless vascular plants
   A. Introductory remarks
   B. Division Pterophyta: The ferns
VII. The seed plants
   A. Introduction
   B. Gymnosperms
      1. General information
      2. Formation of the seed
      3. The conifer leaf
   C. Angiosperms: The flowering plants
      1. General information
      2. The flower
      3. Evolution of the flower
      4. The agents of evolution
      5. Evolution of the fruit
      6. Biochemical evolution
VIII. Summary

MAJOR CONCEPTS

Plants are multicellular photosynthetic organisms that are believed to have evolved from green algae. Both groups have chlorophylls a and b and beta-carotene as their photosynthetic pigments, both store reserve food as starch, and both have cellulose-containing cell walls.

The adaptations of plants to a terrestrial existence include a waxy cuticle, surface pores (stomata) that enable gas exchange, protected reproductive structures, and the retention of the embryonic sporophyte within the female gametophyte.

Plants are divided into two major groups based on the presence (in vascular plants) or absence (in nonvascular plants) of an internal vascular system for transporting water and dissolved particles. The nonvascular plants (bryophytes) include mosses and liverworts. They require a constantly moist environment. Vascular plants include seedless and seed-bearing forms. Ferns are the most numerous seedless vascular plants. The seed plants include naked-seed plants (such as conifers), known as gymnosperms, and flowering plants, the angiosperms.

Flowering plants are the most numerous of the modern plants. Their flowers attract pollinators, resulting in an efficient means of uniting sperm and egg, and their fruits aid in dispersal of seeds.

HOW TO STUDY THE CHAPTER

Read the chapter, focusing on the major concepts.

Reread the chapter, using the questions that follow to help you focus on details as they relate to major concepts. Answer the questions on a separate sheet of paper. Your answers will provide a valuable study aid in preparing for examinations.

FOCUS ON CHAPTER DETAILS

I. **Introduction** (page 454)
   1. In general terms, what are plants?

II. **The transition to land** (pages 454-457)
   A. Introduction
   2. What factors favored the transition of plants from the ocean to the land?
   3. What advantages did life on land offer to photosynthetic organisms?
   B. The ancestral alga
   4. List four reasons why plants are thought to be descendants of the Chlorophyta.
   5. What are the characteristics of the hypothetical ancestor of plants?
   C. The ancestral plant
   6. Identify two characteristics of the ancestral plant that were associated with the transition to land.
   7. How are the reproductive characteristics of the ancestral plant thought to compare with those of modern green algae?
   8. Into what separate lineages did the plants diverge? What is the principal difference between these groups?
   9. What are the four major eras of geologic time? Over what time span does each era extend? In what era and over what time span did vascular plants appear? (See Table 23-1.)

III. **Classification of the plants** (page 458)
   10. What are the common names for each of the ten divisions in the plant kingdom? (See Table 23-2.)
   11. Group these divisions into larger categories according to seed characteristics and the presence or absence of conducting tissues.

IV. **Division Bryophyta: Mosses, hornworts, and liverworts** (pages 458-460)
   A. General information
   12. How must the bryophytes obtain water? What restrictions does this place upon their distribution?
   13. What special structural adaptations to a land environment have the bryophytes made?
   14. Why are bryophytes not believed to be the ancestors of "higher" plants?
   B. Bryophyte reproduction
   15. Describe the life cycle of a moss, paying special attention to the gametophyte and sporophyte, the requirements for fertilization, the protection of the zygote, and the location of the haploid and diploid stages.
   16. How do bryophytes reproduce asexually?

V. **The vascular plants: An introduction** (pages 460-461)
   A. Evolutionary developments in the vascular plants
   17. Describe *Rhynia major*. When did it exist?
   18. What six major evolutionary trends and innovations are seen within vascular plants? Briefly discuss their adaptive value.
   B. Essay: Coal Age plants (pages 466-467)
   19. Explain what peat is and how it is transformed into fossil fuels, such as coal.
   20. Why was such an enormous volume of coal deposited during the Carboniferous period?

VI. **The seedless vascular plants** (pages 462-463)
   A. Introductory remarks
   21. What four divisions represent plants that have vascular tissue but do not produce seeds?
   B. Division Pterophyta: The ferns
   22. What characteristics do the leaves and stems of ferns possess?

23. How are ferns incompletely adapted for life on land? For what types of environments are they best suited?
24. Describe the life cycle of the common fern as shown in Figure 12-4 on page 251. Pay special attention to the nutritional dependence of the sporophyte on the gametophyte, to the process of fertilization, and to the relative importance of the two generations.

VII. **The seed plants** (pages 464–478)

A. Introduction
  25. Describe the climatic conditions favoring the evolutionary development of the seed.
  26. What two major groups of seed plants thrive today?

B. Gymnosperms

General information
  27. What four divisions represent plants having unprotected ("naked") seeds? (See also Table 23-2.)

Formation of the seed
  28. What is the function of a seed?
  29. How does the relative importance of the two generations in seed plants compare to that in the green algae? In the bryophytes? (Refer to page 459, if necessary.) In the ferns?
  30. Describe the two types of spores produced by gymnosperms. In what structures are they formed? What is the ovule? (Locate these structures in Figure 23-15.)
  31. What reproductive events take place within the male cones? Within the female cones? (Refer also to Figure 23-15.)
  32. What special requirements do pine trees have for fertilization?
  33. What events accompany the beginning of the development of the zygote following fertilization?
  34. Explain the meaning of the slogan "three generations under one roof." (See also Figure 23-17.)
  35. Into what two basic structures does the embryo develop?
  36. How do the seeds of pine trees become dispersed, and when do they germinate?

The conifer leaf
  37. What is the function of each of the structures labeled in the cross section of the pine needle in Figure 23-18?
  38. To what climatic conditions is the pine needle adapted?

C. Angiosperms: The flowering plants

General information
  39. Trace the probable sequence of events that established angiosperms as the dominant vegetation of the Cretaceous period.
  40. What improvements over their gymnosperm relatives (and in fact over all other plant groups) do angiosperms exhibit?
  41. How dominant are modern angiosperms?
  42. What are the general characteristics of monocots and dicots? (See Table 23-3.) Cite some examples of each group.

The flower
  43. Identify the structures that make up the female (carpel) and male (stamen) reproductive organs of the flower shown in Figure 23-20.
  44. Briefly describe the process of fertilization in the angiosperms.

Evolution of the flower
  45. What has characterized the evolution of the flower?
  46. What trends are apparent in the evolution of the primitive flower into a highly specialized form?

The agents of evolution
  47. Describe the probable evolution of plants that attract insect pollinators.
  48. What advantage does insect pollination offer over wind pollination?
  49. What refinements did angiosperms contrive

that increased their attractiveness to pollinators?

50. Explain what advantage a flower would gain from being frequented by a constant pollinator rather than a promiscuous one.

51. What structures have evolved that preclude visits from indiscriminate pollinators?

Evolution of the fruit

52. From what flower structure does the fruit develop?

53. How do fruits encourage the dispersal of seeds?

Biochemical evolution

54. List several of the bad-tasting or toxic compounds produced by angiosperms.

55. Are such compounds simply by-products of metabolism? Explain.

VIII. **Summary** (pages 478–479): Read the Summary. If you are familiar with the essential features of the material presented there, you are ready to take the following diagnostic examination.

TESTING YOUR UNDERSTANDING

After you have completed the following examination, compare your answers with the annotated key in the Analysis section.

1. Primitive algae were more successful in coastal waters than on the surface of the open ocean because in coastal waters:
   a. light was more available.
   b. minerals were more abundant.
   c. there was less turbulence.
   d. carbon dioxide was more greatly concentrated.

2. The protistan ancestor of the plants is thought to be a(n):
   a. multicellular brown alga.
   b. colonial volvocine alga.
   c. isomorphic green alga.
   d. oogamous green alga.
   e. photosynthetic euglenoid.

3. The principal difference between the two major lineages of plants lies in the absence or presence of:
   a. vascular tissue.
   b. protected seeds.
   c. specialized photosynthetic organs.
   d. flagellated gametes.
   e. starch as a storage form of carbohydrate.

4. In the bryophytes, the zygote develops inside a multicellular structure called the:
   a. sporangium.
   b. antheridium.
   c. archegonium.
   d. sporophyte.
   e. ovary.

5. Mosses and liverworts are *not* considered to be completely adapted to land because they:
   a. are not anchored to a substrate.
   b. require water for fertilization.
   c. do not absorb minerals from the soil.
   d. provide internal protection for the developing embryo.
   e. Both b and c are correct.

6. In a plant species that is heterosporous, antheridia and archegonia are borne on:
   a. a single gametophyte.
   b. separate gametophytes.
   c. a single sporophyte.
   d. separate sporophytes.
   e. a sporangiophore.

7. Paleobotanists know that seed plants had evolved by the beginning of the Carboniferous period:
   a. because seeds have been found in Devonian deposits.
   b. even though no pre-Mesozoic fossils have ever been found.
   c. because it is known that highly advanced bryophytes existed about 400 million years ago.
   d. because coal is found in Carboniferous strata.
   e. All of the above are correct.

8. Which of the following are seedless vascular plants?
   a. ferns
   b. mosses
   c. liverworts
   d. conifers
   e. ginkgoes

9. Members of the division Pterophyta (ferns) generally exhibit:
   a. extensive secondary growth.

b. very elongated stems.
c. a nutritionally dependent gametophyte.
d. a prominent sporophyte.
e. heterospory.

10. The biggest problem for vascular plants during the Permian period was:
    a. cold climates.
    b. surviving extensive flooding.
    c. water conservation.
    d. competition from angiosperms.
    e. a shortage of insect pollinators.

11. Microsporangia are seed-plant tissues that:
    a. release spores.
    b. contain developing male gametophytes.
    c. contain the ovules.
    d. produce the cones or anthers.

12. Following meiosis in the seed plants, the female gametophyte develops directly from a(n):
    a. megaspore mother cell.
    b. megasporangium.
    c. megaspore.
    d. egg cell.

13. Nonmotile sperm are found in:
    a. all seed plants.
    b. all vascular plants.
    c. all land plants.
    d. angiosperms only.
    e. bryophytes only.

14. Plants that produce seeds also produce:
    a. pollen grains.
    b. a greatly reduced gametophyte.
    c. fruits and flowers.
    d. Both a and b are correct.
    e. All three answers (a, b, and c) are correct.

15. The flowering plants can be divided into two large groups, the:
    a. gymnosperms and angiosperms.
    b. trees and grasses.
    c. monocots and dicots.
    d. wind-pollinated and the insect-pollinated.
    e. Bryophyta and Pterophyta.

16. The female parts of a flower are referred to collectively as the _____ and the male parts as _____.
    a. megasporangia; microsporangia
    b. pistil; anthers
    c. ovary; anthers
    d. carpel; stamens
    e. stigma; pollen grains

17. The female gametophyte of a pine tree is enclosed within:
    a. an ovule.
    b. a cone.
    c. the mother sporophyte.
    d. a seed coat.
    e. All of the above are correct.

18. A primitive flower would possess:
    a. a superior ovary.
    b. fused carpels.
    c. only a small number of floral parts.
    d. bilateral symmetry.
    e. All of the above are correct.

19. A flower is primarily an adaptation by which plants:
    a. disperse their seeds.
    b. protect their seeds.
    c. provide food for beneficial insects.
    d. attract pollinators.
    e. make self-pollination more likely.

20. A flower that _____ is at a decided advantage over other plants.
    a. attracts many different insect pollinators
    b. has specialized landing platforms and nectaries
    c. develops a superior ovary
    d. is small and inconspicuous
    e. Both a and c are correct.

21. Which of these is *not* an adaptation for reproductive success in an angiosperm?
    a. fruits with brightly colored skin
    b. fleshy fruits
    c. sweet-tasting seeds
    d. digestible seed coats
    e. Both b and d are correct.

22. Nicotine, caffeine, and cocaine are products of angiosperms considered by plant biochemists to be:
    a. waste products.
    b. attractants.
    c. worthwhile defenses against predators.
    d. toxic even to plants.
    e. deterrents to competing vegetation.

23. Match each structure with its most appropriate description.

    fruit _____
    anther _____
    ovary _____
    cotyledon _____
    microspore _____
    ovule _____

    a. where a female gametophyte develops
    b. where pollen grains develop
    c. a ripened ovary
    d. an embryonic leaf
    e. develops into a male gametophyte
    f. base of carpel

24. Match each geological period with an event.

    Devonian _____
    Permian _____
    Cretaceous _____
    Carboniferous _____

    a. Angiosperms were abundant.
    b. The climate favored the evolution of the gymnosperms.
    c. The bryophyte fossil record begins.
    d. Ferns dominated the forests.

25. Compare and contrast the first land plants with their hypothetical algal ancestor. Explain differences in terms of adaptation to a terrestrial environment.

26. The history of vascular plants reveals four major evolutionary trends. Point out the adaptive value of each.

27. Beginning with the formation of male and female gametophytes, trace the events necessary for production of the typical conifer seed.

28. Identify the numbered parts of this generalized flower.

1. _____
2. _____
3. _____
4. _____
5. _____
6. _____
7. _____
8. _____
9. _____
10. _____

After you have identified the flower parts, match each part with its description or function listed below.

male reproductive structures (collective term) _____

structure in which the female gametophyte develops _____

contains the microsporangia _____

female reproductive structures (collective term) _____

supports the structure in which pollen grains develop _____

develops into the fruit in angiosperms _____

modified leaves _____

site of attachment of pollen grain to the female structures _____

structure through which the pollen tube must grow to reach the ovary _____

PERFORMANCE ANALYSIS

1. Coastal waters would have been a much more turbulent environment for marine algae than the open ocean. This turbulence would have limited the amount of light useful for photosynthesis but would have been responsible for keeping minerals circulating. Minerals were and are more abundant in coastal waters than on the ocean surface because of input from rivers and streams and because of upwelling due to wave action. Carbon dioxide was not significantly more concentrated in these

environments than in the open ocean. Therefore, choice **b** is correct.

2. The protist ancestor of green plants is thought to have been a green alga that was oogamous (**d**) and heteromorphic as well. Green algae share these two reproductive features as well as other characteristics with the plants.

3. The two major lineages referred to are the bryophytes, which lack conducting or vascular tissue, and the tracheophytes, whose name indicates that they have developed such tissues; therefore, **a** is correct.

4. In bryophytes, such as the mosses and liverworts, the zygote is protected in a reproductive organ of the gametophyte called the archegonium (**c**).

5. Mosses and liverworts are incompletely adapted to land because (1) their flagellated sperm must swim to the egg in order for fertilization to occur, and (2) although they have rootlike structures called rhizoids, these serve only as anchoring devices and do not absorb minerals from the soil. Therefore, choice **e** is correct.

6. A plant that is heterosporous forms two different kinds of spores that develop into separate male and female gametophytes (**b**).

7. Choices b and c are both false statements. Though d is true, ferns, which thrived during the Carboniferous period and are *not* seed plants, could have been the source of the coal. Choice **a** alone is correct, for seeds have been found in Devonian deposits.

8. Mosses and liverworts lack seeds but are not vascular plants. Conifers and ginkgoes are vascular plants but do produce seeds. Only the ferns (**a**) satisfy both conditions stated in the question.

9. The sporophytes of pterophytes (**d**), or ferns, are prominent, leafy structures (see Figure 23-12); the gametophytes are only a few cell layers thick. Ferns do not exhibit secondary growth, although some can grow very tall. Fern gametophytes are nutritionally independent, and the majority of genera are homosporous.

10. Gymnosperms flourished during the dry Permian period because their leaves were well adapted to conserve water (**c**).

11. Inside a microsporangium, microspore mother cells give rise (by meiosis) to microspores, one of which will become the male gametophyte; therefore, **b** is correct.

12. In a process analogous to the development of the male gametophyte, a female gametophyte develops inside a megasporangium directly from a megaspore (**c**) produced by meiosis.

13. Bryophytes are the major exception to choice c, and ferns are the major exception to choice b. However, the seed plants (**a**), the gymnosperms and angiosperms, have no need for motile sperm.

14. Pollen grains are produced in all seed plants and the gametophyte is greatly reduced, but choice c is true only for angiosperms; therefore, **d** is correct.

15. Plants that produce flowers are either monocots or dicots (**c**). Their major points of difference are summarized in Table 23-3.

16. Choice **d** is correct. Refer to Figure 23-20.

17. The female gametophyte of a conifer, such as a pine, develops inside an ovule borne on a scale of one of the sporophyte's female cones. Following fertilization, a seed coat develops around both the embryo and the remaining gametophyte tissue; therefore, **e** is correct.

18. Choices b, c, and d all imply a reduction in the number of floral parts, an evolutionarily advanced condition. The superior ovary (**a**), however, is more primitive than the inferior ovary, the development of which led to the protection of the ovules from foraging insects.

19. Flowers are specialized for attracting pollinators (**d**), which transfer gametes from one plant to another (cross-pollination).

20. Flowers that are specialized to attract specific kinds of pollinators, often a single species, are at an advantage because pollen is less likely to be wasted on plants of other species. Choice **b** is correct.

21. The brightly colored skin and fleshy parts of a fruit are attractive to foragers. The seed coats of some seeds must be digested before the seeds can germinate. But seeds are often bitter so that they will not be eaten; otherwise the embryo would not survive. Therefore, **c** is correct.

22. The substances listed here have evolved as metabolic products for the purpose of defense against herbivores (c).
23. The correct answers in order are **c, b, f, d, e**, and **a**.
24. The correct answers in order are **c, b, a**, and **d**.
25. Similarities between land plants and their algal ancestors are referred to in the analysis for Question 2. The differences are primarily adaptations to the scarcity of water in most terrestrial environments: the retention of the zygote by the parent plant and the development of a cuticle and stomata.
26. The four major trends pointed out in the chapter are (1) the development of vascular tissue; (2) the reduction of the gametophyte stage; (3) the evolution of heterospory; and (4) the evolution of the seed. The development of true roots and leaves could also be included, though the text refers to these adaptations more precisely as innovations rather than as structures that show a gradual evolution.
27. Refer to Figure 23–15, the life cycle of the pine.
28. The answers are

    1. stigma   6. sepal
    2. carpel   7. petal
    3. anther   8. ovule
    4. stamen   9. ovary
    5. filament   10. style

    The matching answers in order are **4; 8; 3; 2; 5; 9; 6, 7; 1**; and **10**.

CHAPTER 24

The Animal Kingdom I: Introducing the Invertebrates

CHAPTER ORGANIZATION

I. Introduction
II. The diversity of animals
III. The origin and classification of animals
IV. Phylum Porifera: Sponges
    A. General information
    B. Reproduction in sponges
V. Phylum Mesozoa: Mesozoans
VI. Radially symmetrical animals
    A. Introduction
    B. Phylum Cnidaria
        1. General information
        2. Class Hydrozoa
        3. Class Scyphozoa
        4. Class Anthozoa
        5. Essay: The coral reef
    C. Phylum Ctenophora
VII. Bilaterally symmetrical animals: An introduction
VIII. Phylum Platyhelminthes: Flatworms
    A. General information
    B. Class Turbellaria
        1. General information
        2. The planarian nervous system
    C. Classes Trematoda and Cestoda
    D. Essay: The politics of schistosomiasis
IX. Other acoelomates
    A. Phylum Gnathostomulida
    B. Phylum Rhynchocoela
X. Pseudocoelomates
    A. General information
    B. Phylum Nematoda
    C. Minor pseudocoelomate phyla
XI. Summary

MAJOR CONCEPTS

Animals are multicellular heterotrophs that are directly or indirectly dependent upon plants or algae for their food energy. Invertebrates (animals without backbones) comprise more than 90 percent of the animal species.

The primary criteria for classifying animals include (1) the number of tissue layers, (2) the basic body plan, (3) the presence or absence of a true coelom, and (4) the pattern of embryonic development.

From the lower animals to the more advanced forms, there is an increasing specialization of cells (from specialized cells to tissues to organs); an increasing complexity of body organization (acoelomate to pseudocoelomate to coelomate); and an increasing complexity of structure and function in the major organ systems (nervous, digestive, circulatory, and excretory).

Sponges (phylum Porifera) and the simple, parasitic mesozoans (phylum Mesozoa) are so different from other animals that they are placed in separate subkingdoms. They possess specialized cells, but there is little coordination of function among the different cells.

Members of the phylum Cnidaria (jellyfish, sea anemones, and corals) and the phylum Ctenophora (sea walnuts and comb jellies) are characterized by radial symmetry, a coelenteron, and two tissue layers (ectoderm and endoderm) separated by the jellylike mesoglea.

Members of the phyla Platyhelminthes (the flatworms), Gnathostomulida (tiny marine worms), and Rhynchocoela (ribbon worms) possess bilateral symmetry, an anterior cluster of nerve cells, and three tissue layers (ectoderm, mesoderm, and endoderm), but no body cavity (they are acoelomates) other than the digestive cavity.

Seven phyla of animals have a body plan based on the pseudocoelom, a fluid-filled cavity between the endoderm and the mesoderm. The pseudocoelom functions as a hydrostatic skeleton. All pseudocoelomates have a one-way digestive tract but lack a circulatory system.

The roundworms (phylum Nematoda) are the largest and most important group of pseudocoelomates. Among the various pseudocoelomates, the entire range of sexual reproductive patterns in animals is seen: separate sexes, simultaneous hermaphrodites, sequential hermaphrodites, external fertilization, internal fertilization, and parthenogenesis.

HOW TO STUDY THE CHAPTER

Read the chapter, focusing on the major concepts.

Reread the chapter, using the questions that follow to help you focus on details as they relate to major concepts. Answer the questions on a separate sheet of paper. Your answers will provide a valuable study aid in preparing for examinations.

FOCUS ON CHAPTER DETAILS

I. **Introduction** (page 481)
  1. Enumerate the general characteristics of an animal.
  2. Invertebrates are both much more numerous and much more diverse than vertebrates. Why do these facts make the invertebrates important in the study of biology?

II. **The diversity of animals** (pages 481–482)
  3. What is the unifying characteristic among all animals? What requirements stem from this?
  4. Why is there such enormous structural and functional diversity in the way animals solve their basic problems?

III. **The origin and classification of animals** (pages 482–483)
  5. Why has it been difficult to trace the evolutionary history of invertebrates?
  6. By what period were most or all of the invertebrate phyla in existence?
  7. What criteria have been used to classify animals into various phyla?

IV. **Phylum Porifera: Sponges** (pages 483–486)
  A. General information
  8. What characteristics warrant the placement of sponges in a subkingdom separate from all other animal phyla?
  9. What areas are commonly occupied by the sponges?
  10. Describe the basic organization of the sponge (see also Figure 24–6).
  11. Why are sponges not considered to be truly multicellular organisms?
  12. What functions do the various types of cells indicated in Figure 24–6 perform in procuring and/or digesting food?
  13. What strategy do large sponges employ for increasing the amount of food they can obtain?
  14. What are the four classes of sponges? What criterion is used to group them into these classes?
  15. By what cells and of what materials are the skeletons of sponges formed?

  B. Reproduction in sponges
  16. In what two ways do sponges reproduce asexually?
  17. Explain how sponges carry out internal fertilization.
  18. Trace the development of the zygote into an adult form.
  19. What is a hermaphrodite? Why is it an advantage for organisms such as sponges to be hermaphroditic?

V. **Phylum Mesozoa: Mesozoans** (page 486)
  20. Describe the mesozoans. What are the

similarities and differences between mesozoans and sponges?

21. What might be the evolutionary significance of this group?

VI. Radially symmetrical animals (pages 486–493)

A. Introduction

22. Give examples of cnidarians and ctenophores.
23. What is radial symmetry?
24. What is the coelenteron? Describe digestion as it occurs in the coelenteron.

B. Phylum Cnidaria

General information

25. Describe the two basic body plans of cnidarians. Define ectoderm, endoderm, and mesoglea.
26. Describe the structure cnidarians use to capture their prey.
27. Briefly explain how asexual and sexual reproductive stages alternate in the typical life cycle of members of this phylum (see Figure 24-12). What type of larva is produced?

Class Hydrozoa

28. Describe the contractile fibers in the ectoderm and endoderm of *Hydra*. What role do these fibers play?
29. Define independent effectors and give two examples found in *Hydra*.
30. How do sensory receptor cells function in *Hydra*?
31. Describe the organization of cells that function as a simple nervous system in *Hydra*.
32. Explain how *Obelia* forms colonies. What differentiation of function is seen among the members of the colony? Why is this type of specialization unique among the lower invertebrates?

Class Scyphozoa

33. What is the dominant life form in the jellyfish?
34. Describe the epitheliomuscular and sensory receptor cells of scyphozoans. How is their nervous system more developed than that of hydrozoans?
35. Indicate the location and function of the specialized multicellular sense organs found within this class.

Class Anthozoa

36. What stage is lacking in the life cycle of anthozoans?
37. Describe asexual and sexual reproduction in anthozoans.
38. How does the basic polyp structural organization within this class differ from that of hydrozoans?
39. What role do anthozoans play in the formation of a coral reef?

Essay: The coral reef (page 492)

40. Describe the reef community, paying particular attention to the role of anthozoans and algae.
41. Where are such reefs typically found?

C. Phylum Ctenophora

42. Describe the ctenophores. What are their two means of locomotion?
43. Describe their body plan. How is food captured?
44. How do ctenophores reproduce?

VII. Bilaterally symmetrical animals: An introduction (pages 493–494)

45. What is bilateral symmetry? What structures tend to develop in the anterior region? In the posterior region?
46. What is characteristic of locomotion and of nervous systems in bilaterally symmetrical animals?
47. What third tissue layer appears in bilaterally symmetrical animals? Where is it located?
48. What is a "germ layer"? What tissues are derived from each of the three germ layers?
49. Study the three basic body plans found in

triploblastic animals in Figure 24-21. Describe each type.

50. What is the difference between a pseudocoelom and a coelom?

VIII. **Phylum Platyhelminthes: Flatworms** (pages 495-498)

A. General information

51. What new organizational level is found within this phylum?

52. How is digestion in the flatworms similar to that in the cnidarians?

53. What basic body plan is found among members of this group?

54. What problems are solved by the flattened bodies and the branching of the digestive cavity exhibited by members of the phylum Platyhelminthes?

55. What controversy is there over the phylogenetic origins of flatworms?

56. Briefly characterize the three classes of flatworms.

B. Class Turbellaria

General information

57. What locomotory specializations do the planarians exhibit?

58. How do planarians procure and ingest their food?

59. Describe the cells that are specialized for maintaining water balance.

60. How do planarians rid themselves of waste?

61. Explain how planarians reproduce sexually and asexually (see Figure 24-25).

The planarian nervous system

62. Describe the basic organization of the planarian nervous system.

63. Where are the three types of sensory receptors located, and what stimuli does each type perceive?

C. Classes Trematoda and Cestoda

64. What adaptations to a parasitic life style are exhibited by the tapeworms and flukes?

65. How do tapeworms cause disease in their hosts?

66. How are parasites thought to have evolved?

D. Essay: The politics of schistosomiasis (page 499)

67. Describe the general life cycle of a trematode worm of the genus *Schistosoma*. What causes the symptoms of schistosomiasis?

68. What factors have been responsible for the rapid spread of this disease in developing countries?

69. What factors inhibit the search for a cure for diseases like schistosomiasis?

IX. **Other acoelomates** (pages 498-500)

A. Phylum Gnathostomulida

70. What is the distinguishing structure of these tiny marine worms, and how is it used?

71. What is the evolutionary significance of the gnathostomulids?

B. Phylum Rhynchocoela

72. What is the distinguishing structure of the ribbon worms, and how is it used?

73. What evolutionary improvements over the flatworms are found among members of this phylum?

74. What are the consequences of a one-way digestive tract?

75. How do ribbon worms reproduce?

X. **Pseudocoelomates** (pages 500-503)

A. General information

76. How does the pseudocoelom of the nematodes provide for more coordinated locomotion than is seen in the acoelomates?

77. What other physiological function does the pseudocoelom serve?

B. Phylum Nematoda

78. Briefly characterize the nematodes.

79. Describe the appearance of a nematode.

80. What type of motion do roundworms exhibit? What muscular action is responsible for this?

81. How do nematodes procure their food?
82. Cite several examples of nematodes that are parasites of humans.
83. What specializations for a parasitic life style have these creatures devised?

C. Minor pseudocoelomate phyla

84. Describe the two phyla of pseudocoelomates whose larvae are parasites of arthropods.
85. Describe three phyla of pseudocoelomates that perform an ecological role similar to that of the gnathostomulids.
86. What is parthenogenesis?
87. What unique digestive and reproductive characteristics are found within the phylum Entoprocta?

XI. **Summary** (pages 504–505): Read the Summary. If you are familiar with the essential features of the material presented there, you are ready to take the following diagnostic examination.

TESTING YOUR UNDERSTANDING

After you have completed the following examination, compare your answers with the annotated key in the Analysis section.

1. For which of these statements do biologists have strong historical evidence?
   a. The invertebrates are descended from a single protist ancestor.
   b. Most of the invertebrate phyla we know today were present during the Cambrian period.
   c. The phylogenetic history of the invertebrates has been constructed accurately from fossil records.
   d. The closest unicellular relative of the animals was a heterotrophic prokaryote.

2. In which of the following ways does digestion in the sponges appear to be severely limited?
   a. Food particles must move between cells by diffusion.
   b. Digestion can only take place within individual cells.
   c. Their water-filtering chambers are essentially two-dimensional.
   d. Their outer body walls cannot recognize harmful substances.
   e. Water currents alone determine the rate at which food particles can be filtered.

3. The sponges are different from other members of the animal kingdom because they:
   a. are not known to reproduce sexually.
   b. have no tissues or organs.
   c. exhibit no motile stages in their life cycle.
   d. do not respond to stimuli in their environment.

4. All of the following are sessile creatures *except*:
   a. sea walnuts.
   b. cnidarian polyps.
   c. entoprocts.
   d. sponges.
   e. All of the above are correct.

5. Choose the pair of phyla that have radially symmetrical adult forms.
   a. Cnidaria and Ctenophora
   b. Rotifera and Porifera
   c. Rhynchocoela and Porifera
   d. Nematoda and Nematomorpha
   e. Trematoda and Cestoda

6. Invertebrates important in the building of coral reefs are members of the:
   a. Scyphozoa.
   b. Hydrozoa.
   c. Mesozoa.
   d. Anthozoa.
   e. Porifera.

7. The adult body form of a jellyfish is referred to as a:
   a. polyp.
   b. colony of polyps.
   c. planula.
   d. zoophyte.
   e. medusa.

8. A coelenteron is a cavity found in cnidarians and their relatives in which:
   a. absorbed materials are passed from cell to cell.
   b. nerve cells interconnect.
   c. extracellular digestion takes place.
   d. internal organs are suspended.
   e. a primitive circulatory system develops.

9. The nervous systems of cnidarians include all of the following *except*:
   a. a simple nerve net.

b. multicellular sense organs.
   c. cells that are both receptors and effectors.
   d. aggregations of nerve cell bodies known as ganglia.
10. In colonial hydrozoans, one type of polyp is specialized for _____ and another for _____.
    a. locomotion; reproduction
    b. defense; feeding
    c. feeding; reproduction
    d. locomotion; defense
    e. filter-feeding; swimming
11. The typical cnidarian life cycle represents an alternation between a _____ form and a _____ form.
    a. motile; sessile
    b. haploid; diploid
    c. asexual; sexual
    d. polyp; medusa
    e. Answers a, c, and d are correct.
12. The ectoderm and endoderm of hydrozoans, scyphozoans, and anthozoans are separated by a:
    a. mesoderm layer.
    b. pseudocoelom.
    c. true coelom.
    d. gelatinous filling.
    e. digestive cavity.
13. In the animal kingdom, bilateral symmetry is most closely associated with:
    a. three germ layers.
    b. a two-ended digestive cavity.
    c. a coelom.
    d. parasitism.
    e. a circulatory system.
14. An invertebrate phylum whose members propel themselves by means of cilia is:
    a. Cestoda.
    b. Porifera.
    c. Nematoda.
    d. Rotifera.
    e. Ciliophora.
15. Internal parasites have little need for:
    a. a digestive system.
    b. a reproductive system.
    c. complex life cycles.
    d. attachment organs.
    e. bilateral symmetry.
16. An example of an acoelomate with a one-way digestive tract is a:
    a. planarian.
    b. gnathostomulid.
    c. ribbon worm.
    d. nematode.
    e. trematode.
17. Which of the following reproductive strategies is found among the pseudocoelomates?
    a. parthenogenesis
    b. simultaneous hermaphroditism
    c. sequential hermaphroditism
    d. separate sexes
    e. All of the above are correct.
18. Evolutionarily speaking, the great improvement in locomotion exhibited by the nematodes is associated with:
    a. muscle tissue.
    b. a pseudocoelom.
    c. development of an extensive circulatory system.
    d. the use of circular muscles.
    e. coordination of muscle activity by a nerve center.
19. In the nematodes and related phyla, a pseudocoelom is located:
    a. within the mesoderm.
    b. within the endoderm.
    c. between mesoderm and ectoderm.
    d. between mesoderm and endoderm.
    e. interior to the endoderm.
20. Hermaphroditism is a reproductive strategy generally found among invertebrates that are:
    a. limited to asexual reproduction.
    b. relatively immobile.
    c. radially symmetrical.
    d. living in high densities.
    e. parasites.

21. Use the terms from the two-column list on the left to identify the various parts of the four basic body plans illustrated. From the two-column list on the right, choose at least one example of each body plan.

coelom	mesentery	annelids	ctenophores
digestive cavity	mesoderm	arthropods	flatworms
ectoderm	mesoglea	chordates	mollusks
endoderm	pseudocoelom	cnidarians	nematodes

(a) TWO-LAYERED

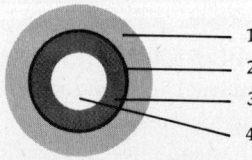

Example(s): _____

(b) THREE-LAYERED

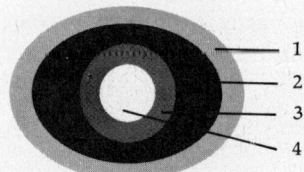

Example(s): _____

(c) THREE-LAYERED

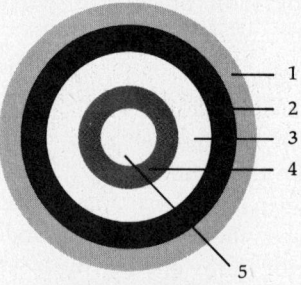

Example(s): _____

(d) THREE-LAYERED

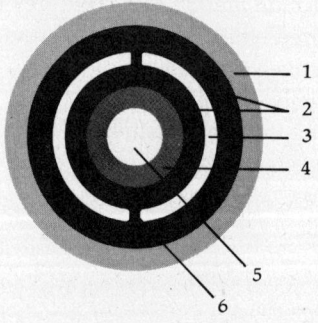

Example (s): _____

22. Match each genus with its distinguishing feature.

Hydra _____
Schistosoma _____
Ascaris _____
Planaria _____
Obelia _____

a. intermediate host is a snail
b. move by cilia
c. colonial marine polyps
d. freshwater solitary polyps
e. intestinal parasites of vertebrates

23. Match each phylum with a short description of its representatives.

Mesozoa _____
Nematomorpha _____
Nematoda _____
Cestoda _____
Ctenophora _____
Kinorhyncha _____
Rotifera _____

a. free-living and parasitic soil inhabitants
b. bioluminescent ocean dwellers
c. "wheel animalcules"
d. marine burrowing forms
e. parasites of arthropod larvae
f. parasites of marine invertebrates
g. parasitic flatworms

24. Animals are conveniently described as multicellular heterotrophs. Formulate a complete set of characteristics that distinguish the animal kingdom from the others you have studied.

25. How do biologists account for the remarkable diversity found among animals?

26. What are the consequences of bilateral symmetry for an animal? How does bilateral symmetry represent a general advantage over radial symmetry?

27. Why is the one-way digestive tract considered an advance over a tract with a single opening?

PERFORMANCE ANALYSIS

1. The Precambrian fossil record is inadequate to reconstruct the phylogenetic history of the invertebrates. Cambrian fossils are plentiful, but by this period, most invertebrate phyla were already in existence (**b** is correct). Evolutionary schemes such as that of Figure 24-4 are based on the study of modern forms.

2. Choices a, c, d, and e are all false statements about the sponges; however, because sponges can only digest food molecules intracellularly, they are limited to feeding on microscopic particles (**b**).

3. Sexual reproduction is quite common among sponges, and the larval forms, the planulae, are free-swimming. Epitheliomuscular cells on the outer surface of the sponge respond to environmental stimuli. But sponges lack the type of cellular organization that is normally associated with tissues or organs; therefore, **b** is correct.

4. Sea walnuts (**a**) are members of the phylum Ctenophora, which move about by the coordinated beating of comblike plates composed of fused cilia (see Figure 24-19). All the others listed here are sessile.

5. The adult forms of both cnidarians and ctenophores (**a**) are radially symmetrical. With the exception of the Porifera, all the other phyla listed are bilaterally symmetrical.

6. Coral reefs are formed by the action of colonial anthozoans (**d**), which secrete a calcium skeleton (see the essay on page 492).

7. The adult jellyfish (a member of the cnidarian class Scyphozoa) is a medusa (**e**), a floating gelatinous mass that is capable of producing gametes.

8. Choices a, b, and e are associated with the mesoglea. Cnidarians have no coelom, but the coelenteron is the site of extracellular digestion; therefore, **c** is correct.

9. Cnidarians have a simple nerve net at the base of the epithelial layer, multicellular sense organs (e.g., ocelli and statocysts), and cells called cnidocysts that function both as receptors and effectors. There is, however, no aggregation of nerve cells (ganglia); therefore, **d** is correct.

10. In a colonial hydrozoan, such as *Obelia* (Figure 24-15a), there are two polyp forms, one specialized for feeding, the other for reproduction; therefore, **c** is correct.

11. Cnidarians typically alternate between a motile adult medusa stage that reproduces sexually and a

sessile polyp stage that reproduces asexually by budding, as in *Hydra*; therefore, **e** is correct.

12. The animals referred to are cnidarians, which have no mesoderm and lack a true coelom. A gastrovascular cavity, the coelenteron, is surrounded by endoderm. The gelatinous filling (**d**) that separates the ectoderm and endoderm is called the mesoglea. See the basic cnidarian body plan in Figure 24-11.

13. Animals that are bilaterally symmetrical are referred to as triploblastic; that is, they have three germ layers (**a**). The flatworms (phylum Platyhelminthes) are examples of bilaterally symmetrical organisms that have neither a two-ended digestive tract, a coelom, nor a circulatory system.

14. Members of Rotifera, the "spinning wheels," propel themselves by means of a crown of cilia located around the mouth; therefore, **d** is correct.

15. In evolving from free-living forms, many parasites have lost their digestive system (**a**).

16. Evolutionarily speaking, the ribbon worms (**c**) (phylum Rhynchocoela) are the first group of animals to exhibit a one-way digestive tract (i.e., one having a mouth and an anus).

17. All the strategies listed here are found in at least some pseudocoelomates (**e**). The rotifers, for example, reproduce by parthenogenesis (**a**), entoprocts are sequential hermaphrodites (**c**), and nematodes exist as separate sexes (**d**). Simultaneous hermaphroditism (**b**) is fairly widespread among the pseudocoelomates.

18. The pseudocoelom, found among the nematodes and related minor phyla, serves as a hydraulic skeleton and allows for effective muscular contractions (**b**).

19. Choice **d** is correct. Refer to Figure 24-21c.

20. Hermaphroditism is most commonly associated with animals that are not active enough to encounter a mate. Sponges, entoprocts, and many flatworms are examples of relatively immobile (**b**) hermaphrodites.

21. Refer to Figure 24-21 in the text for the correct answers to this question.

22. The correct answers in order are **d**, **a**, **e**, **b**, and **c**.

23. The correct answers in order are **f**, **e**, **a**, **g**, **b**, **d**, and **c**.

24. Your set of distinguishing characteristics should include nutritional mode, cellular organization, response to stimuli, and patterns of growth and reproduction.

25. Despite the limitations posed by a heterotrophic mode of nutrition, animals have become remarkably diverse as a result of refinement, reinvention, and elaboration of a few basic strategies.

26. With the evolution of bilateral symmetry, animals developed an anterior/posterior aspect and a dorsal/ventral aspect. Sensory awareness became associated with the head region (cephalization), and digestive, excretory, and reproductive functions, with the posterior region. One great advantage of bilaterally symmetrical organisms over radially symmetrical organisms is a more efficient manner of locomotion.

27. A one-way digestive tract allows for continuous eating and for specialization of function (e.g., excretory functions can be physically separated from digestive ones). You should be able to think of several more reasons on your own.

CHAPTER 25

The Animal Kingdom II: The Protostome Coelomates

CHAPTER ORGANIZATION

I. Introduction
II. Phylum Mollusca: Mollusks
   A. Introduction
   B. Characteristics of the mollusks
      1. General information
      2. Supply systems
   C. Minor classes of mollusks
   D. Class Bivalvia
   E. Class Gastropoda
   F. Class Cephalopoda
      1. General information
      2. Essay: Behavior in the octopus
   G. Evolutionary affinities of the mollusks
III. Phylum Annelida: Segmented worms
   A. General information
   B. Class Oligochaeta: The earthworm
      1. General introduction
      2. Digestion in earthworms
      3. Circulation in earthworms
      4. Respiration in earthworms
      5. Excretion in earthworms
      6. The nervous system of earthworms
      7. Reproduction in earthworms
   C. Class Polychaeta
   D. Class Hirudinea
IV. Minor protostome phyla
V. The lophophorates
VI. Summary

MAJOR CONCEPTS

A coelom functions as a hydrostatic skeleton and provides space for internal organs. Coelomate animals can be divided into protostomes and deuterostomes according to their pattern of embryonic development. Embryonic development in protostomes is characterized by a spiral cleavage pattern, mouth development near the blastopore, and coelom formation by splitting of the mesoderm. Embryonic development in deuterostomes is characterized by a radial cleavage pattern, anus development near the blastopore, and coelom formation by outpocketings of the primitive gut.

The major protostome phyla are Mollusca, Annelida, and Arthropoda. The basic body plan of members of the phylum Mollusca includes a head-foot, a visceral mass, and a mantle (which secretes the shell in those animals that possess shells). Mollusks are also characterized by a toothed tongue, the radula.

Mollusks and annelids (phylum Annelida) are thought to share a common ancestor because both groups possess similar trochophore larvae. The major structural characteristics of annelids include internal and external segmentation, a tubular digestive tract, paired nephridia in each segment, a closed circulatory system, and a well-developed coelom. They also have relatively complex nervous systems.

Of the seven minor phyla of protostomes, four consist of bottom-dwelling marine worms, and the remaining three contain organisms that show both annelid and arthropod features.

The members of three additional phyla, known collectively as lophophorates, are protostomes that exhibit some deuterostome characteristics.

HOW TO STUDY THE CHAPTER

Read the chapter, focusing on the major concepts.

Reread the chapter, using the questions that follow to help you focus on details as they relate to major concepts. Answer the questions on a separate sheet of paper. Your answers will provide a valuable study aid in preparing for examinations.

FOCUS ON CHAPTER DETAILS

I. Introduction (pages 506–507)

1. For what reason is the coelom considered an extremely important evolutionary innovation? What is another useful feature of the coelom?
2. What features of embryonic development are used to classify coelomate animals into two broad phylogenetic groups?
3. What major phyla represent each group?
4. Describe coelom formation in protostomes and deuterostomes.

II. Phylum Mollusca: Mollusks (pages 507–516)

A. Introduction

5. Mollusks are a diverse and abundant (47,000 species) group of animals. Give a general description of mollusks. What kind of diversity does the phylum Mollusca exhibit?
6. What are the three major classes of mollusks? Name several representatives of each class.

B. Characteristics of the mollusks

General information

7. Describe the three distinct body zones of mollusks.
8. In what molluscan functions is the mantle cavity important?
9. What specialized organ is found only in the mollusks? How does it operate?

Supply systems

10. Describe the open circulatory system found in most mollusks.
11. What improvements on this system have been made by the cephalopods?
12. How do aquatic mollusks exchange gases with their environment?
13. Briefly describe the molluscan digestive system. How are digestive wastes excreted?
14. Describe the nephridia. What two functions does the excretory system of mollusks perform?

C. Minor classes of mollusks

15. Into what classes have the following molluscan groups been placed: solenogaster, chitons, *Neopilina*, and tusk shells?
16. What resemblance do these groups bear to other modern mollusks? How are they different?
17. What was the significance of the fossil find known as *Neopilina galatheae*?

D. Class Bivalvia

18. Describe the structure from which this class takes its name.
19. Explain how the bivalves are limited in their range of movement.
20. Explain the filter-feeding system employed by the bivalves.
21. Describe the general layout of a bivalve's nervous system. What specialized cells and tissues provide information about external stimuli?
22. What reproductive strategies are employed within this class?

E. Class Gastropoda

23. What kind of diversity is exhibited by this class, which contains more than three-fourths of the total number of species in the phylum?

24. What anomalies has a spiral type of development produced among the shelled gastropods? Do these same characteristics appear among those that have lost the shell?
25. What respiratory strategy has been adopted by the land-dwelling snails?
26. How is the nervous system organized in the more active gastropod forms? What sensory specializations are found?
27. What type of reproduction do most gastropods exhibit?

F. Class Cephalopoda

General information

28. How has the basic molluscan body plan been modified in the cephalopods?
29. What purposes does the shell serve in cuttlefish and squids?
30. Explain how the mantle cavity in this group is employed in locomotion.
31. What defense against predators have the cephalopods devised?
32. What improvements in the nervous system has the predatory life style of the cephalopods necessitated?
33. How is reproductive behavior within this class more complex than in other mollusks?

Essay: Behavior in the octopus (page 514)

34. How does the octopus procure its prey?
35. What tactic does an octopus use to obtain depth perception?
36. With what environmental stimuli are color changes in the octopus associated?
37. What problems do the arrangement of sensory receptors and the absence of a skeleton pose for this deep-sea creature?

G. Evolutionary affinities of the mollusks

38. Which groups of animals have trochophore larvae (see Figure 25-11), and of what evolutionary significance is this thought to be?
39. What does the degree of segmentation seen among some molluscan forms indicate about the phylogenetic relationship between the mollusks and the annelids?

III. **Phylum Annelida: Segmented worms** (pages 516-521)

A. General information

40. What distinctive body organization is characteristic of annelids?
41. What other set of features characterizes members of this phylum?
42. Describe the three classes of annelids.

B. Class Oligochaeta: The earthworm

General information

43. What structures can be found repeated in each segment of the earthworm?
44. What specializations of body regions are developed?
45. What muscular arrangements account for the style of movement of the earthworm? Describe the coelom found in this class. How does it provide for control over movement?

Digestion in earthworms

46. How does the earthworm acquire its food? What consequences does earthworm digestion have for the areas in which these creatures live?
47. What specialized areas of the digestive tract that we have not thus far encountered in our study of the animal kingdom does this group display? (See also Figure 25-13.)

Circulation in earthworms

48. Provide a description of the internal transport system of the earthworm, and explain how it functions.
49. What is the significance of the development of a closed circulatory system?

Respiration in earthworms

50. How are gases exchanged between individual cells of the earthworm and the medium in which it lives?

Excretion in earthworms

51. How is the excretory system of the earthworm designed?

52. What functions are performed by the nephridia?

The nervous system of earthworms

53. Describe the four kinds of specialized sensory cells that have evolved in this group.
54. What role do the ganglia play in coordinating the movement of each segment? What is the function of the cerebral ganglia?
55. Describe the structure and function of the ventral nerve cords.

Reproduction in earthworms

56. How do earthworms reproduce?
57. What is the role of the clitellum in fertilization?

C. Class Polychaeta

58. What specializations found among these marine annelids distinguish them from the oligochaetes?
59. Define tagmosis. What are the consequences of tagmosis in polychaetes?
60. What feeding strategies are exhibited by this diverse group of worms?
61. How is reproduction among these animals different from that of other annelids?

D. Class Hirudinea

62. Describe the external appearance of a typical leech (see also Figure 25–17).
63. What adaptations to a parasitic mode of existence are seen in this group?

IV. **Minor protostome phyla** (pages 521–523)

64. Briefly characterize each of the four minor phyla of bottom-dwelling marine worms.
65. Describe the proboscis of a peanut worm, a spoon worm, and a priapulid.
66. Briefly characterize each of the remaining three phyla that show combinations of annelid and arthropod features.

V. **The lophophorates** (pages 523–525)

67. Describe the single feature these diverse creatures have in common.
68. What two major functions does this feature serve?
69. How is segmentation within this group more like that of the deuterostomes than that of the protostomes covered in this chapter? What other deuterostome characteristics are apparent?
70. Into what three phyla have the phoronid worms, the bryozoans, and the lamp shells been placed?
71. Describe the appearance of an aggregation of phoronid worms.
72. Why are bryozoans referred to as the "moss animals"? What pseudocoelomates do they resemble?
73. How can the brachiopods be distinguished from bivalves, with whom they were formerly classified?

VI. **Summary** (pages 525–526): Read the Summary. If you are familiar with the essential features of the material presented there, you are ready to take the following diagnostic examination.

TESTING YOUR UNDERSTANDING

After you have completed the following examination, compare your answers with the annotated key in the Analysis section.

1. Which of the following features of embryonic development is found among the protostomes?
   a. a radial cleavage pattern
   b. The coelom develops from a splitting of the mesoderm.
   c. The mouth forms at the blastopore.
   d. Only b and c are correct.
   e. All three answers (a, b, and c) are correct.

2. Which of the following is true regarding the hard, protective shell found in the phylum Mollusca?
   a. All mollusks have a shell, though in some species it may be greatly reduced in size.
   b. In some species, the shell has been internalized and serves as a stiffening support.
   c. Once the shell is secreted, it cannot grow with the organism and needs to be replaced from time to time.

d. In the most advanced species, the shell serves only as a protective casing for the brain.
3. Which of the following invertebrates has a body divided into three distinct body zones?
   a. oligochaetes
   b. clams
   c. onychophorans
   d. leeches
   e. Both a and c are correct.
4. For which of these systems does the mantle cavity of a mollusk provide a supporting function?
   a. respiratory
   b. excretory
   c. reproductive
   d. Only a and b are correct.
   e. All three answers (a, b, and c) are correct.
5. Which of the following is a feature of the circulatory system of a cephalopod?
   a. continuous vessels
   b. a large blood-filled cavity
   c. a three-chambered heart pumping blood to the gills
   d. transport of oxygen and carbon dioxide through the coelom
   e. Both b and c are correct.
6. The nephridia of mollusks and annelids collect fluid from the:
   a. blood vessels.
   b. coelom.
   c. digestive tract.
   d. respiratory tract.
   e. intercellular spaces.
7. Which group of mollusks bears the closest resemblance to the hypothetical ancestor?
   a. bivalves
   b. Gastropoda
   c. Scaphopoda
   d. chitons
   e. Cephalopoda
8. In which of these invertebrate classes is the trend toward cephalization *not* in evidence?
   a. Oligochaeta
   b. Polychaeta
   c. Bivalvia
   d. Gastropoda
   e. Cephalopoda

9. Which of the following molluscan reproductive strategies is considered to be the most primitive?
   a. separate sexes with external fertilization
   b. separate sexes with internal fertilization
   c. sequential hermaphroditism
   d. simultaneous hermaphroditism
   e. specialized "brood pouches"
10. The developmental phenomenon of torsion exhibited by shelled gastropods:
    a. results in an asymmetrical body organization.
    b. has moved the visceral mass to a more posterior position.
    c. produces the characteristically coiled shell.
    d. results from a more rapid growth on one side of the body.
    e. Both a and d are correct.
11. In the most active of the gastropods, the nervous system is more highly evolved. The best evidence for this is the presence of:
    a. a complex image-forming eye.
    b. an anterior aggregation of nerve cells.
    c. a reduction in the numbers of interconnected ganglia.
    d. pairs of elongated nerve cords.
12. Cephalopods are considered to be the most evolutionarily advanced invertebrates because they:
    a. have a closed circulatory system.
    b. engage in internal fertilization.
    c. exhibit a complex larval development.
    d. possess a well-developed brain.
    e. all have a shell that serves as an internal support.
13. The similarity between mollusks and annelids that is strongly suggestive of an evolutionary link is:
    a. the trochophore larva found in marine forms of both groups.
    b. the segmentation of gills and nephridia in some adult forms.
    c. the external segmentation of larval forms.
    d. the homology between setae and head-feet.
14. To what aspects of annelid organization does the phylum owe its name?
    a. distinct external segmentation
    b. internal partitioning by means of septae
    c. the segmented coelom

d. the tubular gut
   e. the rings of circular muscles
15. Which of the following features of the earthworm does *not* exhibit segmentation?
    a. the coelom
    b. the posterior nervous system
    c. the forward part of the digestive tract
    d. the arrangement of circular muscles
    e. the excretory system
16. Which of the following is true regarding the organization of the nervous system of an earthworm?
    a. Nerve fibers run longitudinally inside a dorsal nerve cord.
    b. Movement in one segment does not depend on the activities of adjacent segments.
    c. Muscles are innervated by anterior ganglia.
    d. Cerebral ganglia generally inhibit or modulate the activities of individual segments.
17. Worms that parasitize their hosts by secreting a blood-coagulating chemical are classified as:
    a. polychaetes.
    b. onychophorans.
    c. spoon worms.
    d. hirudineans.
    e. lophophorates.
18. One general similarity between members of the class Oligochaeta and members of the class Polychaeta is:
    a. the use of locomotory appendages as respiratory organs.
    b. a closed circulatory system.
    c. specialization of mouthparts.
    d. hermaphroditic reproduction.
19. Which of the following does *not* support the speculation that onychophorans could be ancestors of the arthropods?
    a. Onychophorans have arthropodlike eyes and antennae.
    b. Onychophorans are protected by a cuticle that is molted periodically.
    c. Onychophorans have jointed appendages.
    d. Onychophorans give birth to live young.
20. Although the lophophorates are a diverse group, they all have:
    a. a number of deuterostome characteristics.
    b. an organ used in filter-feeding.
    c. a superficial resemblance to the bivalves.
    d. a terrestrial mode of existence.
    e. Both a and b are correct.
21. Match each structure or set of structures with the animal group in which it appears.

    radula _____   a. gastropods
    parapodia _____   b. cephalopods
    hatchet foot _____   c. bivalves
    ink sacs _____   d. oligochaetes
    gizzard _____   e. polychaetes
    sucking pharynx   f. leeches
    _____   g. burrowing marine worms
    clitellum _____
    anterior suckers   h. tardigrades
    _____   i. brachiopods
    four pairs of legs _____
    retractable proboscis
    _____
    antennae _____
    lophophore _____

22. Why is the development of a true coelom considered to be such an important event in animal evolution?
23. Distinguish between the protostomes and the deuterostomes, using three aspects of embryological development.
24. Explain how several important life forms suggest that the mollusks, annelids, and arthropods have a common evolutionary lineage.

PERFORMANCE ANALYSIS

1. Choice **d** is correct. Compare the cleavage patterns of protostomes and deuterostomes in Figure 25–1, and the embryonic fate of the blastopore and the development of the coelom in Figure 25–2.
2. A review of the external structure of the octopus should convince you that both a and d are false. Shells of mollusks *can* grow with the organism (c). In squid and cuttlefish the shell is internalized, serving as a stiffening support; therefore, **b** is correct.

3. Choice **b** is correct. The body plan of mollusks (such as a clam) involves three distinct zones: a head-foot, a visceral mass, and a mantle. Oligochaetes, onychophorans, and leeches are not mollusks and do not show a comparable degree of tagmosis.

4. As water passes through the mantle cavity of a mollusk, it aerates the gills and picks up wastes excreted from the digestive tract and gametes produced in the reproductive tract; therefore, **e** is correct.

5. Cephalopods are advanced among the mollusks in having a closed circulatory system, which means that choice **a** is correct. Although cephalopods have a three-chambered heart, blood is pumped to the gills by accessory hearts. Neither choice b nor d would be associated with a closed circulatory system.

6. Nephridia are tubular structures. One end picks up nitrogenous wastes from the coelom, and the other delivers them to the external environment. The correct answer is **b**.

7. The body plan of the hypothetical molluscan ancestor shown in Figure 25-3a is suggestive of the organization of polyplacophorans (chitons) (**d**) and monoplacophorans.

8. The bivalves (**c**), whose bodies are flattened between two halves of a shell, do not exhibit any marked degree of cephalization as do the other four classes listed here.

9. The most primitive reproductive strategy is that exploited by most bivalves, separate sexes with external fertilization (**a**).

10. The torsion found among shelled gastropods is a result of more rapid growth on one side of the body than on the other. This leads to an asymmetry of form (therefore, **e** is correct) and the positioning of the visceral mass directly over the head. The coiling of the gastropod shell is a separate developmental phenomenon.

11. The relatively high degree of activity of many gastropods is associated with an aggregation of nerve cells in the anterior end (**b**).

12. Choices c and e are false statements. Cephalopods have closed circulatory systems, but so do annelids; they have internal fertilization, but this is also found in a number of molluscan groups. It is the well-developed brain (**d**) and the high degree of intelligence that sets the cephalopods apart from other invertebrates.

13. Marine annelids and marine mollusks have in common an unsegmented larval form known as a trochophore (see Figure 25-11). Some segmentation is seen in *Nautilus* and *Neopilina*, but it is considered a later development and is not suggestive of an evolutionary relationship. There is no homology between the bristles and setae of annelids and the fleshy head-feet of mollusks, which contain sensory and motor organs. Choice **a** is therefore correct.

14. Annelid is Latin for "ringed" and refers to the distinct external segmentation (**a**) found among members of the phylum.

15. All these structures exhibit a marked degree of segmentation except for the forward digestive tract (**c**). There, specialization has occurred in the form of a pharynx, esophagus, crop, and gizzard (see Figure 25-13).

16. The nerve cords of the earthworm run ventrally. Muscle movement in a given segment is directed by ganglia within that segment, though it is triggered by movement in adjacent segments and may be inhibited or modulated by cerebral ganglia; therefore, **d** is correct.

17. Leeches get their class name, Hirudinea, from the blood-coagulating chemical hirudin; therefore, **d** is correct.

18. The only similarity between oligochaete and polychaete worms is a closed circulatory system (**b**). Polychaetes show the specializations referred to in a and c. Oligochaetes, but not polychaetes, are characterized by hermaphroditic reproduction (d).

19. Both a and b are true statements and imply an evolutionary link between onychophorans and arthropods. Were c true, it would also be supporting evidence. But the fact that some species of onychophorans bear live young (**d** is correct), while most arthropods lay eggs, throws a wrench into the proposed theory of an evolutionary link between the two groups.

20. The lophophorates, all of which are aquatic

(eliminating d), are protostomes, but they display some deuterostome characteristics (a). They also possess a specialized organ, the lophophore, which they use in filter-feeding (b). The only group that bears any resemblance to the bivalves (c) are the brachiopods, which live inside a shell of two halves. The halves, however, are dorsal and ventral, not right and left. Choice **e** is therefore correct.

21. The correct answers in order are **a**, **b**; **e**; **c**; **b**; **d**; **d**; **d**; **f**; **h**; **g**; **e**; and **i**.

22. Your discussion should include the role of the coelom in locomotion as well as the increase in internal work surface it provides.

23. See the analysis for Question 1.

24. A trochophore larva is considered to be the link between mollusks and annelids; the onychophorans are thought to be intermediate between annelids and arthropods. Your discussion should focus on those characteristics that support evolutionary links between the groups in question.

CHAPTER 26

The Animal Kingdom III: The Arthropods

CHAPTER ORGANIZATION

I. Introduction
II. Characteristics of the arthropods
   A. General information
   B. The exoskeleton
   C. Internal features
   D. The arthropod nervous system
III. Subdivisions of the phylum
   A. The chelicerates
      1. General information
      2. Class Arachnida
   B. The aquatic mandibulates: Class Crustacea
      1. General information
      2. The lobster
      3. Terrestrial crustaceans
   C. The terrestrial mandibulates: Myriapods
   D. The terrestrial mandibulates: Class Insecta
      1. General information
      2. Insect characteristics
      3. Digestive, excretory, and respiratory systems
      4. Insect life histories
IV. Reasons for arthropod success
   A. General information
   B. Arthropod senses and behavior
      1. Vision: The compound eye
      2. Touch receptors
      3. Proprioceptors
      4. Communication by sound
      5. Communication by pheromones
      6. Essay: Firefly light: An advertisement, a warning, a snare
      7. Programmed behavior
V. Summary

MAJOR CONCEPTS

In terms of number of species and number of organisms, the phylum Arthropoda is the largest animal phylum. An arthropod is characterized by a segmented body plan, a jointed exoskeleton, a variety of specialized appendages and sensory organs, an open circulatory system, and a nervous system consisting of a pair of ganglia in each segment connected by a double ventral nerve cord. Gas exchange in arthropods is accomplished by tracheae, book lungs, or book gills. The organs of excretion in terrestrial athropods are Malpighian tubules.

The three major groups of arthropods are the chelicerates, with chelicerae (fangs or pincers) and pedipalps; the aquatic mandibulates, with two pairs of antennae and a pair of mandibles; and the terrestrial mandibulates, with one pair of antennae and a pair of mandibles different from those of aquatic mandibulates.

Terrestrial mandibulates include the class Insecta, the largest single class in the animal kingdom. Ninety percent of the insect species undergo complete metamorphosis and pass through the following forms: egg (nonfeeding), larva (feeding), pupa (nonfeeding), and adult (feeding or nonfeeding).

The success of arthropods can be attributed to several factors, including their exoskeleton, their small size (in general), their specialization with respect to diet and habitat, their diverse sensory organs, and their fairly advanced nervous system (compared to lower animals).

The success of insects in particular may be attributed to complete metamorphosis and their capacity for flight (in most orders).

Among the diverse sensory receptors of arthropods are the compound eye, touch receptors, proprioceptors, and tympanic organs. Arthropod communication is accomplished by sound, by pheromones, and, in fireflies, by luminescence. Arthropod behavior, which is complex and diverse, is genetically programmed, not learned.

HOW TO STUDY THE CHAPTER

Read the chapter, focusing on the major concepts.

Reread the chapter, using the questions that follow to help you focus on details as they relate to major concepts. Answer the questions on a separate sheet of paper. Your answers will provide a valuable study aid in preparing for examinations.

FOCUS ON CHAPTER DETAILS

I. **Introduction** (page 527)
  1. Characterize phylum Arthropoda in terms of numbers of species and total numbers of individuals.

II. **Characteristics of the arthropods** (pages 527–531)
  A. General information
    2. For what characteristic was the phylum named?
    3. What evolutionary trend becomes apparent when the body organization of a highly evolved arthropod is compared to that of an annelid?
    4. What characteristics are used to distinguish the three principal groups of arthropods?
  B. The exoskeleton
    5. Describe the composition of the three layers of the arthropod cuticle.
    6. What functions does the cuticle serve?
    7. How is the exoskeleton capable of precise movements?
    8. Why is the exoskeleton periodically shed? What are the disadvantages of molting?
    9. What were the evolutionary consequences of the exoskeleton?
  C. Internal features
    10. What features of "lower" coelomate phyla are retained among the arthropods?
    11. What functions are served by tracheae, book lungs, and Malpighian tubules?
  D. The arthropod nervous system
    12. What degree of nervous organization and specialized function is seen at the segmental level? How are the nervous functions centrally connected and regulated?

III. **Subdivisions of the phylum** (pages 531–541)
  A. The chelicerates
    General information
    13. For what purposes are the first two pairs of chelicerate appendages suited? The remaining appendages?
    14. To what extent are tagmosis and external segmentation found within this group?
    15. Into what three classes are chelicerates divided? Name a representative organism of each class.
    Class Arachnida
    16. Name several specialized uses of the chelicerae and the pedipalps found within the arachnids.
    17. Describe how spiders obtain and absorb their food.
    18. How do spiders breathe?
    19. What are the functions of the most posterior appendages of spiders? To what special uses has silk been put?
  B. The aquatic mandibulates: Class Crustacea
    General information
    20. List several representative groups of terrestrial and aquatic crustaceans.
    21. What are the principal differences between crustaceans and insects?

22. What reproductive strategies are notable within the class Crustacea?

The lobster

23. Describe the general body plan of the lobster.
24. Describe the antennae and mandibles, and locate them in Figure 26-10. What function do the maxillae and maxillipeds serve?
25. How have the lobster's walking legs been modified?
26. Of what adaptive value is it for a lobster to autotomize its appendages?
27. What are the general designs of the digestive, respiratory, and excretory systems? (See also Figure 26-10.)
28. What reproductive strategies does the lobster employ?

Terrestrial crustaceans

29. Compare and contrast the respiratory strategies of amphibious and true land crabs with those of the land snails.

C. The terrestrial mandibulates: Myriapods

30. What characteristics define the myriapods as a group?
31. From what feature do the myriapods derive their name?
32. What are the principal differences between members of the class Chilopoda and members of the class Diplopoda?
33. What unusual feature is found among the pauropods and symphylans?

D. The terrestrial mandibulates: Class Insecta

General information

34. In terms of both numbers and species, the insects are the dominant terrestrial organisms. How did the evolution of flight provide for this dominance?
35. Describe the four major orders of insects and list representative organisms.

Insect characteristics

36. Study the insect body plan exemplified by the grasshopper in Figure 26-14. What specializations are most notable?
37. How do the mouthparts of grasshoppers compare to those of more highly specialized insects?
38. How many wings do most insects have? How does this characteristic vary among the rest of the insects?

Digestive, excretory, and respiratory systems

39. Describe the insect digestive tract.
40. What excretory and respiratory adaptations help insects to conserve water?

Insect life histories

41. Compare and contrast the immature life stages of insects with those found among marine invertebrates.
42. How has the capacity for flight influenced insect reproductive patterns?
43. Describe the pattern of metamorphosis displayed by the grasshopper.
44. Describe the developmental process in those insects that undergo complete metamorphosis.
45. Describe the interplay of the three different hormones that control development (specifically molting and metamorphosis) in an insect.

IV. **Reasons for arthropod success** (pages 542-548)

A. General information

46. How is the great success of the arthropods related to the nature of the exoskeleton, to body size, to life-history patterns, and to diet?
47. What feature of the terrestrial environment has contributed to the tremendous diversity (750,000 species) of the insects?
48. What features of the nervous system of insects contribute to their great success?

B. Arthropod senses and behavior

Vision: The compound eye

49. Describe the structure of the compound eye (see also Figure 26-20).

50. What organizational aspects of the compound eye account for the fact that it provides poor resolution but far exceeds the vertebrate eye in its ability to detect motion? How has the flicker-fusion test been able to demonstrate this?
51. What other type of photoreceptor do many arthropods utilize?

Touch receptors

52. How do arthropods perceive direct contact or vibrations in their environment?

Proprioceptors

53. How are arthropods able to obtain information about the position of their body parts?

Communication by sound

54. What five distinct types of calls do insects use? To what characteristics of the call do they respond?
55. How do the sensory receptors of arthropods detect sound?
56. Explain how an insect is able to translate a sound into a nerve impulse.
57. How do insects perceive pressure changes in sound waves? How are receptors arranged so that the direction of the sound can be identified?

Communication by pheromones

58. What is a pheromone?
59. What does the example of the gypsy moth demonstrate about the attraction of insects to pheromones?

Essay: Firefly light: An advertisement, a warning, a snare (page 544)

60. Describe the various ways in which the luminescent organs of fireflies are used to procure a mate, to procure prey (a special case), and to protect from predation.

Programmed behavior

61. How are the behavior patterns of arthropods characterized? How does this correlate with the life span of the average insect?

V. Summary (pages 548–549): Read the Summary. If you are familiar with the essential features of the material presented there, you are ready to take the following diagnostic examination.

TESTING YOUR UNDERSTANDING

After you have completed the following examination, compare your answers with the annotated key in the Analysis section.

1. Which of the following statements regarding arthropod diversity is true?
   a. Arthropods constitute the second largest phylum of animals.
   b. There are probably more than a million different arthropod species.
   c. Although insects are the most highly adapted arthropods, they make up less than a quarter of the total arthropod species.
   d. Arthropods are not well suited to aquatic environments.
2. Which of the following most clearly suggests that arthropods are descended from the annelids?
   a. specialization of segments for specific functions
   b. the high degree to which tagmosis has developed
   c. the external segmentation of larval forms
   d. jointed appendages
3. The first pair of appendages in mandibulates are _____, and the first pair in chelicerates are _____.
   a. mandibles; chelicerae
   b. antennae; chelicerae
   c. mandibles; pedipalps
   d. maxillae; antennae
   e. claws; fangs
4. One problem posed by the arthropod exoskeleton is that it:
   a. does not grow with the organism.
   b. does not offer protection from desiccation.
   c. does not allow for finely articulated movements.
   d. offers little protection from predators.
   e. Both b and d are correct.
5. Terrestrial arthropods transport waste materials out of their body cavities and into the hindgut by means of:

a. setae.
   b. tracheae.
   c. Malpighian tubules.
   d. nephridia.
   e. blood vessels.
6. You can get a male praying mantis to copulate with his female partner if you:
   a. stimulate certain mechanoreceptors on the abdomen.
   b. spray him with a pheromone.
   c. inject him with a stimulatory hormone.
   d. cut off his head.
7. In spiders, the chelicerae are modified for:
   a. injecting poison.
   b. chewing food.
   c. copulation.
   d. secreting digestive enzymes.
   e. spinning silk.
8. Which of the following arthropods have book gills?
   a. lobsters
   b. crayfish
   c. amphibious crabs
   d. horseshoe crabs
   e. barnacles
9. Arachnids typically have _____ pairs of appendages.
   a. four
   b. five
   c. six
   d. eight
   e. ten
10. The claws of a lobster:
    a. are of approximately equal size.
    b. can be autotomized, if damaged.
    c. cannot be regenerated.
    d. are modified pedipalps.
    e. are its mandibles.
11. Myriapods are arthropods that have:
    a. retained distinct segments.
    b. paired appendages on each segment.
    c. just a single pair of antennae.
    d. numerous jointed appendages.
    e. All of the above are correct.
12. Reproductive behavior in lobsters and crayfish involves:
    a. the laying of eggs under rocks.
    b. a clitellum.
    c. transfer of semen by the male pedipalps.
    d. separate sexes.
    e. All of the above are correct.
13. The great diversification of insects into previously unoccupied habitats was *primarily* a result of their:
    a. highly specialized feeding appendages.
    b. capacity for flight.
    c. metamorphic development.
    d. waterproof exoskeleton.
14. In the grasshopper both the _____ and _____ are attached at the thorax.
    a. antennae; the first two pairs of legs
    b. forewings; a single pair of legs
    c. mandibles; the first pair of legs
    d. wings; the three pairs of legs
    e. hindwings; the posterior mouthparts
15. Which of the following represents a method of controlling water loss that is associated with the respiratory system of an insect?
    a. formation of uric acid
    b. closing of the spiracles
    c. reabsorption of water from the Malpighian tubules
    d. reabsorption of dilute fluids from the tracheae
16. Which of the following is the correct developmental sequence in an insect that undergoes complete metamorphosis?
    a. egg, nymph, larva, pupa, imago
    b. egg, larva, nymph, adult
    c. egg, larva, pupa, imago
    d. egg, pupa, larva, adult
    e. egg, instar, larva, adult
17. The larval form of most insects neither _____ nor _____.
    a. moves; feeds
    b. molts; grows
    c. feeds; defends itself
    d. flies; reproduces
18. As an insect begins its pupal molt, _____ hormone(s) is (are) released and the secretion of _____ declines.
    a. brain; ecdysone
    b. ecdysone and brain; juvenile hormone
    c. juvenile; ecdysone
    d. molting; chitin
    e. juvenile and brain; ecdysone

19. Ommatidia are arthropod sensory receptors that respond to:
    a. direct contact or air currents.
    b. pressure changes.
    c. movements of various body parts.
    d. airborne chemicals.
    e. visual stimuli.

20. A dragonfly can detect the flight path of a fast-moving mosquito because its photoreceptors:
    a. are capable of high resolution.
    b. are movable.
    c. have separate visual fields.
    d. respond very rapidly.
    e. Both c and d are correct.

21. The sense organ used by an insect to obtain information about the positioning of its legs on a blade of grass is called a(n):
    a. tympanic organ.
    b. rhabdom.
    c. campaniform sensillum.
    d. ocellus.
    e. touch receptor.

22. An insect with a tympanic organ should be able to detect:
    a. low concentrations of a sex pheromone.
    b. a melodic mating call.
    c. differences in light intensity.
    d. how far apart its wings are.
    e. a piece of dirt stuck to its abdomen.

23. Observation of and experimentation with insect behavior have indicated that insects _____ complex behavior patterns.
    a. learn
    b. inherit
    c. do not exhibit
    d. place little emphasis on
    e. ignore

24. Match each structure with its function or description.

    tracheae _____
    book gills _____
    hemocoel _____
    wing veins _____
    sensilla _____
    chelicerae _____

    a. sensory spines or setae
    b. delivery of blood to the tissues
    c. supply oxygen to the tissues
    d. used for biting
    e. chitinous tubules for support
    f. collect and process food

25. Match the structures on the left with the arthropod group (or groups) in which they appear.

    mandibles _____
    two pairs of antennae _____
    three pairs of walking legs _____
    anterior excretory organs _____
    fused head and thorax _____
    one or two pairs of appendages per segment _____

    a. crustaceans
    b. insects
    c. arachnids
    d. myriapods

26. Provide a brief description of the arthropod exoskeleton, and comment on its advantages and disadvantages.

27. Construct a table showing several characteristics that would help differentiate a grasshopper, a spider, and a lobster.

28. Insects, according to number of species, make up more than 90 percent of the animal kingdom. Discuss several factors that have contributed to their overwhelming success.

PERFORMANCE ANALYSIS

1. Arthropods are the largest of the animal phyla, with over 850,000 identified species, 750,000 of which are insects. It is suspected that the total number of arthropod species may be as high as 10 million. Some arthropods (the crustaceans) are aquatic. Therefore, only choice **b** is true.

2. The suggestion that arthropods are descendants of the annelids is supported by the fact that arthropods, most notably the larval forms, show a high degree of external segmentation; therefore, **c** is correct.

3. Choice **b** is correct; this should be apparent to you

from the lobster in Figure 26-10 and the spider in Figure 26-8.

4. The arthropod exoskeleton is advantageous because it offers protection from desiccation and predators, and it allows for finely articulated movements (b, c, and d). However, because the exoskeleton does not grow with the organism (a is correct), it must be periodically shed. Arthropods are particularly vulnerable both to predators and to drying out during these molting periods.

5. Choice c is correct. See Figure 26-4b.

6. Decapitation (d) of the male praying mantis (done by the female prior to copulation) removes the inhibitory effect of the brain.

7. In spiders, the first pair of appendages (the chelicerae) are fangs used to inject a paralyzing poison into their prey (a).

8. All the animals listed here have true gills except the horseshoe crab (d), an aquatic chelicerate. (See Figure 26-6.)

9. As can be seen from the horseshoe crab of Figure 26-6, arachnids have six (c) pairs of appendages: the chelicerae, the pedipalps, and four pairs of walking legs.

10. The claws of a lobster are modified first walking legs, one of which is larger than the other and is used for defense. If damaged, they can be autotomized (dropped off) and regenerated following molting; therefore, b is correct.

11. A quick glance at the centipede, a myriapod, shown in Figure 26-12, should convince you that e is correct.

12. Reproduction in lobsters and crayfish involves external fertilization and separate sexes (d). The eggs, after they are extruded from the female gonopore, become attached to the swimmerets. Transfer of semen by the pedipalps of the male occurs in spiders, and the clitellum is characteristic of annelids.

13. All of the factors listed probably contributed in some degree to the enormous diversification of the insects, but certainly none had more impact on their ability to move into previously unoccupied areas than the capacity for flight (b).

14. Choice d is correct. See the body plan of the grasshopper in Figure 26-14.

15. Insects regulate water loss primarily by opening and closing the spiracles (b) of the respiratory tract.

16. Insects that undergo a complete metamorphosis develop from an egg into the larva (feeding stage), then enter a dormant stage, the pupa. In the pupal stage they form those structures identifiable with the adult, or imago; therefore, c is correct.

17. The larval form is primarily a feeding form that increases in size through molting. Most insect larvae cannot fly and do not reproduce (d).

18. Choice b is correct. Read the caption to Figure 26-18.

19. Ommatidia are the structural units of the arthropod eye (see Figure 26-20) and respond to visual stimuli (e).

20. The ommatidia in the eye of a dragonfly are not particularly high-resolution structures. The dragonfly is able to detect fast-moving objects because each ommatidium has a separate visual field and responds very rapidly to stimuli; therefore, e is correct.

21. A sensory receptor that provides information about the position of various parts of the body is called a proprioceptor. An example is the campaniform sensillum (c), shown in Figure 26-22a.

22. The tympanic organ (see Figure 26-24) responds to pressure changes in sound waves, which would help an insect hear a mating call (b).

23. The complex behavior patterns found among insects are genetically programmed; therefore, b is correct.

24. The correct answers in order are c, c, b, e, a, d, and f.

25. The correct answers in order are a, b, d; a; b; a; a, c; and d.

26. The arthropod exoskeleton or cuticle is secreted by the underlying epidermis. Its outer layer is composed of lipoprotein and is often waxy. The inner and middle layers are chitinous. The advantages and disadvantages of the exoskeleton are summarized in the analysis for Question 4.

27. A number of characteristics used to differentiate insects, spiders, and crustaceans from one another include the number of pairs of antennae, the number of pairs of appendages, the types of mouthparts, external segmentation, respiratory structures, and any notable specializations such as poison, silk, wings, jumping or swimming legs, and so on. The figures in the text should help you construct the table along these lines.

28. Most of the factors that have contributed to the success of the insects are mentioned in Question 13 and its analysis. Additional factors include the degree of refinement of the nervous system and the wide variety of sensory receptors, most notably the compound eye.

CHAPTER 27

The Animal Kingdom IV: The Deuterostomes

CHAPTER ORGANIZATION

I. Introduction
II. Phylum Echinodermata: The "spiny-skinned" animals
   A. Introduction
   B. Class Asteroidea: Starfish
   C. Other echinoderms
III. Phylum Chaetognatha: Arrow worms
IV. Phylum Hemichordata: Acorn worms
V. Phylum Chordata: The cephalochordates and urochordates
VI. Phylum Chordata: The vertebrates
   A. Introduction
   B. Classes Agnatha, Chondrichthyes, and Osteichthyes: Fish
      1. General information
      2. The transition to land
   C. Class Amphibia
   D. Class Reptilia
      1. Introduction
      2. Evolution of the reptiles
   E. Class Aves: Birds
      1. General information
      2. Evolution of flight
   F. Class Mammalia
VII. Summary

MAJOR CONCEPTS

Animals in the following four phyla exhibit deuterostome development: Echinodermata (starfish, sea urchins, and sea lilies), Chaetognatha (arrow worms), Hemichordata (acorn worms), and Chordata (lancelets, tunicates, and vertebrates). Echinoderms are radially symmetrical and are characterized by a water vascular system. The arrow worms have three distinct body regions. The acorn worms show both echinoderm and chordate characteristics.

All chordates possess four specific characteristics at some time during their development: a notochord, a dorsal nerve cord, a pharynx with gill slits, and a tail. In addition to the four chordate characteristics, vertebrates (subphylum Vertebrata) have a vertebral column enclosing the nerve cord and a cranium enclosing the brain.

When vertebrates moved from an aquatic to a terrestrial existence, three major developments were lungs, a strong skeletal support system, and the amniote egg.

HOW TO STUDY THE CHAPTER

Read the chapter, focusing on the major concepts.

Reread the chapter, using the questions that follow to help you focus on details as they relate to major concepts. Answer the questions on a separate sheet of paper. Your answers will provide a valuable study aid in preparing for examinations.

FOCUS ON CHAPTER DETAILS

   I. **Introduction** (page 550)
      1. List the features of embryonic development that are unique to deuterostomes.
      2. Into what four phyla are deuterostomes classified?

II. Phylum Echinodermata: The "spiny-skinned" animals (pages 550–553)

A. Introduction

3. List representatives of each class of echinoderms.
4. What do the general body shapes of echinoderm larvae and adults suggest about their ancestry? How is mobility within this group related to symmetry?

B. Class Asteroidea: Starfish

5. Describe the general organization of a starfish.
6. What type of internal skeleton is found among the starfish?
7. Describe the arrangement of nerves and sensory receptors in the starfish.
8. What reproductive strategies are employed in this group?
9. How are circulation, respiration, and waste removal accomplished in starfish?
10. Describe the action of the water vascular system in starfish locomotion. How does this system assist in the predatory behavior of the starfish?

C. Other echinoderms

11. How are some echinoderms well adapted to a sessile life style?
12. How are tube feet used in brittle stars, sea urchins, and sea cucumbers?
13. What traces of radial symmetry are apparent in sand dollars and sea cucumbers?
14. Describe the feeding habits of sea cucumbers.

III. Phylum Chaetognatha: Arrow worms (page 553)

15. Describe the ecologically important feeding habits of arrow worms.
16. What are the distinguishing characteristics of the arrow worms?

IV. Phylum Hemichordata: Acorn worms (pages 553–554)

17. Describe the general body organization of the acorn worms.
18. Evolutionarily speaking, in what way are the hemichordates intermediate between echinoderms and chordates? What decidedly chordate features do the acorn worms exhibit?

V. Phylum Chordata: The cephalochordates and urochordates (pages 554–555)

19. What are the three groups of chordates?
20. What are the four general chordate features exemplified by *Branchiostoma*? (See Figure 27-8.) What functions does each of these structures perform in the cephalochordates?
21. Why are tunicates (urochordates) rather than *Branchiostoma* thought to be the ancestral chordates?

VI. Phylum Chordata: The vertebrates (pages 555–564)

A. Introduction

22. Describe the principal feature of the vertebrate animal.
23. What major advantage does an endoskeleton have over an exoskeleton?

B. Classes Agnatha, Chondrichthyes, and Osteichthyes: Fish

General information

24. One of the two major features of the Agnatha relates to predatory behavior, the other shows that they are degenerate forms. Explain.
25. Give some examples of chondrichthyans. What are their identifying features?
26. How do the life histories of salmon and eels parallel the evolutionary pattern followed by the bony fish?

The transition to land

27. Under what primitive conditions might freshwater fish have evolved an accessory breathing structure, somewhat like a lung?
28. What three types of bony fish did these lunged fish apparently give rise to? How were the lunglike structures (or their usage) modified in each group?

29. Explain how one of these three forms might have been able to explore a land environment.

C. Class Amphibia
30. From what group did amphibians evolve?
31. What feature of amphibians makes them particularly vulnerable in a terrestrial habitat?
32. What different developmental paths have been taken by the amphibians?

D. Class Reptilia

Introduction

33. What evolutionary advance appeared first in the reptiles and completely freed the vertebrates from water?
34. Make a simple sketch of the amniote egg (see Figure 27-16). How does this structure provide for the needs of the embryo?
35. How do the embryonic stages of reptiles and mammals reflect their aquatic ancestry?
36. What other terrestrial adaptations do reptiles exhibit?
37. List some representatives of the reptiles.

Evolution of the reptiles

38. Trace the evolutionary history of reptiles from the late Carboniferous period through the Permian to the Triassic period.
39. What evolutionary trend was apparent in the early Archosauria?
40. What assumption about dinosaurs as a group is now a matter of controversy?
41. What explanations have been proposed to explain the sudden disappearance of dinosaurs?

E. Class Aves: Birds

General information

42. What secondary adaptations for an existence "on the wing" are found throughout this class?
43. How do birds regulate their body temperatures?

44. In what other ways do birds differ from their reptilian ancestors?

Evolution of flight

45. What popular theory of the origin of flight is now being rethought?
46. What is John Ostrom's hypothesis about the evolution of feathers?
47. For what other purpose might dinosaurs have used feathers? What support has *Archaeopteryx* provided for Ostrum's hypothesis?

F. Class Mammalia
48. What are the three distinguishing characteristics of mammals?
49. Mammals have been split into three groups with respect to development of the embryo and degree of maternal care. Name each group, list its distinguishing characteristics, and name a few representative species.
50. What were the habits of the early placental mammals?
51. What unique skeletal features are exhibited by the mammals?
52. Familiarize yourself with the twelve orders (and representative species) of placental mammals shown in Figure 27-22.
53. What are the distinguishing characteristics of the primate order?
54. Why are human primates considered to be the least specialized of all mammals? What single feature sets our species apart from all others?

VII. **Summary** (page 565): Read the Summary. If you are familiar with the essential features of the material presented there, you are ready to take the following diagnostic examination.

TESTING YOUR UNDERSTANDING

After you have completed the following examination, compare your answers with the annotated key in the Analysis section.

1. The radial symmetry of a starfish reflects a:
   a. probable coelenterate ancestry.

b. radially symmetrical larval form.
   c. probable sessile echinoderm ancestor.
   d. set of adaptations to an active predatory existence.
   e. Both b and d are correct.

2. Which of the following is characteristic of *all* echinoderms?
   a. a downward-directed mouth
   b. reduced importance of the coelom
   c. well-defined radial symmetry
   d. a spiny internal skeleton
   e. a predatory mode of existence

3. The tube feet of echinoderms are specialized:
   a. for locomotion.
   b. for food gathering.
   c. as gills.
   d. for protection from predators.
   e. Both a and b are correct.

4. Which of the following echinoderms exemplify the trend toward bilateral symmetry?
   a. feather stars
   b. sea urchins
   c. sea cucumbers
   d. sand dollars
   e. All of the above are correct.

5. Acorn worms are thought to be close relatives of animals that have a dorsal hollow nerve cord primarily because:
   a. one of their two nerve cords runs dorsally.
   b. they have a notochord.
   c. their pharynx is perforated with gill slits.
   d. they are divided into three distinct body regions.

6. Cartilage represents that part of the endoskeleton of vertebrates that:
   a. is still growing.
   b. is posterior to the anus.
   c. forms the outer protective casing of the brain.
   d. is hollow.
   e. surrounds the central nerve cord.

7. Which of the following is the most likely candidate for the ancestral form of the chordates?
   a. lancelets
   b. tunicates
   c. acorn worms
   d. arrow worms
   e. echinoderms

8. Predatory fish lacking jaws and a bony endoskeleton are classified as:
   a. Agnatha.
   b. Chondrichthyes.
   c. Hemichordata.
   d. Osteichthyes.
   e. invertebrates.

9. The first lunglike structures appeared among the vertebrates in:
   a. primitive fish as a swim bladder.
   b. freshwater fish as accessory respiratory structures.
   c. the cartilagenous fish.
   d. forms similar to modern lungfish.
   e. the first amphibians.

10. Amphibians are incompletely adapted to a terrestrial existence because:
    a. the amphibians' water-permeable skin limits their activity.
    b. in general, they must return to the water in order to reproduce.
    c. the free-living aquatic larval stage appears to be a necessary part of the life cycle.
    d. Both a and b are correct.

11. The real break between vertebrates and a watery existence came with the development of _____ by _____.
    a. the lung; lungfish
    b. a dry, scaly skin; reptiles and some amphibians
    c. the amniote egg; reptiles
    d. wings; birds
    e. a placenta; mammals

12. The explosive increase in the number of reptile species during the Permian period is associated with:
    a. a generally drier climate.
    b. the evolution of endothermy.
    c. lush tropical forests that provided an abundant food supply.
    d. the rise of the dinosaurs.
    e. reduced competition from mammals.

13. The closest reptilian relative of modern birds was a:
    a. pterosaur.
    b. thecodont.
    c. lizard.
    d. dinosaur.
    e. tree-dwelling reptile.

14. *Archaeopteryx* is similar to modern birds only in that it:
    a. had a breastbone to support flight muscles.
    b. developed wings from the bones originally used in claws.
    c. had an insulating layer of feathers.
    d. spent a large part of its time in trees.

15. There is a good deal of controversy occasioned by the proposal that _____ (formerly assumed to be exothermic) might have been endothermic.
    a. the earliest mammals
    b. the thecodonts
    c. some dinosaurs
    d. some primitive birds
    e. some amphibians

16. The first placental mammal might well have resembled a(n):
    a. shrew.
    b. opossum.
    c. theropod.
    d. primate.
    e. dolphin.

17. One characteristic of *all* mammals is that they:
    a. bear live young.
    b. nurse their young.
    c. are endothermic.
    d. Only b and c are correct.
    e. All three answers (a, b, and c) are correct.

18. Humans may be distinguished anatomically from other primates by:
    a. their lack of canine teeth.
    b. low foreheads.
    c. their single pair of mammae.
    d. many highly specialized structures.
    e. their upright posture.

19. In comparison to other animals, humans can be considered extremely:
    a. agile.
    b. fast of foot.
    c. sensitive to external stimuli.
    d. specialized.
    e. intelligent.

20. Which of the following is true of the backbone of vertebrates?
    a. It provides support but remains flexible.
    b. It is not entirely bony.
    c. It encircles the nerve cord.
    d. It exhibits independent, localized movement.
    e. All of the above are correct.

21. Match each description with an appropriate group of animals.

able to autotomize body parts _____	a. acorn worms
have a protective cellulose-containing covering _____	b. tunicates
	c. arrow worms
	d. starfish
employ a water vascular system in locomotion _____	e. brittle stars
possess both dorsal and ventral nerve cords _____	
hermaphroditic planktonivores _____	

22. Match each vertebrate characteristic with the appropriate class.

denticled skin _____	a. birds
moist, water-permeable skin _____	b. mammals
dry, scaly skin _____	c. reptiles
hollow bones _____	d. amphibians
large skull bones _____	e. Chondrichthyes
aquatic larvae _____	f. Agnatha

23. Why are the echinoderms believed to be the invertebrate phylum most closely related to the chordates?

24. What four major characteristics do the cephalochordates and urochordates have in common with the vertebrates? Describe the skeletal features of the vertebrates that distinguish them from all other animals.

25. Explain why certain groups of bony fish are believed to have been the first vertebrates to make the transition onto land.

26. List those features characteristic of the primates. What features set humans apart from the other members of the order?

PERFORMANCE ANALYSIS

1. The bilaterally symmetrical larva of the starfish attests to a bilaterally symmetrical ancestor, which, it is hypothesized, became radially symmetrical as a result of a sessile existence (**c** is correct). Starfish, *despite* their radial symmetry, are active predators and may be evolving toward bilateral symmetry.

2. Echinoderms are named for their spiny internal skeletons (although this is greatly reduced in the sea cucumbers). Not all echinoderms are radially symmetrical (the sea cucumber is the exception), nor are all predatory (sea cucumbers ingest plankton, organic matter in the bottom deposits, or simply the bottom sediment itself). Unlike the starfish and brittle stars, sessile forms do not usually have a downward-directed mouth. Choice **d** therefore is the only characteristic found among all members of the phylum.

3. Tube feet are part of the water vascular system of echinoderms and provide their means of locomotion (**f**). Tube feet are also employed by starfish to open the shells of their bivalve prey and by brittle stars exclusively for the purpose of gathering and handling food (**b**). Choice **e** is therefore correct.

4. You only need to compare the representative species in Figures 27–1, 27–5a, 27–5b, and 27–5c to see that choice **c** alone is correct.

5. The strongest evidence of a relationship between acorn worms (the hemichordates) and animals having a dorsal hollow nerve cord (the chordates) lies in the perforation of the pharynx by gill slits (**c**).

6. Cartilage represents bone that is still growing (**a**) and has not yet been calcified.

7. Among the choices listed, both the tunicates and the lancelets have all four of the chordate characteristics (see analysis for Question 24). However, the lancelets are believed by many biologists to be degenerate forms of fish rather than ancestors of modern chordates. Therefore, choice **b** (tunicates) is correct.

8. Jawless cartilaginous fish, such as the lamprey and the hagfish, are members of the vertebrate class Agnatha (**a**).

9. Freshwater primitive fish are believed to have evolved lunglike respiratory structures, secondary to the gills, that would have been useful in surfacing for air if the water became stagnant. These accessory structures are believed to be the forerunners of the vertebrate lung; therefore, **b** is correct.

10. Amphibians are considered to be incompletely adapted to life on land for the reasons given in a and b. Choice **d** is therefore correct.

11. The reptiles were the first vertebrates to become completely free of an aquatic existence because of the development of the amniote egg (**c**), which provides the growing embryo with a watery environment.

12. The generally drier climate (**a**) of the Permian period is thought to have favored the reptiles, which were well adapted to land. Dinosaurs and mammals did not appear in any abundance until the Triassic period.

13. According to fossil evidence, the closest reptilian relatives of modern birds were small carnivorous descendants of the dinosaurs (**d**) known as therapods.

14. *Archaeopteryx* had feathers (**c**), but it was not a good flyer and may not have flown at all. Instead, it lived on the ground, and its wing feathers, which it used primarily for insulation, were embedded in its skin rather than attached to bones of the forelimb.

15. The proposed use of feathers by *Archaeopteryx* for insulation has led to the supposition that endothermy may have first evolved among the dinosaurs (**c**).

16. The earliest placental mammals were probably small and nocturnal (characteristics that afforded protection from dinosaur predators) and might have resembled modern shrews (**a**).

17. Those characteristics listed apply to all mammals except the monotremes, which lay eggs instead of bearing live young; therefore, **d** is correct.

18. Humans have canine teeth, a high forehead, a single pair of mammary glands (as do all of the primates), and are generally unspecialized. It is our upright posture (**e**) that sets us apart, in an anatomical sense, from primate relatives.

19. Choice **e** is correct. That answers a, b, and c are

incorrect is probably obvious. The versatility of humans is indicative of generalized, not specialized (d), capabilities.

20. A backbone is a flexible support surrounding the nerve cord of vertebrates. It contains cartilaginous disks and muscles associated with its vertebrae that allow some sections to be moved independently of others. Choice **e** is therefore correct.

21. The correct answers in order are **e, b, d, a,** and **c**.

22. The correct answers in order are **e, d, c, a, b,** and **d**.

23. Among the invertebrates, echinoderms are the only deuterostomes. Despite the fact that they are, for the most part, radially symmetrical, they exhibit a number of trends toward the bilateral symmetry found in all chordates.

24. Cephalochordates and urochordates are like vertebrates in that they have a notochord, a dorsal, hollow nerve cord, pharyngeal gill slits, and a tail. Vertebrates may be distinguished from their chordate relatives by the presence of a vertebral column, which encircles the nerve cord, has cartilaginous disks, is associated with muscles, and ends anteriorly in a bony protective enclosure for the brain.

25. A group of bony fish that had lunglike accessory respiratory structures and skeletal supports for the thorax (permitting them to gulp air on land) and that were preadapted in some way for moving on the land were most probably the first vertebrates to make the transition from an aquatic to a terrestrial environment.

26. The primates can be distinguished on the basis of dentition pattern, orientation of the digits of the hand, the number of pectoral mammae, the positioning of the eyes in the head, and features of the brain associated with intelligence. Humans may be set apart from the rest of the primates because they walk upright, have long arms, a high forehead, and little body hair, and are generally unspecialized, omnivorous, and intelligent.

SECTION 5 Biology of Plants

CHAPTER 28

The Plant: An Introduction

CHAPTER ORGANIZATION

I. Introduction
II. The plant body
III. Leaves
   A. Leaf structure
   B. Leaf adaptations and modifications
      1. General information
      2. Essay: Carnivorous plants
IV. The stem
   A. The structure of the stem
      1. Introductory remarks
      2. Ground tissue
      3. Vascular tissues
      4. Stem patterns
   B. Special adaptations of stems
V. Roots
   A. General information
   B. The structure of the root
      1. Introductory remarks
      2. The epidermis
      3. The cortex
      4. The vascular cylinder
   C. Patterns of root growth
   D. Special adaptations of roots
VI. Adaptations to climate change
   A. Introductory remarks
   B. Annuals, biennials, and perennials
VII. Summary

MAJOR CONCEPTS

Angiosperms (flowering plants) are divided into monocots and dicots based on patterns of leaf venation, the arrangement of vascular tissues in the stem and root, the number of cotyledons (seed leaves), and the patterns of root growth.

The plant body has three major regions: roots (for anchoring the plant and for absorption of water and minerals), stems (for structural support and conduction of water and nutrients), and leaves (for photosynthesis). The three tissue systems (dermal, vascular, and ground) are continuous throughout the plant body.

Leaves, stems, and roots have an outer epidermal layer covered by a waxy cuticle, which is very thin in roots. The ground tissue of all three regions is made up largely of parenchyma cells. The two types of vascular (conducting) tissues, which are embedded in the ground tissue, are xylem (for water and dissolved minerals) and phloem (for food molecules). In the stems of angiosperms, the conducting cells of phloem are sieve-tube members, and those of xylem consist of a series of tracheids and vessels.

Dormancy is the chief adaptation of plants to drought. Plants are classified as annuals, biennials, or perennials based on patterns of growth, dormancy, and death.

HOW TO STUDY THE CHAPTER

Read the chapter, focusing on the major concepts.

Reread the chapter, using the questions that follow to help you focus on details as they relate to major concepts. Answer the questions on a separate sheet of paper. Your answers will provide a valuable study aid in preparing for examinations.

FOCUS ON CHAPTER DETAILS

I. Introduction (page 573)
 1. Describe the invasion of land by plants.
 2. To what extent do other organisms depend on photosynthesis?
 3. Describe the differences between monocots and dicots. (Refer to Figure 28-2.)

II. The plant body (pages 573–575)
 4. What are the basic requirements of any plant?
 5. What problems does multicellularity pose for a plant in terms of providing for the requirements of individual cells?
 6. Explain how the basic body plan of a plant (such as the geranium in Figure 28-3) evolved in response to selection pressures.
 7. What three tissue systems are continuous throughout the plant body?

III. Leaves (pages 576–579)
 A. Leaf structure
 8. What conflicting selection pressures played significant roles in determining leaf structure?
 9. Describe the two types of cells that make up the mesophyll.
 10. Describe two important properties of the epidermis.
 11. Describe the two types of tissues that make up the vascular bundles. What materials move into and out of leaves via the vascular bundles?
 12. Describe the structure of a stoma and its function.
 B. Leaf adaptations and modifications
 General information
 13. Give some specific examples of leaf adaptations to wet and dry climates. What consequences do these particular adaptations have for photosynthesis and water conservation?
 14. Cite several examples of leaves that have been modified for food storage.

Essay: Carnivorous plants (page 578)
 15. How do carnivorous plants differ from carnivorous animals? Where are carnivorous plants usually found?
 16. How are the leaves of carnivorous plants modified for attracting, ensnaring, and absorbing unwary prey?

IV. The stem (pages 579–584)
 A. The structure of the stem
 Introductory remarks
 17. What functions are carried out by (or in) the stems of green plants?
 Ground tissue
 18. In addition to the parenchyma cells, what other cells are found in the ground tissue of the stem? Describe their physical appearance, their location in the stem, and their function.
 19. How do sclerenchyma cells differ from collenchyma cells?
 Vascular tissues
 20. What makes up the vascular tissues of stems?
 21. What is the function of phloem tissue? Describe the conducting cells of the phloem in angiosperms.
 22. What substances are found within the sieve-tube members?
 23. Identify the companion cells in Figure 28-9. What functions do these cells perform for the sieve-tube members with which they are associated?
 24. What is the function of the xylem tissue? Describe the conducting cells of the xylem in angiosperms; in lower vascular plants and gymnosperms.
 Stem patterns
 25. Describe the arrangement of vascular tissue in the stems of monocots and young dicots.
 B. Special adaptations of stems
 26. Describe six types of stem modifications, citing examples where possible.

V. **Roots** (pages 584-587)
   A. General information
      27. What general functions are carried out by roots?
      28. What factors influence root growth?
   B. The structure of the root
      Introductory remarks
      29. Describe the arrangement of the different tissue systems in an angiosperm root (see also Figure 28-13).

      The epidermis
      30. What functions are associated with the root epidermis?
      31. Describe the areas of the epidermis that are specialized for absorption.

      The cortex
      32. Describe the cells that make up the cortex. What do these cells contain (see also Figure 28-13), and how are they arranged?
      33. Describe the structure and arrangement of the endodermis and Casparian strip. (Refer to Figure 28-15.) How does this arrangement regulate the movement of substances into the vascular cylinder of the root?

      The vascular cylinder
      34. Compare the arrangement of xylem and phloem in dicots (Figures 28-13 and 28-16) and monocots (Figure 28-17).
      35. Describe the location of the pericycle. What structures arise from this region?
      36. What structural changes occur in the transition region between the root and the stem?
   C. Patterns of root growth
      37. Compare the growth patterns of monocot and dicot roots.
   D. Special adaptations of roots
      38. Cite several examples of aerial-root adaptations.
      39. What type of tissue is present in large taproots, such as those of beet and carrot plants?

VI. **Adaptations to climate change** (pages 587-589)
   A. Introductory remarks
      40. What principal adaptation enables angiosperms to survive in cold climates?
   B. Annuals, biennials, and perennials
      41. Describe the life cycle of a typical annual plant. What is an herbaceous plant?
      42. Describe the life cycle of a beet or carrot plant.
      43. What is a perennial plant? What benefits accrue to plants that are woody perennials?
      44. Describe the type and value of the adaptations made by herbaceous and woody perennials to unfavorable seasonal climates.

VII. **Summary** (pages 589-590): Read the Summary. If you are familiar with the essential features of the material presented there, you are ready to take the following diagnostic examination.

TESTING YOUR UNDERSTANDING

After you have completed the following examination, compare your answers with the annotated key in the Analysis section.

1. Carnivorous plants are not carnivores in the same sense as animals because they do *not*:
   a. make use of digestive enzymes.
   b. bear specialized structures for capturing their prey.
   c. extract energy from their food.
   d. require animal flesh in their diet but use it only as a supplement.
2. The water-conducting tissues of a pine tree consist of cells called:
   a. tracheids.
   b. vessel members.
   c. sclereids.
   d. sieve-tube members.
   e. spongy parenchyma.
3. The primary function of root hairs is to:

a. lend support to the growing roots.
   b. serve as sensory receptors.
   c. store carbohydrates transported from leaves.
   d. provide a large surface area for mineral absorption.
   e. All of the above are correct.
4. Most of the photosynthetic activity of a typical angiosperm takes place in the _____ cells of the leaf.
   a. spongy parenchyma
   b. palisade parenchyma
   c. epidermal
   d. collenchyma
   e. guard
5. A Casparian strip is associated with cells of the _____ and helps to regulate passage of materials into the _____.
   a. phloem; root parenchyma
   b. xylem; stems and leaves
   c. root endodermis; vascular tissues
   d. leaf epidermis; mesophyll
6. Which of these functions is carried out in cortex cells of the root?
   a. production of energy-storage forms
   b. regulation of passage of dissolved substances
   c. photosynthesis
   d. respiration
   e. Both a and d are correct.
7. The components of a sieve tube are actually cells that:
   a. have pores in their end walls.
   b. are alive at maturity.
   c. depend on companion cells to regulate their activities.
   d. transport sap.
   e. All of the above are correct.
8. Rhizomes and tubers are examples of:
   a. roots modified for functions other than water and mineral uptake.
   b. adventitious roots.
   c. modified underground stems.
   d. specialized photosynthetic tissues.
   e. Both a and b are correct.
9. A type of cell that has lignin-impregnated secondary walls and is dead at maturity is a:
   a. sieve-tube member.
   b. collenchyma cell.
   c. sclerenchyma cell.
   d. companion cell.
   e. guard cell.
10. Which of the following is typically found in plants for which water conservation during the growing season is a significant problem?
    a. deciduous leaves
    b. spines
    c. adventitious roots
    d. parallel leaf venation
    e. thick and fleshy petioles
11. Structurally speaking, root hairs are:
    a. extensions of epidermal cells.
    b. secretions of epidermal cells.
    c. tubular parenchyma overlying epidermal cells.
    d. lignin-impregnated epidermal cells.
    e. parts of mycorrhizal associations.
12. Structurally speaking, stomata are:
    a. gas-filled spaces of the leaf mesophyll.
    b. openings in the leaf epidermis.
    c. crescent-shaped epidermal cells.
    d. parts of the epidermis that lack a cuticle.
    e. irregularities in root surfaces through which oxygen is absorbed.
13. Vascular bundles are arranged in a ring surrounding the ground tissue of the stem in:
    a. dicots.
    b. grasses.
    c. corn plants.
    d. monocots.
14. Biennial plants typically _____ during the first year and _____ during the second year.
    a. grow to full height; enter dormancy
    b. are dormant; produce a rosette
    c. store food reserves; flower and form seeds
    d. produce seeds; disperse the seeds
15. A plant that continues to increase in girth year after year would be called a(n):
    a. herbaceous annual.
    b. woody annual.
    c. biennial.
    d. woody perennial.
    e. herbaceous perennial.
16. A woody perennial established in a climate of cool summers and severely cold winters would most likely have:

a. needlelike leaves.
b. broad, deciduous leaves.
c. featherlike leaves.
d. spines.

17. Match each cell or tissue type with its description.

    parenchyma _____
    pericycle _____
    sclerenchyma _____
    collenchyma _____
    pith _____
    companion cells _____

    a. living cells whose primary walls are thickened at the corners
    b. many-sided, thin-walled cells
    c. make up the fibrous tissues of the stem
    d. the central ground tissue of stems and roots
    e. cell layer giving rise to branch roots
    f. secrete substances into phloem cells

18. Match each structure with its description.

    adventitious _____
    sieve plates _____
    cutin _____
    petioles _____

    a. secreted by epidermal cells of leaves
    b. roots that grow from stems or leaves
    c. connect leaf blades to stems
    d. perforated end walls of phloem cells

19. Compare the organization of dermal, ground, and vascular tissues in the leaves, stems, and roots of a young dicot.
20. Explain the differences between the organization of xylem cells and phloem cells in a nonwoody dicot stem.
21. Provide a general comparison of the leaves, stems, and roots of monocots and dicots.
22. Distinguish between annual plants and perennial plants. What are the differences between herbaceous and woody plants? What advantages accrue to plants that are woody perennials?
23. Label the lettered parts of the following diagram of the interior of a leaf.

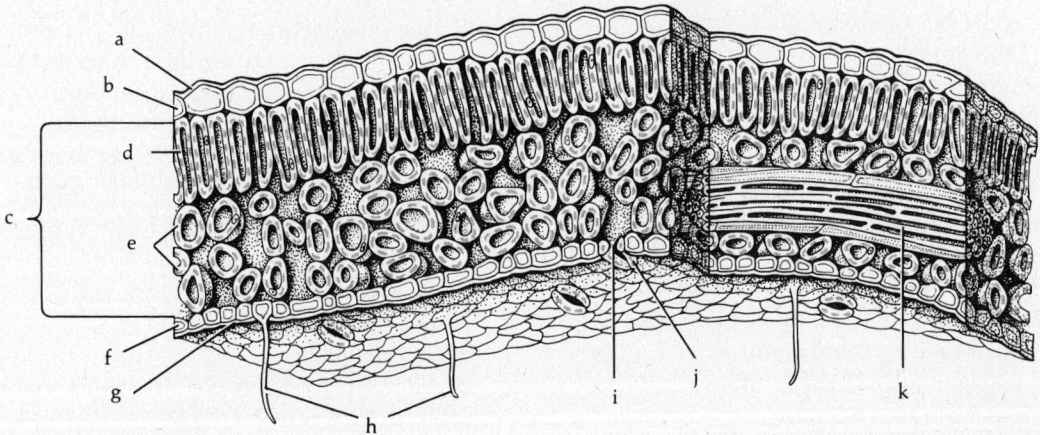

24. The following illustration is a diagrammatic cross section of a root.

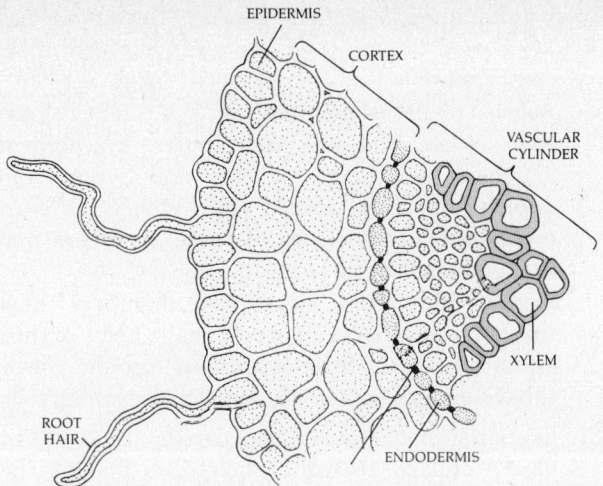

a. Trace the pathway through which most of the solutes and some of the water enter the root.
b. Trace the pathway through which most of the water and some of the solutes enter the root.
c. What structure is a barrier to substances entering the root by the second pathway? How is this barrier circumvented?

PERFORMANCE ANALYSIS

1. Carnivorous plants have a number of specialized structures for capturing the prey from which they derive mineral nutrients. Unlike carnivores, however, they do not derive any caloric value (energy) from their prey (**c**).

2. Tracheids (**a**) are the cells of gymnosperms that make up the long vessels through which water is conducted upward in the stem.

3. Root hairs serve to increase greatly the absorptive surface of the root epidermis; therefore, **d** is correct.

4. Photosynthesis in an angiosperm takes place in both spongy and palisade parenchyma. However, since the palisade parenchyma cells (**b**) are more numerous and are positioned just below the upper-leaf epidermis, they are the primary sites of photosynthesis in the leaf.

5. The Casparian strip (see Figure 28–15) is a waxy band found within the cell walls of the root endodermis. In order for dissolved solutes to enter the vascular bundles, they must first pass through the cell membranes of the endodermal cells, where their passage is regulated. Choice **c** is therefore correct.

6. Cortical cells of the root store starches, as implied in a; since these cells are metabolically active, they engage in respiration (d); therefore, **e** is correct. Dissolved substances are able to pass around and through the cortical cells freely. Photosynthesis does not occur in the root.

7. The sieve-tube members that make up the phloem are living cells that transport sap through pores in their end walls. This transport function is regulated by adjacent companion cells. Therefore, **e** is correct.

8. Rhizomes are underground stems adapted to produce buds; tubers, also underground stems, are adapted for food storage. Choice **c** is therefore correct.

9. Sclerenchyma cells (**c**), which provide support in the stem, often become impregnated with lignin. Their cytoplasm degenerates, so that they are actually dead cells at maturity.

10. Plants for which water loss is a major problem during the growing season exhibit a greatly reduced leaf surface area, since it is through the stomata and cuticle of leaves that transpiration takes place. Spines (**b**) are such an adaptation.

11. Choice **a** is correct. Refer to Figure 28–15.

12. Stomata are the epidermal openings (**b**) found primarily on the undersides of leaves. Their size is controlled by pairs of cells in the leaf epidermis called guard cells (see Figure 28–4).

13. Choice **a** is correct (see the cross section of an alfalfa stem in Figure 28–12a). This same arrangement is seen in all young dicot stems. Corn and grasses are monocots, which are characterized by vascular bundles that are distributed evenly throughout the ground tissue.

14. Biennial plants, such as the beet and carrot, appear during the first year as a rosette of leaves near the ground. The large taproots of these plants store starch during the first year of growth; then, in the second year, stems and flowers shoot up. Choice **c** is therefore correct.

15. A perennial is a plant that lives from one year to the next. Perennials that increase in size do so largely as a result of the accumulation of woody tissue; therefore, **d** is correct.

16. Needlelike leaves are adaptations to dry climates and to cold climates in which water is locked up in ice and not available to plants (**a**). Spines are found among plants living in arid regions. Ferns, which have featherlike leaves or fronds, are adapted to the low light intensity of the forest floor.

17. The correct answers in order are **b**, **e**, **c**, **a**, **d**, and **f**.

18. The correct answers in order are **b**, **d**, **a**, and **c**.

19. Examination of Figures 28-4, 28-6, 28-12a and b, 28-13, 28-15, and 28-16 will reveal the following points of comparison.
    Leaves
    (a) Dermal tissue: Epidermal cells contain no chloroplasts and entirely enclose the leaf; stomata are most common on the lower leaf surface; the epidermis is covered with a waxy cuticle.
    (b) Ground tissue: The interior of the leaf contains spongy and palisade parenchyma with chloroplasts.
    (c) Vascular tissue: Vascular bundles, which are known in the leaf as veins, are embedded in the spongy parenchyma.
    Stems
    (a) Dermal tissue: From one to several layers enclose the entire stem.
    (b) Ground tissue constitutes the bulk of the stem and consists of collenchyma and sclerenchyma cells.
    (c) Vascular tissue: Vascular bundles are embedded in the ground tissue and form a cylinder that encloses the pith.
    Roots
    (a) Dermal tissue encloses the root and is the source of root hairs.
    (b) Ground tissue: The cortex constitutes the bulk of the root and is composed of parenchyma cells.
    (c) Vascular tissue: The vascular cylinder occupies the center of the root interior to the cortex.

20. Examination of Figures 28-9, 28-10, 28-11, and 28-12a and b will reveal the following points of comparison.
    Xylem: The vessel members and the tracheids have thick walls and are surrounded by phloem. Tracheids are long and thin and overlap one another on their tapered ends, where there are pits. Adjoining surfaces of vessel members may lack end walls or have perforations.
    Phloem: Individual sieve-tube members have thinner walls than xylem cells, have holes in their ends (the sieve plate), and contain P-protein. Sieve-tube members are associated with companion cells.

21. Examination of Figures 28-2, 28-12, 28-13, 28-16, 28-17, and 28-18 will be helpful in making this comparison.
    Dicots
    (a) Leaves: veins netted.
    (b) Stems: vascular tissue arranged in or around a central core.
    (c) Roots: taproot with no pith within the vascular cylinder.
    Monocots
    (a) Leaves: veins parallel.
    (b) Stems: vascular bundles scattered throughout the ground tissue.
    (c) Roots: fibrous root system is common and vascular elements surround a large area of pith.

22. Annual plants are those that live for a single year, during which they flower and then overwinter as a seed. They usually show no signs of secondary growth. Perennials have vegetative structures that persist from year to year and increase in size largely through accumulation of woody tissue in the stem or through lengthening of the stem, as in herbaceous perennials. Woody perennials can grow tall, which allows them to compete for light and places their flowers and fruits in positions more attractive to pollinators and seed dispersers, respectively.

23. Check your answers against Figure 28-4 in the text.

24. To check your answer, see Figure 28-15 and read the caption.

CHAPTER 29

Plant Reproduction, Development, and Growth

CHAPTER ORGANIZATION

I. Introduction
II. The flower
III. The pollen grain
IV. Fertilization
V. The embryo
VI. The seed and the fruit
    A. Introductory remarks
    B. Types of fruits
    C. Seed dormancy
    D. Essay: The staff of life
VII. Primary growth
    A. General information
    B. Primary growth of the root
    C. Primary growth of the shoot
        1. General information
        2. Buds, branches, and flowers
        3. Essay: Why the grass grows
VIII. Secondary growth
    A. General information
    B. Essay: The record in the rings
IX. Asexual reproduction
X. Summary

MAJOR CONCEPTS

The flower is the structure of sexual reproduction in the angiosperms. The pollen grains, which contain the male gametophytes, are produced by the anthers of the flower. Within the ovary of the flower are one or more ovules, each of which contains a female gametophyte. A single egg cell is located within the female gametophyte.

Double fertilization, found only among angiosperms, refers to the fertilization of the egg by one sperm nucleus of the pollen grain and the fusion of the other sperm nucleus with two polar nuclei of the female gametophyte. This second fertilization produces the nutritive triploid endosperm.

An angiosperm seed, or mature ovule, is composed of the embryo sporophyte plant, stored food (endosperm), and a seed coat. After essential preparatory stages have been completed, a seed typically enters a dormant period, germinating only when the appropriate environmental conditions exist.

Primary growth takes place at the apical meristems of shoots and roots and involves cell division, cell elongation, and cell differentiation. Secondary growth takes place in woody dicots and results in an increase in girth. Such growth arises primarily from the vascular cambium.

Many plants reproduce asexually, which results in genetically identical offspring.

HOW TO STUDY THE CHAPTER

Read the chapter, focusing on the major concepts.

Reread the chapter, using the questions that follow to help you focus on details as they relate to major concepts. Answer the questions on a separate sheet of paper. Your answers will provide a valuable study aid in preparing for examinations.

FOCUS ON CHAPTER DETAILS

I. Introduction (page 591)
1. In what way are the reproductive structures of plants transitory?

II. The flower (pages 591–592)
2. From what tissues have flower parts evolved?
3. Define all of the terms used to describe the parts of flowers: sepals, calyx, petals, corolla, perianth, stamens, filament, anther, carpels, stigma, style, ovary. See also Figure 29-1.
4. Describe the external arrangement of the male and female reproductive organs of the flower. What important reproductive events take place in each structure prior to fertilization?
5. What tissues form the seed after fertilization occurs?
6. How do complete, staminate, and carpellate flowers differ?
7. Describe and give examples of monoecious and dioecious plants.

III. The pollen grain (pages 592–593)
8. Describe the structure in angiosperms that represents the male gametophyte stage. What three haploid cells are found within it?

IV. Fertilization (page 593)
9. What event takes place under the control of the tube nucleus of each pollen grain?
10. Describe the cells making up the female gametophyte.
11. Why is fusion of gametes in the angiosperms referred to as "double fertilization"? (See also Figure 29-3.) What are the products of double fertilization?

V. The embryo (pages 596–597)
12. What happens to the zygote as it divides mitotically?
13. Follow the sequence of embryonic growth shown in Figure 29-5. Locate the regions that will become apical meristems. What developmental events will take place in these regions as the plant grows?

VI. The seed and the fruit (pages 597–601)
A. Introductory remarks
14. What are the three main parts of the seed, and from what tissues are they derived? From what tissue does the fruit develop?
15. Referring to Figure 29-6, compare and contrast the events that occur within the dicot seed and within the monocot seed as the embryo grows.
16. What happens in the tissues surrounding the seed?

B. Types of fruits
17. What are the distinguishing characteristics of simple, aggregate, and multiple fruits? List several examples of each type.
18. What floral tissues develop into parts of the fruit in a berry, a drupe, and a pome? List several examples of each type of fruit.
19. Distinguish between dehiscent and indehiscent fruits. From what floral tissues are follicles, legumes, achenes, and nuts derived? Give examples of each type.

C. Seed dormancy
20. What is the advantage of seed dormancy?
21. In what two ways might the seed coat maintain dormancy? In each of these cases, what changes must occur in the seed coat before dormancy can be broken?
22. Why do angiosperms exhibit such a variety of mechanisms for maintaining and breaking dormancy?

D. Essay: The staff of life (page 601)
23. What are grains, and why have they been agriculturally important to humans?
24. Describe the three nutritionally significant parts of the wheat kernel. What nutrients does each part supply?

25. How is wheat processed to make white flour? What are the dietary consequences of this processing?
26. How good is wheat as a source of protein?

VII. **Primary growth** (pages 601–606)

A. General information

27. Describe germination.
28. Primary growth originates in apical meristems. What growth processes are considered to be primary?
29. How may plant growth be considered analogous to certain animal behaviors?

B. Primary growth of the root

30. It is convenient to consider the growing root tip as consisting of three zones (see Figure 29-14). What events take place in each of these zones, and how do they contribute to root growth and development?

C. Primary growth of the shoot

General information

31. Compare shoot growth to root growth.
32. How are leaves formed from cells of the apical meristem?
33. What two general types of leaf arrangements are seen in angiosperms? What are nodes and internodes?

Buds, branches, and flowers

34. From what regions of the shoot meristem are lateral buds formed?
35. What structures may form from lateral buds? Is this genetically or environmentally determined?

Essay: Why the grass grows (page 606)

36. How do grasses grow after cutting? How is this growth pattern an adaptation to animal grazing?

VIII. **Secondary growth** (pages 606–608)

A. General information

37. What is secondary growth? What two meristematic regions produce secondary growth? (Locate them in Figure 29-16.)
38. How does the vascular cambium contribute to secondary growth? What is heartwood? Sapwood?
39. How does the cork cambium contribute to secondary growth? What tissues constitute the bark?
40. What traces are left behind by secondary growth processes that make it possible to tell the age of a woody stem?

B. Essay: The record in the rings (page 609)

41. How do seasonal differences in secondary growth produce growth rings?
42. What kinds of information can analysis of growth rings yield other than the age of the tree?

IX. **Asexual reproduction** (pages 610–611)

43. Describe three methods by which plants reproduce asexually.
44. How have humans exploited the natural capacity of plants to reproduce asexually? What benefits may be derived from propagating plants asexually?

X. **Summary** (pages 611–612): Read the Summary. If you are familiar with the essential features of the material presented there, you are ready to take the following diagnostic examination.

TESTING YOUR UNDERSTANDING

After you have completed the following examination, compare your answers with the annotated key in the Analysis section.

1. Prior to fertilization in the angiosperms, pollen grains are released from the _____ of the flower to land upon the _____ of the same or another flower.
   a. male gametophyte; female gametophyte
   b. anther; stigma
   c. carpel; stamen
   d. calyx; ovule
2. A plant that is dioecious has:
   a. complete flowers.
   b. both staminate and carpellate flowers.

c. either staminate or carpellate flowers, but not both.
  d. carpellate flowers only.
  e. staminate flowers only.
3. The pollen grains of angiosperms:
  a. contain three haploid nuclei.
  b. contain two gametes.
  c. are male gametophytes.
  d. direct the formation of the pollen tube.
  e. All of the above are correct.
4. Double fertilization in the angiosperms refers to the fusion of _____ in the ovule of a flower.
  a. two sperm nuclei with the egg cell
  b. two sperm nuclei with two egg cells
  c. one sperm nucleus with an egg cell and a second sperm nucleus with the polar nuclei
  d. two sperm nuclei with the egg and the tube nucleus with the polar nuclei
5. Fertilization in the angiosperms produces:
  a. a diploid zygote and triploid nutritive cells.
  b. two diploid zygotes.
  c. a diploid zygote and diploid nutritive cells.
  d. a diploid zygote and a triploid zygote.
  e. a tetraploid zygote.
6. Reproduction in dicots differs from that in monocots in that:
  a. only dicots undergo double fertilization.
  b. two cotyledons are present in the developing embryos of dicots.
  c. two apical meristems form in the developing embryos of dicots.
  d. only dicots produce fruits.
7. In the angiosperms, a _____ develops from the outer layer of the ovule while _____ develops from the outer wall of the ovary.
  a. female gametophyte; the fruit
  b. pollen tube; a carpel
  c. seed coat; the fruit
  d. seed; an embryo
  e. seed; a female gametophyte
8. A multiple fruit develops from:
  a. a single carpel.
  b. many separate carpels of one flower.
  c. carpels of more than one flower.
  d. many fused carpels of one flower.
  e. Both b and d are correct.
9. Dry fruits may be classified as either _____ or _____.
  a. nuts; legumes
  b. dehiscent; indehiscent
  c. simple; multiple
  d. superior; inferior
  e. berries; pomes
10. The event marking the beginning of germination is:
  a. the release of seeds.
  b. the uptake of water by the seed.
  c. the formation of apical shoot and root meristems.
  d. the elongation of cells of the root cap.
11. A growing root tip is protected by:
  a. the remnants of the seed coat.
  b. the root cap.
  c. a radicle.
  d. a dense cluster of root hairs.
  e. the protoderm.
12. Root hairs arise from cells of the _____, whereas branch roots arise from cells of the _____.
  a. epidermis; endodermis
  b. epidermis; pericycle
  c. cortex; vascular bundles
  d. cortex; endodermis
  e. endodermis; epidermis
13. Primary root growth is chiefly a result of:
  a. cell elongation.
  b. divisions in the apical root meristem.
  c. differentiation of root hairs.
  d. extension of the root cap deeper into the soil.
  e. differentiation of the pericycle.
14. Leaves are produced by:
  a. the apical meristem of the shoot.
  b. the lateral meristem of the shoot.
  c. cells along the internodes of the shoot.
  d. lateral buds.
  e. cells of the shoot epidermis.
15. Which of the following is true of lateral buds?
  a. They remain dormant during one stage of their development.
  b. They develop from the lateral meristems in leaf axils.
  c. They may differentiate to become flowers, branches, or specialized shoots.

d. Only a and c are correct.
e. All three answers (a, b, and c) are correct.

16. The lateral meristems of the stem give rise to:
   a. buds and leaves.
   b. primary growth.
   c. vascular tissues and cork.
   d. branches and twigs.
   e. the vascular and cork cambia.

17. The age of a tree may be determined from growth rings that reflect:
   a. heartwood and sapwood.
   b. early wood and late wood.
   c. densities in secondary xylem.
   d. patterns in both secondary xylem and phloem.
   e. Both b and c are correct.

18. Runners and rhizomes are methods of asexual reproduction involving:
   a. horizontal aboveground stems.
   b. nutritional dependence of clones on the parent plant.
   c. parthenogenesis.
   d. relatively high mortality as compared to sexual reproduction.

19. You can prevent a grass lawn from growing by:
   a. using a spray that selectively kills all monocots.
   b. setting your cutting blade to remove the apical meristem of the grass shoot tips.
   c. allowing hungry herbivores to graze it for a while.
   d. cutting the leaf blades very close to the leaf sheath.
   e. Both a and b are correct.

20. Compared to bread made from whole wheat flour, bread made from white flour:
   a. is a better source of protein.
   b. provides a richer complement of vitamins.
   c. has additional calories.
   d. contains more cellulose.
   e. Both a and c are correct.

21. Match each structure with its description.
   root cap _____
   shoot _____
   node _____
   primordium _____
   vascular cambium _____
   cork _____
   bud _____
   rhizome _____

   a. gives rise to secondary xylem and phloem
   b. point along a stem where leaves and buds develop
   c. tissue in apical meristems that differentiates to form leaves
   d. underground stem forming adventitious roots
   e. replaces epidermis as it is ruptured
   f. forms in leaf axils
   g. a protective layer of cells in the growing root tip
   h. the aboveground parts of a plant

22. Match each floral or embryonic structure with its function.
   ovule _____
   fruit _____
   stigma _____
   perianth _____
   stamen _____
   endosperm _____
   seed coat _____
   cotyledon _____

   a. attachment site for the male gametophyte
   b. an embryonic seed leaf
   c. develops from outer layers of the ovule
   d. encloses the female gametophyte
   e. advertises and protects flower parts
   f. nourishes the growing embryo
   g. a ripened ovary
   h. contains the male reproductive organs

23. How does the seed coat ensure that a seed will not germinate until environmental conditions are favorable?

24. How is primary growth in plants analogous to certain animal behaviors?

25. How have humans exploited the capacity of plants to reproduce asexually in order to propagate desired varieties?

26. Identify the indicated parts of this carpel.

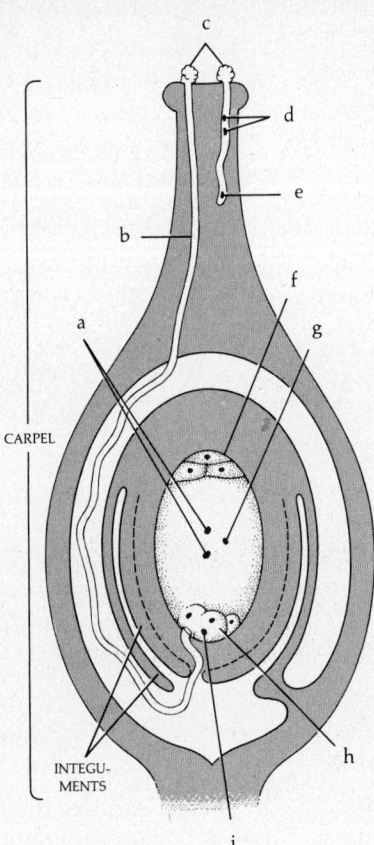

PERFORMANCE ANALYSIS

1. Pollen grains are formed in and released from the anthers of flowers. Pollination is said to occur when the grains land upon the stigma (the sticky surface of the female reproductive organ), as shown in Figure 29-3a. Choice **b** is therefore correct.

2. Dioecious means "in two houses" and refers to the situation in which the male and female flower parts are present on different plants. The flowers of a dioecious plant, therefore, would be either staminate (if they contained only male parts) or carpellate (if they contained only female parts). Choice **c** is correct.

3. Pollen grains contain three haploid nuclei, two of which are sperm nuclei. The mature grain is the male gametophyte, which, after landing upon the stigma of a flower, directs formation of the pollen tube; therefore, **e** is correct.

4. Choice **c** is correct. Refer to Figure 29-3.

5. Fertilization in angiosperms involves the fusion of a sperm nucleus with an egg nucleus to produce a diploid zygote, and the fusion of a second sperm nucleus with two polar nuclei to produce a triploid cell, which becomes the endosperm and nourishes the embryo. Choice **a** is correct.

6. One of the differences between dicots and monocots—in fact, the difference from which they derive their names—is the number of seed leaves, or cotyledons, present in the developing embryo; therefore, **b** is correct.

7. The seed coat develops from the outermost layer or layers of the ovule, and the fruit develops from the wall of the ovary (see Figure 29-7). Choice **c** is therefore correct.

8. A multiple fruit, such as the pineapple, is derived from the carpels of more than one flower (**c**).

9. Dry fruits are classified as either dehiscent (if seeds are released by the rupturing of the mature ovary wall) or indehiscent (if seeds remain in the fruit after the fruit is shed); therefore, **b** is correct.

10. A seed is said to have germinated when changes in the seed coat permit water to enter the tissues of the seed by imbibition. Choice **b** is correct.

11. Choice **b** is correct. Refer to Figure 29-14.

12. Choice **b** is correct. The origin of root hairs was discussed in both this chapter and the previous one. See Figure 29-13 for an illustration of how branch roots arise from the pericycle.

13. Root growth is initiated by cell divisions in the apical meristem. Cells above the meristem then elongate and in doing so increase the length (primary growth) of the root. This elongation is the principal cause of root growth; therefore, **a** is correct. Refer to Figure 29-14.

14. Leaves arise from leaf primordia in the apical meristem of the shoot (**a**). See Figure 29-15.

15. Lateral buds, after breaking dormancy, contribute to primary growth in the form of branches,

specialized shoots, and flowers. Lateral buds, however, develop from the apical (not the lateral) meristem (see Figure 29-15); therefore, choice **d** is correct.

16. The lateral meristems are the vascular cambium, which gives rise to xylem and phloem (vascular, or conducting, tissues), and the cork cambium, which gives rise to cork, a substance that becomes part of the bark. Choice **c** is correct.

17. Rings, made up of secondary xylem, are evident in a trunk cross section because early wood (wood added to the tree in the spring) is less dense than late wood (wood added during the summer). Choice **e** is therefore correct.

18. Runners are aboveground stems, but rhizomes develop underground. Both are a means of vegetative reproduction. They give rise to new plants that are genetically identical to the parent (clones) and for a time are nutritionally dependent on it (**b** is correct).

19. Mowing and grazing, although they may remove the apical meristems, will not prevent a lawn from growing because cells in the leaf sheath (below the blade) can produce more blade. Since grasses are monocots, the selective spray referred to in choice **a** would destroy the lawn.

20. White flour has had both its bran (which contains cellulose and a number of vitamins) and its germ (which contains proteins and vitamins) removed; all that remains is the starchy endosperm. White flour makes available more calories (**c**) than whole wheat flour primarily because its lack of cellulose fiber allows it to remain in the digestive tract for a longer time.

21. The correct answers in order are **g, h, b, c, a, e, f,** and **d**.

22. The correct answers in order are **d, g, a, e, h, f, c,** and **b**.

23. Your discussion should include the fact that in some species seed coats can serve as mechanical barriers that are impermeable to water and gases, and that in other species seed coats contain chemical inhibitors that can be broken down by light or the action of rain water. In these ways, germination can be timed to occur when environmental conditions favor growth and survival.

24. Primary growth in plants involves elongation of shoots and roots. This is analogous to behavior in animals in that plants are able to "seek out" both light, by which to make their food, and minerals and water, which are necessary for maintaining proper ionic concentrations and osmotic pressure.

25. Desirable varieties of plants may be produced by stem cuttings, which can take root, and by grafting stem cuttings onto growing root tissues. Both of these methods exploit the asexual reproductive potential of plants.

26. Check your answers against Figure 29-3 in the text.

CHAPTER 30

Transport Systems in Plants

CHAPTER ORGANIZATION

I. Introduction
II. The movement of water and minerals
   A. Transpiration
   B. The uptake of water
   C. The cohesion-tension theory
   D. Factors influencing transpiration
      1. General information
      2. Mechanism of stomatal movements
      3. Factors affecting stomatal movements
      4. The role of potassium
      5. Crassulacean acid metabolism
   E. Mineral requirements of plants
   F. Uptake of minerals
   G. Functions of minerals
   H. Essay: Halophytes: A future resource?
III. The movement of sugars: Translocation
   A. Introductory remarks
   B. Evidence for the phloem
      1. General information
      2. Essay: Radioactive tracers in plant research
      3. Assistance by aphids
   C. The pressure-flow hypothesis
IV. Soils and plant nutrition
V. Symbioses and plant nutrition
   A. Mycorrhizae
   B. Rhizobia and nitrogen fixation
      1. General information
      2. The symbiotic relationship
      3. Recombinant DNA and nitrogen fixation
VI. Summary

MAJOR CONCEPTS

Transpiration is the process by which water vapor is lost from plants. According to the cohesion-tension theory of water movement in xylem, the evaporation of water from the upper plant parts creates a negative pressure (tension) in the xylem vessels and tracheids. Since water molecules cling together (cohesion), molecules from lower plant regions are pulled up as water molecules are lost through transpiration.

Diffusion of gases, including water vapor, into and out of the leaf is regulated by stomata. The opening and closing of the guard cells of the stomata results from changes in turgor pressure in the cells.

Minerals from the soil enter and move throughout the plant in the transpiration stream of water in the xylem. They are essential components in many important molecules, including chlorophylls and certain enzymes.

Translocation is the process by which organic molecules move from the photosynthetic parts of the plant to other parts of the plant via the phloem. The pressure-flow hypothesis of translocation proposes that molecules are pumped into and out of the sieve tubes of the phloem by active transport and that the movement of molecules is determined by concentration gradients of sugar.

Certain soil characteristics (including the soil's parent rock, pH, composition, and the presence or absence of humus on the surface) affect the availability of minerals to plants.

Symbiotic relationships between plant roots and soil-dwelling fungi (mycorrhizae) and between plants and nitrogen-fixing bacteria are important in providing essential minerals to the plants.

HOW TO STUDY THE CHAPTER

Read the chapter, focusing on the major concepts.

Reread the chapter, using the questions that follow to help you focus on details as they relate to major concepts. Answer the questions on a separate sheet of paper. Your answers will provide a valuable study aid in preparing for examinations.

FOCUS ON CHAPTER DETAILS

I. Introduction (page 614)
1. A plant's success on land depends on what ability?

II. The movement of water and minerals (pages 614–621)

A. Transpiration
2. Explain the role of plants in the movement of water between the soil and the atmosphere.
3. Define transpiration.

B. The uptake of water
4. What roles do diffusion and capillary action play in providing water to roots?
5. Explain how water enters the root cells. What is guttation, and what causes it?
6. What happens if a drop of water is applied to a cut in a xylem vessel? What does this indicate about conditions inside the xylem?
7. Why is atmospheric pressure considered to be an insufficient pulling force on water in the xylem?

C. The cohesion-tension theory
8. What accounts for cohesion between molecules of water? What consequences does this have for a thin column of water such as that of a xylem vessel?
9. How does the cohesion-tension theory explain water movement from the roots into the leaves of a plant? What is the energy source for this process?

D. Factors influencing transpiration

General information
10. How do temperature, humidity, and circulating air affect the rate of transpiration?
11. Why are hairy leaves advantageous to some plants? Which leaf surface (upper or lower) should contain the most hairs? Why?
12. What is the most important factor affecting transpiration?

Mechanism of stomatal movements
13. How does turgor pressure influence the shape of the guard cells?
14. How is turgor maintained or lost?

Factors affecting stomatal movements
15. What is the principal factor affecting stomatal movements?
16. What three lines of evidence suggest that abscisic acid is linked to stomatal movements?
17. How do the rates of photosynthesis and respiration in leaf tissues influence stomatal movement?
18. What is thought to be the significance of the pigment flavin in the functioning of the guard cells?
19. How do temperature changes affect the stomata?
20. What types of plants close their stomata at midday?

The role of potassium
21. In what ways are changes in the K^+ ion concentration associated with changes in turgor in the guard cells? How do these changes occur?
22. What other ion concentrations may also be affected by changes in K^+ ion concentration?

Crassulacean acid metabolism
23. Give several examples of CAM plants. Where do they grow?
24. What pattern of stomatal movement occurs in these plants?

25. Describe Crassulacean acid metabolism. To what photosynthetic pathway is it analogous?

E. Mineral requirements of plants

26. What is a mineral? How do plants acquire minerals?
27. What determines whether a mineral is a macronutrient or a micronutrient?
28. By what two methods are mineral concentrations determined? What problems may arise in using these methods?

F. Uptake of minerals

29. What evidence is there that some minerals enter root cells by active transport?
30. By what mechanism do K^+ ions enter roots against a concentration gradient?
31. What is the electrical consequence of differences in ionic concentrations?

G. Functions of minerals

32. Why do mineral deficiencies often affect structures and functions throughout the entire plant body?
33. Review the functions listed in Table 30-1 for the absorbable forms of various minerals.
34. In terms of minerals (as well as all required substances), it seems logical that plants would make use of those that are readily available in their environments. What are some notable exceptions to this rule of thumb?

H. Essay: Halophytes: A future resource? (page 620)

35. What osmotic problems do plants encounter in soils that contain high Na^+ levels?
36. What mechanisms have the halophytes developed for dealing with high salinity?
37. Explain why irrigated soils are so salty.

III. **The movement of sugars: Translocation** (pages 621-625)

A. Introductory remarks

38. Define translocation.

B. Evidence for the phloem

General information

39. What is "girdling," and what effect does it have on material transport?
40. What plant response makes it difficult to analyze the role of phloem in transporting materials?

Essay: Radioactive tracers in plant research (page 623)

41. What radioactive isotopes can be taken up readily by plants?
42. What factors might determine how long plants need to be exposed to radioactive materials?
43. How are whole-plant and tissue autoradiography used to determine the exact locations at which substances are stored or chemical reactions take place?
44. How have radioactive tracers been used to disclose the role of the phloem in transporting sugars?

Assistance by aphids

45. How are aphids employed as a tool in analyzing the contents of sap?
46. What have aphids revealed about the contents of sap and the rate at which it flows through the stem?

C. The pressure-flow hypothesis

47. What is meant by the source-to-sink pattern of translocation? What plant tissues are examples of sources? Of sinks?
48. Explain the pressure-flow hypothesis, using the apparatus in Figure 30-9 as a model. How does this apparatus resemble a plant stem?
49. What evidence is there for the pressure-flow hypothesis?
50. How can the movement of sugars against a concentration gradient be explained by this hypothesis?
51. What role has been proposed for companion cells? Does the flow of the sugar solution itself require energy?

52. What stages of translocation require the energy the companion cells might provide?

IV. **Soils and plant nutrition** (pages 625-626)

53. Note the general organization of soils (see Figure 30-10). What is the composition of each of the soil horizons?
54. What roles do geological and biological processes play in determining the mineral content of soil?
55. What effect does particle size (clay or sand) have on a soil's ability to support plant growth?
56. In what two ways is the pH of a soil an important factor in determining what kinds of plants can inhabit an area?
57. How do the plants themselves affect the composition of soils?

V. **Symbioses and plant nutrition** (pages 626-629)

A. Mycorrhizae

58. How does a symbiotic relationship with fungi enable plants to grow in soils that would otherwise be unsuitable?

B. Rhizobia and nitrogen fixation

General information

59. Why is nitrogen fixation essential to plant growth?
60. What types of microorganisms carry out nitrogen fixation? What symbiotic relationships are responsible for the bulk of nitrogen fixation?
61. What agricultural method is used to increase quantities of available nitrogen in soils?
62. How are nitrogen-containing fertilizers commercially produced?

The symbiotic relationship

63. How do rhizobia infect a host legume?
64. How does bacterial activity supply the plant with amino acids?
65. Identify the roles of iron and molybdenum in nitrogen fixation.

Recombinant DNA and nitrogen fixation

66. What success have biologists had thus far in transferring nitrogen-fixing capabilities to other organisms? What is the ultimate goal of this research?
67. What nitrogen-fixing symbiotic relationships do not involve legumes? What is the ecological significance of one such relationship?

VI. **Summary** (pages 629-630): Read the Summary. If you are familiar with the essential features of the material presented there, you are ready to take the following diagnostic examination.

TESTING YOUR UNDERSTANDING

After you have completed the following examination, compare your answers with the annotated key in the Analysis section.

1. Root pressure is accounted for by:
   a. diffusion.
   b. capillary action.
   c. osmosis.
   d. negative pressure.
   e. Both a and b are correct.

2. Water moves upward through the stems of plants principally because of the force provided by:
   a. transpiration.
   b. the atmosphere.
   c. simple suction.
   d. adhesion of water and plant tissues.

3. Water losses due to transpiration are *least* important for plants on a day that is:
   a. warm and sunny.
   b. cool and humid.
   c. mild and dry.
   d. hot and windy.

4. As K^+ ions _____ guard cells, stomata _____.
   a. diffuse into; open
   b. diffuse out of; close
   c. are transported into; open
   d. are transported out of; open

5. Which of the following stimulates the opening of the stomata?

a. an increase in respiratory rates in leaf cells
   b. exposing leaves to a strong blue light
   c. the onset of darkness
   d. raising the air temperature
   e. uptake of potassium by root cells

6. Plants that utilize Crassulacean acid metabolism:
   a. take in carbon dioxide only at night.
   b. carry out the energy-capturing reactions of photosynthesis only during the day.
   c. increase respiratory rates at night.
   d. utilize photosynthetic pathways other than the Calvin cycle.
   e. All of the above are correct.

7. Which of the following statements is *false* regarding mineral utilization in plants?
   a. Mineral deficiencies typically affect a number of plant functions.
   b. Minerals are generally present in plant tissues in the same concentration as in the soil.
   c. Methods used in determining the specific requirements of minerals can be misleading.
   d. One mineral can sometimes substitute for another deficient mineral.

8. An ideal soil for plant growth would be one:
   a. whose particles are nearly saturated with H^+ ions.
   b. that is composed mostly of sand and silt.
   c. whose particles are tightly packed around plant roots.
   d. with many negatively charged particles and air spaces.

9. Translocation refers specifically to the movement of:
   a. water and minerals upward through the xylem.
   b. minerals and ions from the soil into root hairs.
   c. any dissolved material through conducting tissues.
   d. photosynthetic products from leaf mesophyll to other tissues.

10. The movement of sugars through the phloem of a plant stem:
    a. does not occur in response to concentration gradients.
    b. takes place as a result of water entering the phloem from the xylem.
    c. proceeds from a sink to a source.
    d. does not involve companion cells.
    e. Both a and d are correct.

11. According to the pressure-flow hypothesis of sugar transport, the overall movement of dissolved solutes within the sieve tubes themselves is by:
    a. diffusion.
    b. capillary action.
    c. bulk flow.
    d. osmosis.
    e. active transport.

12. In a plant that carries out photosynthesis using $^{14}CO_2$ gas, radioactive sugars will:
    a. not be produced.
    b. not be translocated.
    c. be confined to the phloem.
    d. be transported downward through the phloem but upward through the xylem.

13. Nitrogen fixation, as carried out by certain groups of microorganisms associated with plant roots, means:
    a. the incorporation of nitrogen into organic compounds.
    b. the production of nitrogen-containing compounds in a form absorbable by plants.
    c. immobilization of soil nitrogen in the cytoplasm of soil organisms.
    d. the formation of nitrogen gas from dissolved nitrates or ammonium.

14. Which of the following is true concerning symbiotic relationships that lead to the fixation of soil nitrogen?
    a. Bacterial infections are confined to those cells that produce root hairs.
    b. Bacteria provide amino acids directly to the plant.
    c. Bacteria enter the host plant through the tips of its root hairs.
    d. Usually only a single species of microorganism will be able to infect a given plant species.
    e. Both c and d are correct.

15. Agricultural researchers, concerned with maintaining adequate supplies of nitrogen to support plant growth, have been successful in:
    a. stimulating *E. coli* to synthesize an enzyme involved in nitrogen fixation.

b. transferring nitrogen-fixing genes of a legume into the genome of a nonlegume.
c. infecting crops with yeast cells and getting them to fix nitrogen for the crop plants.
d. greatly lowering the expense of producing commercial nitrogen fertilizers.
e. Both b and d are correct.

16. Which of the following represents a symbiotic relationship involving a legume that fixes nitrogen?
   a. actinomycetes and sweet fern
   b. mycorrhizae and white pine
   c. cyanobacteria and corn
   d. rhizobia and alfalfa
   e. Both a and c are correct.

17. The B horizon is defined as the soil layer that contains:
   a. humus.
   b. an accumulation of minerals and inorganic particles.
   c. broken up or loose rock.
   d. the majority of the soil microorganisms.
   e. the root systems of fibrous root plants.

18. Cultivation of halophytes on arid soils could:
   a. provide industry with a new source of salt.
   b. help reduce the salinity of these soils.
   c. be difficult without extensive irrigation.
   d. greatly extend the global total of arable lands.

19. Match each process with an appropriate description.

   transpiration _____   a. a negative pressure in the xylem
   translocation _____   b. responsible for most of a plant's water loss
   turgor _____   c. process by which sugars go from sites of synthesis to storage tissues
   tension _____   d. responsible for the opening of stomata
   cohesion _____   e. allows for photosynthetic activity without substantial water loss
   CAM metabolism _____   f. helps maintain a thin, unbroken water column in vessels
   nitrogen fixation _____ g. requires the action of symbiotic or free-living soil microorganisms

20. Explain how water in the soil surrounding a plant's root system moves into, through, and out of the plant.

21. What are the advantages of using radioactive tracers to learn about plants? Cite an example, explaining the techniques involved.

22. How can our knowledge of the relationship between plants and their soil environments make possible a brighter future for agriculture?

PERFORMANCE ANALYSIS

1. The pressure that draws water and dissolved materials into the root tissues of plants is due to the difference in osmotic pressure (**c**) between the soil and the cells of the root cortex.

2. As water escapes from leaves by transpiration (**a**), an unbroken column of water is pulled upward through the xylem.

3. Choice **b** is correct. Low temperatures and high water-vapor levels are both factors that discourage transpiration.

4. When carbon dioxide concentrations inside the leaf are reduced, K^+ ions are actively transported into the guard cells. This ion movement is followed by osmosis, which causes the guard cells to swell and creates a space between them known as a stoma; therefore, **c** is correct. See Figure 30–6.

5. The conditions in choices a, c, and d would stimulate the closing rather than the opening of the stomata, since photosynthesis proceeds during the day and respiration, encouraged by increases in temperature, generates carbon dioxide that can be used in photosynthesis. Choice e is irrelevant. The illumination of a leaf under a strong blue light (**b**) has been shown to be responsible for the uptake of K^+ ions by the guard cells and, therefore, for the opening of the stomata.

6. Plants adapted to hot, dry climates have their stomata open only at night when conditions are more favorable for avoiding loss of water through the leaves. Because oxygen as well as carbon dioxide is exchanged through the stomata,

respiratory rates increase at night. All plants can carry out the energy-capturing reactions of photosynthesis only when light is available. CAM plants, like C_4 plants, employ a four-carbon cycle (in addition to the Calvin cycle) in which they store carbon dioxide in organic acids. Choice **e** is therefore correct.

7. Choice **b** is the false statement. Many minerals, potassium for example, are much more concentrated in plant tissues than in soils; others, like sodium, are much less concentrated in plants.

8. Ideal soils for plant growth are typically loams made up of clay and silt. Because of their clay content, these soils have an abundance of negatively charged particles that can hold positively charged ions; because of their silt content, they have sufficient air spaces from which roots can take in oxygen for respiration. Therefore, **d** is correct.

9. Translocation is the transport of sugars manufactured in photosynthesis from the leaf mesophyll through the phloem to tissues where they are metabolized or stored (**d**).

10. According to the pressure-flow hypothesis, sugars move from source to sink along a concentration gradient. The process begins when sugar is moved by active transport from the source into a companion cell; from there the sugar moves into a sieve-tube cell of the phloem. Water enters the phloem from the xylem (**b**) by osmosis during translocation.

11. Although osmosis and active transport are both involved in translocation (see Figure 30–9), dissolved substances are moved through the sieve tubes by bulk flow (**c**).

12. By using radioactive carbon dioxide as a tracer, researchers have been able to determine that the role of the phloem is in the transport of sugars (see the essay on page 623). Choice **c** is therefore correct.

13. Nitrogen-fixing microorganisms associated with plant roots (symbiotic nitrogen-fixers) incorporate nitrogen gas in the soil into ammonia, which plants can then incorporate into organic compounds to produce their amino acids; therefore, **b** is correct.

14. Symbiotic nitrogen-fixing bacteria, specific for a given host plant species, enter the plant through the tips of the root hairs and extend infection threads into the cortical cells. They provide ammonia, the nitrogen component that plants use to make amino acids. Therefore, **e** is correct.

15. So far, researchers have been able to incorporate the gene for nitrogenase synthesis into the *E. coli* genome, after which the bacteria was able to synthesize this nitrogen-fixing enzyme (**a**). Choice b refers to another potential method of conferring on plants the ability to fix their own nitrogen. The experiment referred to in choice c was tried but was not successful. In consideration of choice d, it should be pointed out that commercial production of nitrate fertilizers is becoming more expensive as the fossil fuels that supply energy for this process become scarcer.

16. Choice **d** is correct. *Rhizobium* is a symbiotic nitrogen-fixing bacterium; alfalfa is a legume.

17. The B horizon of a soil (see Figure 30–10) is the region in which minerals and inorganic particles accumulate (**b**) as a result of leaching from the A horizon. The A horizon is composed of humus and is where the majority of the soil microorganisms are found. The root systems of fibrous root plants are also found primarily in the A horizon. The C horizon is made up of loose rock.

18. Since halophytes are salt-tolerant plants, they could be grown in saline environments such as marshes and deserts, greatly extending the global total of arable lands (**d**).

19. The correct answers in order are **b, c, d, a, f, e,** and **g**.

20. A complete discussion would include an explanation of root pressure, cohesion, negative tension, transpiration as a pulling force, and the regulatory activity of the stomata.

21. The use of radioactive tracers has allowed plant biologists to describe the location of various metabolic reactions within the plant as well as sites of storage and paths of transport. The role of the phloem was elucidated in this way. Locations of radioactive substances in plants can be determined using the techniques of whole-plant autoradiography and tissue autoradiography. Both of these are aptly described in the essay Radioactive Tracers in Plant Research, page 623.

22. Our agricultural practices can be enhanced by a knowledge of the relationship between plants and the soil. Examples of research that have potential returns are the transfer of nitrogen-fixing genes to important crop species, determination of mineral requirements, and the exploitation of salt-tolerant plants. (See the essay Halophytes: A Future Resource, page 620.)

CHAPTER 31

Hormones and the Regulation of Plant Growth

CHAPTER ORGANIZATION
   I. Introduction
  II. Auxins
      A. General information
      B. Mechanism of action of auxin
      C. Apical dominance and other auxin effects
      D. Synthetic auxins
 III. Cytokinins
      A. Introductory remarks
      B. Responses to cytokinin and auxin combinations
      C. Other cytokinin effects
  IV. The gibberellins
      A. General information
      B. Gibberellins and seed germination
   V. Ethylene
      A. General information
      B. Ethylene and leaf abscission
  VI. Abscisic acid
 VII. Hormonal control of flowering
VIII. Summary

MAJOR CONCEPTS

Hormones are chemicals that are produced in certain tissues of organisms and are transported to other tissues, where they exert specific influences. They are effective in extremely small quantities. In plants, five major groups of hormones have been discovered so far: auxins, cytokinins, gibberellins, ethylene, and abscisic acid.

Auxins, produced mainly in the apical meristems of shoots, are responsible for cell elongation, apical dominance, fruit development, and the retention of leaves and fruits on stems. Low auxin concentrations stimulate the development of branch roots and adventitious roots; high concentrations inhibit root growth.

Cytokinins promote cell division and usually function in combination with auxins to bring about normal plant development.

In many plants, gibberellins promote hyperelongation of the stem. They are also involved in cellular differentiation, and in grasses they stimulate the production of enzymes that make the stored nutrients of the seed available to the embryo.

Ethylene gas is responsible for leaf abscission and ripening of fruit, and it works in conjunction with auxin to promote apical dominance.

Abscisic acid induces and maintains dormancy in seeds and vegetative buds and causes stomata to close under dry environmental conditions.

Evidence exists that flowering is stimulated by a hormone (or hormones), but the specific substance involved has not been isolated.

HOW TO STUDY THE CHAPTER

Read the chapter, focusing on the major concepts.

Reread the chapter, using the questions that follow to help you focus on details as they relate to major concepts. Answer the questions on a separate sheet of paper. Your answers will provide a valuable study aid in preparing for examinations.

FOCUS ON CHAPTER DETAILS

I. Introduction (page 632)
1. Define the term "hormone."
2. What plant functions do hormones affect, and in what quantities are they effective?
3. How do hormones exert their influences?

II. Auxins (pages 632–634)

A. General information
4. What is IAA? What does the term "auxins" denote?
5. Where is IAA produced? Describe the path of transport of this hormone through the plant.
6. How does auxin affect growing shoots?
7. What effect does concentration of the hormone have on shoots? On roots?

B. Mechanism of action of auxin
8. How does auxin alter the plasticity of cell walls? How does this result in cell elongation?
9. By what additional mechanism does auxin act to provide for growth of cell walls?

C. Apical dominance and other auxin effects
10. Cite several examples of how auxin maintains apical dominance.
11. Name several secondary effects of auxin on plant tissues.

D. Synthetic auxins
12. How do synthetic auxins differ from IAA?
13. Why do rooting preparations contain auxin?
14. For what purposes has the synthetic auxin 2,4-D been used?

III. Cytokinins (pages 634–636)

A. Introductory remarks
15. What general stimulatory effect do cytokinins have on cells? From what plant tissues have cytokinins been isolated?

B. Responses to cytokinin and auxin combinations
16. How do auxin and kinetin interact to produce different growth patterns in stems? In roots?
17. How does the presence of Ca^{2+} ions in high concentrations modify the auxin-cytokinin effect? What mechanism may underly this additional interaction?

C. Other cytokinin effects
18. How do cytokinins promote typical growth patterns of lateral branches?
19. What influence do cytokinins have on the condition of plant leaves?
20. What is known about the general mechanism of action of the cytokinins in stimulating cell division?

IV. The gibberellins (pages 636–637)

A. General information
21. How was gibberellin discovered?
22. What is the principal effect of the gibberellins? How do dwarf plants respond to applications of gibberellins?
23. What other effects are attributed to gibberellins?
24. How do interactions between auxins and gibberellins affect differentiation in the vascular cambium?

B. Gibberellins and seed germination
25. What are the general effects of gibberellins on the aleurone layer of immature seeds?
26. What mechanism of action has been proposed to explain these effects?

V. Ethylene (page 638)

A. General information
27. What accounts for the tendency of fruits to ripen when placed in rooms heated by kerosene stoves?
28. What are the effects of ethylene gas on the reproductive tissues of plants? In what part of the cell is ethylene produced?
29. What general relationship exists between auxins and ethylene?

B. Ethylene and leaf abscission
30. What is leaf abscission?

31. Describe the two layers of the abscission zone of a leaf petiole (see also Figure 31-12).
32. What mechanism of action has been proposed for ethylene?
33. Describe the interactive effects of auxins and ethylene on leaf abscission.

VI. **Abscisic acid** (pages 638-639)

34. From what plant tissues were inhibitory hormones first isolated?
35. What effects does abscisic acid have when applied to vegetative buds?
36. Why is abscisic acid considered to be a stress hormone in plants? Cite two examples to support this.
37. What interactions between abscisic acid and other hormones have been observed?

VII. **Hormonal control of flowering** (page 639)

38. Flowering in response to an appropriate light cycle is an example of what phenomenon?
39. What effects on flowering were observed by Chailakhyan when he removed leaves from specific locations on chrysanthemums?
40. What type of hormonal activity did he propose to account for the observed effects?
41. What other observations led to the supposition that flowering hormones are transported through the phloem?
42. Why is the existence of florigen only a supposition?
43. What are the effects of two other previously discussed hormones on flowering?

VIII. **Summary** (pages 640-641): Read the Summary. If you are familiar with the essential features of the material presented there, you are ready to take the following diagnostic examination.

TESTING YOUR UNDERSTANDING

After you have completed the following examination, compare your answers with the annotated key in the Analysis section.

1. Which of the following is generally true of a hormone?
    a. It is produced in the tissue on which it exerts its influence.
    b. It must be present in relatively large quantities to have any significant effect.
    c. It has one or more highly specific effects.
    d. Its effects are independent of other hormones.
    e. Both a and c are correct.
2. Cells in the growing region of the shoot will begin to elongate if the shoot tip is:
    a. removed.
    b. treated with auxin.
    c. removed and the cut surface is treated with auxin.
    d. injected with a concentrated calcium solution.
3. Abscission occurs in the petioles of a deciduous tree when the auxin concentration _____ production.
    a. stimulates abscisic acid
    b. inhibits abscisic acid
    c. stimulates ethylene
    d. inhibits ethylene
    e. Both a and c are correct.
4. Which of the following plant hormones has a general inhibitory effect?
    a. indoleacetic acid
    b. kinetin
    c. gibberellin
    d. abscisic acid
    e. ethylene
5. The lateral buds of a plant will begin to grow:
    a. if the apical meristem is treated with auxin.
    b. if cytokinin is applied to the buds.
    c. from bottom branches when the plant reaches a certain height.
    d. if the plant is sprayed with ethylene gas.
    e. Both b and c are correct.
6. Which of these best describes the effect of cytokinin on plant cells?
    a. It inhibits rapid cell division.
    b. Cells subject to high concentrations of cytokinin remain undifferentiated.
    c. It promotes the rapid enlargement of cells.
    d. It encourages the aging of leaves.
7. Division or enlargement of cells depends upon the relative concentrations of:
    a. gibberellin and auxins.

240 SECTION 5 BIOLOGY OF PLANTS

b. auxins and cytokinins.
   c. auxins and ethylene.
   d. ethylene and abscisic acid.
   e. gibberellin and florigen.

8. Which of the following is a growth effect directly attributable to gibberellins?
   a. growth of dwarf plants to normal height
   b. initiation of embryonic growth
   c. rapid elongation of a flower stem
   d. production of secondary vascular tissue
   e. All of the above are correct.

9. Which of the following is true about the ripening of fruits?
   a. Ripening is initiated by smoke or heat from a fire.
   b. Ripening is principally due to the release of ethylene by symbiotic fungi.
   c. Ripening is promoted by auxins and cytokinins.
   d. Ripening is due to the influence of hormones on the structure of fruit cell walls.

10. Gibberellins were first isolated from:
    a. a rice seedling.
    b. dwarf varieties.
    c. a fungus.
    d. the aleurone layer of immature seeds.
    e. dormant buds.

11. An abscission zone is present at the base of a:
    a. leaf.
    b. petiole.
    c. stem.
    d. root.
    e. root hair.

12. The existence of _____ has been inferred, but the chemical substance itself has never been isolated.
    a. dormin
    b. kinetin
    c. IAA
    d. florigen
    e. ABA

13. Chailakhyan demonstrated that a substance produced in the _____ of a plant causes it to flower.
    a. apical meristems
    b. leaves
    c. lateral buds
    d. flower stems
    e. cortical cells of the root

14. Flowering occurs when a plant is exposed to an appropriate light cycle. This phenomenon is called:
    a. bolting.
    b. breaking dormancy.
    c. abscission.
    d. apical dominance.
    e. photoinduction.

15. Which one of the following hormones has the simplest organic formula?
    a. IAA
    b. ABA
    c. ethylene
    d. kinetin
    e. zeatin

16. Differentiation of secondary tissues in woody plants is stimulated by:
    a. gibberellins.
    b. cytokinins.
    c. ethylene.
    d. abscisic acid.

17. Match each substance with its description or its effect on plant tissues.

    2,4-D _____
    gibberellin _____
    ethylene _____
    abscisic acid _____
    Ca^{2+} ions _____

    a. promotion of dormancy in vegetative buds
    b. a synthetic auxin used to kill weeds
    c. prevention of cell wall expansion
    d. found in high concentrations in immature seeds
    e. regulation of the dropping of senescent leaves

18. For each hormone or class of hormones, indicate its proposed mechanism of action.

    gibberellin _____
    auxin _____
    cytokinin _____
    ethylene _____

    a. promotes synthesis and release of cellulase
    b. stimulates RNA transcription by acting as a derepressor
    c. activates a proton pump in the cell membrane
    d. mechanism unknown, but a clue may be in the hormone's resemblance to adenine

19. Cite several plant processes that result from the interactions of a pair of hormones.
20. Discuss the evidence that has been accumulated to support the proposed existence of a flowering hormone.

PERFORMANCE ANALYSIS

1. A hormone may be defined as a substance that is produced in one tissue and that exerts its highly specific effect, in very small quantities, on a second tissue. Its effect is dependent on other factors in the intercellular environment, including the presence of other hormones. Thus, only choice **c** is correct.

2. Applying auxin to an intact stem (b) has no effect except in high concentrations, where stem growth is inhibited. The presence of CA^{2+} (d) maintains the integrity of cell walls in the shoot tip, thus preventing cells from elongating. Since auxin is produced in the apical meristem, removal of the shoot tip (a) stops shoot growth. However, if the stump is later treated with auxin, normal growth will resume; thus, **c** is correct.

3. Abscission, or the dropping of leaves, is a direct response to the hormone ethylene, whose production is stimulated by auxin; therefore, **c** is correct.

4. Abscisic acid (**d**) is a growth-inhibiting hormone. Among its inhibitory effects is the promotion of bud and seed dormancy.

5. Auxin, produced in the apical meristem, and cytokinin, produced in root cells, have opposite effects on the growth of lateral buds: Auxin inhibits, and cytokinin stimulates. Therefore, buds near the bottom of the plant where auxin concentrations are low and cytokinin concentrations are high will be the first to grow. Application of cytokinin to lateral buds will also promote growth; therefore, **e** is correct.

6. Cytokinin causes cells to divide. As long as cells are dividing, they remain undifferentiated (**b** is correct).

7. Auxins cause cell enlargement, and cytokinins stimulate cell division. The course that undifferentiated tissue will take depends on the relative amounts of these two hormones; therefore, **b** is correct. Refer to Figure 31–7.

8. The general effect of gibberellins is hyperelongation of stems, which accounts for the effects stated in a and c. Gibberellins also cause cell differentiation in the vascular cambium and promote synthesis of hydrolytic enzymes in embryos (**e** is correct).

9. Ethylene causes fruits to ripen by promoting the release of cellulase, which acts to break down cell walls, producing changes in the color, texture, and chemical composition of the fruit (**d**).

10. A Japanese scientist was the first to correlate bolting in diseased rice plants with a substance produced by an infecting fungus (**c**). This substance was named after the fungus *Gibberella*.

11. Choice **b** is correct. See Figure 31–12.

12. Choice **d** is correct. See the analysis for Question 20.

13. Choice **b** is correct. See the analysis for Question 20.

14. Photoinduction (**e**) refers to the initiation of a plant response when the plant is exposed to an appropriate light cycle.

15. Ethylene (**c**), which is a gas at room temperature, has the chemical formula $H_2C{=}CH_2$. The structures of the other hormones introduced in this chapter are given in the figures alongside the text, although it is not likely that your instructor will require you to know them.

16. The relative amounts of secondary xylem and secondary phloem that will be produced from the vascular cambium (a lateral meristem) are determined by gibberellins and their interactions with auxin. Gibberellins act to stimulate the production of phloem. Choice **a** is therefore correct.

17. The correct answers in order are **b, d, e, a,** and **c**.

18. The correct answers in order are **b, c, d,** and **a**.

19. Among the interactive effects discussed in the chapter are the interaction of cytokinin and auxins in growth and differentiation; the interaction of gibberellins and auxins in the differentiation of vascular tissue and the production of flowers; and the interaction of ABA and gibberellins in hyperelongation of stems.

20. Chailakhyan's experiments on flowering, depicted in Figure 31–15, indicated that there is some chemical produced in leaves that initiates flowering. This substance has been given the name florigen, but it has never been isolated.

CHAPTER 32

Plant Responses to Stimuli

CHAPTER ORGANIZATION

I. Introduction
II. Phototropism
III. Geotropism
IV. Photoperiodism
   A. Introductory remarks
   B. Photoperiodism and flowering
      1. Background information
      2. Measuring the dark
   C. Photoperiodism and phytochrome
   D. Other phytochrome responses
   E. Essay: The discovery of photoperiodism
V. Circadian rhythms
   A. Introduction
   B. Biological clocks
      1. Background information
      2. Setting the clock
      3. Clock functions
      4. The nature of the clockwork
VI. Touch responses
   A. Twining and coiling
   B. Rapid movements in the sensitive plant
   C. The effects of touch on plant growth
   D. Responses in carnivorous plants
VII. Summary

MAJOR CONCEPTS

Plants are able to sense and respond to their environment; in addition, they are able to anticipate and prepare for environmental changes. Two responses of high survival value for young plants are phototropism and geotropism. Phototropism, the bending of plants toward a light source, is caused by auxin. Geotropism, the tendency of shoots to grow upward and of roots to grow downward, is apparently accomplished by the action of abscisic acid produced by the root cap.

Photoperiodism involves the changes that organisms undergo in response to the relative lengths of the light and dark periods of a day. Long-day plants flower when the period of light exceeds a critical length; short-day plants flower when the period of light is less than some critical period. Even the briefest interruption of the dark phase of the photoperiod by light reverses the photoperiod effects, indicating that the dark period rather than the light period is the critical factor. Phytochrome is the pigment in plants that detects the transition between light and dark periods.

Circadian rhythms are regular cycles of activity and growth that correlate approximately with a 24-hour day.

All vascular plants respond to touch and other mechanical stimuli with altered growth patterns. Some plants (Venus flytraps and other sensitive plants, for example) exhibit rapid movements in response to tactile stimuli.

HOW TO STUDY THE CHAPTER

Read the chapter, focusing on the major concepts.

Reread the chapter, using the questions that follow to help you focus on details as they relate to major concepts. Answer the questions on a separate sheet of paper. Your answers will provide a valuable study aid in preparing for examinations.

FOCUS ON CHAPTER DETAILS

I. Introduction (page 642)
1. In what ways do plants resemble organisms with nervous systems?

II. Phototropism (pages 642–643)
2. What is phototropism?
3. Describe the experiment Charles Darwin performed to account for the bending of grass shoots in response to light.
4. What technique was used by Frits Went to study phototropism?
5. What results did Went record, and what conclusions did he draw about the phototropic response of stems?
6. How can the behavior of auxin account for the observations made by Darwin and Went?

III. Geotropism (pages 643–644)
7. What is geotropism?
8. In what way was auxin thought to influence the upward growth of the shoot and the downward growth of the root of a seedling?
9. Why is it unlikely that auxins are responsible for either of these phenomena?
10. How is root growth influenced by abscisic acid?
11. What appears to be the role of the root cap in determining the direction of growth of young roots?

IV. Photoperiodism (pages 644–648)
A. Introductory remarks
12. What is photoperiodism? What does it enable plants to do?

B. Photoperiodism and flowering

Background information
13. What is a day-neutral plant?
14. How do short-day and long-day plants differ with respect to flowering?
15. When do plants begin to respond to photoperiodism?

Measuring the dark
16. Under what photoperiodic conditions does the cocklebur flower?
17. What effects do interruptions in the light and dark periods have on the cocklebur's ability to flower?
18. What are the effects of interrupting the dark period of a long-day plant?

C. Photoperiodism and phytochrome
19. Describe the two different forms of phytochrome.
20. How do long-day and short-day plants differ in their responses to concentrations of the active form of phytochrome?
21. How are the relative amounts of the two forms affected by light and dark?

D. Other phytochrome responses
22. What effects do red and far-red light have upon seed dormancy? Explain the adaptive value of these effects.
23. What is meant by etiolated growth? What stimulus must a dark-grown seedling receive before it begins normal growth?
24. What is one hypothesis regarding the mechanism of action of phytochrome? What evidence is there to support it?

E. Essay: The discovery of photoperiodism (pages 646–647)
25. How did the flowering of Maryland Mammoth tobacco and Biloxi soybeans provide agriculturalists with a knowledge of photoperiodism?
26. How is such knowledge useful in explaining plant distributions?
27. How do organisms other than plants make use of photoperiodism?

V. Circadian rhythms (pages 648–650)
A. Introduction
28. Give several examples of activities of plants that exhibit a daily rhythm.
29. What is a circadian rhythm? How is such a rhythm affected by constant environmental conditions?

B. Biological clocks

Background information

30. If plant rhythms are externally controlled, what environmental forces might be responsible?
31. What evidence is there that organisms possess biological clocks—that is, that their rhythmic activities are internally controlled?

Setting the clock

32. What is entrainment? Cite an example. Under what conditions do entrained plants revert to their natural rhythms?

Clock functions

33. What is the primary function of a biological clock?
34. Cite an example of a secondary function attributed to the development of "timekeeping."

The nature of the clockwork

35. Why is it now thought that the reversion of P_{fr} to P_r does *not* serve as a time-measuring system?
36. How might phytochrome conversion serve instead as a clock-setting mechanism?

VI. **Touch responses** (pages 650–653)

A. Twining and coiling

37. What are the functions of tendrils?
38. What is circumnutation, and of what value is it to plants?
39. What stimuli are required in order for a tendril to exhibit a curling response?
40. Why is it thought that some plants have a way of storing tactile information?
41. How is circumnutation advantageous to the parasitic dodder plant?

B. Rapid movements in the sensitive plant

42. Explain how a tactile stimulus causes sudden movements in *Mimosa pudica*.
43. How does the stimulus spread to other parts of the plant?

C. The effects of touch on plant growth

44. How do the growth patterns of plants growing in greenhouses differ from patterns of the same plants growing wild?
45. What external stimuli are responsible for this difference, and what internal effects might these stimuli have on the plants?

D. Responses in carnivorous plants

46. Describe how unwary insects trigger electrical impulses in the modified leaves of the Venus flytrap and the sundew.

VII. **Summary** (pages 653–654): Read the Summary. If you are familiar with the essential features of the material presented there, you are ready to take the following diagnostic examination.

TESTING YOUR UNDERSTANDING

After you have completed the following examination, compare your answers with the annotated key in the Analysis section.

1. The bending of stems toward the light has been shown to be due to the _____ of auxin _____.
   a. migration; away from the light side
   b. migration; away from the dark side
   c. degradation; on the light side
   d. degradation; on the dark side

2. Darwin and his son were among the first to show that plants:
   a. exhibit a phototropic response.
   b. synthesize phytochrome pigments.
   c. respond in specific ways to mechanical stimuli.
   d. are affected by gravity.
   e. respond to auxins by undergoing cellular elongation.

3. The downward growth of roots was at first believed to be due to differing concentrations of an inhibitory substance in the horizontal seedling. This has been difficult to reconcile because:
   a. auxins cannot be transported through cell membranes.
   b. auxins are soluble and would not be affected by gravity.

    c. abscisic acid is present only in leaf and stem tissue.
    d. an inhibitory hormone would prevent elongation at both surfaces of the root.

4. The righting of a seedling is best characterized as a _____ response.
    a. phototropic
    b. photoperiodic
    c. tactile
    d. geotropic
    e. chemotropic

5. Spinach is a long-day plant with a critical photoperiod of 14 hours. This means that spinach will flower if exposed to an uninterrupted_____ period of _____ hours.
    a. dark; 10 or more
    b. dark; 10 or less
    c. light; 14 or more
    d. light; 14 or less
    e. light; 10 or less

6. A plant that is day-neutral will flower:
    a. in its first year of growth.
    b. only during the night.
    c. independent of photoperiods.
    d. when light and dark periods are of equal length.

7. During daylight hours, phytochrome pigment(s):
    a. P_{fr} absorbs far-red light.
    b. P_r is converted to P_{fr}.
    c. P_r and P_{fr} are relatively stable.
    d. P_{fr} is degraded.
    e. Both a and d are correct.

8. Flowering will _____ if a sufficient concentration of _____ has accumulated.
    a. occur in a short-day plant; P_{fr}
    b. not occur in a short-day plant; P_{fr}
    c. occur in a long-day plant; P_r
    d. not occur in a long-day plant; P_{fr}

9. A seedling that is etiolated:
    a. exhibits stunted shoot growth.
    b. produces large, sprawling leaves.
    c. is colorless.
    d. will not survive even if moved into light.
    e. Both a and b are correct.

10. Plants that live on a forest floor beneath a dense canopy of vegetation absorb primarily _____ wavelengths.
    a. red
    b. blue
    c. far-red
    d. green
    e. yellow

11. The evolution of an internal biological clock has enabled organisms to _____ changes in environmental rhythms.
    a. adjust to seasonal changes without having to perceive
    b. adjust their internal rhythms to
    c. make immediate or short-term responses to
    d. avoid entrainment by

12. Which of the following is true about circadian rhythms?
    a. They are approximately 24 hours in duration.
    b. They can be reset by environmental cues.
    c. They persist even under constant environmental conditions.
    d. There is strong evidence that they are endogenous in nature.
    e. All of the above are correct.

13. Circadian rhythms are *not* believed to be controlled by external factors because:
    a. rhythmic patterns differ from individual to individual.
    b. the rhythms persist even when the organism is subjected to constant environmental conditions.
    c. the rhythms do not respond to changes in environmental stimuli.
    d. internal timekeeping mechanisms have been discovered.
    e. Both a and b are correct.

14. A plant that adjusts its daily rhythms when moved to a different environment in a greenhouse is said to exhibit:
    a. entrainment.
    b. photoinduction.
    c. photoperiodism.
    d. circumnutation.
    e. geotropism.

15. Communication among plant cells:
    a. may rely upon the transmission of electrical impulses.

b. must occur without significantly affecting a cell's turgor.
  c. is not moderated by environmental stimuli.
  d. is entirely a hormone-mediated process.
16. Gentle stroking of the tendril of a pea plant will *not* induce curling if:
  a. the tendrils are stroked in the dark.
  b. ATP is unavailable.
  c. auxin is applied to the stroked side of the tendril.
  d. the stroking is discontinuous.
17. The folding of the leaflets of the sensitive plant, *Mimosa pudica*, is a _____ response.
  a. phototropic
  b. tactile
  c. geotropic
  d. predatory
18. Which of the following represents the response of a plant to mechanical stimuli?
  a. etiolation
  b. righting of a seedling
  c. germination
  d. a short, stocky growth pattern
  e. Both a and b are correct.
19. A dodder plant makes use of coiling behavior in order to:
  a. anchor itself to the soil surface.
  b. compete for light.
  c. entrap and digest insects.
  d. set up a feeding network on its host.
20. Match each type of plant response with an appropriate example.

  phototropism   a. leaf folding in the sensitive plant
  _____
  photoperiodism   b. righting of an oat seedling
  _____
  geotropism _____   c. coiling of a pea plant tendril
  circumnutation   d. bending of grass shoots toward the light
  _____
  entrainment
  _____   e. adjustment of sleep movements to an artificial environment

                         f. ragweed flowers when day length is less than 14 hours

  electrically stimu-
  lated rapid response
  _____

21. Discuss those responses in plants known to be associated with the conversion of phytochrome pigments.
22. Why is it thought that circadian rhythms are controlled by a clock mechanism? How is this internal clock affected by environmental changes?
23. Discuss an example of a rapid movement of a plant that is mediated by an electrical impulse.

PERFORMANCE ANALYSIS

1. This question refers to Went's experiment, depicted in Figure 32–2. The result of the experiment was that the stem bent away from the side holding the agar block. Went supposed that this happened because a chemical substance had been transferred from the cut coleoptile tip, via the agar block, to the decapitated seedling, and that this substance caused elongation of the stem on the side on which it was placed. Although Went's experiment did not go so far as to show the mechanism by which light alters auxin concentrations in the stem, it is now known that the bending effect is due to the migration of auxin away from the illuminated side (**a**).
2. Charles Darwin and his son were able to show that the phototropic response (**a**) of plants is due to an influence transmitted from the upper to the lower part of a stem.
3. It was first suspected that auxin inhibited growth in the lower surface of the root, thus causing it to turn downward. But the uneven distribution of any inhibitory hormone (auxin or, as is now suspected, abscisic acid) in the root is not likely to be a simple, direct effect of gravity, since hormones are soluble and would not be affected by gravity (**b**).
4. This type of response is called geotropic (**d**).
5. The determining factor in flowering is the length of the uninterrupted dark period. Thus, if spinach

flowers when the amount of daylight exceeds 14 hours, the dark period should be 10 hours or less (**b**).

6. A day-neutral plant is defined as one that flowers without regard to day length; therefore, choice **c** is correct.

7. During daylight hours, red wavelengths predominate, so that phytochrome P_r is converted to P_{fr} (**b**). During the night, P_{fr} is degraded or converted back to P_r.

8. Phytochrome P_{fr} inhibits flowering in short-day plants and promotes flowering in long-day plants; therefore, **b** is correct.

9. Etiolation—elongation, spindly growth of stems, and growth of small, stunted leaves—occurs when a seedling is kept in the dark. Because the manufacture of chloroplasts is stimulated by light, an etiolated seedling is colorless (**c**). Returning such a plant to the light causes normal growth to resume.

10. Plants absorb more red and far-red than blue, green, and yellow wavelengths (see Figure 10-8, page 216). But plants on the forest floor absorb mainly far-red wavelengths (**c**), since the red wavelengths are absorbed by the overhanging vegetation.

11. Organisms can modify their biological clocks so that their internal rhythms become synchronized to environmental rhythms. This adjustment to an externally imposed rhythm is known as entrainment. Choice **b** is therefore correct.

12. A circadian rhythm is approximately 24 hours long. Because these rhythms appear to persist under constant conditions (although they can be modified by certain environmental cues), they are believed to be endogenous in nature; therefore, **e** is correct.

13. Choices a and b are both true and support the hypothesis that circadian rhythms are not under external control. Choices c and d are false statements. Choice **e** is therefore correct.

14. The adjustment of circadian rhythms to externally imposed rhythms is known as entrainment (**a**).

15. As a result of research on the mechanisms of movement in the sensitive plant and several carnivorous plants, communication among plant cells is no longer considered to be exclusively hormonal in nature. Instead, electrical impulses (**a**) have been shown to play an important role in many responses. Changes in turgor pressure have also been implicated in movement.

16. The tendrils will curl in response to auxin and under the influence of light, even if stroking takes place in the dark or is discontinuous. But if ATP is unavailable (**b**), curling will not take place.

17. When the leaflets of *Mimosa* are touched, turgor-pressure changes occur in the thickenings at the base of the leaflets. Thus, the folding that follows can be categorized as a tactile response (**b**).

18. Etiolation (a) and germination (c) are phototropic responses. Righting (b) is geotropic in nature. Plants respond to tactile stimuli produced by external factors (neighboring plants, the wind, raindrops, etc.) by altering their growth pattern to become short and stocky (**d** is correct).

19. The parasitic dodder plant coils around its host plant, developing haustoria, by which it feeds; therefore, **d** is correct.

20. The correct answers in order are **d**, **f**, **b**, **c**, **e**, and **a**.

21. You should discuss the influence of accumulations of P_r and P_{fr} on phenomena such as flowering, germination, and development.

22. Circadian rhythms are thought to be controlled by an internal clock because plants exhibit consistent, daily rhythms for various activities; these rhythms differ from species to species and from individual to individual; and the rhythms persist even when all environmental conditions are kept constant. The internal clock responds to environmental changes by resetting its own rhythm to become synchronized with external ones.

23. Three examples of rapid movement in plants that are regulated by electrical impulses are leaf folding in *Mimosa*, tentacle bending in the sundew, and leaf shutting in the Venus flytrap.

SECTION 6 — Biology of Animals

CHAPTER 33

The Human Animal: An Introduction

CHAPTER ORGANIZATION

I. Introduction
II. Organization of the human body
III. Cells and tissues
    A. Introductory remarks
    B. Epithelial tissue
    C. Connective tissue
    D. Muscle tissue
        1. General information
        2. Striated muscle
        3. Smooth muscle
    E. Nerve tissue
IV. Functions of the organism
    A. Introductory remarks
    B. Energy and metabolism
    C. Homeostasis
    D. Integration and control
        1. General information
        2. Feedback control
    E. Continuity of life
V. Summary

MAJOR CONCEPTS

As vertebrates, humans have a bony, jointed supporting endoskeleton that includes a skull and a vertebral column enclosing the central nervous system. Human bodies contain a coelom, which is divided into the thoracic and abdominal cavities by the diaphragm.

In multicellular animals above sponges, cells with similar structures and functions are organized into tissues; tissues are functionally integrated to form organs; and organs that are functionally related are grouped into organ systems.

The four major tissue types in the human body are epithelial tissue (forms body coverings and linings and also glands), connective tissue (secretes substances that form the intercellular matrix), muscle tissue (is specialized for contraction), and nerve tissue (conducts electrical impulses).

As an integrated organism, a multicellular animal has the capacity to perform several functions, including the acquisition and processing of nutrients, regulation of its internal environment (homeostasis), integration and control of its life processes, and reproduction.

HOW TO STUDY THE CHAPTER

Read the chapter, focusing on the major concepts.

Reread the chapter, using the questions that follow to help you focus on details as they relate to major concepts. Answer the questions on a separate sheet of paper. Your answers will provide a valuable study aid in preparing for examinations.

FOCUS ON CHAPTER DETAILS

I. Introduction (pages 659–661)

1. List several vertebrate characteristics of humans.
2. Into what parts is the human coelom divided? What major organs are found in the two major body cavities?
3. Humans are mammals. List the major characteristics of mammals.
4. What developmental tendencies found among mammals reach their culmination in humans?

II. Organization of the human body (page 661)

5. How do the cells of a multicellular organism differ from the cells of unicellular organisms with respect to functional organization?
6. Name the four levels of organization.

III. Cells and tissues (pages 661–666)

A. Introductory remarks

7. What are the four types of tissue into which cells are organized?

B. Epithelial tissue

8. What functions are associated with epithelial tissue?
9. What is the basement membrane?
10. Describe an important specialization of epithelial cells.
11. How is epithelial tissue classified? What are the cellular characteristics of the three types of epithelium? (See Figure 33–4.)

C. Connective tissue

12. How is connective tissue organized?
13. What three types of fibers are associated with the intercellular matrix that supports connective tissues?
14. What criteria are used to classify connective tissue? Name some common examples of connective tissue and their characteristic matrix material.

D. Muscle tissue

General information

15. By what process do muscle tissues provide for all locomotory activities? What two proteins are required?
16. List and describe the types of muscle tissue according to three separate classifications.

Striated muscle

17. What two types of muscles are striated in appearance?
18. How are muscles attached to the skeleton?
19. Explain how muscles function in antagonistic pairs to cause both bending and straightening of body parts around a joint. (See Figure 33–7.)
20. How are individual cells organized into muscles?
21. Cardiac muscle is, like skeletal muscle, striated. In what way is it different from skeletal muscle?

Smooth muscle

22. Give a general description of a smooth muscle cell.
23. How do the contractions of smooth muscles differ from those of skeletal muscles?
24. Where are smooth muscles typically found?
25. How are the smooth muscle cells that are found in the linings of hollow organs arranged? What function does this arrangement serve?

E. Nerve tissue

26. What two types of cells constitute nervous tissue? What is the general function of each type?
27. What is the role of each of the three functional classes of neurons?
28. A neuron is strikingly different from other cells because it has numerous extensions, or processes. Describe the functions of these processes, and compare their arrangement in the three classes of neurons (Figure 33–9).

29. Why are neurons often the longest cells found in animals?
30. How is a nerve different from a neuron?

IV. **Functions of the organism** (pages 666–669)

A. Introductory remarks

31. What is the evolutionary significance of organ systems?

B. Energy and metabolism

32. What problem implicit in the second law of thermodynamics must be solved by all organisms?
33. What general requirement must be met in order to solve this problem? What organ systems have evolved to meet this requirement?

C. Homeostasis

34. What problems do the sensitivity of the structure of organic molecules and the chemical reactions in which they participate, pose for organisms?
35. How must organisms deal with these general problems?
36. How does multicellularity and/or increased size enhance an organism's ability to maintain homeostasis?

D. Integration and control

General information

37. A multicellular organism must coordinate the activities of its tissues and organs to meet its changing demands. What two systems share this task? How do they differ in terms of response time?
38. Cite several examples that show the close interrelationship between these two systems.
39. What are the two major divisions of the nervous system? How are these divisions functionally different?
40. How is the autonomic nervous system subdivided? What general role does each of these subdivisions play?

Feedback control

41. How does a negative feedback loop act as a control on an organism's activities?

42. Describe the control of blood glucose levels by the pancreas. What two types of negative feedback loops are involved?
43. Describe an example of negative feedback involving both the nervous and endocrine systems.
44. Describe an example of a negative feedback system that involves an additional loop. How can we modify the principle of negative feedback based on this example?
45. Study Figure 33–13 to aid your understanding of feedback control. What parts of the system represent nervous control? Hormonal control? Which organs or tissues act as effectors? As targets?

E. Continuity of life

46. Outline the essential features of mammalian reproduction.

V. **Summary** (page 670): Read the Summary. If you are familiar with the essential features of the material presented there, you are ready to take the following diagnostic examination.

TESTING YOUR UNDERSTANDING

After you have completed the following examination, compare your answers with the annotated key in the Analysis section.

1. Which of the following is *not* characteristic of mammals?
   a. prolonged juvenile stage
   b. extremely large litters
   c. highly developed brains
   d. homeothermy
   e. high metabolic rates

2. The chief function of epithelial tissue is to:
   a. serve as a protective or absorptive lining.
   b. bind together the cells of other tissues.
   c. provide a supply of substances for the intercellular matrix.
   d. insulate nerve cells.
   e. Both b and c are correct.

3. Fibers, such as collagen, are products of cells of the _____ tissues.

a. epithelial
   b. striated muscle
   c. smooth muscle
   d. nerve
   e. connective

4. The cells of smooth muscle:
   a. contain both actin and myosin.
   b. are generally under involuntary control.
   c. are mononucleate.
   d. are arranged in two layers that contract alternately.
   e. All of the above are correct.

5. Tendons are used to connect _____ to _____.
   a. smooth muscle fibers; epithelial linings
   b. skeletal muscles; bone
   c. nerve fibers; the spinal cord
   d. connective tissues; surrounding cells
   e. internal organs; the lining of the coelom

6. The intercellular matrix of bone is:
   a. called cartilage.
   b. primarily empty space.
   c. calcified.
   d. plasma.
   e. fibrous.

7. A striated muscle fiber is best described as:
   a. a single mononucleate cell.
   b. a single multinucleate cell.
   c. many mononucleate cells working as a unit.
   d. many multinucleate cells and their associated connective tissue.

8. Movement of bones around a joint is caused by the simultaneous _____ of antagonistic groups of muscles.
   a. contraction and relaxation
   b. contraction
   c. expansion
   d. rotation

9. A(n) _____ is a process that carries an electrical impulse _____ the cell body.
   a. axon; away from
   b. dendrite; toward
   c. nerve; away from
   d. interneuron; either toward or away from
   e. Both a and b are correct.

10. The human coelom is divided into its two major cavities by a:
    a. septum.
    b. mesentery.
    c. layer of undifferentiated mesoderm.
    d. muscular diaphragm.
    e. vertebral column.

11. The most important consequence of the second law of thermodynamics for living things is the need:
    a. to extract energy from the inanimate world in order to maintain their organization.
    b. to maintain a constant internal environment.
    c. to dissipate metabolic heat very rapidly.
    d. to create their own sources of energy and raw materials.

12. The ability of an organism to maintain an internal environment distinctly different from its external environment is referred to as:
    a. metabolism.
    b. integration.
    c. homeostasis.
    d. feedback control.
    e. autonomic control.

13. Which of the following animals would have the *fewest* problems maintaining a constant body temperature under stress?
    a. a small aquatic organism
    b. a large aquatic organism
    c. a small terrestrial organism
    d. a large terrestrial organism

14. Which of the following best describes the relationship between the human endocrine and nervous systems?
    a. They act independently in integrating a specific set of activities.
    b. They are anatomically unrelated.
    c. Hormones and neurons interact to produce overall metabolic effects.
    d. Nervous control is primarily stimulatory in nature; endocrine control is primarily inhibitory.

15. Parasympathetic and sympathetic represent subdivisions of the _____ system.
    a. somatic
    b. autonomic
    c. endocrine
    d. skeletomuscular

16. Which of the following effectors is a large ganglion?
    a. the pituitary gland
    b. the adrenal medulla
    c. a sensory neuron
    d. the thyroid gland
    e. a thermoreceptor

17. The somatic nervous system innervates:
    a. skeletal muscle.
    b. smooth and cardiac muscle.
    c. endocrine glands.
    d. Both a and b are correct.
    e. Both b and c are correct.

18. Which of the following would be a result of stimulation (innervation) by the sympathetic nervous system?
    a. waves of contraction moving food along the small intestine
    b. a twitch in an arm muscle
    c. increased heart rate when you are in danger
    d. being able to stand perfectly still for long periods

19. In a negative feedback system, the _____ regulate(s) the production or release of hormones.
    a. secretory tissue itself
    b. concentration of another hormone
    c. response of the target tissue
    d. concentrations of the hormones themselves
    e. Both c and d may be true, depending on the system.

20. All mammals:
    a. have hair, fur, or feathers.
    b. have one or more pairs of mammary glands.
    c. reproduce by sexual means.
    d. can regulate their internal temperature.
    e. Only b, c, and d are correct.
    f. All four answers (a, b, c, and d) are correct.

21. Match each entry listed on the left with the applicable term or terms from the list on the right.

    nerve fibers _____
    muscle protein _____
    extracellular parts of connective tissue _____
    stimulate effector organs _____
    receive/relay information to the central nervous system _____
    striated cells _____

    a. sensory receptors
    b. ground substance
    c. motor neurons
    d. skeletal muscle
    e. axons
    f. hormones
    g. smooth muscle
    h. actin
    i. fibers
    j. dendrites
    k. myosin
    l. sensory neurons
    m. cardiac muscle

22. Cite a biological example of negative feedback control. Use a diagram to show the interactions.

23. Relate as many mammalian characteristics as you can to the development of homeothermy.

24. Compare the nervous and endocrine systems. Consider both organizational and functional aspects. How are the two systems structurally interrelated?

25. What is the primary role of interneurons? Where are they usually found? Label the indicated parts of this interneuron.

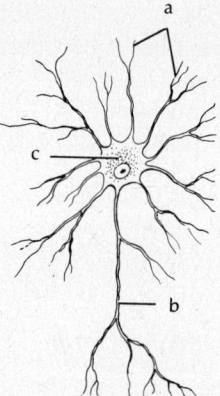

PERFORMANCE ANALYSIS

1. Choice **b** is correct. See the last two paragraphs of the chapter introduction.

2. Gland cells in epithelial linings secrete mucus, which helps to protect the linings. Any substance entering or leaving the body passes through an epithelial cell. Therefore, **a** is correct.

3. Collagen, a fiber found in skin and bone, is secreted by cells of connective tissues (**e**).

4. Smooth muscle cells contain the contractile proteins actin and myosin and are mononucleate. They are usually under involuntary control. Smooth muscle is arranged in two layers—an inner circular layer and an outer longitudinal layer—that contract alternately. Choice **e**, then, is correct.

5. Tendons, which are made up of supporting fibers of connective tissues, attach skeletal muscles to bone (**b**).

6. Young bone cells (see Figure 33-5) secrete calcium, which hardens to become part of the intercellular matrix of mature bone; therefore, **c** is correct.

7. Choice **b** is correct. A striated muscle fiber is a single cell with numerous nuclei that is formed by the fusion of many small mononucleate cells during embryonic development.

8. Choice **a** is correct. Note how the biceps and triceps contract and relax in opposition to one another in the raising of the forearm in Figure 33-7.

9. Choice **e** is correct. Dendrites are processes that receive stimuli from other cells and transmit them to the nerve cell body; axons are processes that carry the nerve impulse away from the cell body.

10. The diaphragm (**d**), a large, dome-shaped muscle, divides the human coelom into the thoracic and abdominal cavities.

11. Since the second law of thermodynamics states that all natural processes tend to proceed in the direction of increasing disorder (entropy), organisms must be efficient at extracting and using energy, the source of which is ultimately the inanimate world (e.g., energy from the sun), in order to maintain their own high degree of internal organization (**a**).

12. The capacity of an organism to control its internal environment is known as homeostasis; choice **c** is correct.

13. Since water serves as a temperature buffer for aquatic organisms, and since rate of heat loss depends on the surface-to-volume ratio of the organism, the large aquatic organism (**b**) would be most capable of maintaining homeostasis under temperature stress.

14. Choices a, b, and d are all false statements. The interaction of the nervous and endocrine systems (choice **c** is correct) is exemplified by control of body temperature in mammals (Figure 33-13).

15. The autonomic nervous system (**b**), which innervates smooth and cardiac muscle as well as glands, exercises involuntary control over body functions. It is divided into the parasympathetic and sympathetic subdivisions.

16. The adrenal medulla (**b**), generally thought of as a gland, is in reality an aggregation of nerve cell bodies (a ganglion).

17. Only choice **a** applies to the somatic nervous system. Choices b and c apply to the autonomic nervous system.

18. The sympathetic division of the nervous system exercises control over functions associated with emotion, stress, and danger. Thus, an increased heart rate in response to danger (**c**) would be due to sympathetic innervation.

19. In a negative feedback system, the production and release of a hormone can be regulated by altered levels of the hormone itself or by the response to the hormone by the target tissue (this response might also be hormonal in nature). (See the discussion of the influence of thyroxine on cellular metabolism on page 668.) Choice **e** is therefore correct.

20. All the choices listed except a (feathers are not found in mammals) are characteristic of mammals. Again, see the last two paragraphs in the introduction to the chapter. Choice **e** is therefore correct.

21. The correct answers in order are **e, j; h, k; b, i; c, e, f; a, j, l;** and **d, m**.

22. Any of the following would serve nicely as examples of negative feedback control: pancreatic control of blood glucose levels, hormonal and neural control of blood pressure, TSH-thyroxine regulation of metabolic rate. Your diagram should resemble Figure 33-13.

23. Homeothermy must have evolved concomitantly with high rates of metabolism and activity so that mammals could generate sufficient heat to warm themselves. They would have needed insulation in

the form of fat or fur to keep in the heat. The relatively high and constant body temperature would have allowed for increased physical and mental activity, which in turn would have required more refined sensory apparati and a more complex system to integrate and analyze information from the environment.

24. Basically, the endocrine system is a slow-response system. The messengers of the endocrine system, hormones, must travel through the bloodstream to effector tissues or organs. The nervous system, conveying messages by electrical impulses, is a rapid-response system. The two systems are closely interrelated: endocrine glands are often the effectors of neurons, and certain neurotransmitters at the synapses of neurons may act as hormones.

25. Interneurons transmit signals between neurons within the central nervous system (see Figure 33-11). Consult Figure 33-9b on page 665 to check your labels on the diagram.

CHAPTER 34

Energy and Metabolism I: The Digestive System

CHAPTER ORGANIZATION

I. Introduction
II. **Digestive tract in vertebrates**
   A. Introduction
   B. The oral cavity
   C. The pharynx and esophagus: Swallowing
      1. General information
      2. Essay: The Heimlich maneuver
   D. The stomach: Storage and liquefaction
   E. The small intestine: Final digestion and absorption
      1. General information
      2. Absorption of nutrients
   F. Essay: Mother's milk—It's the real thing
   G. Essay: Aids to digestion
   H. The large intestine: Absorption of water and elimination
III. **Major accessory glands**
   A. The pancreas
   B. The liver
IV. **Regulation of blood glucose**
V. **Some nutritional requirements**
   A. General information
   B. Essay: Amino acids and nitrogen
   C. The price of affluence
VI. **Summary**

MAJOR CONCEPTS

Digestion, which is under both hormonal and neural control, is the process by which food is broken down into molecules that can be absorbed into and used by the body.

Digestion begins in the mouth (breakdown of starches), continues in the stomach (mechanical churning, chemical liquefaction, and beginning of protein breakdown), and is completed in the small intestine (breakdown of carbohydrates, fats, and proteins). Absorption of monosaccharides, amino acids, and dipeptides into the blood vessels and of large fatty acids into lymph vessels occurs in the small intestine. Hormones secreted from duodenal cells stimulate the digestive functions of the liver and pancreas. The liver secretes bile, which emulsifies fats, and the pancreas secretes digestive enzymes.

Water and some minerals are reabsorbed from the food mass in the large intestine. Symbiotic bacteria in the large intestine are the source of certain vitamins.

Glucose is the primary source of energy to mammalian cells, and the liver, under hormonal control, is the main regulator of blood glucose levels.

Nutritional requirements of humans include calories, essential amino acids, essential fatty acids, vitamins, and minerals.

HOW TO STUDY THE CHAPTER

Read the chapter, focusing on the major concepts.

Reread the chapter, using the questions that follow to help you focus on details as they relate to major concepts. Answer the questions on a separate sheet of paper. Your answers will provide a valuable study aid in preparing for examinations.

FOCUS ON CHAPTER DETAILS

I. Introduction (page 671)

1. What is digestion? How are digested molecules used by organisms?
2. Trace the evolution of digestive systems.

II. Digestive tract in vertebrates (pages 671–682)

A. Introduction

3. Describe the general organization of the vertebrate digestive tract. What two digestive processes take place there?
4. What are the four layers of digestive tissue?
5. What digestive functions are performed by cells and tissues of the intestinal epithelium?
6. How does the organization of the muscularis externa help move food along the tract as well as control the food's passage from one area to another?

B. The oral cavity

7. Describe the arrangement of teeth in the adult human. How are teeth specialized according to function?
8. Describe the external and internal structure of a tooth.
9. What is the basic function of the tongue? Cite examples of how the tongues of some animals have become specialized for other functions.
10. Describe the content of saliva. What digestive roles do salivary secretions play?
11. How is the secretion of saliva controlled? What factors stimulate and inhibit this secretion?

C. The pharynx and esophagus: Swallowing

General information

12. Trace the path of food from the mouth to the stomach. To what extent is the movement of food under voluntary control? Under involuntary control?
13. Describe the peritoneum and its folds, the mesenteries. What function do they serve?

Essay: The Heimlich maneuver (page 675)

14. How does food strangulation occur? How can this condition be distinguished from a heart attack?
15. Describe the procedure (called the Heimlich maneuver) you would use to assist someone who was choking on a piece of food. How would you modify the maneuver if the victim was lying on his or her back?
16. Why should the victim see a physician even after the food has been dislodged?

D. The stomach: Storage and liquefaction

17. What is the appearance of the stomach when it is not full?
18. How is the size of the stomach related to an animal's feeding habits?
19. Describe the tissue in the stomach's mucosal lining whose secretions serve a digestive function. What are the functions of these secretions?
20. Explain how the stomach is under both hormonal and neural control.
21. Describe the processing of food in the stomach. How is food passed on to the small intestine?

E. The small intestine: Final digestion and absorption

General information

22. Describe the structural features of the lining of the small intestine. How do these features provide for a large surface area across which food molecules may enter the blood?
23. Name and describe the action of those enzymes that break down food molecules in the small intestine (see Table 34-1). Where is each of these enzymes produced, and how do they arrive at their sites of action?
24. Describe two other digestive secretions, their sites of production, and the roles they perform in the small intestine.
25. Describe the activities of the two hormones produced in the duodenum that regulate

digestive activities (see Table 34–2). To what extent is digestion under the control of the nervous system?

Absorption of nutrients

26. What food molecules enter the bloodstream by facilitated diffusion? By active transport?
27. How do large fatty acids and cholesterol enter the bloodstream?
28. What is the fate of fats and cholesterol once they have entered the blood?

F. Essay: Mother's milk—It's the real thing (page 678)

29. How does the content of human milk compare with that of cow's milk?
30. What other substances are present in the milk of mammals? How are these substances critical to the welfare of the suckling infant?
31. Why do milk formulas pose a hazard to the health of infants in developing countries?
32. Why is the consumption of lactose a problem for many children? What specific problems arise? Why?
33. What factor is apparently associated with the ability of the adult human to produce lactase?

G. Essay: Aids to digestion (page 681)

34. What benefits do herbivores derive from symbiotic microorganisms? How do the microorganisms profit?
35. What digestive organs are associated with gut symbionts?
36. What curious adaptation that exploits bacterial fermentation is exhibited by the lagomorphs?
37. What are the special adaptations of ruminants that have made them the most successful of the terrestrial herbivores?
38. What benefits do we receive from our own gut bacteria?

H. The large intestine: Absorption of water and elimination

39. What is the chief function of the large intestine?
40. How does water enter the digestive tract during the course of digestion?
41. What causes diarrhea? Why is diarrhea sometimes fatal?
42. How do symbiotic bacteria of the large intestine serve us?
43. Locate the appendix in Figure 34–5. What happens if it becomes inflamed? What important activity occurs in the appendix?
44. Describe the content of the feces. What happens to digestive wastes that remain in the large intestine after the water has been absorbed into the body?

III. **Major accessory glands** (pages 682–683)

A. The pancreas

45. What is the developmental origin of the pancreas?
46. List the digestive enzymes produced by the pancreas, and describe their functions. How do they travel to the small intestine?
47. What cells of the pancreas serve an endocrine function? What substances are produced there?

B. The liver

48. The liver, the body's largest internal organ, carries out a variety of functions. List the liver's functions in the following categories: synthesis, storage, regulation, and degradation.

IV. **Regulation of blood glucose** (page 684)

49. The blood sugar levels of vertebrates are maintained at fairly constant levels. Describe the role of the liver in this process.
50. How are these activities regulated hormonally?

V. **Some nutritional requirements** (pages 684–686)

A. General information

51. What are the relative caloric values of the three principal types of food molecules that supply humans with energy?
52. What requirements do vertebrates have with regard to the manufacture of proteins and fats? What is an essential amino acid?

53. What is a vitamin? What are the dietary sources and the metabolic roles of the various vitamins? (See Table 34–3.)
54. What benefit do people derive from ingesting large amounts of vitamins?

B. Essay: Amino acids and nitrogen (page 685)

55. How does the biosynthesis of nitrogen-containing compounds such as amino acids differ from that of carbohydrates and fats?
56. What nitrogen sources are used by plants in the manufacture of amino acids?
57. Which amino acids are considered to be essential for humans?
58. What risk might you assume in being a vegetarian? How could you adjust your diet to reduce this risk? How are agricultural scientists assisting you in this regard?

C. The price of affluence

59. What dangers are associated with the typical affluent American's diet with regard to intake of excess calories, excess salt, eating of animal fat, and nutritional fads?

VI. **Summary** (pages 686–687): Read the Summary. If you are familiar with the essential features of the material presented there, you are ready to take the following diagnostic examination.

TESTING YOUR UNDERSTANDING

After you have completed the following examination, compare your answers with the annotated key in the Analysis section.

1. Which of the following hormones is produced in cells of an organ that also contains the hormone's target tissues?
   a. cholecystokinin
   b. gastrin
   c. secretin
   d. trypsin
2. In each jaw of the adult human, there are _____ incisors, _____ canines, _____ premolars, and _____ molars.
   a. two, two, six, six
   b. four, two, four, six
   c. six, four, three, three
   d. four, no, six, six
   e. two, two, eight, four
3. Peristalsis refers to the movement of food along the digestive tract due to:
   a. wavelike contractions of smooth muscle.
   b. mucus secretions.
   c. the contraction and relaxation of sphincters.
   d. the entry of water into the tract by osmosis.
4. In humans, the breakdown of starches begins in the:
   a. stomach.
   b. small intestine.
   c. pancreas.
   d. mouth.
   e. large intestine.
5. The digestive tracts of carnivores differ from omnivores like ourselves in that carnivores have no:
   a. smooth muscle in the lining of the esophagus.
   b. sphincter muscles.
   c. digestive enzymes in their saliva.
   d. Both b and c are correct.
   e. Both a and c are correct.
6. The upper part of the esophagus is striated muscle and the lower part is smooth muscle. This implies that:
   a. peristalsis is under voluntary control.
   b. swallowing is under both somatic and autonomic control.
   c. peristalsis occurs in the upper part of the esophagus only.
   d. the muscle of the diaphragm is continuous with that of the upper esophagus.
7. Gastric glands are responsible for the production of:
   a. hydrochloric acid.
   b. pepsinogen.
   c. pepsin.
   d. gastrin.
   e. Both a and b are correct.
8. Which of the following breakdown products of digestion *cannot* enter directly into blood vessels of the villi?
   a. amino acids
   b. dipeptides
   c. simple sugars
   d. large fatty acids
9. Which of the following is a pair of hormones produced by the pancreas whose effects are antagonistic?

a. insulin and glucagon
   b. secretin and cholecystokinin
   c. somatostatin and bile
   d. gastrin and secretin
   e. cholecystokinin and bilirubin

10. What role does the pair of hormones referred to in the previous question perform?
    a. They regulate blood glucose levels.
    b. They stimulate or inhibit the production of sodium bicarbonate by the pancreas and liver.
    c. They determine the amount of water reabsorbed during digestion.
    d. They influence the rate of conversion of pepsinogen to pepsin.

11. The site of action for those enzymes secreted into the pancreatic duct is the:
    a. liver.
    b. pancreas.
    c. small intestine.
    d. gallbladder.
    e. stomach.

12. An essential amino acid is one that:
    a. is necessary for proper protein functioning.
    b. is present at the active site of the enzyme of which it is a part.
    c. can only be obtained by an organism from its diet.
    d. must be synthesized by an organism.
    e. Both a and b are correct.

13. Stomach acid is neutralized in the lower digestive tract as a result of the secretion of an alkaline fluid into the tract by the:
    a. pancreas.
    b. gallbladder.
    c. liver.
    d. duodenum.
    e. Both a and b are correct.

14. Colon cancer may result from eating a diet high in:
    a. fibrous foods.
    b. animal fat and protein.
    c. carbohydrates.
    d. vitamins.
    e. calories.

15. The chief source of vitamin K for humans is:
    a. green, leafy vegetables.
    b. symbiotic bacteria living in the large intestine.
    c. certain animal tissues.
    d. the liver, where it is synthesized from other vitamins.

16. Which of the following events is stimulated by a low pH?
    a. secretion of pancreatic enzymes
    b. conversion of pepsinogen to pepsin
    c. the release of gastrin from stomach cells
    d. the release of enzymes into the saliva
    e. the absorption of the breakdown products of proteins

17. In the human digestive tract, most food molecules are absorbed through the epithelium of the:
    a. liver.
    b. stomach and mouth.
    c. stomach only.
    d. small intestine.
    e. large intestine.

18. Which of the following is necessary for proper digestive functioning?
    a. The pH of the small intestine must be closely regulated.
    b. Following digestion, water must be reabsorbed from the tract.
    c. Mucus must be secreted by epithelial tissues of the stomach to protect it from self-digestion.
    d. The movement of food through the tract must be controlled by muscle movements.
    e. All of the above are correct.

19. Encouraging African mothers to adopt the use of milk substitutes instead of nursing is:
    a. praiseworthy because these substitutes are rich in calories, vitamins, and minerals.
    b. unwise because their infants usually are unable to break down lactose.
    c. unwise because mother's milk provides immunity against many diseases.
    d. probably neither really helpful nor harmful to the health of their infants.
    e. Both b and c are correct.

20. The Heimlich maneuver is employed in cases where:
    a. eating certain foods has brought on a heart attack.
    b. food has become lodged in the trachea.
    c. bones are caught in the esophagus.
    d. vomiting must be induced.

21. Animals having an enlarged cecum _____ in order to obtain additional nutrients.
    a. chew regurgitated food
    b. must produce vast quantities of digestive enzymes
    c. eat their feces
    d. can digest cellulose

22. Match each hormone or enzyme listed at the left with its site of synthesis.

    lactase _____   a. pancreas
    glucagon _____   b. stomach
    secretin _____   c. small intestine
    amylase _____   d. duodenum
    dipeptidase _____   e. liver
    gastrin _____
    lipase _____
    cholecystokinin _____

23. Match each structure with its function.

    incisors _____   a. greatly increase the absorptive surface of the digestive tract
    mesenteries _____
    microvilli _____   b. manipulates and mixes food with saliva
    mucosa _____
    muscularis externa _____   c. layers of smooth muscle that help to move ingested food
    peritoneum _____   d. a moist, protective lining of the abdominal cavity
    premolars _____
    salivary glands _____ e. folds of connective tissue used for support
    tongue _____   f. used for cutting food
    villi _____   g. epithelial tissue containing secretory cells
                             h. used for grinding food

24. Using a diagram of the human digestive tract, indicate what happens to a food particle containing carbohydrate, fat, and protein from the time it is ingested to the time its absorbable components enter the bloodstream.

25. How has our affluence contributed to serious health problems associated with our diet?

26. What risk do people take in becoming strict vegetarians? How can they plan their meals to obtain a well-balanced diet?

27. On the following diagram of the human digestive system, show where each of the events listed below takes place. Place the identifying letters after the part.

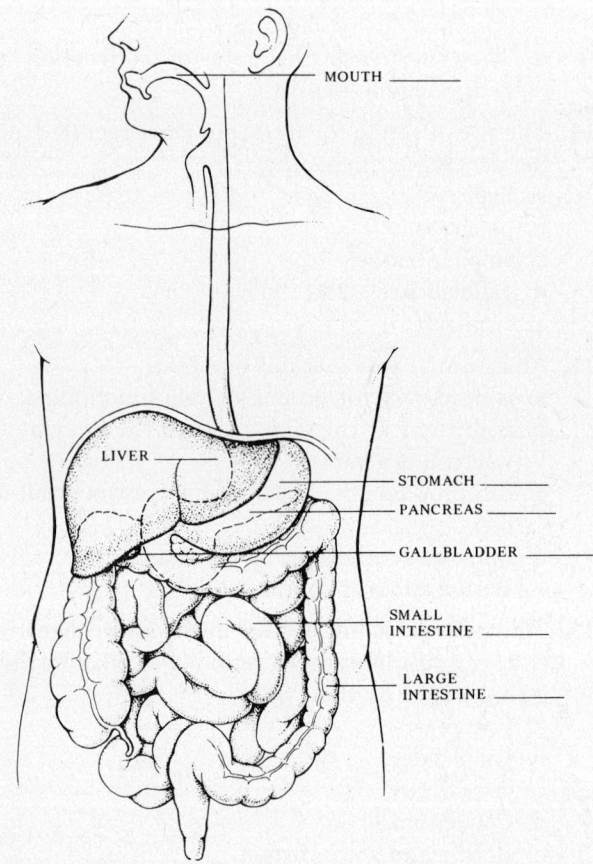

a. Lipases hydrolyze fats into glycerol and fatty acids.
b. Bile is produced.
c. The pyloric sphincter opens.
d. Most of the food molecules are absorbed.
e. Pepsinogen is converted to pepsin.
f. Proteins are broken down into amino acids.

g. The presence of food stimulates the secretion of amylases.
h. Alkaline fluid is secreted.
i. The stomach acids are neutralized.
j. HCl loosens the fibrous components of food.
k. Indigestible food remains are broken down by bacteria.
l. Bile is stored.
m. Mucus is secreted.
n. Gastrin is released into the bloodstream.
o. Glucose is converted to glycogen and stored.

PERFORMANCE ANALYSIS

1. Gastrin (**b**) is produced in pits of the mucosal layer of the stomach. Its target tissues are the chief cells and parietal cells of the stomach, which respond by producing pepsinogen and hydrochloric acid, respectively.
2. Choice **b** is correct.
3. The lower part of the esophagus is lined with smooth muscle whose wavelike contractions provide for movement of the food mass to the stomach (**a**).
4. Saliva, secreted into the mouth by the salivary glands, contains the enzyme amylase, which begins to break down starches; therefore, **d** is correct.
5. Carnivores, such as dogs, that tear and gulp their foods, have striated muscle along the length of the esophagus and no digestive enzymes in their saliva; therefore, **e** is correct.
6. Choice **b** is correct, since striated muscle is innervated by the somatic nervous system and smooth muscle by the autonomic nervous system.
7. Choice **e** is correct. See the analysis for Question 1.
8. Large fatty acids (**d**) diffuse through mucosal cells of the intestine where they are packaged as chylomicrons. These are then secreted into lymph vessels, which eventually carry them to the blood.
9. Insulin stimulates the uptake of glucose by cells, thus decreasing blood sugar levels. Glucagon, another pancreatic hormone, promotes glycogen breakdown in the liver and so increases blood sugar levels; therefore, **a** is correct.
10. As implied above, insulin and glucagon regulate blood sugar levels (**a**).
11. The pancreas, an organ whose evolutionary and embryological origin is as an outpouching of the small intestine, secretes enzymes (see Table 34–1) into the small intestine (**c**) via the pancreatic duct.
12. Amino acids that cannot be synthesized from other amino acids and so must be obtained in the diet (**c**) are referred to as essential amino acids.
13. Stomach acid that enters the small intestine is neutralized by alkaline fluid released by the pancreas (a) through the pancreatic duct and by bile (which contains sodium bicarbonate), secreted by the liver and released into the tract from the gallbladder (b). When these alkaline solutions reach the small intestine, its pH is maintained between 7 and 8. Choice **e** is therefore correct.
14. Colon cancer has been associated with the amount of time food sits in the lower digestive tract. Because diets high in animal fat and protein (**b**) lack fibrous materials that are essential for moving food through the digestive tract, people who have such diets are more susceptible to cancer of the colon.
15. Symbiotic bacteria living in the large intestine (**b**), such as *E. coli*, provide us with our chief source of vitamin K as a result of their metabolic activities.
16. When hydrochloric acid is secreted by glands in the stomach lining, the stomach pH drops to between 1.5 and 2.5, favoring the conversion of pepsinogen to its active form, pepsin (**b**), an enzyme that begins the breakdown of proteins in the stomach.
17. Although the digestion of starches begins in the mouth and the digestion of proteins begins in the stomach, the digestion of all food molecules is completed in the small intestine (**d**), where most of them are absorbed into the blood.
18. All four answers (**e**) are correct. With reference to choices a and d, see the analysis for Questions 3, 6, 13, and 16. Water entering the digestive tract by osmosis must be recovered in the intestine, or dehydration will occur. The mucus secreted by tissues in the stomach epithelium not only helps move food along the digestive tract but also prevents the digestive enzymes in the stomach from digesting its walls.

19. Giving infants lactose-rich milk substitutes instead of breast milk deprives them of antibodies normally contained in their mother's milk, leaving them more vulnerable to diseases (**c** is correct). Most infants do have enzymes that break down lactose. However, the production of lactase decreases after the age of four, especially in humans not of northern European descent.

20. The Heimlich maneuver is a procedure used to dislodge food that is threatening strangulation by blocking the trachea (**b**).

21. The lagomorphs, a group of mammals that includes the rabbit, have an enlarged sac at the junction of the small and large intestine (the cecum) in which bacterial fermentation occurs. These animals eat their feces (**c**), which are rich in bacteria and nutrients metabolized by these bacteria.

22. The correct answers in order are **c, a, d, a, c, b, a,** and **d**. (Refer to Tables 34–1 and 34–2.)

23. The correct answers in order are **f, e, a, g, c, d, h, g, b,** and **a**.

24. The carbohydrate component of the food particle is first broken down in the mouth by amylase into oligosaccharides. These are later digested in the small intestine by the enzymes shown in Table 34–1. Proteins are broken down into peptide fragments in the stomach by pepsin. These fragments are digested by various peptidases in the small intestine (again see Table 34–1). Fats are degraded to fatty acids in the small intestine by the action of lipases. Cholesterol and fatty acids too large to be absorbed into the blood are packaged as chylomicrons which first enter the lymphatic system.

25. Because Americans are affluent, they tend to overeat (becoming obese makes one more susceptible to circulatory disease and diabetes), use too much salt (encouraging high blood pressure), eat a diet overly concentrated in animal fats (responsible for atherosclerosis, heart attacks, and colon cancer), and subject themselves to a number of liabilities by experimenting with fad diets.

26. A number of vegetables are low in essential amino acids. Protein deficiencies may be avoided by combining the right vegetables in a single meal. For example, combining rice, which is deficient in isoleucine and lysine but contains adequate amounts of the other essential amino acids, with beans, which are high in both of these amino acids but deficient in tryptophan, provides all of the essential amino acids in adequate quantities.

27. Answers: **mouth**: g; **liver**: b, o; **stomach**: c— **between stomach and small intestine**, e, j, m, n; **pancreas**: h; **gallbladder**: l; small intestine: a, d, f, i, m; **large intestine**: k.

CHAPTER 35

Energy and Metabolism II: Respiration

CHAPTER ORGANIZATION

I. Introduction
II. Diffusion and air pressure
III. Evolution of respiratory systems
   A. Background
   B. Evolution of gills
   C. Evolution of lungs
IV. Respiration in large animals: Some principles
V. Human respiratory system
VI. Mechanics of respiration
VII. Essay: Cancer of the lung
VIII. Transport and exchange of gases
   A. Hemoglobin and its function
   B. Myoglobin and its function
IX. Control of respiration
X. Essay: Diving mammals
XI. Summary

MAJOR CONCEPTS

Respiration is the process by which an organism obtains the oxygen needed in cellular respiration and releases carbon dioxide.

Oxygen enters into cells and body tissues by diffusion, moving from areas of higher partial pressure to areas of lower partial pressure. Selection pressures for efficient gas exchange led to the evolution in vertebrates of gills and lungs, which provide the large surface areas required for oxygen diffusion. Respiration in large animals involves both diffusion and bulk flow.

In humans, air enters the nose or mouth and passes from the pharynx through the larynx and into the trachea. The trachea branches into two bronchi, which subdivide into bronchioles that terminate in alveoli. Gas exchange between the air and the blood in humans takes place in alveoli.

The respiratory pigment hemoglobin, which is a component of the red blood cells of vertebrates, transports oxygen in the blood. A single hemoglobin molecule has four subunits, each of which can combine with one molecule of oxygen.

The control center for respiration is in the brainstem. The respiratory neurons located there respond to very small changes in the blood concentrations of carbon dioxide or hydrogen ions and to larger changes in oxygen concentration.

HOW TO STUDY THE CHAPTER

Read the chapter, focusing on the major concepts.

Reread the chapter, using the questions that follow to help you focus on details as they relate to major concepts. Answer the questions on a separate sheet of paper. Your answers will provide a valuable study aid in preparing for examinations.

FOCUS ON CHAPTER DETAILS

I. Introduction (page 688)

1. How do heterotrophs obtain energy?
2. Distinguish between the two meanings of the term "respiration."
3. What is meant by basal metabolic rate? How may an organism's metabolic rate be measured? What effect does activity have on metabolic rate?

II. Diffusion and air pressure (pages 689–690)

4. In every organism, gas exchange takes place by diffusion. What is diffusion?
5. How may the partial pressure of specific gases in the atmosphere be determined? How is this calculated for oxygen in dry air?
6. What is meant by a blood oxygen partial pressure of 40 mm Hg?
7. How is equilibrium established between a liquid and gases in the air surrounding it?
8. How does altitude affect the P_{O_2} of the atmosphere?
9. What adaptations are seen in people who live at high altitudes?
10. Explain what is meant by the "bends." What damage can they cause?

III. Evolution of respiratory systems (pages 690–693)

A. Background

11. What limitations did an oxygenless atmosphere impose on life processes? Why was the accumulation of atmospheric oxygen such an important evolutionary step? (Refer to page 210 if necessary.)
12. Why is diffusion adequate for the supply of oxygen to small organisms but inadequate for large ones?
13. How do large organisms obtain oxygen efficiently?
14. How do earthworms and arthropods increase the effectiveness of diffusion for gas exchange?
15. How do these strategies influence the shape or size of organisms within these groups?

B. Evolution of gills

16. How are gills organized so as to provide for maximum gas exchange? (Consult Figure 35–5.)
17. Trace the evolution of the gill in primitive vertebrates.
18. Explain the functioning of the countercurrent arrangement of circulatory vessels. How is this beneficial for organisms with gills?

C. Evolution of lungs

19. What are lungs? What is the disadvantage of lungs in comparison with gills?
20. Why is it so much easier for a land-dwelling organism to obtain oxygen than one that lives in water?
21. What advantage do terrestrial organisms with lungs have over those without lungs?
22. What organisms besides vertebrates breathe by means of lungs?
23. How were lungs used among the primitive fishes? What structure gave rise to the lungs of the lungfish?
24. What respiratory strategies are employed by amphibians and reptiles? What is the purpose of the trachea, glottis, and nostrils?
25. What is a ventilation lung? How do frogs differ from other land vertebrates in their method of lung ventilation?
26. Study the respiratory systems of the various groups illustrated in Figure 35–8. What major similarities and differences do you see?

IV. Respiration in large animals: Some principles (page 694)

27. Describe the four stages by which large animals move oxygen from their external environment to their tissues.
28. How is carbon dioxide removed from the body?

V. Human respiratory system (pages 694–696)

29. Trace the passage of air from the nasal passages to the point at which oxygen

diffuses into the bloodstream (see Figure 35-9).

30. What four specializations are found within the nasal cavity? What is the purpose of each?
31. For what purpose is the larynx adapted? What happens to the human voice when changes in the vocal cords occur?
32. Where do smooth muscle and cartilage surround the respiratory tract? What functions do these structures serve?
33. How are foreign particles removed from the tract?
34. How do the alveoli provide for the entry of gases into the bloodstream?
35. How do the alveoli increase the respiratory surface area?
36. What are the functions of the pleura? What is pleurisy?

VI. **Mechanics of respiration** (pages 696–697)

37. How do the differences in atmospheric pressure and alveolar pressure determine the direction of air flow?
38. How do muscle movements accomplish inhalation and exhalation?
39. What percent of inspired air can actually be exchanged?
40. Why do large aquatic mammals suffocate on land?

VII. **Essay: Cancer of the lung** (page 697)

41. What are the alarming statistics regarding the incidence of lung cancer in the United States?
42. What protective functions are cancer cells unable to perform?
43. Speculate as to why looking at photographs of cancerous lung tissues does not always elicit responsible fear in the beholder and why concern often comes only when one is diagnosed as having the disease.

VIII. **Transport and exchange of gases** (pages 698–701)

A. Hemoglobin and its function

44. Why must active animals employ respiratory pigments?
45. What respiratory pigment is found in vertebrates? What respiratory pigments are found among invertebrates? How are these transported within the circulatory system?
46. Review the structure of hemoglobin (see Figure 3-27 on page 71). How does the molecule bind oxygen?
47. In what manner do individual oxygen atoms occupy the heme sites? What influence on the beta-chain structure does this binding have?
48. Relate changes in hemoglobin affinity for oxygen to the characteristic shape of the association-dissociation curve (Figure 35-13).
49. What determines whether oxygen combines with hemoglobin or is released from it?
50. Explain the unloading of oxygen from hemoglobin as determined by the P_{O_2} at various locations in the oxygen delivery system. (See also Figure 35-14.)
51. How is carbon dioxide carried in the blood?
52. According to what chemical equation does carbonic acid dissociate in the blood? How is this dissociation affected by changes in the P_{CO_2}? How does the P_{CO_2} in the blood affect the loading or unloading of oxygen?

B. Myoglobin and its function

53. Where are myoglobin pigments found? How do they resemble hemoglobin pigments?
54. How does myoglobin function in supplying a reserve of oxygen when the organism is engaged in strenuous exercise?

IX. **Control of respiration** (pages 701–703)

55. How are the rate and depth of respiration controlled?
56. How do respiratory neurons influence inhalation and exhalation?
57. How are changes in concentrations of blood gases detected?
58. What is hyperventilation? How are blood gas levels returned to normal following hyperventilation?

59. Why is it impossible to commit suicide by holding your breath?
60. Why are there chemoreceptors that monitor changes in oxygen concentration?

X. **Essay: Diving mammals** (page 702)
61. How do the circulatory systems of marine mammals that dive to great depths differ from our own?
62. What is the "diving reflex"? How does it allow diving mammals to survive despite a discontinuity in their oxygen supply?
63. How can you tell that humans have also inherited a "diving reflex"? Of what adaptive value is this reflex?
64. How did Nemiroff account for the recovery of people who had nearly drowned and the fact that some survived without brain damage?
65. If a person who has been submerged in cold water appears to be dead, what action should be taken?

XI. **Summary** (page 703): Read the Summary. If you are familiar with the essential features of the material presented there, you are ready to take the following diagnostic examination.

TESTING YOUR UNDERSTANDING

After you have completed the following examination, compare your answers with the annotated key in the Analysis section.

1. Which of the following pertains to the role of diffusion in the exchange of oxygen between organisms and their environment?
   a. It occurs more rapidly in a liquid medium than in dry air.
   b. Oxygen enters individual cells by diffusion in both large and small organisms.
   c. Oxygen is unable to diffuse through the moist external membranes of most animals.
   d. Only a and b are true.
   e. All three answers (a, b, and c) are true.
2. Gases in the lung will diffuse into the capillaries of the alveoli if the partial pressure of these gases is:
   a. greater in the capillaries than in the lung.
   b. greater in the lung than in the capillaries.
   c. equal on both sides of the alveolar membrane.
   d. less than in the atmosphere.
   e. Choice a is true only for oxygen; b is true only for carbon dioxide.
3. Divers who have surfaced too rapidly from the ocean floor will have _____ in their blood.
   a. little oxygen
   b. bubbles of nitrogen gas
   c. too high a concentration of carbon dioxide
   d. too high a concentration of H^+ ions
   e. Both c and d are correct.
4. In large animals, oxygen enters by _____ across a membrane and is transported through the circulatory system by _____.
   a. active transport; facilitated diffusion
   b. bulk flow; osmosis
   c. passive diffusion; bulk flow
   d. facilitated diffusion; active transport
5. Which of the following parts of the respiratory tract is specialized to produce sounds using expired air?
   a. the bronchioles
   b. the nasal passages
   c. the pharynx
   d. the larynx
   e. the trachea
6. In the human lung, gases are exchanged across a greatly increased surface area provided for by:
   a. alveoli.
   b. convolutions of the bronchus.
   c. the pleural membranes.
   d. the high concentration of respiratory pigments.
   e. enlarged blood vessels below the epithelium of the lung.
7. The amount of air taken into the lungs is determined by the:
   a. endocrine system.
   b. somatic nervous system.
   c. diaphragm and intercostal muscles.
   d. smooth muscles that control the width of the bronchus.
8. In the circulatory systems of vertebrates and echinoderms, oxygen is _____, but carbon dioxide is primarily _____.

a. dissolved in plasma; bound to hemoglobin
   b. bound to respiratory pigments; present in the form of carbonic acid
   c. dissolved in plasma; carried inside red blood cells
   d. readily soluble; present in an insoluble form

9. The characteristic shape of the oxygen-hemoglobin association-dissociation curve is due to the fact that:
   a. the unbinding of oxygen from hemoglobin is not dependent on P_{O_2}.
   b. the structure of the beta chain remains constant if pH and temperature do not change.
   c. the affinity of hemoglobin for oxygen increases as more and more oxygen molecules are bound.
   d. the entire hemoglobin molecule has its greatest affinity for oxygen when it holds just a single oxygen atom.

10. Which of the following takes place in oxygen-rich capillaries that supply metabolically active tissues?
    a. Hemoglobin becomes saturated with oxygen.
    b. Oxygen dissociates completely from hemoglobin.
    c. Enough oxygen is released from hemoglobin to supply metabolic demands.
    d. The heme units exchange their oxygen atoms for carbon dioxide atoms.
    e. Both b and d are correct.

11. All active animals that have a circulatory system must make use of respiratory pigments because:
    a. oxygen cannot diffuse through the walls of capillaries unless it is bound to protein.
    b. oxygen diffuses so slowly through a watery medium that without the pigments animals could not afford to be active.
    c. oxygen is relatively insoluble in plasma.
    d. neither oxygen nor carbon dioxide is soluble in plasma to any significant degree.

12. Hemoglobin contains _____ heme units, each of which can combine with _____ of oxygen.
    a. four; one molecule
    b. eight; one molecule
    c. two; one molecule
    d. four; one atom
    e. two; one atom

13. Blood that is in capillaries surrounding tissues that are using oxygen is:
    a. saturated with molecular carbon dioxide.
    b. depleted of hemoglobin.
    c. acidic.
    d. depleted of red blood cells.
    e. Both b and d are correct.

14. The respiratory pigment myoglobin:
    a. is found in skeletal muscle.
    b. gives up most of its oxygen below a P_{O_2} of 20 mm Hg.
    c. provides a reserve supply of oxygen during periods of heavy exercise.
    d. has an association-dissociation curve that lies to the left of that for hemoglobin.
    e. All of the above are correct.

15. The lung is believed to have evolved:
    a. from the swim bladder of saltwater fish.
    b. as an accessory respiratory structure in freshwater fish.
    c. from the gills of a freshwater fish.
    d. from the gills of a "higher" invertebrate.

16. Which of the following organisms requires forced ventilation of its lungs?
    a. an alligator   d. a whale
    b. a mouse   e. a frog
    c. an eagle

17. Diving mammals can remain submerged for relatively long periods because:
    a. they possess a special respiratory pigment that releases oxygen only at very low partial pressures of oxygen.
    b. blood is circulated only to essential tissues.
    c. they do not require oxygen as an electron acceptor in cellular respiration.
    d. lactic acid serves as a reserve source of energy.

18. In order for exhalation to occur:
    a. chemoreceptors located in blood vessels must signal the brain.
    b. the diaphragm and intercostal muscles must contract.
    c. the volume of the thoracic cavity must be increased.
    d. respiratory neurons of the brainstem must be inhibited.

19. The countercurrent arrangement of blood vessels in the gills of fish ensures that:
    a. blood will flow to and from the gills.

b. gas exchange with the environment occurs at an optimum rate.
c. capillaries lie directly beneath the thin integument of the gill.
d. very active fish can breathe when motionless.

20. The vertebrate gill is believed to have originated from a(n) _____ of primitive vertebrates.
    a. lung
    b. water-filtering organ
    c. outgrowth of the circulatory system
    d. region of the integument specialized for respiration

21. Match each structure with its description.

    tracheae _____
    tracheal cilia _____
    alveoli _____
    bronchi _____
    pleura _____
    carotid arteries _____

    a. branch from the trachea to the right and left lungs
    b. saclike terminations of the bronchioles
    c. help expel foreign particles from the respiratory tract
    d. sites of chemoreceptors detecting changes in blood gas levels
    e. lubricating membrane
    f. respiratory network of arthropods

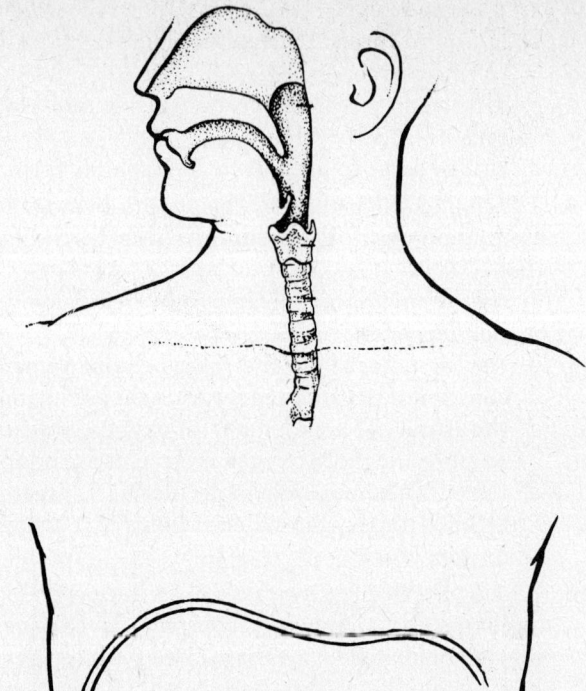

22. What respiratory problem would a person vacationing at a high altitude encounter? Explain how the local inhabitants are physiologically adapted to these conditions.
23. What are the advantages of the ventilation lung over gas exchange through external membranes? Over gills? What is its primary disadvantage?
24. Why is a person more likely to survive prolonged submersion in cold water than in warm water?
25. Complete this drawing of the human respiratory system, and label these parts: bronchi, bronchiole, diaphragm, heart, larynx, lungs, pharynx, trachea.

PERFORMANCE ANALYSIS

1. Although the rate of diffusion of a substance through water is much slower than through dry air, oxygen enters cells by diffusion in both large and small organisms (**b** is correct) across moist membranes, either externally, as in the gill, or internally, as in the lining of the ventilation lung.
2. The partial pressure of a gas in a mixed sample of gases is proportional to its concentration. Gases diffuse from regions of higher concentration to regions of lower concentration; therefore, **b** is correct.
3. Nitrogen is more soluble in the blood at increased pressures, such as in the sea. Divers who surface

too rapidly get the bends, in which the nitrogen in their blood becomes insoluble and forms bubbles (**b**).

4. In large animals, such as terrestrial mammals, oxygen diffuses across the alveolar membrane and into capillaries in the lung. The pumping of the blood by the heart is an example of movement by bulk flow. Choice **c** is therefore correct.

5. Air expelled from the lungs passes over ligaments stretched across the larynx (**d**). These ligaments, known as the vocal cords, produce sound when vibrated by the expired air.

6. The alveoli (**a**), shown in the electron micrograph of Figure 35-10a, are small air sacs clustered around the ends of bronchioles. They greatly increase the surface area across which gas exchange takes place.

7. Depth of inspiration is determined by the diaphragm and intercostal muscles (**c**).

8. Choice **b** is correct. Vertebrates and echinoderms use respiratory pigments such as hemoglobin to transport oxygen. Carbon dioxide dissolves in the blood plasma to form carbonic acid. (See the equation on page 699.)

9. Refer to the association-dissociation curve of Figure 35-13. The affinity of hemoglobin for oxygen increases as oxygen is bound (**c**) and decreases as oxygen is released. Oxygen will be given up more rapidly at lower partial pressures than at higher ones; hence, the characteristic sigmoid shape of the curve. Answer d is incorrect because the affinity for oxygen increases in a stepwise fashion and because oxygen molecules, not atoms, are bound.

10. Tissues that are metabolically active are consuming oxygen in chemical reactions. Therefore, the partial pressure of oxygen in these tissues is kept continually low. Hemoglobin circulating through these tissues will give up its oxygen (**c**). (Again refer to the curve in Figure 35-13.) However, the hemoglobin will not release all of the oxygen that it is carrying.

11. Oxygen is relatively insoluble in water at body temperatures. Blood plasma is largely water; therefore, **c** is correct.

12. See Figure 3-27 on page 71 to convince yourself that **a** is correct.

13. Tissues using oxygen are also releasing carbon dioxide into the blood. Since carbon dioxide and water react according to the equation on page 699, blood containing a high concentration of carbon dioxide will be acidic (**c**).

14. The myoglobin curve in Figure 35-16 shows that choices b and d are correct. Owing to the properties demonstrated by the curve, myoglobin (which is abundant in skeletal muscle) accepts oxygen from hemoglobin and unloads it at very low partial pressures, such as those present in muscle tissue during heavy exercise; therefore, **e** is correct.

15. Some groups of freshwater fish developed primitive lungs as accessory structures to the gill. This was probably of adaptive value because fresh water often becomes stagnant. Therefore, choice **b** is correct.

16. Frogs (**e**), like many amphibians, are unable to create the negative pressure in their lungs that is required to bring in air. Instead, they must force air into their lungs by gulping.

17. Choices a, c, and d are all false statements. By constricting peripheral blood vessels, a diving mammal (such as a whale) allows blood to flow only to essential tissues such as the heart and brain (**b**). In this way, these mammals can conserve the oxygen in their blood and can remain submerged for an hour or more.

18. Inhalation is due to the expansion of the lungs inside the thoracic cavity by the action of the diaphragm and intercostal muscles. This contraction is innervated by neurons in the respiratory center (in the brainstem). In order for exhalation (caused by the relaxation of these muscles) to occur, these neurons must be momentarily inhibited. Choice **d** is therefore correct.

19. The countercurrent arrangement of blood vessels in the gills of fish is shown in Figure 35-5. It provides for an optimum rate of exchange of blood gases with the environment (**b**). You may want to review the principle of countercurrent exchange discussed on page 129.

20. Primitive aquatic vertebrates, which exchanged

gases through their integuments, became subject to selection pressures that resulted in a trend toward the development of a thick, scaly, water-impermeable skin. It is hypothesized that a water-filtering organ (**b**) used in feeding became modified for the purpose of respiration.

21. The correct answers in order are **f**, **c**, **b**, **a**, **e**, and **d**.
22. Since the partial pressure of oxygen decreases with altitude, the normal hemoglobin of a person adapted to sea level conditions would not be saturated with oxygen at a higher altitude. In time, he or she would acclimate to conditions by producing more red blood cells, breathing more deeply, and pumping blood more rapidly as the size of the heart increased. All these adaptations are found among people who live at high altitudes.
23. The advantages of a ventilation lung over integumental exchange is primarily that of reduced water loss by evaporation. Organisms with lungs have an energetic advantage over those with gills since gill-breathers must actively move the oxygen-containing medium over the gill surface. A distinct disadvantage of the ventilation lung is that the movement of air over the exchange surface occurs in two stages—inhalation and exhalation—and so there is not a continuous flow of oxygenated air.
24. Cold water activates the diving reflex: The heart rate slows and the reduced supply of oxygenated blood is shunted to the heart and brain, which would begin to die within 4 minutes without oxygen.
25. Compare your completed sketch with Figure 35-9 on page 695.

CHAPTER 36

Energy and Metabolism III: Circulation of the Blood

CHAPTER ORGANIZATION

I. Introduction
II. Composition of the blood
   A. Blood plasma
   B. Blood cells
      1. Introductory remarks
      2. Red blood cells (erythrocytes)
      3. White blood cells (leukocytes)
      4. Platelets
      5. Blood clotting
III. The heart
   A. Evolution of the heart
   B. The human heart
IV. The blood vessels
   A. Structure and blood flow
   B. The vascular circuits
V. Lymphatic system
VI. Cardiovascular dynamics
   A. Cardiac output
   B. Blood pressure
      1. General information
      2. Cardiovascular regulating center
VII. Diseases of the heart and blood vessels
VIII. Summary

MAJOR CONCEPTS

Oxygen, nutrients, other essential substances, and waste products are transported throughout the body in the blood. The blood is composed of a liquid portion (plasma) in which nutrients, hormones, enzymes, and wastes, among other compounds, are suspended. The solid portion of the blood is composed of red blood cells (erythrocytes), which contain oxygen-bearing hemoglobin, white blood cells (leukocytes), which attack foreign particles in the body, and platelets, which function in the formation of blood clots.

Heart structure in vertebrates varies. Fish have a two-chambered heart; the amphibian heart has three chambers; and birds and mammals are characterized by a four-chambered heart. The mammalian heart pumps blood into a network of vessels, consisting of arteries, arterioles, capillaries, venules, and veins through which the blood passes before it returns to the heart. Exchange of substances between the blood and cells occurs in the capillaries.

The synchronization of heart muscle contractions is regulated primarily by the sinoatrial node (the pacemaker) and by the atrioventricular node. Secondary regulation of the heartbeat rate is accomplished by neural and hormonal means. Parasympathetic stimulation slows the heartbeat; sympathetic stimulation and adrenaline accelerate it.

HOW TO STUDY THE CHAPTER

Read the chapter, focusing on the major concepts.

Reread the chapter, using the questions that follow to help you focus on details as they relate to major concepts. Answer the questions on a separate sheet of paper. Your answers will provide a valuable study aid in preparing for examinations.

FOCUS ON CHAPTER DETAILS

I. Introduction (page 705)

1. What important materials are carried in the blood?
2. Describe the functional organization of the vertebrate circulatory system.

II. Composition of the blood (pages 705–707)

A. Blood plasma

3. What is the composition of blood plasma?
4. What are the functions of plasma proteins in general? What are the major functions of the three different classes of specific plasma proteins?
5. What is blood serum?

B. Blood cells

Introductory remarks

6. What elements of the blood are not dissolved in plasma?

Red blood cells (erythrocytes)

7. What are the major characteristics of mature red blood cells?
8. What is the turnover rate of erythrocytes?

White blood cells (leukocytes)

9. How do the numbers of red and white blood cells compare?
10. What are the major characteristics of white blood cells?

Platelets

11. Describe a platelet and its function. Is it also a cell? Explain.

Blood clotting

12. Describe the sequence of events by which a blood clot forms. What is the purpose of a clot?
13. What happens when a sample of blood is placed in a glass test tube?
14. What deficiency in the blood-clotting process is responsible for hemophilia?

III. The heart (pages 707–710)

A. Evolution of the heart

15. What form of a simple heart is found in an invertebrate such as the earthworm?
16. Trace the evolution of the vertebrate heart and its role in the circulatory system from fish to mammals (see Figure 36–4). What is the advantage of the four-chambered heart?

B. The human heart

17. Use Figure 36–5 to trace the flow of blood through the heart.
18. What are the roles of the atria, ventricles, and valves in the pumping of blood?
19. What evidence is there that the heart does not require external stimulation in order to function?
20. Describe the sinoatrial node. What does it stimulate?
21. How does the impulse spread to cause a delayed, simultaneous contraction of the ventricles?
22. Why is it possible to measure the heart's activity by placing electrodes on the skin surface?
23. What do the three types of waves in the electrocardiogram of Figure 36–7 represent?
24. What accounts for the characteristic sound of the beating heart?
25. How does the sound of a heart that has been damaged by rheumatic fever differ from the normal sound?
26. How do the parasympathetic and sympathetic divisions of the autonomic nervous system exercise control over the heartbeat?
27. How does the hormone adrenaline influence the heartbeat rate?

IV. The blood vessels (pages 710–714)

A. Structure and blood flow

28. Compare the structure of artery, vein, and capillary walls. (Refer to Figure 36–8).
29. Blood leaves the heart in spurts. What mechanism is responsible for the smooth flow of blood in the vessels after it leaves the arteries? What accounts for the pulse?
30. How does the difference between hydro-

static and osmotic pressure change along the length of a capillary bed? What exchanges occur between capillaries and interstitial fluid as a result of this?

31. How is excess interstitial fluid returned to the cardiovascular system?
32. What is edema?
33. How does the structure of a capillary make it well suited for supplying surrounding cells with materials?
34. How do various materials (oxygen, glucose, hormones, dissolved substances, etc.) move out of or into capillaries?
35. Relate the structure of veins and venules to their function in returning blood to the heart. What assistance do the veins obtain from muscle movement? What is the function of the valves within the veins?

B. The vascular circuits

36. Describe the pulmonary circuit with respect to its function in oxygenating the blood.
37. What is unique about the blood carried by pulmonary arteries? Pulmonary veins?
38. What structures constitute the systemic circuit? What is the role of the aorta in this circuit?
39. How sensitive is the brain to oxygen deprivation?
40. Describe the hepatic portal system. What is its purpose?

V. **Lymphatic system** (page 714)

41. What are two main functions of the lymphatic system of higher vertebrates?
42. Compare the lymphatic system to the blood vascular system.
43. How is lymph moved through the vessels of the lymphatic system?

VI. **Cardiovascular dynamics** (pages 714–716)

A. Cardiac output

44. What is meant by cardiac output? Write the definition as a formula.

B. Blood pressure

General information

45. How is blood pressure defined? What does a pressure of 120/80 mean?
46. How is blood pressure influenced by the heart rate? How is it related to blood flow?
47. How does the organization of smooth muscle in the walls of arterioles rely on a simple physical principle in regulating blood flow to capillaries?
48. How is the activity of these muscles controlled?
49. What is the adaptive value of fainting?
50. What psychological events are known to influence the distribution of blood flow?

Cardiovascular regulating center

51. What is the general function of the cardiovascular regulating center? Where is it located?
52. How does the cardiovascular regulating center provide for the blood-pressure reflex? How is this reflex an example of negative-feedback control?

VII. **Diseases of the heart and blood vessels** (pages 716–718)

53. What is the incidence of cardiovascular disease in the United States?
54. Thrombus and embolus are key words in the discussion of heart disease. Define and distinguish between these terms.
55. What is a heart attack? What conditions can cause a heart attack? What determines whether or not a victim can recover?
56. The symptoms of angina pectoris are similar to those of a heart attack. What distinguishes angina physiologically?
57. What is a stroke? What conditions may result in a stroke?
58. What is the condition of the cardiovascular system of a person with atherosclerosis? Why is such a person susceptible to heart attacks and strokes?
59. What do doctors suspect are the causes of atherosclerosis? How does the risk of this disease differ between men and women?

60. What is surprising about the change in mortality rates due to ischemic disease?
61. What is hypertension? Why is it so dangerous? How common is hypertension in the United States?
62. What are the mechanisms by which drugs control hypertension?

VIII. **Summary** (page 718): Read the Summary. If you are familiar with the essential features of the material presented there, you are ready to take the following diagnostic examination.

TESTING YOUR UNDERSTANDING

After you have completed the following examination, compare your answers with the annotated key in the Analysis section.

1. Which of the following is *not* dissolved in blood plasma?
   a. carbon dioxide
   b. hemoglobin
   c. hormones
   d. platelets
   e. Both b and d are correct.
2. Which of the following accurately describes the composition of human blood?
   a. Blood, by weight, is about 90 percent water.
   b. Plasma proteins of blood are so large that they do not contribute to its osmotic pressure.
   c. Substances dissolved in blood are either nutrients or waste products.
   d. Cells or cell fragments make up about 40 percent of the blood.
3. Which of the following elements of the blood are actually living cells containing a nucleus?
   a. erythrocytes
   b. leukocytes
   c. platelets
   d. globulins
4. The conversion of fibrinogen to its active form, fibrin, is necessary for:
   a. blood clotting.
   b. circulation of blood through the capillaries.
   c. conduction of an electrical impulse through the heart.
   d. maintaining proper osmotic pressure of the blood.
   e. Both b and d are correct.
5. A woman who suffers from hemophilia cannot form a blood clot because she:
   a. has an insufficient quantity of thromboplastin.
   b. lacks a factor necessary to produce the active enzyme thrombin.
   c. does not have platelets in her blood.
   d. produces a nonfunctional form of fibrin.
   e. lacks the ability to manufacture immunoglobulin.
6. Deoxygenated blood returning from the systemic circulation enters the _____ of the heart through the _____.
   a. right atrium; venae cavae
   b. left atrium; venae cavae
   c. right atrium; aorta
   d. right atrium; pulmonary veins
   e. left atrium; pulmonary veins
7. Oxygenated blood returning from the lungs enters the _____ of the heart through the _____.
   a. right atrium; venae cavae
   b. left atrium; venae cavae
   c. right atrium; aorta
   d. right atrium; pulmonary veins
   e. left atrium; pulmonary veins
8. The beat of the vertebrate heart is initiated:
   a. by the autonomic nervous system.
   b. by the hormone epinephrine.
   c. within the sinoatrial node.
   d. at the atrioventricular node.
   e. Both a and b are correct.
9. The 100-millisecond delay between atrial and ventricular excitation is due to the activity of the:
   a. sinoatrial node.
   b. atrioventricular node.
   c. bundle of His.
   d. vagus nerve.
   e. sympathetic nervous system.
10. The "lubb" sound of the heart is due to the _____, and the "dup" sound is a result of the _____.
    a. contraction of the atria; relaxation of the atria
    b. contraction of the atria; contraction of the ventricles

    c. contraction of the ventricles; the closing of valves leading to the aorta
    d. closing of valves between the atria and the ventricles; closing of valves between ventricles and arteries

11. Fluid leaves the capillaries at their _____ end because hydrostatic pressure of the blood is _____ its osmotic pressure.
    a. venous; greater than
    b. venous; less than
    c. arteriolar; greater than
    d. arteriolar; less than

12. Substances moving into or out of the blood must pass through _____ of a capillary.
    a. a single endothelial cell
    b. the lumen
    c. a junction in the endothelium
    d. three cell layers constituting the wall
    e. Either a or c may be correct.

13. Because of the arrangement of the pulmonary and systemic circulations, the _____ are the only vessels of their type that carry fully deoxygenated blood.
    a. pulmonary veins
    b. pulmonary arteries
    c. inferior and superior venae cavae
    d. pulmonary capillaries
    e. coronary arteries

14. Which of these represents the correct direction of blood flow from the left ventricle to the right atrium?
    a. aorta, arteries, capillaries, veins, vena cava
    b. arteries, aorta, capillaries, vena cava, veins
    c. arterioles, arteries, capillaries, veins, venules
    d. vena cava, arteries, venules, capillaries, veins
    e. arteries, arterioles, capillaries, vena cava, veins

15. After receiving blood from the left ventricle, the aorta first branches to supply blood to the:
    a. brain.
    b. liver.
    c. left arm.
    d. heart.
    e. lungs.

16. Because of the hepatic portal system, blood from the _____ is transported first to the _____ before being sent to the heart.
    a. brain; extremities
    b. liver; kidneys
    c. lungs; systemic circulation
    d. small intestine; liver
    e. kidneys; liver

17. Which of the following correctly describes the human lymphatic system?
    a. Like the cardiovascular system, it is a closed network of vessels.
    b. Tiny lymph hearts positioned at critical points propel the lymph.
    c. It serves as a shuttle between interstitial fluid and the vena cava.
    d. The lymphatic vessels resemble arteries and arterioles in their structure but not in their functions.
    e. All of the above are correct.

18. A blood pressure given as 120/80 means that:
    a. cardiac output is 120 liters per minute with a pulse of 80.
    b. arterial pressure is 120 mm Hg at ventricular contraction and 80 mm Hg at relaxation.
    c. diastolic pressure is 120 mm Hg and systolic pressure is 80 mm Hg.
    d. the left ventricle moves 120 milliliters of blood per beat and the right moves 80 milliliters.
    e. hypertension has reached a critical level.

19. Blood flow to a capillary bed is regulated chiefly by:
    a. valves at the arteriolar end of the bed.
    b. rings of smooth muscles in the walls of arterioles.
    c. the force with which the heart muscles contract and expand.
    d. differences in osmotic pressure between capillaries and interstitial fluid.

20. The cardiovascular regulating center responds to a lowered blood pressure by stimulating the autonomic nervous system to:
    a. increase the heart rate.
    b. decrease the heart rate.
    c. cause arterioles to constrict.
    d. cause arterioles to dilate.
    e. Both a and c are correct.

21. Match each disease or condition with its description.

atherosclerosis _____
stroke _____
heart attack _____
hypertension _____
angina pectoris _____

a. increases the risk of an embolus
b. caused by ischemia in the brain
c. thickening of the linings of the arteries
d. chest pain resulting from narrowing of coronary blood vessels
e. results in the death of cardiac muscle cells

22. Match each term with its function or description.

stretch receptor _____
medulla _____
lymph duct _____
bundle of His _____
right atrium _____
plasma protein _____
cardiac output _____

a. transports absorbed fats into the bloodstream
b. site of the cardiovascular regulating center
c. indicates the amount of work done by the heart in a given time period
d. site of the sinoatrial node
e. responsible for maintaining the osmotic pressure of the blood
f. responds to changes in blood pressure by signaling brain centers
g. conducts a nerve impulse to the ventricles

23. Compare the circulatory systems of a fish, an amphibian, and a mammal.
24. Explain how nerve impulses initiated within the heart cause simultaneous contractions of the atria and nearly simultaneous contractions of the ventricles.
25. Explain why atherosclerosis and hypertension increase the risk of heart attacks and strokes.
26. Label the indicated parts in the following diagram of the heart.

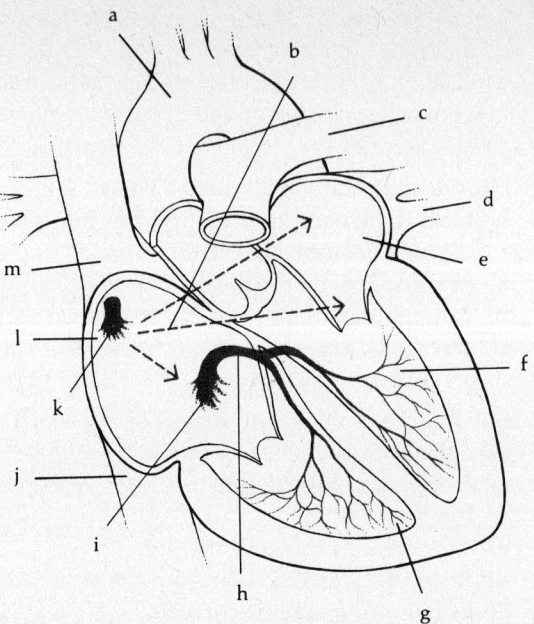

PERFORMANCE ANALYSIS

1. Platelets (d) are cytoplasmic fragments of large cells called megakaryocytes and therefore are not soluble in the plasma portion of the blood. Hemoglobin (b) is carried inside red blood cells and is therefore not dissolved in the plasma. The correct answer is **e**.

2. By weight, 60 percent of human blood is plasma, and 40 percent is composed of cells (erythrocytes and leukocytes) and cell fragments (platelets). Choice **d** is correct.

3. Of the four choices listed only a and b are cells. The nuclei of erythrocytes, however, are extruded at maturity; therefore, **b** is correct.

4. Fibrin is a blood protein formed whenever plasma encounters a rough surface and is produced from fibrinogen in the series of reactions shown on pages 706 and 707. Fibrin molecules form an insoluble network around red blood cells and platelets to produce a clot (see Figure 36–3). Choice **a** is therefore correct.

5. A hemophiliac has defective Factor VIII, a substance necessary for the activation of prothrombin in the chain of clotting reactions; therefore, **b** is correct.

6. Choice **a** is correct; examine Figure 36-12, the diagram of the human circulatory system.

7. Again see Figure 36-12; choice **e** is correct.

8. The heartbeat is initiated by the sinoatrial node (**c**), a group of specialized cardiac muscle cells lying in the wall of the right atrium of the heart. (See Figure 36-6.) The impulse generated by this pacemaker spreads to the atrioventricular node, which is an electrical bridge between the atria and the ventricles. The autonomic nervous system (**a**) and the hormone epinephrine (**b**) are responsible for the control of the *rate* at which the heart beats.

9. Because the atrioventricular node (**b**) is composed of slow-conducting fibers, it imposes a delay between atrial and ventricular contractions.

10. The characteristic sounds of the heart are produced as the valves between the atria and the ventricles close at ventricular contraction ("lubb") and as the valves between the ventricles and arteries close at ventricular relaxation ("dup") (**d**).

11. Whenever the hydrostatic pressure of the blood is greater than its osmotic pressure, fluid will leave the blood. This happens at the arteriolar end of a capillary bed. When osmotic pressure is greater than hydrostatic pressure (which drops at the venous end of a capillary bed), fluid will be taken up by the blood from interstitial spaces; therefore, **c** is correct.

12. As can be seen from Figure 36-10, substances in the blood flowing through the lumen of a capillary must pass through either the capillary endothelium (composed of a single cell layer) or a junction in this endothelium in order to enter the interstitial space; therefore, **e** is correct.

13. Choice **b** is correct; see Figure 36-12.

14. Choice **a** is correct; see Figure 36-12.

15. The first two branches of the aorta are the right and left coronary arteries, which supply blood to the heart (**d**).

16. Note in Figure 36-12 that the capillary bed in the small intestine is connected to a capillary bed in the liver (**d**) by the hepatic portal vein. Venous blood filtered by the liver is taken to the heart via the inferior vena cava.

17. The human lymphatic system (Figure 36-13) is an open-ended circulatory system composed of vessels (similar to the veins that carry blood) that take up materials from interstitial spaces and return them to the blood via the superior vena cava (**c**). In some nonmammalian organisms, the lymph is propelled by tiny hearts, but in humans and other mammals it is propelled by contractions of the body muscles.

18. Blood pressure is usually measured in an artery of the upper left arm. In a healthy young adult the pressure, when the ventricles of the heart are contracting, should be approximately 120 mm Hg, and when the ventricles are relaxing, it should be about 80 mm Hg (**b**).

19. Sphincter muscles, rings of smooth muscles surrounding the outer walls of arterioles (**b**), can contract or expand to determine the amount of blood that flows through a capillary bed.

20. Since blood pressure is proportional to blood flow, factors that increase the flow also increase the pressure. An example is an increase in heart rate (a). But blood pressure also increases if flow remains the same while the tube through which the blood is moving narrows in diameter, as is the case when arterioles constrict (c); therefore, **e** is correct.

21. The correct answers in order are **a**, **c**; **b**; **e**; **a**; and **d**.

22. The correct answers in order are **f**, **b**, **a**, **g**, **d**, **e**, and **c**.

23. Your discussion should reflect the anatomical arrangements depicted in Figure 36-4.

24. Nerve excitation within the heart spreads from the sinoatrial node (the pacemaker) throughout the atria, causing almost simultaneous contractions of both atria. The wave of excitation is transmitted to the atrioventricular node; from there it passes to the bundle of His, which causes the two ventricles to contract almost simultaneously. Because the atrioventricular node fibers conduct relatively slowly, the ventricles do not contract until after the atria have completed their contractions.

25. Because atherosclerosis is associated with thickening of the arteries, it increases the chances of

formation of an embolus, a wandering clot. Emboli are one of the chief causes of heart attacks and strokes. Hypertension, which is associated with narrowing of arterioles, results in an increased strain on arteriole walls, which also increases the chances of an embolus formation.

26.
a. aorta
b. spread of atrial excitation
c. pulmonary arteries
d. pulmonary veins
e. left atrium
f. left ventricle
g. right ventricle
h. bundle of His
i. atrioventricular node
j. inferior vena cava
k. sinoatrial node
l. right atrium
m. superior vena cava

CHAPTER 37

Homeostasis I: Excretion and Water Balance

CHAPTER ORGANIZATION

I. Introduction
II. **Regulating the chemical environment**
   A. Introduction
   B. Excretion of metabolic wastes
   C. Controlling concentrations of chemicals
   D. Maintaining water balance
III. **Water balance**
   A. An evolutionary perspective
   B. Sources of water gain and loss in terrestrial animals
   C. Water compartments
IV. **The kidney**
   A. Introduction
   B. Function of the kidney
      1. General information
      2. Water conservation: The loop of Henle
   C. Control of kidney function: Aldosterone and ADH
V. Summary

MAJOR CONCEPTS

The excretory system contributes to the maintenance of homeostasis by removing metabolic by-products, by controlling ion concentration of body fluids and helping to regulate blood pH, and by maintaining water balance in the body. Nitrogenous wastes (from the deamination of amino acids) are excreted from mammals chiefly in the form of urea.

Equalizing water loss and water gain is the key factor in maintaining water balance. Water sources for mammals include water in the diet and the oxidation of food molecules. Water is lost from mammals in feces and urine, by respiration, and through the skin.

In mammals, the organ primarily responsible for regulating chemical composition of the blood is the kidney. The functional unit of the kidney is the nephron, which consists of a long tubule with a closed front end, known as Bowman's capsule, that encloses a cluster of capillaries, the glomerulus. Fluid from the blood known as filtrate (plasma minus large proteins) enters the capsule. As the filtrate moves through the tubule, almost all the water, ions, and other useful substances are reabsorbed into the bloodstream through the peritubular capillaries. Other substances are secreted from these capillaries into the tubules. The remaining water and solutes, along with most of the urea, are excreted as urine.

The secretion of urine that is hypertonic with respect to the blood is a principal mechanism of water conservation in mammals. The formation of hypertonic urine is made possible by the portion of the nephron known as the loop of Henle.

The function of the nephron is under the control of hormones, chiefly antidiuretic hormone (ADH) and aldosterone.

HOW TO STUDY THE CHAPTER

Read the chapter, focusing on the major concepts.

Reread the chapter, using the questions that follow to help you focus on details as they relate to major concepts. Answer the questions on a separate sheet of paper. Your answers will provide a valuable study aid in preparing for examinations.

FOCUS ON CHAPTER DETAILS

I. Introduction (page 720)

1. One way a multicellular organism can maintain homeostasis is by controlling the composition of its blood. How is this accomplished in terrestrial vertebrates?
2. Differentiate between the mechanism for removing wastes from the bloodstream and the elimination of feces.

II. Regulating the chemical environment (pages 720–721)

A. Introduction

3. What three problems must be solved in order for an organism to control its internal chemical environment?

B. Excretion of metabolic wastes

4. What are the two chief metabolic waste products of animals?
5. In what form are nitrogenous wastes eliminated in aquatic organisms? Why do terrestrial organisms differ in this regard?
6. Explain how nitrogenous wastes are eliminated in birds, terrestrial reptiles, and insects.
7. In what form are nitrogenous wastes excreted in mammals?

C. Controlling concentrations of chemicals

8. Cite evidence to support the assertion that kidneys are more accurately regarded as regulatory organs than as excretory organs.
9. Are all substances that enter the kidneys excreted? Explain.
10. Under what conditions is glucose excreted by the kidneys?
11. Name some ions whose concentrations are regulated by the kidneys. List some of the functions these ions perform.

D. Maintaining water balance

12. Solutions to the problem of water balance vary widely among animals. For any organism, what single factor determines the particular solution to this problem?

III. Water balance (pages 722–724)

A. An evolutionary perspective

13. What problems would have been encountered by organisms that evolved in an isotonic environment and later moved to fresh water?
14. How have freshwater fish solved the problem of living in a hypotonic environment?
15. How have cartilaginous fish solved the problem of living in a saltwater environment? How does their unusual tolerance for urea demonstrate the "opportunism of evolution"?
16. How have the bony fish solved the problems associated with the move from fresh water to salt water?

B. Sources of water gain and loss in terrestrial animals

17. What are three ways terrestrial organisms can obtain water?
18. Explain why the desert-dwelling kangaroo rat must have a diet high in fat.
19. What are the four routes of water loss for humans?

C. Water compartments

20. What are the three principal body compartments in which water is found? What proportion of the total body water is held in each one?
21. How does water enter the body, and in what ways is it exchanged among the principal compartments? How can such rates of exchange be experimentally determined?
22. How do dehydration and profuse sweating affect the movement of water from compartment to compartment?
23. What malfunctions can lead to edema?

IV. The kidney (pages 725–730)

A. Introduction

24. Where are the human kidneys located? Describe their general appearance.
25. Describe the overall structure of the human

kidney (Figure 37-7a). What is the functional unit of the kidney?

B. Function of the kidney

General information

26. How is blood from the renal artery routed when it reaches the vicinity of a nephron?
27. Describe filtration, the first step in the formation of urine. What structures are involved in filtration?
28. How does the filtrate compare in volume and in chemical composition to the blood plasma entering the kidney?
29. How does the secretion process contribute to renal function? What structures are involved in secretion?
30. What occurs during the reabsorption process? In what part of the nephron does it take place? Why is it expensive in energy terms?
31. What happens to the urine after it leaves the nephron?

Water conservation: The loop of Henle

32. What is the primary means by which animals regulate their body water?
33. How does the urine of animals that have free access to water differ in composition from the urine of animals that do not?
34. What structure in the kidneys of mammals gives them the ability to produce hypertonic urine?
35. Trace the passage of filtrate through the nephron (see Figure 37-12).
36. Explain how the walls of the following structures differ in their permeability to salts and water: the proximal convoluted tubule; the descending branch of the loop of Henle; the ascending branch of the loop of Henle; the distal tubule; the collecting duct.
37. In what regions does the filtrate become isotonic, hypotonic, and hypertonic with respect to the blood plasma?
38. How is the hypertonicity of the urine controlled?
39. Why is the loop of Henle able to act as a countercurrent multiplier? On what features of the loop does this multiplying effect depend?

C. Control of kidney function: Aldosterone and ADH

40. How does the secretion of aldosterone affect the solute concentration of the urine?
41. On what does the secretion of aldosterone depend?
42. Explain how Addison's disease affects urine formation.
43. How does the hypothalamus control the production and release of ADH?
44. Under what conditions is ADH secretion inhibited? How does this help to restore water balance?

V. **Summary** (pages 730-731): Read the Summary. If you are familiar with the essential features of the material presented there, you are ready to take the following diagnostic examination.

TESTING YOUR UNDERSTANDING

After you have completed the following examination, compare your answers with the annotated key in the Analysis section.

1. Desert animals that require a high-fat diet rely on _____ in order to maintain proper water balance.
   a. oxidative processes
   b. water-containing food
   c. living close to a stream or pond
   d. making a hypotonic urine
   e. Both b and d are correct.

2. Cartilaginous saltwater fish limit their water losses by:
   a. forming a highly concentrated urine.
   b. keeping their body fluids isotonic with respect to their environment.
   c. drinking a sufficient quantity of water to offset the losses.
   d. excreting wastes through specialized glands instead of as dissolved material.
   e. eating prey with a high water content.

3. In which of the following organisms does the

kidney serve the same purpose as the contractile vacuole of *Paramecium*?
a. a shark
b. a frog
c. a lake trout
d. an ocean fish
e. a lizard

4. Most of the water content of the human body may be accounted for by water held in:
a. the circulatory system.
b. the interstitial fluid.
c. the intracellular fluid.
d. the digestive tract.
e. the upper respiratory tract.

5. In a dehydrated person, a significant quantity of water moves out of the _____ and into the _____.
a. interstitial fluid; tissues
b. blood; interstitial fluid
c. fluid surrounding the nephron; collecting duct of the kidney
d. intracellular; interstitial fluid
e. interstitial fluid; blood

6. During the absorption of food molecules in the small intestine:
a. water moves into capillaries by osmosis.
b. salts move into capillaries by diffusion.
c. small organic molecules move into capillaries by diffusion.
d. water enters capillaries due to an increase in hydrostatic pressure.
e. Both b and c are correct.

7. By drinking sea water, a cod:
a. replaces water lost by osmosis from its body fluids.
b. ingests excessive quantities of salt, which must then be eliminated.
c. will eventually become dehydrated.
d. dilutes its body fluids, thus interfering with nervous function.
e. Both a and b are correct.

8. Solutes that are filtered from the blood by the kidney reenter the bloodstream:
a. by way of the lymphatic system.
b. via active transport from the proximal tubule.
c. if ADH is present in cells lining the collecting duct.
d. by passive diffusion through walls of the tubules of nephrons.

9. Filtrate in the descending loop of Henle is _____ with respect to blood plasma, since the walls of the descending loop are _____.
a. hypotonic; freely permeable to water
b. hypotonic; provided with a chloride ion pump
c. hypertonic; freely permeable to water
d. hypertonic; freely permeable to salts
e. isotonic; freely permeable to both water and salts

10. The loop of Henle is constantly bathed in a salty fluid because:
a. the entire loop of Henle is impermeable to water.
b. sodium ions are actively transported from its descending branch.
c. chloride ions are actively transported from its ascending branch.
d. sodium ions passively diffuse through the walls of the ascending branch.
e. Both c and d are correct.

11. Fluid in the ascending branch of the loop of Henle is passed directly to the:
a. collecting duct.
b. renal pelvis.
c. proximal convoluted tubule.
d. distal convoluted tubule.
e. peritubular capillaries.

12. The presence of ADH in the walls of the collecting duct ensures that:
a. the collecting duct will be impermeable to water.
b. urine will be hypertonic.
c. blood pressure will be lowered.
d. sodium will be reabsorbed from the distal tubule.
e. All of the above are correct.

13. ADH is produced by the _____ if it receives a signal from _____ that blood volume is low.
a. adrenal gland; pressure receptors
b. adrenal gland; osmotic receptors
c. hypothalamus; pressure receptors
d. hypothalamus; osmotic receptors
e. pituitary gland; osmotic receptors

14. Consuming alcoholic beverages will cause the:
a. pituitary gland to release ADH.

b. walls of the collecting duct to become less permeable to water.
   c. secretion of aldosterone to be inhibited.
   d. blood pressure to increase.
   e. Both a and d are correct.

15. The loop of Henle is a countercurrent multiplier and thus:
    a. speeds the flow of urine in a direction opposite to that of the blood.
    b. transports out chloride ions while pumping in sodium ions.
    c. maintains a difference in solute concentration between opposite ends of the tubule and between the nephron and the interstitial fluid.
    d. keeps the temperature of the urine roughly equal to that of the blood.

16. Terrestrial organisms excrete nitrogenous wastes in the form of uric acid or urea because:
    a. uric acid cannot be excreted as crystals.
    b. ammonia, the immediate breakdown product of proteins, is toxic.
    c. urea is crystalline and requires very little water for its excretion.
    d. Only a and c are correct.
    e. All three answers (a, b, and c) are correct.

17. Deamination refers to the:
    a. breakdown of ammonia.
    b. removal of an —NH_2 group from an amino acid.
    c. filtering of amino acids by Bowman's capsule.
    d. failure of the kidney to maintain a proper amino acid balance.

18. Match each structure with its description.
    loop of Henle _____
    glomerulus _____
    Bowman's capsule _____
    collecting duct _____
    peritubular capillaries _____
    urethra _____

    a. a blind end of the renal tubule
    b. tube through which urine is excreted from the body
    c. a hairpin-shaped structure in the renal tubule
    d. a cluster of capillaries delivering blood to a nephron
    e. site of reabsorption of dissolved substances from the renal tubule
    f. its permeability is under hormonal influence

19. Match each process or condition with its description.
    edema _____
    dehydration _____
    oxidation _____
    reabsorption _____
    filtration _____
    countercurrent exchange _____

    a. maintains concentration differences between nephron and interstitial fluid
    b. accumulation of interstitial fluid
    c. provides water as a metabolic end product
    d. return of dissolved solutes to the bloodstream
    e. takes place between Bowman's capsule and glomerular capillaries
    f. loss of water from cells

20. Contrast the strategies for maintaining fluid homeostasis used by a saltwater bony fish and a freshwater bony fish.

21. Discuss the principal means by which terrestrial vertebrates gain or lose water.

22. Describe the three stages by which urine is formed in the nephron of the kidney.

23. Match each nephron structure listed below with the appropriate letter in the diagram.

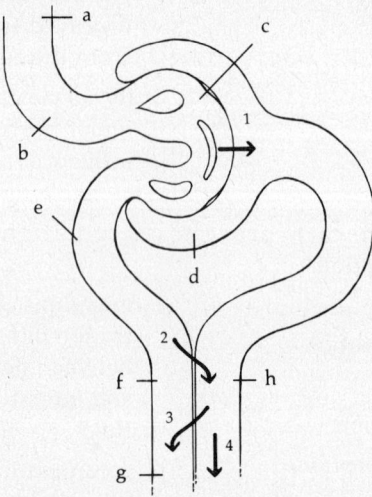

Bowman's capsule _____
efferent arteriole _____
artery _____
vein _____
afferent arteriole _____
tubule _____
peritubular capillary _____
glomerulus _____

Match each of the four basic renal functions listed below with the number in the diagram that indicates the site of the function.

excretion _____
filtration _____
reabsorption _____
secretion _____

PERFORMANCE ANALYSIS

1. Some animals that live in environments where drinking water is not readily available gain water by oxidizing organic compounds with a high hydrogen content, such as fats; therefore, **a** is correct.

2. Cartilaginous saltwater fish, such as sharks and skates, have body fluids isotonic with sea water (**b**), owing to their high tolerance for and retention of urea.

3. Freshwater fish, such as the lake trout (**c**), excrete a hypotonic urine in order to rid themselves of water taken up through their integument (skin) by osmosis. *Paramecium*, which also lives in fresh water, achieves the same results by expelling water with its contractile vacuole.

4. Choice **c** is correct; 65 percent of body fluid is intracellular (within the cell) fluid.

5. In dehydration, water moves out of cells and into interstitial fluids in response to an increased solute concentration in the extracellular fluid; therefore, **d** is correct.

6. During digestion, salts and small organic molecules are actively transported into capillaries of the digestive tract. Water follows by osmosis (**a**).

7. The body fluids of marine bony fish, like the cod, are not as salty as the sea water in which the fish live, and so they face the problem of replacing water lost by diffusion. They do this primarily by drinking sea water and excreting the salts contained in the water through specialized glands; therefore, **e** is correct.

8. Reabsorption of glucose, amino acids, and ions takes place in the peritubular capillaries by active transport from the proximal tubule of the nephron (**b**).

9. Since the walls of the descending loop of Henle are freely permeable to water, large quantities of water move out of the tubule into the surrounding zone of high salt concentration, and the filtrate becomes hypertonic; therefore, **c** is correct.

10. Choice **e** is correct; see Figure 37–12. The continual movement of Na^+ and Cl^- ions out of the ascending branch of the loop of Henle establishes a salt gradient in the interstitial fluid surrounding the tubules of the nephron. Chloride ions are pumped out (**c**), and sodium ions follow by diffusion (**d**).

11. Choice **d** is correct; again refer to Figure 37–12.

12. ADH, antidiuretic hormone, acts on the walls of

the collecting duct to make them more permeable to water. As a result, the fluid remaining in the tube (now more properly called the urine) becomes hypertonic (**b**).

13. Pressure receptors in the walls of the heart, aorta, and carotid arteries transmit information about blood volume to the hypothalamus, where ADH is produced. Osmotic receptors in the hypothalamus also influence ADH secretion, but these monitor the solute content of the blood, not changes in blood volume. Choice **c** is therefore correct.

14. Alcoholic beverages, by suppressing ADH secretion, make the walls of the collecting duct less permeable to water (**b**), increasing urinary flow and decreasing blood pressure.

15. Because the loop of Henle works as a countercurrent system associated with active transport (see Figure 37–13), it creates a difference in solute concentration between opposite ends of the tubule and between the nephron and the interstitial fluid (**c**). The longer the loop, the greater the concentration difference that can be established.

16. Nitrogen removed from amino acids in the form of amino groups is incorporated into ammonia (see Figure 37–1). Since ammonia is toxic to many organisms, they must incorporate it either into urea, a water-soluble compound, or, if water is unavailable, into crystalline uric acid, which they then excrete as a dry, semisolid paste; **b** is correct.

17. Deamination means the removal of an amino group ($-NH_2$) from amino acids (**b**).

18. The correct answers in order are **c**, **d**, **a**, **f**, **e**, and **b**.

19. The correct answers in order are **b**, **f**, **c**, **d**, **e**, and **a**.

20. The strategies are discussed in the analyses for Questions 3 and 7.

21. Terrestrial vertebrates gain water by drinking water, by ingesting foods that contain water, and as a result of oxidation of nutrient molecules. They lose water during respiration, by evaporation from the skin, and in urine and feces.

22. **Filtration** involves movement of blood plasma out of the glomerulus into Bowman's capsule. The constriction of the efferent arterioles creates the pressure in the glomerulus that is needed to maintain this movement. **Secretion** involves the selective removal, by the cells of the tubular walls, of molecules that remain in the plasma after filtration. During **reabsorption**, most of the water and solutes that entered the tubules during filtration are transported back into the bloodstream by the cells of the tubular walls.

23. Answers: **d**, **e**, **a**, **g**, **b**, **h**, **f**, **c**; 4, 1, 3, 2.

CHAPTER 38

Homeostasis II: Temperature Regulation

CHAPTER ORGANIZATION

I. Introduction
II. Principles of heat balance
   A. Background
   B. Heat transfer
   C. Size and temperature
   D. "Cold-blooded" and "warm-blooded"
III. Poikilotherms
IV. Homeotherms
   A. General information
   B. The mammalian thermostat
      1. General information
      2. Regulating as body temperature rises
      3. Regulating as body temperature falls
   C. Meeting energy costs
   D. Cutting energy costs
V. Adaptations to extreme temperatures
   A. Introductory remarks
   B. Adaptations to extreme cold
      1. General information
      2. Countercurrent exchange
   C. Adaptations to extreme heat
VI. Summary

MAJOR CONCEPTS

The complex life processes of an animal require that the organism's body temperature be maintained within a narrow temperature range. Heat balance requires that net heat loss from an organism equal heat gain.

The two major sources of heat gain available to animals are the radiant energy of the sun and cellular metabolism. Heat is lost from animals by conduction, radiation, and evaporation.

Ectotherms are animals that maintain their body heat by using external energy sources. Endotherms maintain their body heat by producing and conserving energy internally. Most ectotherms are poikilotherms, animals whose internal temperature varies with that of the environment. Most endotherms are homeotherms, animals that maintain a constant internal temperature.

The mammalian temperature-regulation center is located in the hypothalamus of the brain. The thermostat detects temperature changes in the circulating blood.

Adaptations to extremely cold climates include countercurrent blood circulation and insulation (fat and fur or feathers). Adaptations to extremely hot environments are primarily water-conservation techniques and include a variety of behavioral responses.

HOW TO STUDY THE CHAPTER

Read the chapter, focusing on the major concepts.

Reread the chapter, using the questions that follow to help you focus on details as they relate to major concepts. Answer the questions on a separate sheet of paper. Your answers will provide a valuable study aid in preparing for examinations.

FOCUS ON CHAPTER DETAILS

I. Introduction (pages 732–733)

1. What dictates the narrow temperature range within which life can exist?
2. How does freezing affect chemical reactions that take place in living tissue?
3. How are biochemical reactions affected by increases in temperature?
4. As a consequence of these factors, what is the temperature range of external environments within which life usually exists?
5. What experiment demonstrates how efficient mammals are in regulating their internal body temperatures?

II. Principles of heat balance (pages 733–736)

A. Background

6. What are the two primary sources of heat gain for an organism?

B. Heat transfer

7. What are the three major ways in which organisms lose heat?
8. What is conduction? How do water and air compare as conductors?
9. What kinds of insulators do animals use to conserve heat?
10. How does convection influence heat loss by conduction?
11. How is heat exchanged between a body and its surroundings by radiation?
12. How does the color of a body influence its ability to absorb radiant energy?
13. How is evaporation of water useful in maintaining heat balance?

C. Size and temperature

14. Explain how maintenance of heat balance is dependent upon the size of an organism.

D. "Cold-blooded" and "warm-blooded"

15. Why are the terms "cold-blooded" and "warm-blooded" often misleading?
16. By what criterion are organisms classified as either ectotherms or endotherms?
17. What is the difference between a poikilotherm and a homeotherm?

III. Poikilotherms (pages 736–738)

18. Why are most aquatic animals necessarily poikilotherms?
19. Why is a fish able to maintain a constant temperature even though it is an ectotherm? To what conditions is a fish particularly sensitive, and why?
20. To what extent can a lizard, generally considered a poikilotherm, maintain a constant temperature? What types of behavior account for this?

IV. Homeotherms (pages 738–741)

A. General information

21. What is characteristic of homeotherms? What groups of animals are true homeotherms?
22. Why is the cost of being a homeotherm considered high? How does size affect this cost?
23. How does a mammal exchange heat between its body core and its periphery?
24. In what two ways does a mammal regulate its body temperature?

B. The mammalian thermostat

General information

25. Compare the hypothalamus to a thermostat that regulates a furnace.
26. From what sources does the hypothalamus receive its information about changes in body temperature? In what cases do some sources take priority over others?
27. What has happened to the thermostat when one has a fever? Why are chills often experienced at the onset of fever?
28. What may be the adaptive value of a fever?

Regulating as body temperature rises

29. List two physiological responses to increased body temperature in mammals.
30. Why do dogs pant and cats lick themselves when it is hot?

Regulating as body temperature falls

31. What are the physiological responses to a decrease in body temperature?

32. How do homeotherms trap the heat they generate internally?

C. Meeting energy costs

33. What demands are routinely placed on a bird's metabolic rate?

34. Why do birds migrate?

D. Cutting energy costs

35. What are some adaptive strategies used by animals when they cannot get enough food to meet their normal energy requirements?

36. What animals are considered to be true hibernators? (Note that bears, popularly believed to hibernate, are not included.)

37. What physiological changes accompany hibernation?

38. What types of stimuli are likely to arouse a hibernator? What changes accompany arousal from hibernation?

V. **Adaptations to extreme temperatures** (pages 742–744)

A. Introductory remarks

39. How do humans remain comfortable under extreme temperature conditions?

B. Adaptations to extreme cold

General information

40. How do mammals generally cope with extreme cold?

41. How are aquatic mammals able to maintain core temperatures as high as those of terrestrial mammals? What role does fat play?

42. How do the extremities of the Arctic fox contribute to the regulation of its body temperature?

Countercurrent exchange

43. How does the arrangement of blood vessels leading to and away from the body core of many Arctic animals contribute to heat conservation?

C. Adaptations to extreme heat

44. In what three ways do camels have a distinct advantage over humans in a desert environment?

45. Of what advantage is the hump on a camel's back?

46. What heat-unloading strategy is used by humans under heat stress? By some other mammals?

47. Why can't small desert animals employ a similar strategy? How do these animals maintain both their temperature and their water balance?

VI. **Summary** (pages 744–745): Read the Summary. If you are familiar with the essential features of the material presented there, you are ready to take the following diagnostic examination.

TESTING YOUR UNDERSTANDING

After you have completed the following examination, compare your answers with the annotated key in the Analysis section.

1. The transfer of heat between two objects that are in direct contact occurs by:
   a. conduction.
   b. convection.
   c. radiation.
   d. absorption.
   e. oxidation.

2. The upper temperature limit for life is determined by the point at which:
   a. intracellular fluid begins to vaporize.
   b. proteins become denatured.
   c. the metabolic rate becomes so high it can no longer be regulated.
   d. core temperature has increased by 10°C above normal.

3. An animal that depends on the sun's radiant energy and heat conduction to maintain a relatively constant body temperature is most accurately described as:
   a. a homeotherm.
   b. an endotherm.

c. an ectotherm.
  d. warm-blooded.

4. Small animals lose heat more rapidly than do large ones because they:
   a. have higher metabolic rates.
   b. have a greater surface-to-volume ratio.
   c. contain less total body water.
   d. are generally ectotherms and so cannot regulate their internal temperatures effectively.

5. Which of the following homeotherms would have the greatest difficulty remaining cool if placed in the open sun on a hot, humid day?
   a. a coyote with a light-colored coat
   b. a fox with a pair of long ears
   c. a black Labrador retriever
   d. a person wearing thin summer clothing

6. A terrestrial poikilotherm can maintain a nearly constant internal temperature under heat stress by:
   a. sweating or panting.
   b. lowering its metabolic rate.
   c. remaining active.
   d. orienting its body to expose minimum surface area to the sun.

7. Endotherms keep warm chiefly as a result of:
   a. heat generated in oxidative reactions.
   b. radiative heat gain from the environment.
   c. insulating layers of fat or fur.
   d. constricting peripheral blood vessels.
   e. Both c and d are correct.

8. Heat is transported from the body core of a mammal to its surface primarily by:
   a. conduction through body tissues.
   b. the nervous system.
   c. circulation of the blood.
   d. radiation.

9. In mammals, information about external temperature changes is usually detected by _____, which signal the _____.
   a. arterial receptors; pacemaker of the heart
   b. arterial receptors; arteriolar sphincters
   c. skin receptors; pituitary gland
   d. skin receptors; hypothalamus

10. A homeotherm maintains homeostasis at a temperature that is generally _____ that of its environment.
    a. a few degrees cooler than
    b. a few degrees warmer than
    c. nearly the same as
    d. substantially higher than
    e. substantially lower than

11. Which of the following explains why a camel can survive in the desert even without water?
    a. It can tolerate a water loss equal to 25 percent of its body weight.
    b. It is very efficient at maintaining a constant body temperature.
    c. It can produce a highly concentrated urine.
    d. Only a and c are correct.
    e. All three answers (a, b, and c) are correct.

12. Unlike a winter sleeper, a true hibernator:
    a. becomes aroused only when environmental temperatures become favorable.
    b. exhibits a marked decrease in metabolic rate.
    c. resets its thermostat to below normal.
    d. lives on stored food reserves.

13. The primary reason that animals hibernate is to:
    a. stay out of the cold.
    b. lower their energy requirements.
    c. avoid predators.
    d. conserve body water.

14. The most effective form of insulation for a mammal that lives in water is:
    a. short-cropped fur.
    b. a layer of scaly skin.
    c. body fat.
    d. a dark-colored outer surface.

15. Which of these animals uses fat deposits to keep itself *cool*?
    a. a bear
    b. a penguin
    c. a camel
    d. a whale
    e. none of these animals

16. When the air temperature is higher than 98.6°F, humans can regulate their internal temperature *only by*:
    a. dilating peripheral blood vessels.
    b. convectional heat loss.
    c. sweating profusely.
    d. lowering the metabolic rate.
    e. Both a and b are correct.

17. Which of the following events takes place when the body temperature of a mammal begins to fall?
    a. Blood vessels near the skin surface constrict.
    b. Insulating layers increase in thickness.
    c. Involuntary muscles contract.
    d. Adrenaline and/or thyroxine secretion is stimulated.
    e. All of the above are correct.

18. The countercurrent arrangement of blood vessels in the extremities of an Arctic mammal works to:
    a. warm venous blood.
    b. cool venous blood.
    c. warm arterial blood.
    d. cool arterial blood.
    e. Both a and d are correct.

19. The evaporation of water from the respiratory tract of an animal:
    a. takes place most rapidly when humidity is high.
    b. removes 540 calories of heat per gram of water evaporated.
    c. is the chief means of water loss in all mammals.
    d. is the principal means by which desert mammals remain cool during the heat of the day.
    e. Both b and c are correct.

20. Match each group of organisms with the description that fits it best.

    poikilotherms _____
    endotherms _____
    hibernators _____
    small desert mammals _____
    marine mammals _____
    birds _____

    a. have high energy requirements that necessitate constant feeding
    b. have temperatures that fluctuate with changes in environmental temperatures
    c. can tolerate a great drop in skin surface temperature
    d. are active nocturnally to avoid heat
    e. warm themselves with metabolic heat
    f. minimize energy requirements when food is scarce

21. Match each means of heat exchange on the left with an appropriate example (or examples) from the list on the right.

    radiation _____
    conduction _____
    convection _____
    evaporation _____

    a. between a snake and a warm rock it is lying on
    b. between a squirrel and its burrow
    c. from a human to the air moving over his or her skin on a dry day
    d. between a basking lizard and the sun overhead
    e. from a pig to a pond it is wallowing in

22. How does the circulatory system of a mammal contribute to the maintenance of a constant internal temperature?

23. What is meant by "fever"? Why is it often accompanied by chills? What might be its adaptive value?

24. Why are the temperature limits within which life exists relatively narrow?

PERFORMANCE ANALYSIS

1. When two objects are in direct contact, kinetic energy of molecular motion is transferred from the hotter object to the colder one. This phenomenon is referred to as conduction (**a**).

2. Since the function of proteins, and enzymes in particular, depends on their three-dimensional shapes, these molecules become nonfunctional when temperatures rise too far above the optimum. It is apparently this feature of our biochemistry that sets the upper temperature limit for life. Choice **b**, then, is correct.

3. Loosely speaking, ectotherms are organisms that must be warmed from the outside. They are unable to generate heat from metabolic processes to any great degree—the tuna is a notable exception—and must gain heat by radiation (as from the sun) or by conduction (when in direct contact with a hot object, such as a rock). Choice **c**, therefore, is correct.

4. Small endotherms, such as rodents, must have

very high metabolic rates in order to offset the great amount of heat lost across their body surfaces, which are large in relation to their volume; the smaller the animal, the larger its surface-to-volume ratio. Therefore, **b** is correct.

5. The Labrador retriever (**c**) would have the most difficulty because its black coat would gain heat quickly from the sun's radiation. Its problems would be complicated by the high humidity, which would lessen the effectiveness of evaporative heat loss. The man in thin summer clothing would not be insulated by a thick coat of fur, so evaporation, even though slow on a humid day, would be more effective. The long ears of the fox would tend to radiate heat away from its body.

6. Terrestrial poikilotherms—reptiles, for the most part—may maintain remarkably stable internal temperatures by orienting their bodies (**d**) so as to encourage or reduce radiative and conductive heat gain. Sweating and panting (a) are used by homeotherms, not poikilotherms. Choice b is incorrect because poikilotherms do not typically regulate body temperature metabolically. Remaining active (c) is related to the creation of heat metabolically; it is a means of generating heat in homeotherms but it is not used by poikilotherms.

7. Endotherms, by definition, are those organisms that are able to generate and retain heat from metabolic reactions, primarily oxidation (**a**).

8. Although it may be argued that heat produced in the body core moves slowly through tissues by conduction, the great portion of the heat exchanged at the body surface of a mammal is brought there by circulation of the blood (**c**).

9. Sensory receptors in the skin transmit electrical impulses to the hypothalamus (**d**), which acts as a thermostat for mammals.

10. Choice **d** is correct. It is easier to generate heat than it is to release it. For many homeotherms, heat loss involves loss of body water, which is often not immediately replaceable.

11. Camels survive desert conditions by conserving water (they produce a hypertonic urine) or, when this is not fully possible, by tolerating a water loss often as high as 25 percent of body weight. Since camels store heat during the day and release it at night, they are literally poikilotherms, their body temperatures vary by as much as 5° or 6°C during the course of a day. Both choices a and c are correct; therefore, the answer is **d**.

12. Hibernators, such as the ground squirrel, and winter sleepers, such as the bear, both drop their thermostats so as to reduce energy requirements at a time of year when food is unavailable. But true hibernators exhibit marked decreases in metabolic rates (**b**) (often 20 to 100 times lower than normal), whereas winter sleepers lower their body temperatures only slightly and maintain this temperature by living off stored food reserves.

13. Choice **b** is correct. See the analysis for Question 12.

14. Because fur and feathers are poor insulators when wet, aquatic mammals, such as seals and whales, make use of a fat layer (**c**) to retain their body heat.

15. Whereas aquatic mammals, as mentioned above, make use of fat deposits to stay warm, the desert camel (**c**) stores fat in its hump and thereby limits radiative heat gain from its environment.

16. Radiative heat loss (implied in choice a) and convectional heat loss work only when the temperature of the environment is lower than body temperature. Only by sweating (**c**) can people cool themselves when the air temperature is above 98.6°F. For each gram of water that is evaporated from the skin surface, a human loses approximately 540 calories of heat.

17. When body temperatures fall, mammals may constrict peripheral blood vessels and increase the thickness of their insulating layers to reduce radiative heat losses. More heat can be produced by involuntary contractions of the muscles (such as in shivering) or by increased metabolic rate, which is stimulated by adrenaline or thyroxine secretions. Therefore, **e** is correct.

18. Choice **e** is correct. The principle of this countercurrent exchange is illustrated in Figure 38–12.

19. Evaporative heat loss is possible only if the surrounding air is not saturated with water vapor and is employed only by animals that carry enough body water to tolerate some water loss. For each gram of water evaporated, 540 calories of heat are removed (**b**). If you think that c is correct, reread page 723.

20. The correct answers in order are **b**, **e**, **f**, **d**, **c**, and **a**.
21. The correct answers in order are **b**, **d**; **a**, **b**, **e**; **c**; and **c**.
22. Your discussion should include control of the size of peripheral blood vessels as well as heat conservation made possible in Arctic mammals by the countercurrent arrangement of arteries and veins (see Figure 38–12).
23. A fever occurs when the hypothalamus resets the internal thermostat to a temperature that is above normal. Until your body temperature reaches the new thermostat setting, you will feel chills. Fever may be of adaptive value in creating an unfavorable environment for pathogens.
24. The upper temperature limit for life is discussed in the analysis for Question 2. The lower limit is the environmental temperature at which body water freezes, thus immobilizing or locally concentrating solutes.

CHAPTER 39

Homeostasis III: The Immune Response and Other Defenses

CHAPTER ORGANIZATION

I. **Introduction**
II. **Nonspecific defenses**
   A. Anatomic barriers
   B. The inflammatory response
III. **Interferon**
IV. **The immune response**
   A. Introductory remarks
   B. The immune system
   C. B lymphocyte responses and the formation of antibodies
      1. General information
      2. The B lymphocyte: A life history
      3. The action of antibodies
      4. The structure of antibodies
      5. Monoclonal antibodies
      6. The clonal theory of antibody formation
      7. The genetics of antibody formation
   D. Essay: Death certificate for smallpox
   E. T lymphocytes and their functions
      1. Introductory remarks
      2. Discovery of the T lymphocytes
      3. The functions of T lymphocytes
V. **Disorders associated with the immune system**
   A. Allergies
   B. Autoimmune diseases
   C. The Rh factor
VI. **Tissue and organ transplants**
   A. Blood transfusion
      1. General information
      2. Inheritance of blood types
   B. Organ transplants
      1. General information
      2. The major histocompatibility complex
VII. **Immunity and cancer**
VIII. **Summary**

MAJOR CONCEPTS

There are three levels of defense against foreign-particle invasion in vertebrates: (1) nonspecific defenses (the skin and mucous membranes and also the inflammatory response); (2) a semispecific response to viral invasions (interferon); and (3) a highly specific response to individual antigens (the immune response).

The two types of white blood cells involved in the immune response are B lymphocytes, which are responsible for antibody production, and T lymphocytes, which function in cell-mediated immunity.

The general structure of an antibody (a protein molecule) consists of two identical light chains and two identical heavy chains. Each chain has regions that are common to all antibodies of the same class and regions that vary with specific antibodies.

The most widely accepted theory of antibody formation is the clonal selection model, which proposes that each one of an organism's B lymphocytes is capable of producing antibodies against one specific type of antigen. When a B lymphocyte encounters its specific antigen, it produces clones of plasma cells, each of which produces antibodies that act upon the original antigen. Memory cells are also produced. They persist in the bloodstream following infection and are capable of immediate synthesis of antibodies upon subsequent exposures to the antigen. This memory-cell response is the cause of the rapid, enhanced immunity that develops following vaccination or infection.

T lymphocytes destroy cells by cell-to-cell contact (cell-mediated immunity). They destroy the body's own cells that are harboring intracellular viruses or other parasites.

Disorders associated with the immune system include allergic reactions, Rh disease, autoimmune diseases, blood-transfusion reactions, and tissue-transplant rejections.

HOW TO STUDY THE CHAPTER

Read the chapter, focusing on the major concepts.

Reread the chapter, using the questions that follow to help you focus on details as they relate to major concepts. Answer the questions on a separate sheet of paper. Your answers will provide a valuable study aid in preparing for examinations.

FOCUS ON CHAPTER DETAILS

  I. **Introduction** (page 746)
  1. List the three basic ways in which vertebrates protect themselves.
 II. **Nonspecific defenses** (pages 746–747)
  A. Anatomic barriers
  2. How do the various external membranes of the body help prevent infection by microorganisms?
  B. The inflammatory response
  3. What events trigger an inflammatory response?
  4. What are the roles of the three types of granulocytes and of the monocytes in an inflammatory reaction?
  5. What symptoms resulting from inflammatory response are systemic (system wide)? What is a local symptom?
III. **Interferon** (pages 747–748)
  6. How does interferon differ from other defense mechanisms?
  7. What are interferons? How were they discovered?
  8. How does interferon protect against viral infection?
  9. Why has interferon become a valuable research tool? How is its high cost currently being reduced?
 IV. **The immune response** (pages 748–756)
  A. Introductory remarks
  10. What accounts for the high degree of specificity of the immune response?
  B. The immune system
  11. Describe a lymph node. What general functions do lymph nodes serve?
  12. Describe what happens as lymph passes through a lymph node; as blood passes through a lymph node (see Figure 39–4).
  13. Describe the various organs and tissues, in addition to lymph vessels and nodes, that constitute the immune system. What are their functions?
  C. B lymphocyte responses and the formation of antibodies
  General information
  14. What are antigens and antibodies? What is their relationship to one another?
  15. List the five classes of antibodies and their specific locations and functions, where known.
  The B lymphocyte: A life history
  16. Where are B lymphocytes found in the body?
  17. What happens when a B lymphocyte recognizes an antigen?
  18. What two types of cells result from the division of a B lymphocyte? What functions do these cell types carry out?
  19. What accounts for the body's occasional inability to respond immediately to infection by microorganisms? Why, on the other hand, is full-scale antibody production sometimes immediate?
  20. Describe four general ways in which vaccines are prepared.
  The action of antibodies
  21. In what three ways can antibodies act against foreign substances?
  22. What is complement? Describe the actions of complement proteins.

The structure of antibodies

23. How did Gerald Edelman use multiple myeloma to determine the structure of an immunoglobulin?
24. Review the structure of an antibody as shown on page 341, paying special attention to the C and V regions.
25. What region of an antibody is responsible for its antigen-specific properties?

Monoclonal antibodies

26. Describe the technique that has made antibodies available in quantities sufficient for research and therapy. What types of cells are involved?

The clonal theory of antibody formation

27. Estimate the number of different antigens against which an organism can form antibodies. Can antigens be synthetic? Explain.
28. Describe the clonal theory of antibody selection. What evidence exists for this theory?

The genetics of antibody formation

29. Why was the clonal theory initially discounted as an adequate explanation for antibody formation?
30. How can the amount of genetic material in the cell actually account for the great diversity of antibody structure?

D. Essay: Death certificate for smallpox (page 754)

31. What did people suffering from smallpox come to discover about the disease that later would become important in preventing it?
32. What practices were used to combat smallpox in the Middle and Far East?
33. What experiment did Edward Jenner perform that led to the development of a vaccine against smallpox? From the knowledge of the immune system you have gained thus far, explain why Jenner succeeded.
34. What efforts were made by the World Health Organization to eradicate smallpox from the disease's last stronghold?

E. T lymphocytes and their functions

Introductory remarks

35. Describe the other highly specific immune response that exists besides antibody production.

Discovery of the T lymphocytes

36. How was it determined that certain lymphocytes are produced by the thymus gland?
37. What is the role of the T lymphocytes in the immune response?

The functions of T lymphocytes

38. What are the secondary effects of T lymphocyte production?
39. What other types of T lymphocytes have been discovered, and what are their functions?
40. Why is current immunological research so important?

V. **Disorders associated with the immune system** (pages 756–757)

A. Allergies

41. Describe the immune response that takes place in the presence of allergens (environmental antigens). What is anaphylactic shock?
42. Why is the immune response sometimes considered to be maladaptive?
43. How do histamines and corticosteroids affect the immune response?

B. Autoimmune diseases

44. What is autoimmunity?
45. Why are some substances recognized as "self" and thus not attacked by the body's immune system?
46. What diseases are the result of an autoimmune response? What other diseases are suspected?

C. The Rh factor

47. What is the Rh factor?
48. What problem arises during pregnancy in a

woman lacking the Rh factor if she carries a child that has the Rh factor?

49. How can Rh disease be prevented?

VI. Tissue and organ transplants (pages 757–759)

A. Blood transfusion

General information

50. What reason did Landsteiner give to explain that blood transfusions are often unsuccessful?

51. What are the four major blood groups? From whom can people belonging to each of the four groups receive blood? (See Figure 39–10.) (It should be noted that with modern blood-banking and blood-typing techniques, transfusions of blood of a different type or subtype from the recipient's would be considered only in extreme emergencies.)

52. Why are certain blood groups incompatible?

Inheritance of blood types

53. Explain the relationship between the inheritance of blood groups and the inheritance of multiple alleles.

54. Write the possible genotypes for people in each of the four blood groups.

B. Organ transplants

General information

55. Why are grafts of one's own skin usually successful while donor grafts are not? What immune responses take place during graft rejection?

56. What problems can arise when drugs that suppress the immune response are administered following transplants?

57. In what less dangerous ways might the immune response be circumvented in a person who has received a transplant?

The major histocompatibility complex

58. What is the major histocompatibility complex (MHC)?

59. How does this complex provide for accurate discrimination between "self" and "not-self"?

60. By what mechanisms do Class I and Class II MHC molecules initiate or moderate immune responses?

VII. Immunity and cancer (page 760)

61. Why do cancer cells usually induce an immune response? What does this indicate about the immune system of a person who develops cancer? About the evolutionary origin of cell-mediated immunity?

VIII. Summary (pages 760–761): Read the Summary. If you are familiar with the essential features of the material presented there, you are ready to take the following diagnostic examination.

TESTING YOUR UNDERSTANDING

After you have completed the following examination, compare your answers with the annotated key in the Analysis section.

1. Which of the following cells are the most numerous of those that move to the site of infection to phagocytize microorganisms?
    a. lymphocytes
    b. basophils
    c. neutrophils
    d. erythrocytes
    e. antibodies

2. Lupus has been identified as an autoimmune disease. This means that:
    a. the immune system makes antibodies against cells that are "not-self."
    b. antibodies are made that attack the affected individual's own cells.
    c. the immune system cannot produce its own antibodies to foreign substances.
    d. antibodies are produced that are defective and may form cancerous clones.

3. Which of the following classes of immunocompounds is effective only against viral pathogens?
    a. interferons
    b. histamines
    c. immunoglobulins
    d. complement
    e. immunosuppressives

4. The specific action of antibody molecules is

accounted for by the amino acid sequence of:
   a. the C region.
   b. the V region.
   c. the light chains.
   d. the heavy chains.

5. Memory cells produced from B lymphocytes allow an organism to respond immediately to a second infection by a disease because they:
   a. can rapidly produce antibodies selected during the first infection.
   b. contain copies of antigens associated with the disease.
   c. divide immediately after "remembering" a particular antigen.
   d. can quickly initiate a cell-mediated response.

6. A B lymphocyte responds to an encounter with the antigen to which it is specific by *first*:
   a. producing a substance that coats the foreign particle bearing the antigen.
   b. producing a particular antibody.
   c. stimulating the migration of macrophages.
   d. dividing to form plasma and memory cells.
   e. producing a plasma cell that differentiates to form a memory cell.

7. If a woman who is _____, gives birth to an _____ baby as her first child, red blood cells of the next fetus she carries may be destroyed by maternal antibodies.
   a. Rh^+; Rh^+
   b. Rh^+; Rh^-
   c. Rh^-; Rh^+
   d. Rh^-; Rh^-
   e. Both b and c are correct.

8. A person whose ABO blood group is determined by two codominant alleles has a genotype of:
   a. AO.
   b. BO.
   c. AB.
   d. AA.
   e. BB.

9. A person who has type B blood cannot receive a transfusion from a person who has type _____ blood.
   a. A
   b. B
   c. AB
   d. O
   e. Both a and c are correct.

10. Skin grafts from which of the following donors would be most likely to be accepted without complications?
    a. a sibling of the same blood type
    b. either one of the parents
    c. an identical twin
    d. any healthy donor, provided that an immunosuppressive drug is given
    e. Grafts from any of the four donors would be readily accepted.

11. The saying that "no two individuals are identical" is best exemplified by:
    a. ABO blood groups.
    b. MHC cell-surface molecules.
    c. the clonal theory of antibody formation.
    d. the cell-mediated response.
    e. the Rh factor.

12. Antibodies may actually bring about the lysis of foreign cells when they act in conjunction with:
    a. memory cells.
    b. interferon.
    c. MHC cell-surface molecules.
    d. complement.
    e. white blood cells.

13. Which of the following is an organ of the immune system whose job is to filter the interstitial fluid?
    a. the thymus gland
    b. the bone marrow
    c. the spleen
    d. a lymph node
    e. Peyer's patches

14. Which of the following is a type of nonspecific response to foreign particles?
    a. inflammation
    b. the immune response
    c. the cell-mediated response
    d. the release of interferon
    e. autoimmune responses

15. Which of these represents a systemic manifestation of the inflammatory response?
    a. phagocytosis by white blood cells
    b. increased blood flow at the site of invasion
    c. fever
    d. production of specific antibodies
    e. release of histamines

16. IgE antibodies are produced by _____ in response to the presence of a(n) _____.
    a. B lymphocytes; infectious bacterium
    b. B lymphocytes; allergen
    c. T lymphocytes; virus
    d. T lymphocytes; transplanted tissue
    e. neutrophils; foreign substance

17. The clonal theory of antibody formation:
    a. is unable to account for the great diversity of antibodies produced by a single organism.
    b. has yet to be confirmed by substantial evidence.
    c. predicts that a single plasma cell forms only one type of antibody.
    d. depends upon the stable arrangement of C and V genes in the genome.

18. Large quantities of antibodies for research can be produced by forming a hybrid between a B lymphocyte and a(n):
    a. T lymphocyte.
    b. memory cell.
    c. granulocyte.
    d. cancer cell.
    e. active cell.

19. Which of the following is a type of white blood cell that participates in the inflammatory response?
    a. neutrophils
    b. eosinophils
    c. basophils
    d. monocytes
    e. All of the above are correct.

20. The *initiation* of the cell-mediated response requires that:
    a. lymphocytes release chemicals that will destroy invading cells directly.
    b. lymphocytes divide and bind to antigens of invading cells.
    c. antibodies produced by lymphocytes combine with antigens.
    d. an inflammatory response occurs first.
    e. Both c and d are correct.

21. Match each substance with its definition or its role in defending the body against foreign invaders.

    granulocytes _____
    gamma globulins _____
    IgA molecules _____
    complement _____
    lymphokines _____
    B lymphocytes _____
    "killer cells" _____
    histamines _____

    a. antibodies present in external secretions
    b. produce antibodies in response to bacteria in the blood
    c. chemicals released by T lymphocytes that attract macrophages
    d. T lymphocytes that secrete cytotoxins
    e. circulating white blood cells
    f. circulating antibodies
    g. blood proteins that function as lytic enzymes
    h. released by mast cells in response to allergens

22. What anatomical barriers provide a first line of defense against invading microorganisms?

23. Describe the immune system as an organ system, and indicate the individual functions of its parts.

24. Explain how the clonal selection model of antibody formation accounts for the great diversity of antibodies produced by the body.

25. Match the lettered parts of the following diagram of a lymph node with the terms listed below. (Structures may have more than one location.)

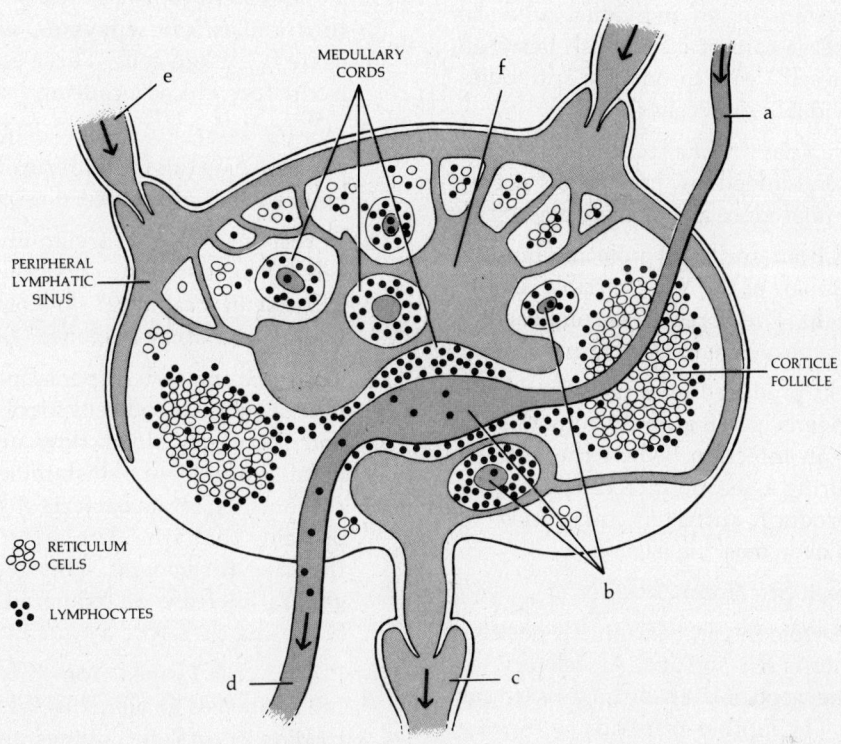

afferent lymphatic vessel _____
arteriole _____
central lymph cistern _____
efferent lymphatic vessel _____
venule _____

PERFORMANCE ANALYSIS

1. Neutrophils (**c**) are the most numerous of the white blood cells that move through the blood to the site of infection, where they phagocytize microorganisms.

2. The immune system of an individual with an autoimmune disease cannot distinguish between "self" and "not-self" and produces antibodies against the individual's own cells (**b**).

3. Interferons (**a**) are a class of small proteins produced in response to viral infections. Interferons act by stimulating cells to produce antiviral enzymes.

4. The site at which an antibody molecule binds a specific antigen is part of the V, or variable region, of the molecule (**b**). The V region is composed of both a light and a heavy chain (see Figure 39-6).

5. The memory cells produced by B lymphocytes in response to antigens (see Figure 39-8) persist in the blood after an infection. When the antigen reappears, as during a second infection, memory cells rapidly produce sufficient quantities of antibodies (**a**) to overcome the pathogen.

6. Choice **d** is correct. See Figure 39-8.

7. An Rh-negative woman can bear an Rh-positive baby if the father is Rh-positive. At delivery, or just prior to it, the blood of the baby mixes with the mother's blood. The mother responds by making antibodies to the Rh factor. In subsequent pregnancies, these antibodies are passed into the fetal bloodstream along with other antibodies during the last month of pregnancy and cause destruction of the red blood cells of an Rh-positive fetus. Choice **c** is therefore correct.

8. Since A and B are codominant alleles, this person's blood type must be AB (**c**).

9. Since persons with type B blood have antibodies in their blood plasma to antigen A, they could not receive blood from those whose blood cells carry this antigen, that is, from those with either type A or type AB blood. Therefore, **e** is correct.

10. A patient will exhibit an immune response to a skin graft from all donors other than his or her identical twin (**c**). The immune response may be suppressed by drugs, but this often presents complications, mainly lowered resistance to infection.

11. No two individuals should ever have the same major histocompatibility complex (**b**), since the antigens that make up the complex are coded for by at least 20 different genes, each of which has from 8 to 10 alleles.

12. Complement (**d**) is a group of blood proteins that function as lytic enzymes, aiding antibodies to destroy foreign cells. These enzymes digest holes in the foreign cells, causing them to rupture.

13. Lymph vessels, which end blindly in interstitial fluid, take up fluid for filtering by lymph nodes (see Figure 39-4); therefore, **d** is correct.

14. In response to any foreign invader, infected cells release histamines, chemicals that trigger an increase in local blood flow. This is referred to as the inflammatory response (**a**).

15. The inflammatory response includes the mobilization of white blood cells involved in phagocytosis and increased blood flow at the invasion site because of release of histamine, but these are local responses. Certain bacteria or host cells can release proteins that stimulate a resetting of the hypothalamic thermostat. This results in fever (**c**), a general increase of temperature throughout the body, and therefore a systemic response.

16. Plasma cells formed from B lymphocytes produce IgE antibodies to allergens. When antigen-antibody binding occurs, histamines are released, and the inflammatory response begins. Choice **b** is correct.

17. Choice **c** is correct. See the analysis for Question 24.

18. Cancer cells (**d**) divide continuously; by hybridizing them with the B lymphocytes that produce desired antibodies, great quantities of these antibodies can be obtained.

19. During an inflammatory response, all four types of white blood cells listed here (**e**) move to the site of infection, where they phagocytize foreign particles.

20. When exposed to specific antigens, T lymphocytes divide into active cells and memory cells. Receptor sites (not antibodies) on the active T cells bind to these antigens, activating the lymphocytes; therefore, **b** is correct. Some, but not all, T lymphocytes do release cytolytic chemicals (**a**). However, this

occurs only after the cell-mediated response has been initiated.

21. The correct answers in order are **e, f, a, g, c, b, d,** and **h**.

22. The anatomical barriers that provide the first line of defense are the skin and mucous membranes.

23. The organs and tissues of the immune system are depicted in Figure 39-3. Lymph nodes (Figure 39-4) serve as filters of the lymph. The spleen filters foreign material from the blood, and the tonsils filter out airborne particles entering through the nose and mouth. Peyer's patches filter microorganisms from the intestinal tract. Bone marrow is the site of production for B lymphocytes; the thymus produces T lymphocytes.

24. According to Burnet's clonal selection model, each B lymphocyte is genetically different, producing a specific type of antibody. When a given antigen binds with the B lymphocyte able to produce an antibody for that antigen, clones of plasma cells are produced that turn out the specific antibody in great quantities (see Figure 39-8).

25. Answers: **e; a; f; c;** and **b, d**.

CHAPTER 40

Integration and Control I: The Nervous System

CHAPTER ORGANIZATION

I. Introduction
II. **Organization of the vertebrate nervous system**
   A. Introduction
   B. The central nervous system
   C. The peripheral nervous system
      1. General information
      2. Divisions of the peripheral nervous system: Somatic and autonomic
      3. Divisions of the autonomic nervous system: Sympathetic and parasympathetic
III. **The nerve impulse**
   A. Background
   B. The ionic basis of the action potential
      1. General information
      2. Propagation of the impulse
   C. The myelin sheath
      1. Introductory remarks
      2. The function of the myelin sheath
IV. **The synapse**
V. **Summary**

MAJOR CONCEPTS

The integration and control of an organism's life processes are accomplished by the nervous system in conjunction with the endocrine system.

The neuron is the functional unit of the nervous system. It consists of dendrites, which receive impulses; a cell body, which contains the nucleus and metabolic machinery; and an axon, which relays the impulses to other cells.

The vertebrate nervous system is divided into two major sections: the central nervous system (the brain and spinal cord) and the peripheral nervous system (all nerves outside the central nervous system). The two main divisions of the peripheral nervous system are the somatic nervous system, which is responsible for control of the skeletal muscles, and the autonomic nervous system, which is involved in the control of cardiac muscle, smooth muscle, and glands. The autonomic nervous system is composed of the sympathetic division (reacts to the external environment) and the parasympathetic division (controls restorative activities).

Nerve impulses are transmitted by sequential changes in the electrical charges on the outside and inside of nerve fiber membranes. Nerve impulses are transmitted across synapses by chemical neurotransmitters that excite or inhibit the production of an action potential in the target cell.

HOW TO STUDY THE CHAPTER

Read the chapter, focusing on the major concepts.

Reread the chapter, using the questions that follow to help you focus on details as they relate to major concepts. Answer the questions on a separate sheet of paper. Your answers will provide a valuable study aid in preparing for examinations.

FOCUS ON CHAPTER DETAILS

I. **Introduction** (pages 762–763)
   1. How do nervous systems differ from endocrine systems?
   2. Review the general organization of the nervous systems of cnidarians, planarians, annelids, arthropods, and vertebrates.
   3. What major trend is apparent in the evolution of vertebrate nervous systems?

II. **Organization of the vertebrate nervous system** (pages 763–769)

   A. Introduction
   4. Draw a diagram to show the various subdivisions of the vertebrate nervous system.
   5. Describe a characteristic neuron.
   6. What terms are used to designate (1) insulating cells, (2) clusters of nerve cell bodies, and (3) bundles of nerve fibers when these are located inside the central nervous system? In the peripheral nervous system?
   7. What is a synapse?

   B. The central nervous system
   8. Distinguish between the gray matter and the white matter of the spinal cord.
   9. Describe the structural link between the spinal cord and the brain. What functions are controlled from this region?

   C. The peripheral nervous system

   General information
   10. Describe the organization of the sensory and motor neurons that lead to and from the central nervous system (see Figure 40–3).
   11. Describe a typical reflex arc (see Figure 40–4). Cite one example from everyday experience.

   Divisions of the peripheral nervous system: Somatic and autonomic
   12. What is the functional difference between the autonomic and the somatic nervous systems?
   13. Why is it inaccurate to distinguish these divisions in terms of involuntary and voluntary control?
   14. How does the arrangement of motor neurons differ in the two divisions?
   15. How is the control of motor activity by the autonomic and somatic divisions different?
   16. How do the divisions differ in terms of sensory input?
   17. How do autonomic and somatic reflex arcs differ?

   Divisions of the autonomic nervous system: Sympathetic and parasympathetic
   18. Briefly characterize the sympathetic and parasympathetic divisions.
   19. List the three major physical differences between the two divisions.
   20. How do the two divisions differ in their influence over bodily functions? Cite some specific examples for each division.

III. **The nerve impulse** (pages 769–773)

   A. Background
   21. How do electrically charged particles behave?
   22. What is a conductor? An insulator?
   23. What is electric potential? How is an electric potential converted to electrical energy?
   24. For what three reasons did biologists abandon the idea of the nerve impulse as a flow of electrons?
   25. Why did the squid become a valuable tool in neurological research?
   26. Explain how a voltmeter is connected to determine the resting potential of a neuron membrane. What is the resting potential?
   27. What is an action potential? What is a nerve impulse? Compare a nerve impulse to an electric current.
   28. What is meant by the all-or-nothing response of a neuron to a stimulus?

B. The ionic basis of the action potential

General information

29. How are Na⁺ and K⁺ distributed on both sides of the membrane of the resting nerve cell?
30. What factors determine these ionic distributions?
31. How does the movement of K⁺ ions account for the establishment of the resting potential?
32. How does stimulation of the membrane account for the production of an action potential?
33. How is the resting potential restored after stimulation?

Propagation of the impulse

34. How is a nerve impulse self-propagated along an axon?
35. What prevents the nerve impulse from flowing in the reverse direction?

C. The myelin sheath

Introductory remarks

36. Besides neurons, what types of cells are found in nervous tissues? Describe Schwann cells.

The function of the myelin sheath

37. Why are impulses conducted more rapidly in vertebrates, many of whose neurons are surrounded by a myelin sheath, than in invertebrates, which lack such a structure?
38. How may the myelin sheath be considered economical?

IV. **The synapse** (pages 773-774)

39. What is the function of a synapse? Describe an electrical synapse.
40. Describe impulse transmission across a chemical synapse. Name two of the most important neurotransmitters.
41. What happens to a neurotransmitter once it has influenced a postsynaptic membrane?
42. What accounts for the excitatory or inhibitory nature of a synapse?
43. How are synapses able to function as relay and control points?

V. **Summary** (pages 774-775): Read the Summary. If you are familiar with the essential features of the material presented there, you are ready to take the following diagnostic examination.

TESTING YOUR UNDERSTANDING

After you have completed the following examination, compare your answers with the annotated key in the Analysis section.

1. An evolutionary trend in the nervous systems of invertebrates that becomes highly refined in vertebrates is the:
    a. aggregation of ganglia in the head region.
    b. development of a dorsal nerve cord.
    c. interconnection of ganglia from different regions of the body.
    d. Only a and c are correct.
    e. All three answers (a, b, and c) are correct.

2. Which of the following gives an accurate comparison of the sympathetic and parasympathetic divisions of the autonomic nervous system?
    a. Each influences a completely different set of effectors.
    b. Each uses a different neurotransmitter at postganglionic nerve endings.
    c. The sympathetic slows the heartbeat, while the parasympathetic speeds it.
    d. Preganglionic and postganglionic fibers in the two divisions are of the same relative length.

3. Nerves running from the ventral side of the spinal cord in the thoracic region contain:
    a. sympathetic motor fibers.
    b. sympathetic sensory fibers.
    c. parasympathetic motor fibers.
    d. parasympathetic sensory fibers.
    e. Both b and d are correct.

4. The sensory neurons of a reflex arc synapse with _____ on the _____ side of the spinal cord.
    a. motor neurons; ventral
    b. other sensory neurons; dorsal
    c. interneurons; ventral
    d. interneurons; dorsal

5. How is the anatomy of the autonomic nervous system different from that of the somatic system?
   a. The autonomic division has both preganglionic and postganglionic motor fibers.
   b. Sensory and motor fibers of the autonomic division run through the spinal cord as tracts.
   c. Motor neurons of the autonomic division exit the spinal cord on its ventral side.
   d. Motor neurons of the autonomic division synapse with an effector in the peripheral nervous system.
   e. Autonomic synapses are electrical, whereas somatic synapses are chemical.
6. Which of these activities is under the control of the parasympathetic division?
   a. secretion of adrenaline
   b. contraction of peripheral blood vessels
   c. increased smooth muscle contractions in the intestinal wall
   d. formation of goose bumps
   e. relaxation of digestive sphincters
7. When a neuron is *not* conducting an impulse, a(n) _____ is present across its cell membrane.
   a. action potential
   b. depolarizing current
   c. electric potential
   d. insulator
   e. impermeability to K^+ ions
8. When a resting potential is established across a neuron membrane:
   a. there is no net movement of K^+ ions through the membrane.
   b. the membrane is impermeable to Na^+ ions.
   c. there is a chemical gradient across the membrane.
   d. there is an electrical gradient across the membrane.
   e. All of the above are correct.
9. When a stimulus arrives at the cell membrane of a neuron, the membrane becomes _____ permeable to _____ ions.
   a. less; Na^+
   b. more; Na^+
   c. less; K^+
   d. more; K^+
10. A nerve impulse moves along an unmyelinated axon toward a synapse:
    a. much like an electric current flows through a wire.
    b. as the axon responds to numerous stimuli along its length.
    c. by self-propagation.
    d. by saltatory movement.
11. Which of these events *initiates* the transmission of a nerve impulse across a synapse?
    a. Vesicles at the synaptic knobs release their contents.
    b. Neurotransmitters diffuse across the synaptic cleft.
    c. Permeability of the postsynaptic membrane is altered.
    d. Neurotransmitters combine with receptor molecules.
    e. Neurotransmitters are enzymatically degraded.
12. Which of the following describes the effect produced by the simultaneous arrival of a number of chemical signals at the postsynaptic membrane?
    a. propagation
    b. depolarization
    c. postganglionic excitation
    d. summation of excitatory and inhibitory signals
    e. inhibition
13. Saltatory transmission refers to:
    a. movement of Na^+ and K^+ ions from one neuron to the next.
    b. the rapid conduction of a nerve impulse from node to node along a myelinated axon.
    c. the active transport of Na^+ ions into interstitial fluid surrounding the nerve.
    d. the transmission of the nerve impulse across the synaptic cleft.
14. Cells of the nervous tissue referred to as Schwann cells:
    a. are neurons.
    b. make up the myelin sheath.
    c. are found only in the central nervous system.
    d. form a continuous layer along the length of a nerve fiber.
    e. Both b and d are correct.
15. The reversal of polarity that occurs when a neuron membrane becomes permeable to Na^+ ions is called:
    a. conduction.
    b. an action potential.

c. synaptic transmission.
d. a monosynaptic reflex.

16. If there were no sodium-potassium pumps in the membranes of neurons, the concentration of _____ inside the membrane would _____.
    a. Na^+; slowly diminish
    b. Na^+; rapidly diminish
    c. K^+; slowly diminish
    d. K^+; rapidly diminish

17. The all-or-nothing principle as applied to nervous control of muscular activity implies that:
    a. the membrane of any given neuron is always depolarized to the same degree.
    b. all receptor molecules on a postsynaptic membrane are bound to molecules of a neurotransmitter.
    c. all of the positive ions inside the neuron exit immediately following an action potential.
    d. if a stimulus is strong enough, all neurons of the peripheral nervous system will become depolarized.

18. Order the following regions of the central nervous system from anterior to posterior.
    a. cranial, thoracic, lumbar, cervical, sacral
    b. cranial, sacral, lumbar, thoracic, cervical
    c. lumbar, cranial, cervical, sacral, thoracic
    d. cranial, cervical, thoracic, lumbar, sacral
    e. lumbar, cervical, cranial, thoracic, sacral

19. The somatic division of the peripheral nervous system is responsible for the innervation of:
    a. heart muscle.
    b. skeletal muscle.
    c. smooth muscle.
    d. glands.
    e. Both a and b are correct.

20. Identify the structures of the polysynaptic reflex arc by matching each term listed below with the appropriate letter in the diagram.

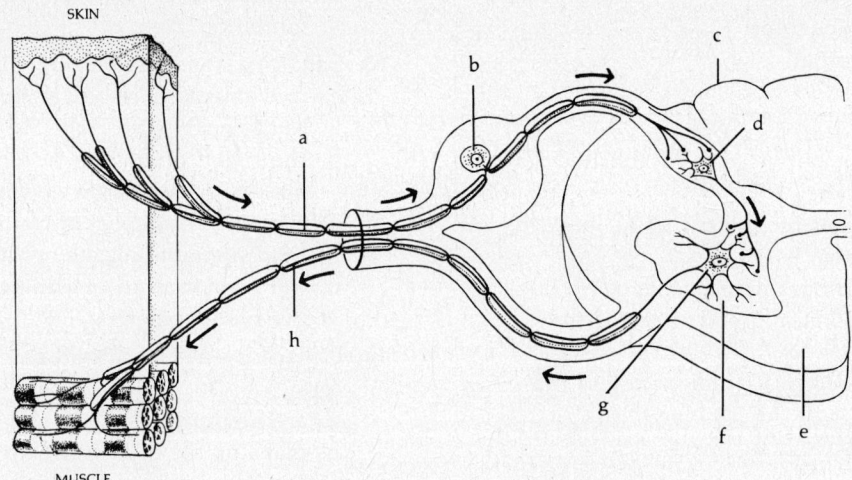

sensory neuron _____
motor neuron _____
motor neuron cell body _____
gray matter _____

interneuron cell body _____
spinal cord _____
white matter _____
sensory neuron cell body _____

21. Match each structure with its description.

    tracts _____
    nerves _____
    ganglia _____
    nuclei _____
    gray matter _____
    white matter _____

    a. bundles of nerve fibers in the central nervous system
    b. bundles of nerve fibers in the peripheral nervous system
    c. clusters of nerve cell bodies in the central nervous system
    d. clusters of nerve cell bodies in the peripheral nervous system

22. Match each structure with its description.

    preganglionic fibers _____
    postganglionic fibers _____
    motor axons _____
    sensory dendrites _____
    myelinated fibers _____
    synaptic clefts _____

    a. link neurons in the same pathway
    b. fibers carrying messages away from the central nervous system
    c. fibers conveying messages directly from receptors
    d. insulated, rapid-conducting fibers
    e. nerves closest to effector organs
    f. nerves closest to the central nervous system

23. Describe a polysynaptic reflex arc.

PERFORMANCE ANALYSIS

1. The trends toward aggregation of ganglia in the head region (cephalization) and the interconnection of ganglia from different regions of the body found among invertebrates become more refined in vertebrates. Choice **d** is therefore correct. (Note that in vertebrates, clusters of nerve cell bodies are called nuclei when they occur in the central nervous system and ganglia only when they occur outside the central nervous system.) Vertebrates, instead of having a ventral nerve cord as do invertebrates, had evolved a dorsal cord.

2. As you can tell from Figure 40-5, the parasympathetic and sympathetic divisions of the autonomic nervous system innervate the same effectors, though their effects are, for the most part, antagonistic (choice c, however, is incorrect—the effects should be reversed). Postganglionic fibers are short in the parasympathetic division and long in the sympathetic. But there is a difference in the neurotransmitters used at postganglionic synapses in the two divisions: acetylcholine in the parasympathetic division and noradrenaline in the sympathetic (therefore, **b** is correct). These differences are summarized in Table 40-1.

3. Motor fibers exit the spinal cord on the ventral side (see Figure 40-3) and sensory fibers exit on the dorsal side. Parasympathetic fibers exit only in the cranial and sacral regions, while sympathetic fibers exit in the cervical, thoracic, and lumbar regions; therefore, **a** is correct.

4. Look at Figures 40-3 and 40-4 to convince yourself that **d** is correct.

5. Choices b and c are true of both the autonomic and somatic divisions, while choice d is generally more true of the somatic division than of the autonomic. Synapses of both divisions are chemical, employing neurotransmitters. But somatic fibers travel all the way from cell bodies in the central nervous system to skeletal muscles without interruption, while autonomic fibers synapse at ganglia located outside the central nervous system, as implied in choice **a**.

6. Only choice **c** is under parasympathetic control. The other choices are stimulated by the sympathetic division and prepare the body for "fight or flight."

7. Because the ionic concentration is more positive outside the neuron membrane than inside, an electric potential (**c**) of about 70 millivolts is present across the membrane. This potential is referred to as the resting potential, the potential that exists when the neuron is not conducting a nerve impulse.

8. At rest, a neuron's membrane is impermeable to Na^+ ions but slightly permeable to K^+ ions, which diffuse out of the cell until equilibrium is reached.

Because negatively charged organic ions inside the neuron are too large to diffuse out, an electrical gradient is established: the inside of the cell is negative with respect to the outside. The chemical gradient results from the impermeability of the membrane to Na^+ ions, which means that there are many more positive ions outside the membrane than inside. (All this, of course, means that choice **e** is correct.)

9. A stimulus causes the nerve cell membrane to become more permeable to Na^+ ions (**b**), which rush into the cell. This brings about a depolarization and an action potential, or firing of the neuron.

10. The nerve impulse is self-propagated (**c**) along an axon (read the caption to Figure 40-12). If the axon is myelinated, the nerve impulse jumps from node to node along the axon in a manner referred to as saltatory movement.

11. When a nerve impulse reaches the end of an axon (the synaptic knob), small vesicles containing neurotransmitters release their contents (**a**). Then these neurotransmitters diffuse across the space between two neurons (the synaptic cleft) and combine with receptor molecules at the postsynaptic membrane, altering its permeability. Immediately after they are used, the neurotransmitters are enzymatically degraded.

12. Depolarization need not always occur at the postsynaptic membrane, but depends on the strength of the summation (**d**) of the electrical signals that arrive. Although individual signals may be excitatory or inhibitory, the overall effect is excitatory or else an action potential is not produced.

13. Choice **b** is correct. The phenomenon is discussed in the analysis for Question 10 and is illustrated in Figure 40-14.

14. Choice **b** is correct. See Figure 40-13.

15. Choice **b** is correct. See the analysis for Question 9.

16. The sodium-potassium pump restores the resting potential by pumping Na^+ ions back out of the nerve cell and K^+ ions back in. If this active transport process did not take place, the concentration of K^+ ions inside the membrane would diminish slowly (**c**), since only a small amount of K^+ ions leave the cell during generation of an action potential.

17. When a stimulus arrives at the neuron membrane, an action potential is either generated or it is not. The voltage by which the membrane is depolarized is always the same (**a**). This is what is meant by the all-or-nothing principle.

18. Choice **d** is correct. Refer to Figure 40-5.

19. Only skeletal muscle (**b**) is under somatic control. Heart muscle, smooth muscle, and glands are stimulated by the autonomic division.

20. Answers: **a**, **h**, **g**, **f**, **d**, **c**, **e**, **b**.

21. The correct answers in order are **a**, **b**, **d**, **c**, **c**, and **a**.

22. The correct answers in order are **f**, **e**, **b**, **c**, **d**, and **a**.

23. In a mammalian polysynaptic reflex arc, nerve impulses are transmitted from sensory receptors along sensory neurons that enter the spinal cord on the dorsal side. Inside the spinal cord, the sensory neurons synapse with interneurons or motor neurons (a simple reflex arc), or turn toward the brain, or both. From the interneurons, the impulses travel along motor neurons (which exit from the spinal cord on the ventral side) to their effector organs, where the reflex action is carried out. (See also Figure 40-4.)

CHAPTER 41

Integration and Control II: The Endocrine System

CHAPTER ORGANIZATION

I. Introduction
II. Some characteristics of hormones
III. Androgens: The male sex hormones
   A. Introduction
   B. Regulation of testosterone production
IV. Estrogens: The female sex hormones
V. Hormones of the adrenal cortex
VI. Thyroid hormone
VII. The pituitary hormones
VIII. The pituitary-hypothalamus axis
IX. Adrenaline and noradrenaline
X. Pancreatic hormones
XI. Parathyroid hormone
XII. Melatonin: The pineal hormone
XIII. Prostaglandins
XIV. Essay: Circadian rhythms
XV. Mechanisms of action of hormones
   A. Background
   B. Intracellular receptors
   C. Membrane receptors
XVI. Summary

MAJOR CONCEPTS

In vertebrates, hormones are organic molecules that are secreted from one part of the animal and are carried in the blood to another part, where they exert specific influences. A hormone is effective in extremely small quantities and only exerts an influence in target cells that have receptors for that specific hormone.

The hypothalamus of the brain and the anterior pituitary gland influence the production of many hormones in other parts of the body through complex negative feedback interactions.

Prostaglandins are fatty acids that resemble hormones in exerting effects on specific target tissues but are unlike hormones in that they often act directly on the tissues that produce them or, in some cases, on the tissues of another individual.

Hormones act by at least two mechanisms: by directly influencing the transcription of RNA and by triggering the release of a "second messenger" inside the target cell.

HOW TO STUDY THE CHAPTER

Read the chapter, focusing on the major concepts.

Reread the chapter, using the questions that follow to help you focus on details as they relate to major concepts. Answer the questions on a separate sheet of paper. Your answers will provide a valuable study aid in preparing for examinations.

FOCUS ON CHAPTER DETAILS

I. **Introduction** (pages 776–777)
   1. What practices made scientists aware of the influence of endocrine glands?

2. Describe Berthold's experiment with cockerels. Why were his results significant?
3. Why is the original Greek meaning for the word "hormone" not really accurate? What is a better definition of a hormone?
4. What is the difference between an endocrine gland and an exocrine gland?
5. How do messages conveyed by hormones differ from neuron messages?
6. What determines whether or not a hormone produces a response in a tissue?

II. **Some characteristics of hormones** (page 777)
7. Give a vertebrate-oriented definition of a hormone.
8. What are the three general chemical types of hormones?
9. In what two important ways are the concentrations of hormones controlled?

III. **Androgens: The male sex hormones** (pages 777–780)
A. Introduction
10. Where is the male hormone testosterone produced? When is it active, and what are its influences at those times? Distinguish the primary effects of androgens from the secondary sex characteristics they produce.
11. What are some effects of testosterone in animals other than humans?
B. Regulation of testosterone production
12. Describe the negative feedback system that regulates the production of testosterone.
13. How is the maturation of sperm in Sertoli cells regulated?
14. Why is inhibin a likely candidate for a male contraceptive?
15. Cite some examples of the effect of external events on testosterone levels in animals.

IV. **Estrogens: The female sex hormones** (page 780)
16. What are some of the effects of estrogens?

V. **Hormones of the adrenal cortex** (pages 780–781)
17. Where is the adrenal cortex located? What types of hormones are produced there?
18. What is the general effect of cortisol? When is it most likely to be produced?
19. How is the action of glucocorticoids similar to the action of the sympathetic nervous system?
20. How are glucocorticoids used medically? What are their undesirable side effects?
21. How is the concentration of glucocorticoids regulated?
22. What is the general effect of mineralocorticoids?
23. What is the usual result of a tumor in the female adrenal cortex?

VI. **Thyroid hormone** (page 781)
24. How is the production of thyroxine regulated?
25. How does thyroxine resemble an amino acid?
26. What is the general effect of thyroxine on metabolism?
27. What are the causes and symptoms of hyperthyroidism? Of hypothyroidism?
28. What other hormone is released from the thyroid gland? What is its effect?

VII. **The pituitary hormones** (page 782)
29. Why was the pituitary once considered to be the master gland? How has this notion proved to be inaccurate?
30. How is the production of the tropic hormones in the anterior pituitary controlled?
31. What is the role of somatotropin in normal growth? What happens when there is a deficiency of this hormone? When is it produced in excess?
32. How is the production of prolactin controlled? What are its effects?
33. What is the role of the intermediate lobe of the pituitary in reptiles and amphibians? In humans?
34. What is the function of the posterior lobe of the pituitary?

VIII. **The pituitary-hypothalamus axis** (pages 782–785)
- 35. What is unusual about the peptide hormones produced by the hypothalamus? What is their target tissue? What are their effects?
- 36. What are the roles of TRH, GnRH, and somatostatin?
- 37. How does the relationship between the pituitary and the hypothalamus exemplify a dual feedback control system?
- 38. What other hormones are produced in the hypothalamus, and what are their effects?
- 39. How is the release of oxytocin regulated?
- 40. Why do ADH and oxytocin exhibit a cross action?

IX. **Adrenaline and noradrenaline** (page 785)
- 41. Why does the adrenal "gland" not really qualify as a gland?
- 42. What hormones are produced in the adrenal medulla? What are their effects? How are these effects related to those of the sympathetic division of the autonomic nervous system?

X. **Pancreatic hormones** (pages 785–786)
- 43. What are the effects of insulin?
- 44. What is diabetes, and how may it be diagnosed?
- 45. Why are untreated diabetics susceptible to death from dehydration?
- 46. What is the role of glucagon in the regulation of blood sugar? What is the role of somatostatin?
- 47. Why is the level of blood glucose under such tight control? Name at least five hormones that are involved.

XI. **Parathyroid hormone** (page 786)
- 48. Where are the parathyroid glands located?
- 49. What is the role of parathormone in calcium metabolism? How is its production regulated?
- 50. How are the effects of parathormone related to those of calcitonin?
- 51. What are the results of tumors in or surgical removal of the parathyroid glands?

XII. **Melatonin: The pineal hormone** (pages 786–787)
- 52. Where is the pineal gland located?
- 53. What is the function of the pineal gland in lower vertebrates?
- 54. How does the action of the pineal gland resemble photoperiodic responses in plants?
- 55. What are the effects of melatonin in humans? In other animals?
- 56. What must be demonstrated before the pineal gland can be conclusively considered a timekeeping device?

XIII. **Prostaglandins** (pages 787–789)
- 57. Why are the prostaglandins so named?
- 58. What are five ways in which prostaglandins differ from hormones?
- 59. What is the most striking effect of prostaglandins? How is this effect related to their suspected role in reproduction?
- 60. How are prostaglandins thought to cause menstrual cramps (dysmenorrhea)?
- 61. What role do prostaglandins play in inflammatory and immune responses? What is the effect of aspirin on prostaglandins?

XIV. **Essay: Circadian rhythms** (page 788)
- 62. What are circadian rhythms?
- 63. Cite several examples of human physiological changes that follow circadian schedules.
- 64. Cite two situations in which it might be very important to consider a person's chronobiology before making certain decisions about them.

XV. **Mechanisms of action of hormones** (pages 789–791)
- A. Background
  - 65. Describe the two mechanisms underlying hormone specificity, and cite examples of each mechanism.
- B. Intracellular receptors
  - 66. How do steroid hormones enter a cell?

67. Explain the action of the steroid hormone-receptor complex.
68. What accounts for the specificity of action of the steroid hormones?

C. Membrane receptors

69. What happens when an amine, peptide, or protein hormone arrives at its target cell?
70. How is the form of diabetes found in the young different from the type found in the old and the obese?
71. Carefully study the mechanism of action of the hormone adrenaline in releasing glucose from liver cells (see Figure 41-15). Describe the role of cyclic AMP as a second messenger.
72. How does adrenaline demonstrate the effectiveness of hormones in minute quantities?
73. What evidence is there that so-called "mammalian" hormones are not unique to mammals?

XVI. **Summary** (pages 791-793): Read the Summary. If you are familiar with the essential features of the material presented there, you are ready to take the following diagnostic examination.

TESTING YOUR UNDERSTANDING

After you have completed the following examination, compare your answers with the annotated key in the Analysis section.

1. Which of the following organs or tissues is both an endocrine and an exocrine gland?
   a. adrenal cortex
   b. pancreas
   c. sebaceous glands
   d. pituitary gland
   e. hypothalamus
2. In animals, hormones are chemical messengers that are transported through _____ and that have _____ effects.
   a. the bloodstream; excitatory or inhibitory
   b. the lymphatic system; excitatory
   c. ducts; excitatory
   d. the digestive tract; stimulatory
   e. interstitial fluids; inhibitory
3. Which of the following classes of chemical messengers is often produced in their own target tissues?
   a. gonadotropic hormones
   b. gonadotropin-releasing hormones
   c. glucocorticoids
   d. prostaglandins
   e. peptide hormones of the hypothalamus
4. A possible male contraceptive is a hormone that is produced in the _____ of males.
   a. pituitary gland
   b. Sertoli cells
   c. islets of Langerhans
   d. semen
   e. adrenal medulla
5. Female sex hormones are known collectively as _____, and male sex hormones are classified as _____.
   a. estrogens; androgens
   b. progesterones; testosterones
   c. estrogens; prostaglandins
   d. prostaglandins; testosterones
   e. steroids; peptides or proteins
6. Which of these pairs of hormones regulates the metabolism of calcium?
   a. thyroxine and parathormone
   b. parathormone and calcitonin
   c. ACTH and TSH
   d. prolactin and oxytocin
7. Which of the following is true of the hormones adrenaline and noradrenaline?
   a. They are produced in the hypothalamus and stored in the pituitary gland.
   b. Their effects are usually antagonistic.
   c. They are secreted by nerve endings rather than by glands.
   d. Their action resembles that of the parasympathetic division.
   e. Both b and c are correct.
8. The pineal gland has attracted more than passing curiosity because:
   a. it is light-sensitive in some organisms.
   b. it appears to assume the functions of the pituitary gland if the latter is damaged.
   c. it is present in all vertebrates except humans.
   d. it is not really a gland but a large ganglion that secretes hormones.

9. Steroid hormones combine with specific _____ molecules (receptors) that are located _____.
   a. steroid; in the cell membrane
   b. steroid; in the cytoplasm or nucleus
   c. protein; in the cell membrane
   d. protein; in the cytoplasm or nucleus
   e. cyclic AMP; inside the cell

10. Which of the following is true with regard to the effects of adrenaline on liver cells storing glycogen?
    a. An enzyme is activated by cyclic AMP.
    b. ATP is converted to cyclic AMP.
    c. Glycogen is broken down to glucose.
    d. Adrenaline binds with receptors on the outer surfaces of cell membranes.
    e. All of the above are correct.

11. Research into the mechanism of action of protein hormones has revealed that cyclic AMP:
    a. is a membrane-bound hormone receptor.
    b. acts as a second messenger.
    c. is responsible for degrading protein hormones.
    d. promotes RNA transcription.
    e. is always present in the target cells of steroid hormones.

12. Which of the following is true regarding the relationship between the hypothalamus and the pituitary?
    a. The pituitary is the master of the hypothalamus.
    b. The pituitary may be overridden by the hypothalamus.
    c. The hypothalamus stores and releases hormones manufactured by the pituitary.
    d. There is no circulatory connection between the hypothalamus and the pituitary gland.

13. Pain resulting from the action of prostaglandins may be relieved by:
    a. an injection of estrogen.
    b. removal of the prostate gland.
    c. aspirin.
    d. electrical stimulation of the autonomic nervous system.

14. Releasing hormones that could have either a stimulatory or an inhibitory effect were first isolated from the _____ of mammals.
    a. adrenal glands
    b. brains
    c. pancreases
    d. testes and ovaries
    e. mammary glands

15. When blood testosterone levels fall below normal:
    a. LH production in the pituitary ceases.
    b. LH is released from the pituitary.
    c. sperm production is inhibited.
    d. sperm production is stimulated.
    e. Both b and c are correct.

16. When cortisol is secreted into the bloodstream, the uptake of glucose by:
    a. all cells is inhibited.
    b. all cells is stimulated.
    c. brain and heart muscle is inhibited.
    d. brain and heart muscle is favored.
    e. storage cells in the liver is stimulated.

17. A group of hormones of prime importance in maintaining water balance and proper ion concentrations are the:
    a. androgens.
    b. prostaglandins.
    c. mineralocorticoids.
    d. luteinizing hormones.
    e. pancreatic hormones.

18. Match each hormone or class of hormones with one of its medical applications.

    glucocorticoids _____
    somatotropin _____
    oxytocin _____
    prostaglandins _____
    testosterone _____
    ADH _____

    a. promotion of normal growth in dwarfs
    b. related to cramping during menstruation
    c. elevation of blood pressure in some vertebrates
    d. reduction of inflammation in arthritis
    e. promotion of contractions during labor
    f. castration removes its influence

19. Match each hormone or class of hormones with the appropriate description of its chemical structure.

prostaglandins _____
_____
mineralocorticoids _____
ACTH _____
thyroxine _____
insulin _____
progesterone _____
adrenaline _____

a. peptide
b. protein
c. modified amino acid

prolactin _____
somatotropin _____

d. fatty acid
e. steroid

20. Compare the mechanism of action of adrenaline with that of estradiol.

21. Explain how the relationship between the hypothalamus and the pituitary gland allows an organism to respond to internal changes in hormone levels.

22. Complete the following table.

Condition	Hormone Directly Involved	Source of Hormone	Hormone Is Controlled by
a. Dwarfism			
b. Growth of body hair in men			
c. Diabetes			Concentration of sugar in blood
d. Development of female breasts			
e. Hypothyroidism			
f. Increased rate of heartbeat			
g. Milk production in mammals			

23. Identify the hormone-producing organs in the following diagram.

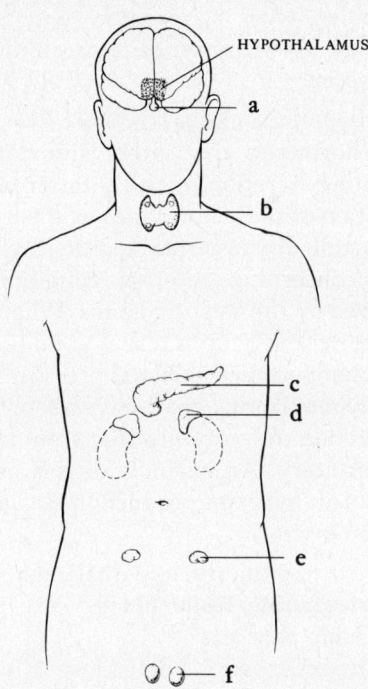

PERFORMANCE ANALYSIS

1. The products of exocrine glands are secreted into ducts, whereas the products of endocrine glands (hormones) are secreted into the bloodstream. The pancreas (**b**) qualifies as both an exocrine gland, because it secretes an alkaline fluid containing digestive enzymes into the pancreatic duct, and an endocrine gland, because it secretes the hormones insulin, glucagon, and somatostatin into the bloodstream.

2. Hormones, which are transported through the bloodstreams of animals, may be excitatory or inhibitory (**a**).

3. Prostaglandins (**d**) differ from hormones in that they often exert their influence on the same tissues in which they are produced. The prostaglandins present in menstrual fluid, for example, cause muscle cramps in the walls of the uterus.

4. There is evidence that inhibin, a hormone produced in the Sertoli cells (**b**) of the testes, inhibits FSH production. This substance, although not yet isolated, is a candidate for a male contraceptive.

5. Choice **a** is correct. The female hormones are referred to collectively as estrogens; there is also a specific hormone known as estrogen. Male hormones are classified as androgens.

6. Parathormone from the parathyroid glands and calcitonin (**b**) from the thyroid gland act in an antagonistic fashion to regulate blood calcium levels. Parathormone causes calcium to be released from bone and to be reabsorbed in the kidney, and calcitonin inhibits the release of calcium from bones.

7. The hormones adrenaline and noradrenaline are produced in the adrenal medulla and are secreted from nerve endings (**c**) in the central portion of the adrenal gland, which is actually a giant ganglion. Their effects are similar and resemble those of the sympathetic division.

8. In lower vertebrates, the pineal gland contains light-sensitive (**a**) cells and is involved with photoperiodism. The production of a pineal hormone, melatonin, varies with light intensities and causes widespread physiological effects in many animals. For this reason, it is tempting to suggest that the biological-clock mechanism may lie within this gland.

9. The mechanism of action of steroid hormones involves protein receptors that are present in either the cytoplasm or the nucleus of the target cell; therefore, **d** is correct.

10. The mechanism of action of adrenaline on liver cells that store glycogen is typical of protein hormones. Review Figure 41–15 to see that **e** is correct.

11. When many of the protein hormones bind to their receptor molecules on or in their cell membranes, ATP is converted to cyclic AMP in the cytoplasm of the target cell. Cyclic AMP serves to activate an enzyme in a reaction sequence that brings about the effect we associate with the hormone. Because of this role, cyclic AMP is often referred to as a second messenger (**b**).

12. The circulatory and nervous connections between

the hypothalamus and the pituitary gland are shown in Figure 41-9. Though it was once thought to be the master gland, the pituitary may be overridden by the hypothalamus (b), which can signal it to release stored hormones.

13. It is now known that aspirin (c) provides relief from the pain of inflammation by inhibiting the synthesis of prostaglandins.

14. The releasing hormones, TRH, which stimulates the release of thyrotropin, and GnRH, which controls the release of gonadotropins, were first isolated from the brains (b) of pigs and sheep. These hormones are produced by the mammalian hypothalamus and are stored in the pituitary gland.

15. Since testosterone acts on the cells of the testes to produce sperm, lowered levels would inhibit sperm production (c). The hypothalamus responds to these lowered levels of testosterone by signaling the pituitary gland to release LH (b), luteinizing hormone, which acts on the male gonads to stimulate testosterone production; therefore, choice e is correct.

16. The hormone cortisol acts by inhibiting the uptake of glucose by all cells in the body except those of the brain and heart muscle (d). The activity of these organs is therefore favored by cortisol.

17. The mineralocorticoids (c), such as aldosterone, which stimulates the reabsorption of sodium ions from renal tubules, are of general importance in water and ion balance.

18. The correct answers in order are d, a, e, b, f, and c.

19. The correct answers in order are d, e, a, c, a, e, c, b, and b. Refer to Table 41-1 on page 792.

20. The mechanism of action of adrenaline is shown in Figure 41-15. Estradiol is currently thought to enter the target cell by diffusion. Once inside, it combines with a receptor, and the complex initiates RNA transcription, resulting in protein synthesis.

21. The hypothalamus is known to secrete at least nine hormones that either stimulate or inhibit hormone secretion by the anterior pituitary. The secretion of these hormones by the hypothalamus is controlled by negative feedback processes and by data concerning internal conditions that are received by the hypothalamus. When a change in hormone levels or internal conditions (such as body temperature or blood pressure) is detected, the hypothalamus may increase or decrease its production of hormones that stimulate or inhibit the pituitary. An example of this process is the control of thyroxine production discussed on pages 668–669.

22. a. growth, pituitary, hypothalamus
    b. testosterone, testis, LH
    c. insulin, pancreas
    d. estrogens, ovary, FSH
    e. thyroxine, thyroid, TSH
    f. epinephrine, adrenal medulla, sympathetic division
    g. prolactin, pituitary, sucking stimulation

23. a. pituitary gland
    b. thyroid gland
    c. pancreas
    d. adrenal gland
    e. ovary
    f. testis

CHAPTER 42

Integration and Control III: Sensory Receptors and Skeletal Muscle

CHAPTER ORGANIZATION

I. Introduction
II. **Sensory receptors**
  A. Introduction
  B. Types of sensory receptors
  C. Light reception: Vision
   1. Introduction
   2. The retina
   3. Rods and cones
   4. Visual pigments
   5. Vision and behavior
   6. Essay: What the frog's eye tells the frog's brain
  D. Chemoreception: Taste and smell
   1. Background
   2. Taste
   3. Smell
   4. Chemical communication in mammals
  E. Sound reception: Hearing
III. **Skeletal muscle**
  A. Introduction
  B. Structure and function of skeletal muscle
   1. General information
   2. Essay: Twitch now, pay later
   3. Actin and myosin
   4. Regulation of muscle contraction
   5. Initiation of muscle contraction: The neuromuscular junction
   6. The motor unit
IV. **The knee-jerk revisited**
V. **Summary**

MAJOR CONCEPTS

Sensory receptors are cells or groups of cells adapted to detect changes in an organism's internal or external environment and to transmit the information to sensory neurons. Receptors may be classified by the type of stimuli they detect (mechanoreceptors, photoreceptors, chemoreceptors, temperature receptors, and pain receptors) or by the type of information they provide (interoceptors, proprioceptors, and exteroceptors).

The retina of the human eye contains two types of photoreceptor cells: rods, which are light-sensitive cells responsible for night vision, and cones, which provide high-resolution and color vision. Taste and smell are accomplished through the process of chemoreception, the detection of specific molecules.

Hearing involves the mechanical transmission of sound waves from the tympanic membrane (which vibrates when struck by sound waves), through the bones in the middle ear, to the fluid in the cochlea of the inner ear. The resulting movements of the basilar membrane stimulate the hair cells of the organ of Corti; the pattern of stimulation is then relayed to the brain.

Skeletal muscles are composed of muscle fibers, each of which is a single muscle cell that contains 1,000 to 2,000 myofibrils. Myofibrils are divided into functional units called sarcomeres. The thick and thin filaments in sarcomeres slide past one another, causing the muscle to contract. Muscle fiber contraction is stimulated by motor neurons, and the energy for contraction is supplied by ATP.

HOW TO STUDY THE CHAPTER

Read the chapter, focusing on the major concepts.

Reread the chapter, using the questions that follow to help you focus on details as they relate to major concepts. Answer the questions on a separate sheet of paper. Your answers will provide a valuable study aid in preparing for examinations.

FOCUS ON CHAPTER DETAILS

I. **Introduction** (page 794)
   1. Cite several examples of ways in which our senses serve us.
   2. Why are skeletal muscles studied along with sensory receptors?

II. **Sensory receptors** (pages 794–808)
   A. Introduction
       3. Describe how human sense organs have been tailored to our specific requirements by adaptation and evolution.
   B. Types of sensory receptors
       4. What types of sensory receptors do humans possess?
       5. Describe the action of the various mechanoreceptors that are found in the human skin (consult Figure 42–1).
       6. Sensory receptors may be classified as interoceptors, exteroceptors, and proprioceptors. What kind of information is received by each of these types? Cite a number of examples for each group.
       7. What is a transducer?
       8. How is the information from sensory receptors conveyed to the nervous system?
       9. What is the basis for the differences among the senses?

   C. Light reception: Vision
   Introduction
       10. Name several animals that possess highly developed light-receptor systems.
       11. How is the vertebrate eye like a camera?
       12. Describe how light enters the mammalian eye and becomes focused on the retina. How is this different from focusing in the eye of a fish?
       13. What is the function of the iris?
       14. How does our stereoscopic vision provide us with the ability to judge distance?
       15. How do other animal groups employ stereoscopic vision?
       16. Why do some animals have an eye positioned on each side of the head?

   The retina
       17. Why is the vertebrate eye described as "anatomically inside out"?
       18. How much of the light that strikes the cornea reaches the retina? What happens to the portion of this light that is not captured by the photoreceptors?
       19. What feature in the eyes of some nocturnal vertebrates maximizes their light-detecting capabilities?
       20. Describe how signals from the photoreceptors of the retina are carried to the optic nerve. (See Figure 42–6. Note, however, that this simplified diagram does not show the true degree of convergence that exists in the neural pathways of the eye.)
       21. What accounts for the existence of the blind spot? What effect does it have on vision?

   Rods and cones
       22. How does vision provided by rod cells differ from that of the cone cells?
       23. How does the proportion of rods and cones differ in the eyes of nocturnal and diurnal animals?
       24. What is the significance of the fovea of the retina?

25. How does the primate eye compare to that of a bird?
26. How is the arrangement of photoreceptors in the human eye associated with the use of the hands?

Visual pigments

27. Describe the basic structure of the visual pigments found in photoreceptors.
28. Why are animals dependent on plants for proper visual perception?
29. What is the visual pigment found in rod cells? How does it respond to a light stimulus? How do changes in pigment structure generate a nerve impulse? How is the light-absorptive form of the pigment restored?
30. What evidence exists to support a three-receptor hypothesis of color vision? How does this hypothesis account for the perception of colors other than the three fundamental colors?
31. To what portion of the electromagnetic spectrum do our photoreceptors respond? Even though human photoreceptors can detect ultraviolet light, why are most people unable to see by ultraviolet light? Who can see by ultraviolet light?

Vision and behavior

32. What different roles does color play in animal behavior?
33. How does the vision of humans differ from that of most other mammals?

Essay: What the frog's eye tells the frog's brain (page 802)

34. Compare the retina of the frog eye to the human retina.
35. What did Lettvin's work demonstrate about the way a frog's eye functions? What was Lettvin's method for studying the frog eye?
36. How does his finding account for the well-known feeding behavior of frogs?

D. Chemoreception: Taste and smell

Background

37. What animals rely primarily on the sense of smell to obtain information about their environment? Why did a reliance on smell not develop in primates?

Taste

38. What adaptations do fish have for taste perception?
39. How is the location of human taste cells related to their function?
40. What are the four primary tastes perceived by human taste buds? How do taste buds respond to these primary tastes?
41. Cite some examples of differences in taste discrimination and preference in animals.

Smell

42. How are the senses of taste and smell similar in fish and higher vertebrates?
43. Define the sense of smell as it applies to terrestrial animals.
44. How do humans perceive smells?
45. Explain John Amoore's theory of odor discrimination.

Chemical communication in mammals

46. Cite some examples of the role of smell in animal behavior.

E. Sound reception: Hearing

47. Relate the structure of the human ear to the mechanism by which sound waves are amplified and transmitted to the inner ear. (Refer to Figure 42-14.)
48. Why must sounds received in the outer ear be amplified? In what two ways is this accomplished?
49. Locate the auditory tube in Figure 42-14. What is its function? How may we suffer from this anatomical arrangement?
50. Note the structure of the inner ear in Figure 42-15, particularly the cochlea and sensory receptors (hair cells). Explain how sound travels through the cochlea. What is the function of the round window?
51. How are pressure waves transmitted from cochlear fluid to the organ of Corti?

52. How does the structure of the basilar membrane allow for the discrimination of different frequencies of sound waves?
53. What frequencies can be detected by the human ear?
54. Why is it fortunate that we cannot hear sounds at very low frequencies?
55. What is the range of frequencies audible to mammals as a group?

III. **Skeletal muscle** (pages 808–814)

A. Introduction

56. Why is the study of skeletal muscle important?

B. Structure and function of skeletal muscle

General information

57. Describe the structure of a skeletal muscle.
58. Describe the organization of myofibrils in a muscle cell. What is the sarcoplasmic reticulum? How are the T tubules arranged in relation to the myofibrils?
59. Describe the structural unit of the myofibrils, the sarcomere (see Figure 42–18). What is the composition of each of the filaments?
60. According to the sliding filament hypothesis, what changes are induced in the sarcomeres when a muscle is stimulated?

Essay: Twitch now, pay later (page 811)

61. How is the color of muscle fibers related to their degree of activity? What accounts for these colors?
62. Compare red and white muscle fibers with respect to their structure and metabolism.
63. How would the proportions of red to white fibers differ in the muscles of a long-distance runner and a sprinter? What determines these proportions?

Actin and myosin

64. Describe the structures of the thin (actin) and the thick (myosin) filaments (Figure 42–19).
65. What are the functions of the globular "heads" of a myosin molecule?
66. How does the action of the globular heads account for the shortening of the sarcomere?
67. What is the role of ATP in muscle contraction?

Regulation of muscle contraction

68. Describe the relationship between troponin and tropomyosin molecules in the thin filaments of an extended muscle fiber.
69. What changes occur in the sarcoplasmic reticulum of a muscle fiber when the muscle is stimulated?
70. What is the function of Ca^{2+} ions in muscle contraction?

Initiation of muscle contraction: The neuromuscular junction

71. Note the anatomical arrangement of a motor neuron and the muscle fiber it innervates (see Figure 42–22). What happens at the neuromuscular junction in response to a signal traveling along the motor axon?
72. What effect does this activity have on the sarcolemma of the muscle fiber?
73. How do curare and the toxin of botulism induce paralysis?

The motor unit

74. What is a motor unit?
75. What determines the number of muscle fibers that a given motor unit will contain?
76. In terms of the motor unit, what factors influence the strength of contraction?
77. How may we interpret highly coordinated activity in terms of muscle contractions?

IV. **The knee-jerk revisited** (pages 814–815)

78. Describe the knee-jerk reflex, including all the steps involved in the transmission of a nerve impulse and the stimulation of muscle contraction. (Note that the receptor in this case is the muscle spindle, a proprioceptor, and the effector is the muscle that the spindles surround.)

V. **Summary** (pages 815–816): Read the Summary.

If you are familiar with the essential features of the material presented there, you are ready to take the following diagnostic examination.

TESTING YOUR UNDERSTANDING

After you have completed the following examination, compare your answers with the annotated key in the Analysis section.

1. Which of the following is (are) the simplest kind of exteroceptors?
   a. the olfactory epithelium
   b. chemoreceptors in the carotid artery
   c. pain receptors in the skin
   d. hair follicles
   e. Meissner's corpuscles

2. The semicircular canals of the ear are an example of a(n):
   a. chemoreceptor.
   b. exteroceptor.
   c. interoceptor.
   d. proprioceptor.
   e. Both a and c are correct.

3. Animals that have stereoscopic vision:
   a. always have eyes placed on each side of the head.
   b. can calculate the distance to nearby objects.
   c. rely on rod cells for color vision.
   d. generally have an excellent sense of smell.

4. The capture of light by the retina of the human eye:
   a. causes the dissociation of a visual pigment known as rhodopsin.
   b. stimulates an action potential in the bipolar cells.
   c. is enhanced by the presence of rod cells.
   d. is not 100 percent efficient.
   e. All of the above are correct.

5. The 125:1 ratio of photoreceptor cells to ganglion cells in the human eye:
   a. applies only to the region known as the fovea.
   b. greatly increases resolution.
   c. is the result of the convergence of many receptors on comparatively few bipolar cells.
   d. means that little processing of visual information actually takes place in the retina.

6. The photoreceptor cells known as rods _____ than the photoreceptor cells called cones.
   a. provide greater resolution
   b. function better in dim light
   c. are more numerous in the center of the retina
   d. are more abundant in animals requiring a high degree of visual acuity
   e. contain a greater variety of visual pigments

7. The retina of a nocturnal mammal would probably have:
   a. a reflecting layer behind its photoreceptors.
   b. a higher proportion of rods than cones.
   c. an enlarged fovea.
   d. more than one blind spot.
   e. Both a and b are correct.

8. An object is focused on the retina of the human eye:
   a. as a result of changes in the shape of the lens.
   b. in the form of an inverted image.
   c. only in the area known as the fovea.
   d. due to the refractive powers of the iris.
   e. Both a and b are correct.

9. The three-receptor hypothesis of color vision proposes that:
   a. in animals that can distinguish colors, there is a third type of photoreceptor in addition to rods and cones.
   b. there are three different kinds of pigments in the retina, each responding to a different fundamental color.
   c. cones are organized so that light stimulates the three types of differently pigmented cones equally.
   d. rod cells assist cone cells in the perception of colors.

10. In humans, the chemoreception of airborne substances takes place *primarily*:
    a. at the tympanic membrane.
    b. at free nerve endings embedded in the skin.
    c. in the olfactory epithelium.
    d. in the taste buds and associated cells.
    e. in chemoreceptors of the pharynx and upper trachea.

11. Mechanical vibrations of the air are converted into nerve impulses by a group of sensory receptors known as:

a. the olfactory epithelium.
b. hair cells.
c. bipolar cells.
d. interoceptors.
e. semicircular canals.

12. Sound waves entering the ear set up vibrations in (give the correct chronological order):
    a. the tympanic membrane, the bones of the middle ear, the oval window membrane, the cochlear fluid, and the basilar membrane.
    b. the oval window membrane, the tympanic membrane, the bones of the middle ear, the basilar membrane, and the cochlear fluid.
    c. the basilar membrane, the cochlear fluid, the bones of the middle ear, the tympanic membrane, and the oval window membrane.
    d. the bones of the middle ear, the tympanic membrane, the cochlear fluid, the oval window membrane, and the basilar membrane.
    e. None of the above is correct.

13. Because different sensory receptors of the ear respond to different _____ of sound waves, we are able to perceive _____ of the sound.
    a. frequencies; the pitch
    b. frequencies; changes in the intensity
    c. intensities; the pitch
    d. intensities; changes in the intensity

14. A myofibril is:
    a. an individual muscle cell.
    b. a group of protein filaments inside a muscle cell.
    c. composed of a single sarcomere.
    d. a group of muscle fibers that contract together.
    e. a motor axon innervating a muscle fiber.

15. When a muscle fiber is stimulated to contract, Ca^{2+} ions are released from _____ and bind to molecules of _____.
    a. the sarcoplasmic reticulum; troponin
    b. the synaptic knobs of a motor neuron; tropomyosin
    c. myofibrils; actin
    d. the sarcolemma; tropomyosin
    e. the T system; troponin

16. Contraction of a muscle fiber occurs when cross-bridge binding takes place between:
    a. troponin and tropomyosin molecules.
    b. Ca^{2+} ions and troponin.
    c. globular heads of myosin and the thin filaments.
    d. complementary chains of the thin filaments.
    e. the myosin molecules of adjacent sarcomeres.

17. The sliding filament model refers to the movement of _____ past one another.
    a. sarcomeres of adjacent myofibrils
    b. myosin and actin filaments
    c. troponin and tropomyosin molecules
    d. myosin filaments
    e. muscle fibers

18. The role of ATP in muscle contraction includes a close biochemical association with:
    a. the sarcoplasmic reticulum of myofibrils.
    b. the globular heads of myosin.
    c. Ca^{2+} ion binding sites of tropomyosin.
    d. cross-bridge binding sites of actin.
    e. Both b and d are correct.

19. A neuromuscular junction is, in reality, a synapse between:
    a. a neuron and a myofibril.
    b. a neuron and a muscle fiber.
    c. a motor axon and the sarcolemma.
    d. a motor axon and the sarcoplasmic reticulum.
    e. Both b and c are correct.

20. A muscle that is under fine control:
    a. is stimulated to contract by a single motor unit.
    b. has each of its fibers stimulated by several motor axons.
    c. has two or three fibers stimulated by a single motor axon.
    d. has each of its fibers stimulated by a large number of motor axons.

21. Match each muscle region with its description.

    Z line _____
    A band _____
    I band _____
    H zone _____

    a. a light band in the center of the sarcomere
    b. separates adjoining sarcomeres
    c. region extending the length of the thick filament
    d. an electrical connection between a muscle cell membrane and its sarcoplasmic reticulum

T system _____

e. region containing only thin filaments

22. Match each sensory receptor with its description.

stretch receptors _____

Pacinian corpuscles _____

ciliary body _____

cochlea _____

bones in the middle ear _____

whiskers of a catfish _____

iris _____

a. proprioceptors in muscle tissue
b. action is similar to the shutter of a camera
c. smooth muscles controlling the focusing of visual stimuli
d. fluid-filled chamber of the inner ear
e. pressure receptors in the skin
f. transmission of vibrations to the oval window
g. chemoreceptors

23. Match each of the membranes with its function.

sarcolemma _____

sarcoplasmic reticulum _____

basilar membrane _____

round window _____

tympanic membrane _____

a. equalizes pressure in the cochlea
b. propagates an action potential in a muscle fiber
c. amplifies intensity of pressure waves in air
d. furnishes calcium ions for muscle contraction
e. supports the sensory receptors that transduce sound to electricity

24. Describe in detail the nerve and muscle activity involved in the reflex of the knee.

25. Cite at least two examples of how sense organs fulfill their roles as transducers.

26. Differentiate between red and white muscle fibers. For what kind of activity is each of these muscle types suited?

27. Correlate the changes in the appearance of a sarcomere (as you would see it in a longitudinal section) with the biochemical events accompanying muscle contraction.

(Chapter 42 continues on page 326)

28. Identify the labeled parts in the following set of diagrams.

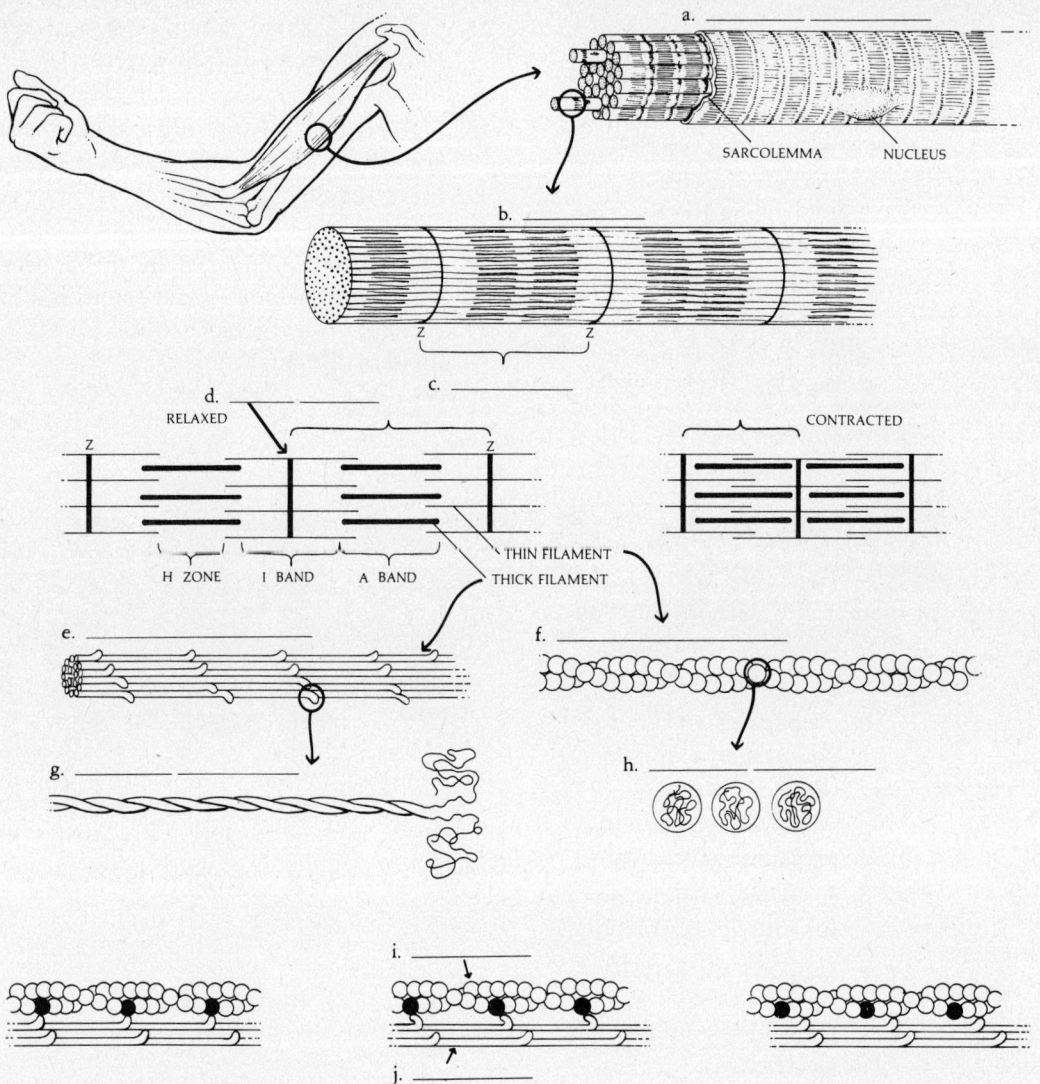

PERFORMANCE ANALYSIS

1. Choices a, c, d, and e are all exteroceptors, but the pain receptors in the skin (**c**) are the least complex—they are simply free nerve endings.

2. Proprioceptors (**d**) are those sensory receptors that provide information about the orientation of the body in space. The semicircular canals of the ear are the major proprioceptive organs in many animals, including humans.

3. For animals with stereoscopic vision, nearby objects present slightly different images to each eye, a phenomenon known as parallax. Because of this disparity, stereoscopic vision allows an animal to calculate the distance between itself and a nearby object (**b**).

4. The light-sensitive cells of the retina, the rods, capture only about 10 percent of the light that strikes the cornea of the eye. These cells contain a pigment, rhodopsin, that dissociates upon absorbing light, causing a change in the resting potential of the cell membrane. This causes a release of neurotransmitter that generates an action potential in the bipolar cells (see Figure 42-6). Therefore, **e** is correct.

5. On the average, there are about 125 photoreceptor cells in the retina for each ganglion cell that leads to the optic nerve. This type of arrangement results in the convergence of many receptors on comparatively few bipolar cells (**c**), which feed into even fewer ganglion cells. Because of the interconnections and interactions that occur, visual information is processed before it leaves the retina. In the fovea, where resolution is greatly increased, there is a 1:1 relationship between photoreceptors and ganglion cells.

6. Because rods are more light-sensitive than cones, they function better in dim light (**b**). Cones, which contain three different kinds of rhodopsin, are closely packed in the fovea (the center of the retina), providing greater resolution and therefore greater visual acuity.

7. A reflecting layer behind the photoreceptors increases the chance that dim light will stimulate them. Because rods are extremely light-sensitive, a higher proportion of them in the retina of a nocturnal animal allows it to see much better in the dark than an animal whose retina contains mostly cones. Choice **e** is correct.

8. Focusing of an object by the human eye takes place as ciliary muscles contract or expand to control the shape of the lens (see Figure 42-4). The image of the object on the retina is inverted. Choice **e** is correct. The iris simply controls the amount of light entering the eye, and the image is not confined to the fovea.

9. There are three different kinds of cone cells in the retina, each containing a different visual pigment. According to the three-receptor hypothesis, each pigment is most sensitive to the wavelength of one of the three fundamental colors (**b**). Because different shades of color stimulate different combinations of these cones, we are able to perceive all the colors (wavelengths) of the visible light spectrum.

10. Chemoreceptors in the olfactory epithelium (**c**) of the nose (see Figure 42-13) bind to airborne molecules that become dissolved in a watery layer of mucus overlying the epithelium. This binding stimulates the receptors to generate an action potential in afferent neurons.

11. Mechanical vibrations are received by the human ear, amplified by its membranes, and transmitted to fluid in the inner ear. Vibrations inside the fluid set up vibrations in sensory receptors called hair cells within the organ of Corti (**b**). The hair cells transmit information to the nervous system regarding the strength and frequencies of the vibrations, which are then interpreted as volume and pitch.

12. Choice **a** is correct. Review Figure 42-15.

13. Because the basilar membrane on which the hair cells of the inner ear lie does not vibrate uniformly along its length, each hair cell responds to a different range of frequencies. This allows us to discriminate pitch; therefore, **a** is correct.

14. A myofibril is a strand of protein filaments (**b**) embedded in the cytoplasm of a skeletal muscle cell and running along the length of the cell. Refer to Figure 42-19.

15. When a muscle fiber is stimulated to contract, the electrical impulse passes along the sarcolemma through the T system to the sarcoplasmic

reticulum. Sacs of the sarcoplasmic reticulum release Ca^{2+} ions that bind to troponin molecules attached to the thin filament, which is composed of tropomyosin and actin molecules; therefore, **a** is correct.

16. When a muscle is stimulated to contract, calcium ions are released from the sarcoplasmic reticulum and combine with troponin molecules of the thin filament (see Figure 42-21). As a result, these molecules undergo changes in conformation that expose cross-bridge binding sites, where the globular heads of the thick filaments (myosin) can bind to the actin molecules of the thin filaments (see Figure 42-19f); therefore, **c** is correct.

17. The sliding filament model proposes that a muscle fiber contracts when the thin (actin) filaments of the cell move past the thick (myosin) filaments (again, see Figure 42-19f). Choice **b** is correct.

18. The globular heads of myosin (**b**) act as enzymes that hydrolyze ATP, releasing the energy for contraction. As ATP is produced, it binds with myosin in such a way as to release the myosin head from its cross-bridge binding site with the thin (actin) filaments.

19. A look at Figure 42-22c should convince you that both choices b and c are correct. The answer is **e**. A motor axon is a fiber extending from the cell body of a neuron, and a muscle fiber is the cell that the motor axon stimulates. The membrane of this cell, whose electric potential is changed by a nerve impulse, is called the sarcolemma.

20. A motor unit is defined as a motor axon and all the muscle fibers that it innervates. In muscles under fine control, such as the muscle that moves the eyeball, the motor unit is composed of two or three fibers. Choice **c** is correct.

21. The correct answers in order are **b**, **c**, **e**, **a**, and **d**.

22. The correct answers in order are **a**, **e**, **c**, **d**, **f**, **g**, and **b**.

23. The correct answers in order are **b**, **d**, **e**, **a**, and **c**.

24. Refer to the Knee-Jerk Revisited section, pages 814-815, for a detailed review of the reflex arc in terms of nervous and muscular action. Study this section carefully to be sure you have not omitted any of the essential steps.

25. You should make use of Figure 42-6 to discuss the eye, Figure 42-15 to discuss the ear, and Figure 42-1 to discuss sensory receptors in the skin. John Amoore's theory of odor discrimination (see page 805) explains how the olfactory epithelium acts as a transducer.

26. Red fibers have many capillaries and mitochondria and large amounts of myoglobin. They depend on oxidative phosphorylation for their energy. White fibers have fewer capillaries and mitochondria and less myoglobin. They depend on glucose or glycogen anaerobically metabolized for energy. Lactic acid accumulates and an oxygen debt occurs if white muscle undergoes frequent repeated contractions. Red muscle is best suited for continuous contractions that require endurance. White muscle is best suited for quick spurts of power.

27. In the contracted sarcomere, the H zone and I band are greatly reduced in width—compare the electron micrograph of Figure 42-18 (uncontracted) with that of Figure 42-20 (contracted). In contraction, the thick and thin filaments slide past each other (see Figure 42-19c). In the contracted state, calcium ions released from the sarcoplasmic reticulum bind to troponin, which causes tropomyosin molecules to shift positions. The cross-bridge binding sites are thereby exposed and bind to the myosin heads.

28. The correct answers are:
    a. **muscle fiber**
    b. **myofibril**
    c. **sarcomere**
    d. **Z line**
    e. **myosin (filament)**
    f. **actin (filament)**
    g. **myosin molecule**
    h. **actin molecules**
    i. **actin (filament)**
    j. **myosin (filament)**

CHAPTER 43

Integration and Control IV: The Brain

CHAPTER ORGANIZATION

I. Introduction
II. The structural organization of the brain
   A. Introductory remarks
   B. Hindbrain and midbrain
   C. Forebrain
   D. Brain circuits
      1. Introductory remarks
      2. The reticular activating system
      3. The limbic system
   E. The cortex
   F. Left brain/right brain
   G. Split brain
III. Chemical activity in the brain
   A. Neurotransmitters
   B. The endorphins
IV. Electrical activity of the brain
   A. The electroencephalogram
      1. Introduction
      2. Alpha, beta, and delta waves
      3. Sleep, dreams, and the EEG
   B. Neuromagnetism
   C. Single-cell recordings
V. Sex and the brain
VI. Learning and memory
VII. Summary

MAJOR CONCEPTS

The major regions of the human brain are the cerebellum (fine motor coordination); the brainstem, composed of the medulla, pons, and midbrain (control of vital functions and consciousness); the diencephalon, which includes the thalamus and hypothalamus (basic drives and emotions and integration of nervous and endocrine systems); and the cerebrum (processing of sensory information, movement, and organization of ideas).

The motor and sensory cortices of the right cerebral hemisphere control motor and sensory functions in the left side of the body. Those in the left hemisphere control the right side of the body. Specific areas of the cerebral cortex have been correlated with particular functions. The control of some faculties, such as speech and spatial orientation, appears to be restricted to one hemisphere or the other.

There are several neurotransmitters in the brain, including acetylcholine, noradrenaline, seratonin, dopamine, and gamma-aminobutyric acid. The observation that brain neurons contain opiate receptors led to the discovery of internally produced opiates known as endorphins.

Three major types of brain-wave patterns that exist in normal humans are alpha waves (associated with relaxation), beta waves (associated with alertness), and delta waves (associated with sleep). Electrical activity in the brain is studied by electroencephalography, by the detection of neuromagnetism, and by the use of microelectrodes.

Several studies have revealed that there are differences in the brains of male and female members of a species.

The two types of memory are short-term memory and long-term memory.

HOW TO STUDY THE CHAPTER

Read the chapter, focusing on the major concepts.

Reread the chapter, using the questions that follow to help you focus on details as they relate to major concepts. Answer the questions on a separate sheet of paper. Your answers will provide a valuable study aid in preparing for examinations.

FOCUS ON CHAPTER DETAILS

I. Introduction (page 817)
1. Describe the structure of the white and the gray matter of the brain. What is the function of the glial cells?

II. The structural organization of the brain (pages 817–825)

A. Introductory remarks
2. What are the three principal regions of the vertebrate brain? What is the evolutionary and embryological origin of these regions?

B. Hindbrain and midbrain
3. What two structures are associated with the hindbrain and midbrain in mammals?
4. Describe the nervous organization of the brainstem. The control centers for what functions reside in the brainstem?
5. How does the location at which visual information is processed differ in higher and lower vertebrates?
6. What is the function of the cerebellum?

C. Forebrain
7. What are the important structures of the diencephalon, and what are their functions?
8. Describe the telencephalon. Explain the evolutionary significance of the rhinencephalon, the corpus striatum, and the cerebrum.
9. What accounts for the greatly increased size of the human brain?
10. Where is the corpus callosum located?

D. Brain circuits

Introductory remarks
11. Proper brain function requires that information in various brain centers be integrated. Name two of these integrating circuits.

The reticular activating system
12. Describe the organization of the reticular activating system.
13. What is the role of the reticular activating system in sensory awareness? How was this determined experimentally? Cite some situations in which the reticular activating system functions.

The limbic system
14. Describe the organization of the limbic system. What is its role?
15. Why is the limbic system considered to be a primitive structure?

E. The cortex
16. Describe the human cerebral cortex (refer to Figure 43-7).
17. How may the functions of various regions of the brain be determined experimentally?
18. What functional areas of the cortex have been mapped? (See Figure 43-8.) Identify the lobes in which each of these cortices is located. Explain how each region is organized for response to sensory receptors (or motor effectors, in the case of the motor cortex).
19. The unmapped cortex has been referred to as the "silent" cortex and as the "association" cortex. Why are both of these designations misleading?
20. The relative size of the association cortex differs among mammals. How has this finding been interpreted? How does the size of the association cortex influence the shape of the skull?
21. What general brain functions are attributed to the association cortex?

F. Left brain/right brain
22. Locate Broca's area and Wernicke's area (see Figure 43-10). What motor functions are associated with these regions?
23. What abnormalities occur when damage occurs to these areas?
24. Why has the left hemisphere of the cortex been considered dominant over the right?
25. Describe the types of disorders that result from damage to the right hemisphere.
26. What anatomical constraint may be responsible for the lateralization of function seen in the cortex?
27. What evidence is there that the functional relationship between the left and right hemispheres is related to developmental processes?
28. How may variability in the size of the song centers of birds be accounted for?

G. Split brain
29. What is a split-brain patient?
30. How did Roger Sperry's patients respond when asked to identify objects they had touched? How did they respond to visual images?
31. How can you account for Sperry's results in terms of the way information is received and channeled to the speech and visual centers of the brain?

III. **Chemical activity in the brain** (pages 825–827)

A. Neurotransmitters
32. Describe the role of each of the following compounds in the central nervous system: noradrenaline, dopamine, serotonin, and GABA. How are these compounds structurally similar?
33. How are severe depression and Parkinson's disease related to concentrations of certain neurotransmitters? How does an antidepressant drug work?
34. How do mood-altering drugs exert their effects?

B. The endorphins
35. What is the natural source of opiates? What effects do they produce in humans? Why are pharmaceutically produced opiates (such as Demerol) as dangerous as natural opiates?
36. What is the mechanism of action of opiates and other structurally similar substances?
37. What conclusion has been reached about why the human nervous system has receptors for opiates? What substances, recently isolated from the brain, support this conclusion?
38. What two types of endorphins have been isolated? What are their general structures? Where in the central nervous system have they been found?
39. Why have endorphins attracted medical interest?
40. What functional relationship is suspected between the pituitary hormone ACTH and the endorphins? What evidence is there for this?
41. What hypothesis has been suggested to explain why users of external opiates become addicted to them?

IV. **Electrical activity of the brain** (pages 827–831)

A. The electroencephalogram

Introduction
42. What is an electroencephalogram? How is it recorded?
43. What is actually measured by an electroencephalogram? What medical diagnoses may be made based on EEG recordings?

Alpha, beta, and delta waves
44. Describe the three types of brain waves that can appear in an electroencephalogram. With what mental states is each of these wave patterns associated?

Sleep, dreams, and the EEG
45. What is paradoxical sleep?
46. How does REM sleep differ from the other sleep stages?

47. What is the apparent relationship between sleep and dreaming?
48. What are some of the proposed functions of REM sleep?

B. Neuromagnetism
49. Upon what principle is the neuromagnetic study of the brain based?
50. What types of information can be obtained by using neuromagnetism to study the brain?
51. What are some practical applications of neuromagnetic studies?

C. Single-cell recordings
52. How are microelectrodes used to study brain activity?
53. Why do invertebrates, such as the squid, make ideal research subjects?
54. How was Kandel able to analyze habituation in the sea hare? What factor was shown to be associated with habituation?
55. Explain how Hubel and Wiesel examined the electrical responses in the visual cortex of the cat to a range of visual stimuli.
56. What principle, introduced in Chapter 42, did their work demonstrate?

V. **Sex and the brain** (pages 831–832)
57. Explain how studies of (a) the influence of testosterone on the song centers of birds and (b) sexual behavior in rats have revealed that there are differences in the brains of male and female members of a species.
58. Summarize the experimental evidence that demonstrates the effect of sex hormones on the brains of rats.
59. What type of behavior in female rhesus monkeys was induced by administration of testosterone during a critical period of their development?

VI. **Learning and memory** (pages 832–833)
60. What is an engram? How did Karl Lashley attempt to find particular engrams? What results forced him to admit that memory is "nowhere and everywhere present"?
61. What is short-term memory? Give an example of something that you might use it for.
62. What analogy is useful in describing how short-term memory is transferred to long-term memory?
63. What structural changes are thought to take place in the brain when a memory pathway is established?
64. What evidence is there that the two types of memory are functionally separate?

VII. **Summary** (pages 833–835): Read the Summary. If you are familiar with the essential features of the material presented there, you are ready to take the following diagnostic examination.

TESTING YOUR UNDERSTANDING

After you have completed the following examination, compare your answers with the annotated key in the Analysis section.

1. The hollow bulges observed at the anterior end of the spinal cord in a human embryo develop to become the:
   a. brainstem and cerebellum.
   b. diencephalon and telencephalon.
   c. medulla, pons, and midbrain.
   d. forebrain, midbrain, and hindbrain.
   e. frontal, parietal, and occipital lobes of the cerebrum.

2. The crossing over of fiber tracts coming from the left side of the body into the right side of the brain is seen in the:
   a. brainstem.
   b. corpus callosum.
   c. cerebellum.
   d. spinal cord.
   e. cerebral cortex.

3. The part of the brain that would be most different in a fish and a human is called the:
   a. cerebellum.
   b. diencephalon.
   c. telencephalon.
   d. midbrain.
   e. thalamus.

4. Information coming to the brain from sensory receptors in the peripheral nervous system is filtered through the:
   a. limbic system.
   b. reticular activating system.
   c. sensory cortex.
   d. association cortex.
5. A deep fissure in the cerebral cortex called the central sulcus divides the _____ from the _____.
   a. left hemisphere; right hemisphere
   b. frontal lobe; parietal lobe
   c. sensory cortex; motor cortex
   d. visual cortex; auditory cortex
   e. Both b and c are correct.
6. On a map of the human sensory cortex, which of these regions would be the largest?
   a. hand and mouth
   b. legs and arms
   c. eyes, nose, and ears
   d. genitalia
   e. All of these sensory regions are of about the same size.
7. The association, or "silent," cortex:
   a. connects the sensory and motor cortices.
   b. is confined to the area posterior to the central sulcus.
   c. shows a reduction in size in higher mammals.
   d. generally does not respond in a specific way to electrical stimulation.
8. Damage to the right hemisphere of the brain often results in:
   a. impairment or loss of speech.
   b. spatial disorientation or failure to recognize close friends.
   c. inability to identify an object held in the left hand.
   d. Parkinson's disease.
   e. Both a and c are correct.
9. A split-brain patient is one in whom the:
   a. frontal lobes have been surgically removed.
   b. right hemisphere controls the right side of the body and the left hemisphere controls the left side.
   c. right and left hemispheres are functionally separate.
   d. corpus callosum has been severed.
   e. Both c and d are correct.
10. Which of the folowing is a neurotransmitter that stimulates major portions of the central nervous system?
    a. acetylcholine
    b. GABA
    c. noradrenaline
    d. beta-endorphin
    e. dopamine
11. Which of the following clues might provide an explanation for the presence of opiate receptors in vertebrate brains?
    a. drug addiction
    b. habituation
    c. delta waves
    d. natural analgesics
12. The role of endorphins was suspected as a result of their relationship to:
    a. adrenaline.
    b. dopamine.
    c. peptide hormones.
    d. ACTH.
13. Which of the following supports the proposal that lateralization of function in the brain is a developmental process?
    a. A young child whose left cerebral hemisphere has been damaged can speak normally.
    b. A split-brain patient with extensive right-hemisphere damage can identify an object held in the right hand.
    c. A person with amnesia can still use his or her short-term memory.
    d. All of the above indicate lateralization.
14. A sleeping person whose EEG shows a period of rapid, low-amplitude waves similar to those observed in an alert person:
    a. is in deep sleep.
    b. has entered REM sleep.
    c. is awakening from sleep.
    d. is tossing and turning in sleep.
15. What type of brain waves would show up on the EEG of a student taking an exam?
    a. alpha
    b. beta
    c. delta
    d. gamma
    e. All of the above would be present.

16. A person deprived of REM sleep probably will *not*:
    a. feel rested when awakened.
    b. remember things he or she has learned.
    c. dream.
    d. talk in his or her sleep.
    e. respond when disturbed.

17. Hubel and Wiesel, in single-cell recordings of the brains of cats, showed that certain kinds of visual stimuli elicited:
    a. responses from specific classes of neurons.
    b. a general response from the entire visual cortex.
    c. a set of specific responses from a single neuron.
    d. a habituation response.

18. Which of the following observations supports the hypothesis that there are functional differences in the brains of male and female mammals?
    a. Rats, even if castrated, still attempt to mate.
    b. Normal male rats injected with estrogens exhibit lordosis.
    c. Electrical stimulation of the hypothalamus of a castrated rat induces increased sexual activity.
    d. Females whose ovaries are removed exhibit lordosis if testosterone is injected.

19. Lashley used the term "engram" to denote:
    a. where electrodes should be placed for an EEG.
    b. the binding sites of endorphins with specific receptors.
    c. the physical change in brain tissue associated with memory for a specific task.
    d. the brain area in which electrical activity is most intense.
    e. the part of the brain where sensory and motor responses are integrated.

20. Which of the following is true with respect to the two kinds of memory humans appear to have?
    a. Short-term memories may be transferred to long-term memory.
    b. The amount of information that can be held in short-term memory is quite limited.
    c. The chemical nature of short-term memory has been explained.
    d. Loss of short-term memory is always accompanied by long-term memory loss.
    e. Both a and b are correct.

21. Match each region of the brain with its description.

    cerebellum _____
    diencephalon _____
    limbic system _____
    visual cortex _____
    auditory cortex _____
    Wernicke's area _____

    a. links the hypothalamus to the cerebral cortex
    b. occupies the occipital lobe of the cerebrum
    c. located in parts of the temporal and parietal lobes of the cerebrum
    d. coordinates complex muscle movements
    e. region of the forebrain containing the hypothalamus

22. Match each term with its description.

    gray matter _____
    enkephalins _____
    alpha waves _____
    dopamine _____
    exogenous opiates _____

    a. highly addictive
    b. densely packed with nerve cell bodies
    c. small peptides that have an opiumlike effect
    d. a neurotransmitter in the central nervous system
    e. produced during periods of relaxation

23. Compare the roles of the limbic and reticular activating systems in relaying and integrating information in the brain. Provide several examples of the types of behaviors associated with each system.

24. Match each numbered brain area in the diagram with the correct term from the list below.
    a. visual cortex _____
    b. lateral sulcus _____
    c. cerebrum _____
    d. frontal lobe _____
    e. central sulcus _____
    f. motor cortex _____
    g. speech areas _____
    h. thalamus _____
    i. occipital lobe _____
    j. auditory cortex _____
    k. cerebellum _____
    l. parietal lobe _____
    m. sensory cortex _____
    n. medulla _____
    o. temporal lobe _____

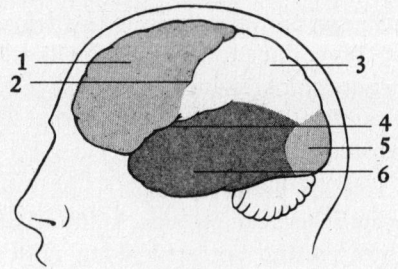

DIAGRAM A

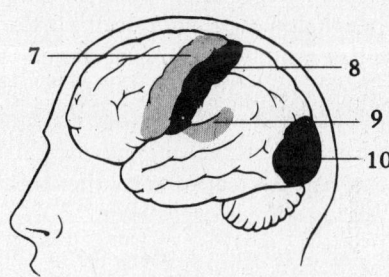

DIAGRAM B

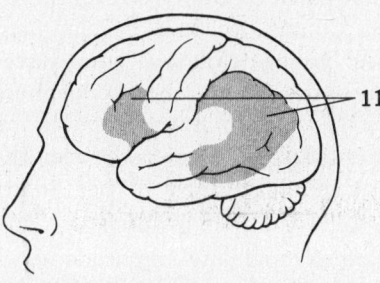

DIAGRAM C

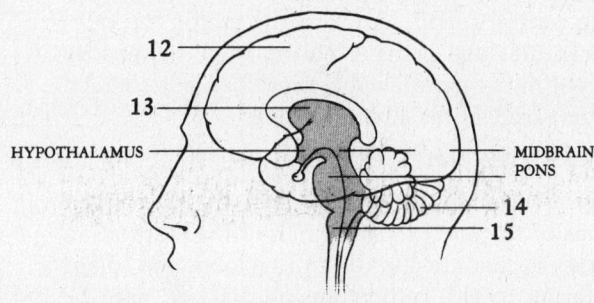

DIAGRAM D

25. Cite several examples of evidence demonstrating that hormones influence the brains of male and female animals differently.

PERFORMANCE ANALYSIS

1. The three hollow bulges, or ventricles, at the anterior end of the embryonic spinal cord develop into the major areas of the brain: forebrain, midbrain, and hindbrain (**d**).

2. The brainstem (**a**) contains fiber tracts running from the spinal cord to higher nerve centers in the brain. It is at this point that signals from one side of the body are channeled to the opposite side of the cerebral cortex. The corpus callosum is a mass of fiber tracts connecting the two hemispheres of the cortex.

3. The telencephalon (**c**), the anterior portion of the forebrain (see Figure 43-2) has changed the most in the course of vertebrate evolution. In a fish, it is primarily concerned with olfactory function and is called the smell brain, or rhinencephalon. In humans, it has developed into the highly convoluted, highly complex cerebrum.

4. The reticular activating system (**b**) of the brainstem apparently filters incoming stimuli from the sensory systems and relays the "important" information to higher brain centers.

5. The central sulcus separates the motor cortex in the frontal lobe of the cerebrum from the sensory cortex in the parietal lobe; therefore, **e** is correct. (Compare Figures 43-7 and 43-8.)

6. Choice **a** is correct. See Figure 43-9.

7. The association cortex refers to the unmapped areas of the cortex. About half of the association cortex is contained in the frontal lobes and so lies anterior to the central sulcus. Its size may be correlated with intelligence. Although it was originally thought to be a horizontal connection between sensory and motor cortices, it now appears that its organization is vertical. Only **d** is correct; it explains why the function of the association cortex has been so difficult to ascertain.

8. Since speech centers are located in the left hemisphere in most individuals, choice a could not be correct. The situation described in c occurs in split-brain patients. Parkinson's disease is associated with low levels of dopamine. Choice **b** alone is correct.

9. In a split-brain patient, the hemispheres of the cerebrum are functionally separate (c), due to the severing of the neural connections (the corpus callosum) between them (d). This is sometimes done to relieve the symptoms of epilepsy. Choice **e** is correct.

10. All choices except d (which is an endogenous opiate) are neurotransmitters in the central nervous system. However, acetylcholine and dopamine are found at only a small number of synapses, and GABA is inhibitory in its effects. Noradrenaline (**c**) is a major neurotransmitter in the reticular activating and limbic systems.

11. The presence of opiate receptors in the brain has been at least partially explained by the isolation of a group of endogenous morphinelike substances, the endorphins, which are believed to act as natural pain relievers (analgesics). Choice **d** is therefore correct.

12. The role of endorphins as pain relievers was suspected because it was found that they are synthesized and secreted along with the stress hormone ACTH (**d**).

13. Lateralization of function means differentiation of function between the two cerebral hemispheres. This is thought to be a developmental process because young children who have sustained damage to their left cerebral hemisphere have been known to develop normal speech patterns (**a**). There is a close correlation between the age of the child at injury and his or her ability to speak normally.

14. A period of rapid, low-amplitude waves from the brain of a sleeping person similar to the waves observed in an alert person is associated with rapid eye movements (REMs). The subject is said to be in REM sleep (**b**).

15. Beta waves (**b**), low-voltage waves 18 to 32 cycles per second in frequency, are associated with mental activity and excitement.

16. Laboratory rats will forget tasks they have learned if they are deprived of REM sleep; it is believed that the same is true for humans (**b**). REM sleep is thought to be an information-processing period.

17. By placing microelectrodes at single-neuron locations in the brain of a cat, Hubel and Wiesel (see Figure 43-22) were able to identify classes of

neurons (**a**), each of which is associated with a different type of visual stimulus.

18. Experimentation on the sexual behavior of male rats revealed that normal sexual behavior can be stimulated even in castrated males by administrations of male hormones or by electrical stimulation of that area of the brain that responds to male hormones (**c**). None of the behaviors implied in choices a, b, or d were observed.

19. Lashley coined the term "engram" to describe the physical change in the brain that is associated with memory for a specific task or problem; therefore, **c** is correct. After failing to locate the engram, Lashley concluded that memory is "nowhere and everywhere present."

20. The chemical nature of memory has not been determined for either short-term or long-term memory (**c**). Long-term memory loss, as in amnesia, need not be accompanied by short-term loss, nor must short-term memory loss, as in patients with damage to the hippocampus of the brain, be accompanied by long-term loss. However, the limited amount of information (**b**) that can be held in short-term memory (usually no more than the seven digits of a phone number) can be transferred to long-term memory (**a**). Choice **e**, then, is correct.

21. The correct answers in order are **d**, **e**, **a**, **b**, **c**, and **c**.

22. The correct answers in order are **b**, **c**, **e**, **d**, and **a**.

23. Basically, the reticular activating system appears to filter incoming stimuli in such a way as to allow a person to discriminate between important and unimportant information. The limbic system is associated with the translation of drives and emotions into complex behaviors.

24. The correct answers are:
    a. **10**
    b. **4**
    c. **12**
    d. **1**
    e. **2**
    f. **7**
    g. **11**
    h. **13**
    i. **5**
    j. **9**
    k. **14**
    l. **3**
    m. **8**
    n. **15**
    o. **6**

25. You could draw on the following areas of research to demonstrate that hormones influence the brains of male and female animals differently: (1) the relationship between the size of song centers of birds and levels of testosterone in their blood, and (2) sexual behavior in castrated male rats in response to injections of testosterone or electrical stimulation of the brain. (See the analysis for Question 18.)

CHAPTER 44

The Continuity of Life I: Reproduction

CHAPTER ORGANIZATION

I. Introduction
II. The male reproductive system
   A. Background
   B. Spermatogenesis
      1. Introduction
      2. Differentiation of spermatids
   C. Pathway of the sperm
      1. General information
      2. The penis and orgasm in the male
III. The female reproductive system
   A. Background
   B. Oogenesis
   C. The menstrual cycle
   D. Pathway of the egg
      1. General information
      2. Orgasm in the female
IV. Contraceptive techniques
V. Estrus
VI. Summary

MAJOR CONCEPTS

In vertebrates, reproduction is characteristically sexual and involves two parents, one producing sperm and the other producing eggs.

After the onset of puberty, gamete production is continuous in human males but occurs cyclically in females (the menstrual cycle). Male gametes (sperm) are produced by meiosis in the seminiferous tubules of the testes. The production of primary oocytes occurs in the ovaries; meiosis in the female is completed at the time of fertilization. Gamete production in both males and females is regulated by hormones involved in complex negative feedback interactions.

The male reproductive system includes the testes, which produce sperm and testosterone; the epididymis, in which sperm gain motility; the penis, which delivers sperm to the female; various glands that produce fluids to nourish and support the sperm; and a system of ducts to carry the semen components from their organs of origin to the outside.

The female reproductive system includes the ovaries, which produce oocytes, progesterone, and estrogen; the oviducts or uterine tubes, which transport the egg cell to the uterus and serve as the usual site of fertilization; the uterus, which serves as the normal site of zygote implantation and houses and nourishes the developing fetus; the cervix, a muscular passageway between the uterus and the vagina; and the vagina, which receives the male penis and is the channel through which the fetus passes at birth.

The female menstrual cycle is under hormonal control and normally includes the following events: maturation of an ovarian follicle, development of the endometrium, ovulation, development of a corpus luteum which maintains the endometrium, and shedding of the endometrium if implantation does not occur.

HOW TO STUDY THE CHAPTER

Read the chapter, focusing on the major concepts.

Reread the chapter, using the questions that follow to help you focus on details as they relate to major concepts. Answer the questions on a separate sheet of paper. Your answers will provide a valuable study aid in preparing for examinations.

FOCUS ON CHAPTER DETAILS

I. Introduction (page 836)
1. What two basic events are involved in sexual reproduction?
2. How are the gametes of vertebrates specialized?
3. What is the evolutionary trend found among vertebrates that involves the relationship between parents and their young?
4. In what organisms is fertilization external? Internal?

II. The male reproductive system (pages 836–841)

A. Background
5. Where are gametes formed in the male?
6. How is the proper temperature for sperm production maintained?

B. Spermatogenesis

Introduction
7. Describe the internal structure of the testes.
8. Outline the events of spermatogenesis. What role do Sertoli cells play?

Differentiation of spermatids
9. What is differentiation?
10. How is the acrosome formed? What is its function during fertilization?
11. How does the flagellum of the sperm develop? What organelles are associated with its function?
12. What events account for the final shape of a sperm cell?
13. Describe the structure of a mature sperm cell. For what activities is it adapted?

C. Pathway of the sperm

General information
14. Trace the movement of sperm from the testes to the tip of the penis. What assistance do the sperm receive from muscle contractions?
15. Describe the surgical procedure of a vasectomy.

The penis and orgasm in the male
16. What is the function of the penis? How common is this organ in the animal kingdom?
17. Describe the internal and external structure of the penis.
18. What causes the penis to become erect?
19. Describe the fluids secreted by the bulbo-urethral glands, the seminal vesicles, and the prostate gland. What is the purpose of each fluid?
20. How are sperm ejaculated?
21. What is semen? How much of the semen is sperm?
22. Only one sperm is able to fertilize the egg cell of the female. What is the suspected role of the other sperm cells?

III. The female reproductive system (pages 842–847)

A. Background
23. What are the functions of the uterus, the endometrium, the cervix, and the vagina?
24. What accounts for the pH of the vagina?
25. Describe the external genitalia (the vulva) of the female.

B. Oogenesis
26. Describe the location of the organs in which female gametes are produced.
27. Describe the initial development of oocytes in terms of meiotic events.
28. What is responsible for the increase in size of a primary oocyte as it undergoes meiosis?
29. Describe the completion of meiosis in an oocyte. How are the cellular materials distributed at the end of meiosis? What is the fate of a polar body?
30. What is the function of an ovarian follicle? What is its role in ovulation?
31. How can you account for the fact that a human oocyte is an unusually large cell?

C. The menstrual cycle
32. What external factors may influence the menstrual cycle?

33. Name the hormones (and their site of production) that function in the feedback system regulating the menstrual cycle. How do these hormones interact?
34. Describe the events of the menstrual cycle (see also Figure 44-12). How do the levels of the gonadotropins and estrogens vary during the several phases of the cycle? What are the effects of these hormones on the ovaries (ovarian phase) and the endometrium (uterine phase)?
35. How much variability is there in the female menstrual cycle?

D. Pathway of the egg

General information

36. Trace the path of the oocyte (or fertilized egg cell) through the female reproductive tract (see also Figure 44-13).
37. What time intervals separate ovulation, fertilization, and implantation?
38. What happens to an egg cell if it is not fertilized?

Orgasm in the female

39. What physiological changes take place in the genitalia of the female during sexual arousal?
40. What appears to be a function of the female orgasm?

IV. **Contraceptive techniques** (page 847)

41. Birth-control methods are ranked according to effectiveness in Table 44-1 on page 848. How do you account for the high success rate of some contraceptive techniques and the low success rate of others?

V. **Estrus** (page 847)

42. What is estrus? Describe some estrus cycles in the animal kingdom. How do human mating habits differ from those of other animals?
43. What nongenetic factors may account for the variety of siblings in the litters of some animals?

44. Why is it economical for females of some species to ovulate only when stimulated by copulation?
45. What explanation has been suggested for the receptivity of human females to mating during infertile periods?

VI. **Summary** (page 849): Read the Summary. If you are familiar with the essential features of the material presented there, you are ready to take the following diagnostic examination.

TESTING YOUR UNDERSTANDING

After you have completed the following examination, compare your answers with the annotated key in the Analysis section.

1. The penis of the male becomes erect as a result of:
   a. contraction of smooth muscles surrounding the urethra.
   b. secretions from the seminal vesicles and prostate gland.
   c. accumulation of blood in spongy tissues.
   d. spermatogenesis.
2. The acidity of the female reproductive tract is neutralized by fluid from the:
   a. vagina during sexual stimulation.
   b. seminal vesicles.
   c. testes.
   d. prostate gland.
   e. Both a and d are correct.
3. A vasectomy involves cutting and tying of the:
   a. seminiferous tubules.
   b. epididymis.
   c. vas deferens.
   d. spermatic cord.
   e. urethra.
4. The leading portion of a sperm cell is its _____, and the trailing portion is its _____.
   a. nucleus; cytoplasm
   b. nucleus; flagellum
   c. acrosome; flagellum
   d. cytoplasm; mitochondrion
   e. seminal vesicles; Golgi body
5. A sperm cell is actually a(n):
   a. differentiated spermatid.

b. spermatocyte.
   c. acrosome.
   d. flagellum.
   e. spermatid.
6. All of the following are packaged inside a sperm cell *except*:
   a. DNA.
   b. mitochondria.
   c. lytic enzymes.
   d. microtubules.
   e. a nutrient source.
7. In humans, male gametes are produced by cells in the _____, and female gametes are produced in the _____.
   a. testes; ovaries
   b. seminiferous tubules; oviducts
   c. vas deferens; the uterus
   d. seminal vesicles; corpus luteum
8. Which of the following indicates the sequence of developmental stages for sperm cells?
   a. spermatogonia, spermatids, sperm, spermatocytes
   b. spermatogonia, spermatocytes, spermatids, sperm
   c. sperm, spermatids, spermatocytes, spermatogonia
   d. spermatids, spermatocytes, spermatogonia, sperm
9. Contractions of smooth muscles in the walls of the uterus are strongest during:
   a. menstruation.
   b. intercourse.
   c. orgasm.
   d. labor.
   e. ovulation.
10. Which of the following is *not* considered to be a part of the female external genitalia?
    a. clitoris
    b. labia majora
    c. labia minora
    d. vagina
11. Just before puberty, the gonads of the female contain oocytes that have:
    a. completed meiosis I.
    b. completed meiosis II.
    c. entered meiosis I.
    d. entered meiosis II.
    e. not yet undergone meiosis.
12. The second meiotic division in human female gamete development does not take place until:
    a. just before birth of the female fetus.
    b. the onset of puberty.
    c. ovulation.
    d. after fertilization.
    e. menstruation.
13. As a primary oocyte completes meiosis, _____ of its cytoplasm and _____ of its chromosomes are conserved in the ovum.
    a. one-fourth; one-fourth
    b. one-half; one-half
    c. one-half; one-fourth
    d. all; one-half
    e. nearly all; one-fourth
14. Which of the following cells (in the human gonads) are haploid?
    a. spermatogonia
    b. primary spermatocytes
    c. spermatids
    d. primary oocytes
    e. All of the above are haploid.
15. Indicate the sequence of events that can occur in the female reproductive tract during a single menstrual cycle.
    a. fertilization, ovulation, menstruation, implantation
    b. ovulation, implantation, fertilization, menstruation
    c. menstruation, ovulation, fertilization, implantation
    d. ovulation, fertilization, menstruation, implantation
16. Indicate the average time intervals between the events in the answer to the preceding question.
    a. 14 days, 7 days, 7 days
    b. 1 day, 14 days, 4 days
    c. 14 days, 1 day, 7 days
    d. 7 days, 7 days, 14 days
17. Beating of cilia and smooth muscle contractions in the oviduct are thought to encourage:
    a. the movement of sperm toward the oocyte.
    b. the movement of the oocyte following its release from the ovary.

c. the sloughing off of the lining of the uterus.
   d. implantation of the fertilized egg.
   e. Both a and b are correct.
18. When blood levels of estrogen decrease, the:
   a. endometrium is sloughed off.
   b. pituitary gland increases its secretion of LH.
   c. corpus luteum is formed.
   d. production of GnRH is stimulated.
   e. Both c and d are correct.
19. Ovulation occurs in response to _____ in the blood levels of _____.
   a. decreases; estrogen
   b. increases; progesterone
   c. decreases; FSH
   d. increases; LH
   e. decreases; LH
20. The _____ is unusual among mammals in that the female of the species is receptive to copulation during infertile periods.
   a. dog
   b. human
   c. horse
   d. cow
   e. rat
21. Match each birth-control method with its mode of action.

   rhythm _____
   vaginal foam _____
   intrauterine device (IUD) _____
   tubal ligation _____
   morning-after pill _____
   condom _____

   a. involves surgical blockage of the uterine ducts
   b. abstinence from intercourse when conception is likely to occur
   c. prevents sperm from entering the female
   d. kills sperm before they enter the uterus
   e. prevents implantation

22. Match each reproductive structure with its definition or function.

   uterus _____
   cervix _____
   epididymis _____
   scrotum _____
   urethra _____
   seminal vesicle _____
   ovarian follicle _____

   a. organ in which a fertilized egg implants
   b. serves as a pathway for both urine and semen in the male
   c. adds prostaglandins to the semen
   d. helps maintain proper temperature for spermatogenesis
   e. region of male reproductive tract in which sperm gain motility
   f. a sphincter muscle between the uterus and the vagina
   g. provides the nutrients needed by the growing oocyte

23. Match each process with its definition or description.

   oogenesis _____
   spermatogenesis _____
   menstruation _____
   estrus _____
   circumcision _____
   implantation _____

   a. begins before birth in the female
   b. removal of a thin layer of skin encircling the glans penis
   c. occurs continually in a fertile male
   d. rupture of the endometrial lining by the fertilized egg
   e. period during which a female animal is receptive to intercourse
   f. occurs monthly in a nonpregnant human female

24. What is the origin of the corpus luteum? What is its role in the menstrual cycle?
25. Explain how "the pill" prevents conception.
26. What is estrus? What is a possible explanation for the evolution in humans of female receptivity to mating during nonestrus periods?
27. Match each of the following events with the uterine phase (or phases) during which it occurs.

implantation _____
increase in LH _____
ovulation _____
low progesterone _____
fertilization _____
increase in progesterone _____
corpus luteum _____
follicle _____
estrogen at maximum concentration _____

a. menstrual flow
b. proliferative phase
c. secretory phase

28. Identify the parts of the human male and female reproductive and urinary systems shown, and give the function of each structure.

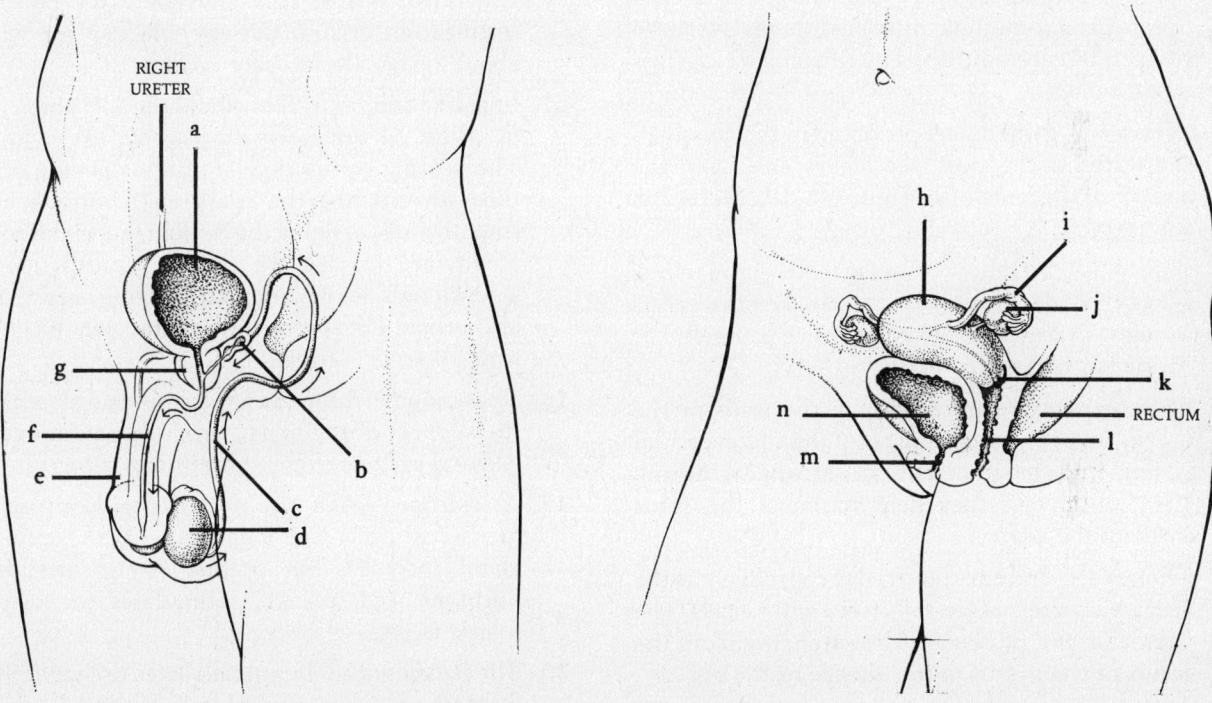

PERFORMANCE ANALYSIS

1. Choice **c** is correct. See Figure 44-7.

2. The vagina is normally mildly acidic, but its pH must be raised if sperm are to survive on their way to the egg. Sexual stimulation of the clitoris causes secretion of an acid-neutralizing fluid into the vagina. The prostate gland of the male adds an alkaline fluid to the semen, which also neutralizes the acidic environment of the vagina; therefore, **e** is correct.

3. Choice **c** is correct. See Figure 44-6.

4. The acrosome, a vesicle containing enzymes that digest the protective layer surrounding the egg cell, is enclosed in the head of the sperm cell. The motile tail consists of a long, powerful flagellum; therefore, **c** is correct.

5. Spermatids, the final products of meiotic divisions in the male, are spherical or polygonal cells that undergo a number of striking changes (differentiation) to become sperm cells (see Figure 44-5); therefore, **a** is correct.

6. Mitochondria and microtubules (b and d) function in the motility of the sperm. Lytic enzymes (c) from the acrosome digest the protective layer around the egg. The delivery of the genetic material, DNA (a), is of course the main function of the sperm. Sperm cells contain no nutrients themselves (**e** is correct) but are nourished by a fructose-rich fluid in the semen.

7. In humans, gametes are produced in the gonads—the testes of the male (see Figure 44-3) and the ovaries of the female (see Figure 44-10); therefore, **a** is correct.

8. Figure 44-3b shows sperm cells in various stages of development in the seminiferous tubules. Take a close look to convince yourself that choice **b** is correct.

9. Smooth muscle contractions in the walls of the uterus assist the sloughing off of the endometrium during menstruation but are strongest during labor (**d**), when they help to move the fetus through the vagina.

10. Choices a, b, and c are referred to collectively as the vulva, the external genitalia (see Figure 44-9). The vagina (**d**) is a muscular tube extending from the cervix of the uterus to the outside of the body.

11. Prophase I of meiosis begins in the human female during fetal development. The primary oocytes remain in prophase until sexual maturity (puberty) is reached (**c** is correct); a single oocyte completes its first division at the time of ovulation. The second meiotic division does not occur until after fertilization.

12. Choice **d** is correct. See the analysis for Question 11.

13. Choice **e** is correct. Although three-fourths of the genetic material of an oogonium is lost as the polar bodies disintegrate, most of its cytoplasm is donated to the cell that becomes the ovum. See Figure 12-12.

14. Spermatids (**c**) are the products of meiosis and so are haploid.

15. The sequences in choices a, b, and d are all impossible—implantation cannot occur immediately after menstruation, and fertilization must occur before implantation. You should also be able to see from Figure 44-12 why choice **c** is correct.

16. Menstruation marks the beginning (day 1) of the menstrual cycle. Ovulation, on the average, occurs within 14 days, at the midpoint of the cycle. The egg cell is viable for only about 24 hours. If fertilization occurs, the egg will implant within about 7 days; therefore, **c** is correct.

17. Implantation and the sloughing of the endometrium do not normally involve the oviducts. The beating of the cilia lining the opening of the tube surrounding the ovary and contractions of smooth muscles lining the oviduct propel the oocyte downward toward the uterus. Contractions of smooth muscles in the oviduct during orgasm may also propel the sperm toward the oocyte. Therefore, choice **e** is correct.

18. The endometrium cannot be sustained without the support of the gonadotropic hormones estrogen and progesterone; therefore, **a** is correct.

19. As estrogen levels rise near the midpoint of the menstrual cycle (see Figure 44-12) the pituitary gland increases its production of luteinizing hormone, LH (**d**). This stimulates the ovarian follicle to release its egg.

20. The females of all the animals listed except humans (**b**) mate only during periods of fertility, known as estrus. The human female is one of the few female animals receptive to copulation when infertile.

21. The correct answers in order are **b, d, e, a, e,** and **c**.

22. The correct answers in order are **a, f, e, d, b, c,** and **g**.

23. The correct answers in order are **a, c, f, e, b,** and **d**.

24. Use Figure 44-12 to explain the origin and hormonal role of the corpus luteum in the menstrual cycle.

25. Women taking "the pill" do not ovulate. Taken daily at the beginning of the cycle, it keeps levels of estrogen and progesterone artificially high, inhibiting the production of FSH and LH by the pituitary. In the absence of FSH, follicles do not ripen; when LH is not produced, they do not rupture to release the egg.
26. Estrus is a period of fertility, during which females are receptive to copulation. The receptivity of the human female during nonestrus periods may have coevolved with the strengthening of pair-bonds, allowing for the establishment of the family as the basic unit of human society.
27. The correct answers in order are c; b; **end of b, start of c**; a, b; b; c; c; c; a, b; b. See Figure 44–12.
28. The correct answers are:
    a. **urinary bladder: storage of urine**
    b. **seminal vesicles: secretion of some of the fluid portion of semen**
    c. **vas deferens: storage of sperm cells**
    d. **testis: production of sperm cells and male sex hormones**
    e. **penis: depositing of sperm in the female reproductive tract**
    f. **urethra: passageway for urine and semen**
    g. **prostate gland: secretion of fluid portion of semen**
    h. **uterus: housing and nourishment of embryo**
    i. **oviduct: passageway for ovum to uterus**
    j. **ovary: production of ova and female sex hormones**
    k. **cervix: opening of the uterus**
    l. **vagina: reception of the penis during copulation; birth canal**
    m. **urethra: passageway for urine**
    n. **urinary bladder: storage of urine**

CHAPTER 45

The Continuity of Life II: Development

CHAPTER ORGANIZATION

I. Introduction
II. Development of the sea urchin
   A. Background
   B. The influence of the cytoplasm
III. Development of the amphibian
   A. General information
   B. The amphibian egg: Cleavage and gastrulation
   C. The amphibian: Neural tube formation
   D. The organizer
   E. Induction and inducers
IV. Development of the chick
   A. General information
   B. Extraembryonic membranes of the chick
   C. Organogenesis
     1. Introductory remarks
     2. Ectoderm
     3. Mesoderm
     4. Endoderm
V. Development of the human embryo
   A. Background
   B. Human embryonic membranes
   C. The placenta
   D. Hormones and pregnancy
   E. The first trimester
   F. The second trimester
   G. The final trimester
   H. Birth
VI. Epilogue
VII. Summary

MAJOR CONCEPTS

Development, the process whereby a fertilized egg becomes a complete organism, involves three overlapping processes: growth, differentiation, and morphogenesis.

The three stages of development are cleavage, involving division of the zygote with little or no change in total volume; gastrulation, resulting in establishment of three tissue layers; and organogenesis, the formation of organs and organ systems. Embryonic induction is the process by which one tissue induces changes in the development of an adjacent tissue.

Each organ or organ system in the body develops from one of the three primary tissue layers (endoderm, mesoderm, and ectoderm) formed in gastrulation.

In amniote eggs, such as those of reptiles, birds, and mammals, four extraembryonic membranes—the yolk sac, amnion, chorion, and allantois—form during the development of the embryo.

Early embryonic development of the human embryo involves the formation of a blastodisc (the embryo) enclosed within a trophoblast. The trophoblast develops into the chorion and later becomes the fetal component of the placenta. By the end of the third month, all the organ systems in a human fetus have been laid down. In humans, birth occurs an average of 266 days after conception.

HOW TO STUDY THE CHAPTER

Read the chapter, focusing on the major concepts.

Reread the chapter, using the questions that follow to help you focus on details as they relate to major concepts. Answer the questions on a separate sheet of paper. Your answers will provide a valuable study aid in preparing for examinations.

FOCUS ON CHAPTER DETAILS

I. Introduction (page 851)
1. Define development with respect to the multicellular organism.
2. Describe the three general stages by which organisms develop.
3. Why are we able to learn about our own development from studying that of other animals?

II. Development of the sea urchin (pages 851–855)

A. Background
4. Why is the sea urchin a favorite experimental tool of embryologists?
5. What event marks the beginning of development?
6. Describe the gametes of the sea urchin.
7. What are four important consequences of fertilization?
8. What events can stimulate development of an unfertilized egg?
9. What does activation of an enucleated egg cell demonstrate?
10. What happens to an embryo during cleavage? How does this affect the ratios of surface area to volume and nuclear volume to cytoplasmic volume in the blastula?
11. Outline the process of gastrulation, paying special attention to the role of the blastopore (see also Figure 45-4). Why are sea urchins considered to be deuterostomes?
12. Identify the three layers that are present in the gastrula (see Figure 45-5).
13. What degree of differentiation is achieved during gastrulation?

B. The influence of the cytoplasm
14. Describe the two halves of a sea urchin egg prior to cleavage and the planes along which the first three cleavage divisions occur.
15. How does separation along the third cleavage line of the embryo provide evidence concerning the role of the cytoplasm in development?
16. How is development affected if cells of the blastula stage are kept separated? If they are separated and then allowed to reaggregate? What does such an experiment show?

III. Development of the amphibian (pages 855–859)

A. General information
17. How does the appearance of an amphibian egg differ from that of the sea urchin egg?
18. How is the cytoplasm of a frog's egg reorganized following fertilization?
19. Describe the experiment in which Spemann demonstrated the importance of the gray crescent in embryonic development (see also Figure 45-7).
20. What hypothesis did Spemann's research support?

B. The amphibian egg: Cleavage and gastrulation
21. How does the cleavage pattern of amphibians differ from that of sea urchins?
22. Compare gastrulation in the frog's egg with that of the sea urchin egg. What determines the future axis of the organism?
23. What is chordamesoderm?

C. The amphibian: Neural tube formation
24. Explain how the neural tube of an amphibian embryo is formed.
25. How does migration of tissues give rise to somites and to the coelom? (See Figure 45-10b.) Identify the endoderm in Figure 45-10c.

D. The organizer
26. What did Spemann demonstrate about the role of the dorsal lip of the blastopore? What is the relationship between the dorsal lip of a

blastopore and the position of the gray crescent in the egg?

27. Explain how Mangold was able to show that the dorsal lip is an organizer of embryonic tissues.
28. What general statement can be made about the role of the chordamesoderm in chordate development?

E. Induction and inducers

29. What is embryonic induction?
30. What is the role of the inducing substance in induction? What experiments have revealed this role?

IV. **Development of the chick** (pages 859–864)

A. General information

31. What adaptations to a terrestrial environment are exhibited by the egg of a chicken?
32. What is a blastodisc? What groups of animals develop from a blastodisc?
33. Describe the blastodisc of a fertilized chick egg.
34. What is the primitive streak? What is its role in gastrulation?

B. Extraembryonic membranes of the chick

35. What is the origin of the extraembryonic membranes of the chick embryo?
36. Describe how the four extraembryonic membranes are formed. (Locate them in Figure 45-17.) What is the function of each of these membranes?
37. How is the chorioallantoic membrane formed? What purpose does it serve?

C. Organogenesis

Introductory remarks

38. What is organogenesis? How does it begin?

Ectoderm

39. Explain how motor neurons differentiate from neural tube cells.
40. What happens to ectodermal cells at the neural crest?

41. Explain how the vertebral column is formed from mesoderm.
42. From what embryonic tissue is the brain derived?
43. Explain how tissues of the vertebrate eye are formed from protrusions of the brain (the optic vesicles).
44. Describe secondary induction as it occurs in the formation of the eye lens.
45. What are the origins of auditory and olfactory organs?

Mesoderm

46. What three types of tissues are differentiated from the somites?
47. What is the embryonic fate of the mesodermal cells lateral to the somites? Of the ventral mesoderm?

Endoderm

48. What tissues are formed in (a) lower vertebrates and (b) higher vertebrates when endodermal pouches of the anterior gut fuse with overlying ectoderm?
49. Explain how the major respiratory and digestive structures form as outpocketings of endoderm.
50. Cite an example of an organ derived from all three tissue types.

V. **Development of the human embryo** (pages 865–873)

A. Background

51. When (and where) does cleavage of a fertilized human egg cell begin?
52. Describe the relationship between the early blastula and the mother.
53. Describe the structure of the blastocyst (consult Figure 45-24c).
54. To what will the inner cell mass and trophoblast eventually give rise?
55. How does the formation of extraembryonic membranes in the human embryo differ from that in the chick?
56. What events take place once the blastocyst

makes contact with the uterus of the mother?

B. Human embryonic membranes

57. Why is the yolk sac found in embryos of placental mammals considered to be a secondary evolutionary development?

58. How does the allantois function in excretion and in the exchange of gases and nutrients?

59. Where is the amniotic cavity formed? What is its significance in genetic testing procedures?

60. What tissues form the chorion?

61. What events mark the maturation of the placenta?

C. The placenta

62. From what fetal and maternal tissues does the placenta develop?

63. Explain how the placenta provides for the supply of nutrients and oxygen to the fetus and for the disposal of its wastes.

64. What nutritional changes occur in the mother as the placenta develops?

65. Describe the physical relation of the fetus to the placenta and the amniotic cavity.

D. Hormones and pregnancy

66. Why can the presence of chorionic gonadotropin be used to determine if a woman is pregnant?

67. What changes in hormone production take place after the placenta has developed?

E. The first trimester

68. Detail the developmental processes that are initiated during the first, second, and third months of embryonic development. From what point is the embryo referred to as a fetus?

69. When do germ cells originate, and how do they arrive at their proper site? What may be the reason that this process occurs so early in the embryo's development?

70. What influences do the germ cells have on sex determination? What particular effect does the presence or absence of a Y chromosome have?

71. What is a teratogen? How might certain substances ingested by the mother adversely affect normal embryonic development? Give examples. What is one reason the embryo is at greatest risk during the first two months of pregnancy?

F. The second trimester

72. What changes have occurred in the fetus after the fourth, fifth, and sixth months?

G. The final trimester

73. What important neurological developments take place within the final three months of pregnancy?

74. How might the physiological status of the mother at this time affect fetal development?

75. How does the fetus acquire immunity from the mother?

76. What is the most significant factor in infant mortality? Why do black children suffer higher mortality rates than do white children?

77. What happens to the placenta during the final month of pregnancy?

H. Birth

78. How long after conception can a mother expect to wait before bearing her child?

79. Describe the changes that occur in the female reproductive tract during each of the three stages of labor.

80. When does a baby begin to breathe on its own?

VI. **Epilogue** (pages 873–874)

81. Evaluate from an evolutionary perspective the controversy over the exact moment at which human life begins.

VII. **Summary** (pages 874–876): Read the Summary. If you are familiar with the essential features of the material presented there, you are ready to take the following diagnostic examination.

TESTING YOUR UNDERSTANDING

After you have completed the following examination, compare your answers with the annotated key in the Analysis section.

1. In eggs that contain a relatively large amount of yolk, cleavage results in:
   a. uniform divisions throughout the egg.
   b. divisions primarily in the vegetal hemisphere.
   c. divisions primarily in the animal hemisphere.
   d. divisions only in the region of the gray crescent.

2. If the early sea urchin embryo is divided in half along its equator, _____ half (halves) will develop normally.
   a. both
   b. neither
   c. only the vegetal
   d. only the animal

3. The gray crescent of an amphibian egg:
   a. appears on the side of the egg where the sperm enters.
   b. gives rise to the dorsal lip of the blastopore.
   c. is the egg nucleus visible through the transparent membrane of the egg cell.
   d. is part of the vegetal hemisphere of the egg.

4. In all chordate embryos, the primary organizer is a tissue called the:
   a. neural plate.
   b. primitive streak.
   c. chordamesoderm.
   d. chorion.
   e. blastula.

5. The fact that an enucleated egg can be activated without penetration by sperm indicates that genetic information is _____ prior to fertilization.
   a. not present in the nucleus
   b. replicated
   c. translated
   d. transcribed

6. The three basic tissue layers—endoderm, mesoderm, and ectoderm—are laid down during:
   a. cleavage.
   b. blastulation.
   c. gastrulation.
   d. embryonic induction.
   e. organogenesis.

7. The portion of the chick egg that undergoes cleavage is called the:
   a. blastopore.
   b. blastula.
   c. blastodisc.
   d. blastocyst.
   e. blastocoel.

8. The lens of the vertebrate eye is formed by _____ induction of the _____ by the optic vesicle.
   a. primary; brain
   b. primary; mesoderm
   c. secondary; epidermis
   d. secondary; myotome

9. Sclerotome, dermatome, and myotome cells arise from _____.
   a. somites
   b. neural folds
   c. gut endoderm
   d. ventral mesoderm
   e. epidermis

10. Organs of the respiratory and digestive tract develop from _____ of _____.
    a. infoldings; mesoderm
    b. infoldings; ectoderm
    c. outpouchings; endoderm
    d. outpouchings; mesoderm
    e. fusion; ectoderm with mesoderm

11. Human egg cells are usually fertilized in the _____ of the female reproductive tract.
    a. vagina
    b. oviducts
    c. uterus
    d. ovaries
    e. ovarian follicle

12. In the human embryo, the yolk sac:
    a. is not present.
    b. is discarded early in development.
    c. contains no yolk.
    d. serves a nutritional function.
    e. completely surrounds the embryo.

13. A human blastocyst resembles a ring with a stone inside. The ring is referred to as the _____ and the stone as the _____.
    a. trophoblast; inner cell mass
    b. chorionic membrane; embryo
    c. placenta; blastocoel

    d. neural tube; notochord
    e. amnion; yolk

14. Which of the following is true regarding the placenta?
    a. It is formed from the chorion of the fetus.
    b. It is formed from the endometrium of the mother.
    c. It allows passage of waste products from fetal to maternal blood.
    d. It allows passage of oxygen and nutrients from maternal to fetal blood.
    e. All of the above are correct.

15. A fetus is connected to the placenta by means of:
    a. the uterine artery and vein.
    b. an umbilical cord.
    c. the chorionic membrane.
    d. the allantoic sac.

16. Early in pregnancy, chorionic gonadotropin is produced by the:
    a. corpus luteum.
    b. maternal gonads.
    c. placenta.
    d. trophoblast.
    e. uterus.

17. During its development inside the mother, a fetus floats in a fluid immediately surrounded by the _____ membrane.
    a. amniotic
    b. yolk sac
    c. allantoic
    d. chorionic
    e. chorioallantoic

18. The rudimentary gonads of the fetus develop into testes or ovaries under the influence of the:
    a. germ cells.
    b. Y chromosome.
    c. hormone chorionic gonadotropin.
    d. androgens.
    e. Both c and d are correct.

19. A human embryo begins to take on distinctly human features during its _____ week of development.
    a. second
    b. third
    c. sixth
    d. eighth
    e. twelfth

20. During labor, just prior to crowning:
    a. uterine muscles exhibit a sustained contraction.
    b. the umbilical cord becomes detached from the placenta.
    c. the amniotic sac usually ruptures.
    d. the cervix has not yet begun to dilate.
    e. Both a and c are correct.

21. Match each embryonic structure with its definition or description.

    morula _____
    blastocoel _____
    gastrula _____
    archenteron _____
    chordamesoderm _____
    neural crest _____
    blastodisc _____
    primitive streak _____

    a. an elongated blastopore
    b. cells along the future axis of the organism
    c. a multicellular, undifferentiated sphere
    d. the fluid-filled cavity of a blastula
    e. forerunner of the gut
    f. formed as a result of cell migrations
    g. ectodermal ridges
    h. the blastula of birds, reptiles, and mammals

22. Match each tissue, organ, or system with the germ layer(s) from which it develops.

    muscle _____
    liver _____
    bone _____
    circulatory system _____
    skin _____
    brain _____
    intestine _____
    coelomic lining _____
    optic vesicle _____
    vertebral column _____

    a. mesoderm
    b. ectoderm
    c. endoderm

23. Match each developmental event with the period of pregnancy during which it occurs.

    fetus doubles in size _____ organs begins _____ a. first month
    limb abnormalities may occur _____ fetal movements begin _____ b. second month
    teeth form dentine _____ migration of germ cells toward rudimentary gonads _____ c. third month
    beginning of the major organ systems _____ cell replacement begins _____ d. second trimester
    development of external sexual _____ brain waves can be detected _____ e. third trimester

24. What are the four consequences of fertilization?
25. Describe two experiments that show the importance of the cytoplasm in influencing development.
26. Make a general comparison of amphibian and avian embryonic development.
27. Explain how Hilde Mangold demonstrated that the dorsal lip of an amphibian embryo's blastopore is an organizer.
28. Identify the indicated structures in the following diagram.

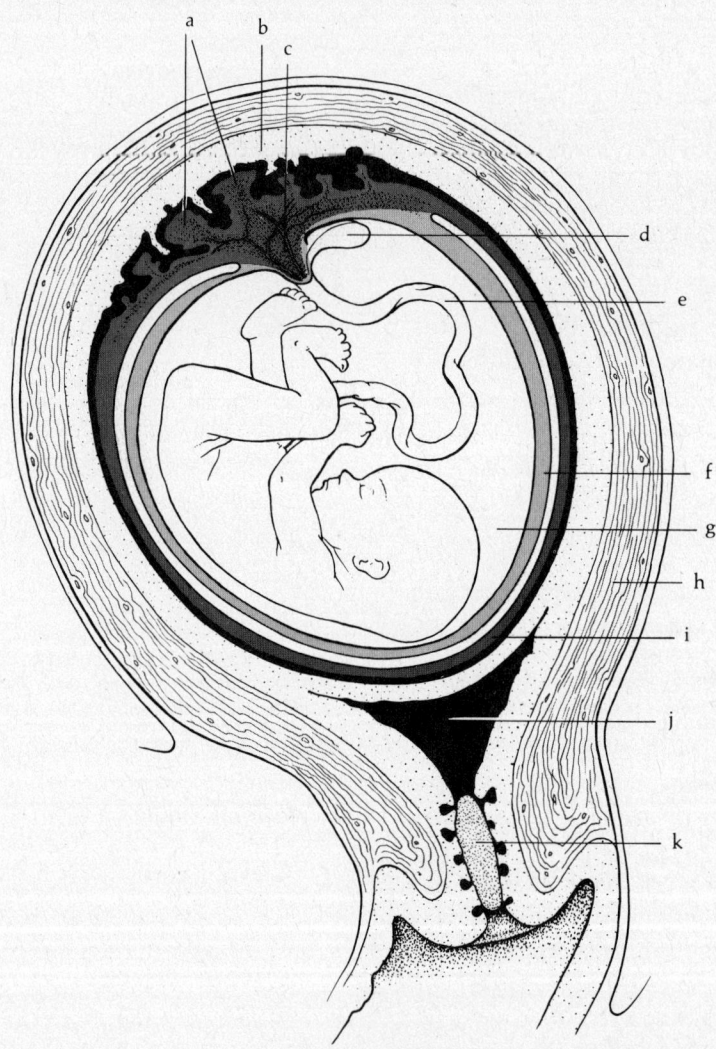

PERFORMANCE ANALYSIS

1. Eggs containing a relatively large amount of yolk, such as that of the frog (see Figure 45-8), cleave unequally, with divisions occurring much more slowly in the vegetal hemisphere than in the animal hemisphere; therefore, choice **c** is correct.

2. If the early sea urchin embryo is divided along its equator into two halves, one that is largely animal and the other that is largely vegetal, neither half (**b**) will develop normally. This indicates the importance of the cytoplasm in controlling development.

3. After fertilization, the pigment cap of a frog's egg rotates so that the gray crescent is opposite the point of sperm entry (see Figure 45-6). During gastrulation (see Figure 45-9b), cells at the boundary between the gray crescent and the vegetal hemisphere give rise to a crescent-shaped slit called the blastopore (**b**).

4. The chordamesoderm (**c**) refers to those mesodermal cells that lie along the longitudinal axis of the embryo. The chordamesoderm serves as the primary organizer in all chordates, inducing differentiation of the overlying ectoderm and setting other developmental events in motion.

5. Egg activation without benefit of a nucleus must mean that genetic instructions are present in the cytoplasm. This is made possible by the transcription (**d**) of mRNA from DNA in the nucleus prior to fertilization. It is possible that translation of genetic instructions into proteins may also have occurred, but this has no direct relationship to the removal of the nucleus.

6. By the end of gastrulation (**c**), the three tissue layers have been formed. Refer to Figure 45-9c.

7. Cleavage in the chick embryo (see Figure 45-15c) results in a disc-shaped layer of cells, the blastodisc (**c**), which sits on the surface of the yolk.

8. The vertebrate eye lens is differentiated from the epidermis by the optic vesicle. This is an example of secondary induction; the optic vesicle itself is the result of a primary induction of ectoderm by underlying mesoderm. Choice **c**, then, is correct.

9. Somites (**a**) are formed from mesoderm and differentiate into sclerotome cells, which will form skeletal elements, dermatome cells, which will develop into skin, and myotome cells, which will become muscle tissue.

10. Outpouchings of endodermal tissue (**c**) in the anterior end of the gut are the origins of the digestive and respiratory tracts.

11. Choice **b** is correct. See Figure 44-13.

12. There is no yolk (**c**) in the eggs of placental mammals, although there is a prominent cavity, the yolk sac, indicating that the lack of yolk is a secondary evolutionary development.

13. In cross section, a human blastocyst (Figure 45-24c) resembles a ring with a stone inside. The ring is a double cell layer called the trophoblast, which gives rise to the placenta. The stone is the inner cell mass, a ball of cells that develop into the embryo. Choice **a**, then, is correct.

14. The placenta is formed from both the chorionic membrane of the embryo and the endometrium of the mother. The placenta allows for the exchange of materials between the fetus and the maternal bloodstream. Waste products pass from fetal blood through the placental tissue and into maternal blood, and oxygen and nutrients diffuse from maternal blood through placental tissue to fetal blood. Therefore, choice **e** is correct.

15. The fetal connection to the placenta is through the umbilical artery and vein, which run through the umbilical cord (**b**). See Figure 45-27.

16. After the blastocyst makes contact with the endometrium, cells of the outer membrane of the trophoblast (**d**) begin production of chorionic gonadotropin, a hormone that stimulates the corpus luteum to continue its secretion of gonadotropic hormones.

17. Choice **a** is correct. This is apparent in Figure 45-35.

18. Whether or not a fetus develops to be male or female depends on the presence of an unidentified chemical whose production is controlled by a gene or genes on the Y chromosome (**b**). If the chemical is present, the rudimentary gonads become testes.

19. By the eighth week (**d**) of embryonic development, the major steps in the formation of organ systems are complete, and the fetus can be recognized as human.

20. By the end of the first stage of labor, dilation, the cervix has opened and strong contractions of the uterus occur every one to two minutes. During this stage, the amniotic sac usually ruptures (c). The second stage, expulsion, begins with crowning, the appearance of the head in the cervix. The umbilical cord is still attached to the placenta at delivery and is surgically separated.
21. The correct answers in order are **c**, **d**, **f**, **e**, **b**, **g**, **h**, and **a**.
22. The correct answers in order are **a**; **c**; **a**; **a**; **b**; **b**; **a**, **b**, **c**; **a**; **b**; and **a**. Refer to Figure 45-22.
23. The correct answers in order are **e**, **b**, **d**, **a**, **c**, **c**, **a**, **d**, and **e**.
24. The four consequences of fertilization are listed on pages 851 and 852.
25. One such experiment is referred to in Question 2 and its analysis. Another is Spemann's separation of amphibian egg cells at the two-cell stage (see Figure 45-7).
26. To discuss amphibian development, make use of Figures 45-6, 45-7, and 45-10. Avian development is presented in Figures 45-15c, 45-16, and 45-17. The major points of difference you should emphasize are the relative amounts of yolk present in the egg and the influence of this factor on cleavage and gastrulation.
27. The Mangold experiment is outlined in Figure 45-13.
28. The correct answers are:
    a. **chorionic villi**
    b. **uterine lining**
    c. **placenta**
    d. **remnant of yolk sac**
    e. **umbilical cord**
    f. **amnion**
    g. **amniotic cavity**
    h. **uterine muscle**
    i. **chorion**
    j. **uterine cavity**
    k. **cervical plug**

PART III   Biology of Populations

SECTION 7   Evolution

CHAPTER 46

The Evidence for Evolution

CHAPTER ORGANIZATION
I. Introduction
II. Lines of evidence
   A. The number of species
   B. Biogeography
   C. The fossil record
   D. Homology
   E. Essay: The record in the rocks
   F. Adaptation
III. Darwin's theory
IV. Evolution in action
   A. Introductory remarks
   B. The peppered moth
   C. Drug resistance in bacteria
V. The theory today
VI. Summary

MAJOR CONCEPTS

Darwin's theory of evolution was based on five categories of evidence: the number and diversity of species, the distribution of specific types of organisms, the fossil record, homologous anatomical structures, and adaptations of organisms to specific environments.

The two main components of Darwin's theory are (1) the production of inheritable characteristics by chance and (2) natural selection.

Natural selection has been observed in the development of drug resistance in bacteria and in the increase in the black form of the peppered moth in industrial England.

HOW TO STUDY THE CHAPTER

Read the chapter, focusing on the major concepts.

Reread the chapter, using the questions that follow to help you focus on details as they relate to major concepts. Answer the questions on a separate sheet of paper. Your answers will provide a valuable study aid in preparing for examinations.

FOCUS ON CHAPTER DETAILS

I. **Introduction** (page 881)
   1. What two reasons are given for the intellectual revolution that took place in biology during the middle of the nineteenth century?

II. **Lines of evidence** (pages 881–888)
   A. The number of species
     2. Why did Darwin and Wallace come to doubt the doctrine of special creation?
     3. How did the creationists explain the number of species?
   B. Biogeography
     4. Define biogeography.
     5. How was the distribution of species in geographically similar regions inconsistent with creationist doctrine?
     6. What speculation did Darwin make about

the diversity found among Galapagos finches?

7. What central theme of biogeography challenged (but did not disprove) the doctrine of special creation?

C. The fossil record

8. What correlations did Darwin make between fossils and the strata in which they were found?
9. Why was the giant armadillo of South American plains of interest to Darwin?
10. What evidence did Darwin hope would emerge from the fossil record?
11. What piece of fossil evidence convinced Thomas Huxley that evolution may have occurred as Darwin speculated?
12. What aspect of the fossil record, as we know it today, provides further evidence for Darwin's theory?

D. Homology

13. Describe some homologous structures among vertebrates that provided evidence for evolution.
14. Why did the existence of homologous structures cause Darwin to question creationism?

E. Essay: The record in the rocks (pages 886–887)

15. How have various forces shaped the geological strata at or near the earth's surface?
16. How can individual strata be distinguished from one another? How does a knowledge of these strata allow us to piece together bits of fossil evidence into a coherent story?
17. What are the names of the geologic eras? How were many of the geologic periods named?
18. How did Lord Kelvin arrive at an estimate for the age of the earth?
19. What new dating methods have indicated that the earth is older than Lord Kelvin thought?
20. How does the presence of radioactive elements in a rock sample allow one to date the stratum from which the sample is taken?
21. Why are radioactive clocks important to students of evolution?

F. Adaptation

22. What are three biological meanings of the word "adaptation"?
23. How did creationists use the idea of adaptation of specific features to deny the possibility of evolution?
24. How did Darwin regard the notion of a perfect adaptation?
25. How did Darwin believe that organisms become adapted to their environment?

III. **Darwin's theory** (page 888)

26. What made Darwin more than just "another traveler with an odd tale"?
27. Outline the four premises that formed the foundation of Darwin's theory.
28. What did Darwin mean by natural selection?

IV. **Evolution in action** (pages 888–890)

A. Introductory remarks

29. Darwin believed that evolution occurs too slowly to be observed. Why is this presumption no longer valid?

B. The peppered moth

30. Why was a light-colored body of adaptive value to the peppered moth, *Biston betularia*, prior to the Industrial Revolution?
31. How did industrialization affect the habitat of the moths? What changes occurred in the local moth population at that time? How did Kettlewell show that such changes were due to natural selection?
32. What changes have been occurring in English moth populations in recent years? Why?

C. Drug resistance in bacteria

33. What bacteriological phenomenon accompanied the widespread use of antibiotics?

CHAPTER 46 *The Evidence for Evolution* 357

34. Describe the Lederberg experiment.
35. What did the Lederbergs conclude?

V. **The theory today** (pages 890–891)

36. Newer evidence supports what view of the relatedness of all living organisms?
37. What contributions have cytology and biochemistry made to the theory of evolution?
38. What is the central question biologists now ask concerning evolution?

VI. **Summary** (page 891): Read the Summary. If you are familiar with the essential features of the material presented there, you are ready to take the following diagnostic examination.

TESTING YOUR UNDERSTANDING

After you have completed the following examination, compare your answers with the annotated key in the Analysis section.

1. Darwin referred to evolution as:
   a. "the striving of nature for absolute perfection."
   b. "a process that should readily lend itself to demonstration."
   c. "descent with modification."
   d. "not at all at odds with the idea of special creation."

2. The first serious challenge to the theory of special creation came about as a result of observations in the field of:
   a. taxonomy.
   b. cytology.
   c. vertebrate paleontology.
   d. geology.
   e. biochemistry.

3. Which of the following arguments might have been offered by a nineteenth-century creationist as a rebuttal to Charles Darwin?
   a. There are no new species, only "altered forms" of the originally created species.
   b. God did not create species; he created genera.
   c. The exquisite structures found in nature could not have arisen as adaptations to environments, since intermediate stages of such structures would have been useless and nonsensical.
   d. All three arguments might have been offered.

4. The study of the worldwide distribution of plants and animals is known as:
   a. paleontology.
   b. biogeography.
   c. natural selection.
   d. adaptation.
   e. creationism.

5. While serving as a naturalist aboard the H.M.S. *Beagle*, what observations did Charles Darwin make that were inconsistent with the concept of special creation?
   a. There were 57 separate species of kangaroos that live only in Australia.
   b. Polar bears did not live in the tropics.
   c. The 13 species of finches on the Galapagos Islands were similar to mainland species.
   d. Only a and c are correct.
   e. All three answers (a, b, and c) are correct.

6. According to Darwin, the inheritable variations present in a population are the results of:
   a. natural selection.
   b. chance.
   c. mutation.
   d. adaptation.
   e. Both a and d are correct.

7. Which of the following proved difficult for Darwin to acknowledge?
   a. Organisms sometimes experience abrupt changes in their genetic material.
   b. The earth is relatively young, probably about 100 million years old.
   c. Distantly related organisms have a number of homologous structures.
   d. Most islands have many fewer species than do continents.
   e. Most of the offspring of many species die before reaching reproductive age.

8. On an evolutionary time scale, it is correct to say that _____ become adapted to their environments.
   a. individual organisms
   b. genotypes
   c. phenotypes
   d. populations
   e. physiological processes

9. In explaining how such an intricate structure as

the vertebrate eye could be an adaptation, Darwin would have said that such a refinement:
   a. is the achievement of perfection by nature.
   b. resulted from gradual improvement.
   c. evolved rapidly as a one-in-a-million possibility.
   d. admittedly could only have been the work of an ingenious creator.

10. Kettlewell explained the case of industrial melanism in the peppered moth by saying that:
    a. light-colored moths underwent a physiological adaptation in becoming darker.
    b. predatory birds could detect the darker moths but maintained a preference for the lighter ones.
    c. a rare, recurring mutation was selected for by environmental changes.
    d. black moths were more resistant to the effects of air pollution than light-colored ones.

11. Colonies of the Lederbergs' bacteria that would grow when plated onto a medium containing penicillin:
    a. had developed resistance as a result of exposure to the drug.
    b. had been penicillin-resistant all along.
    c. had not been able to grow in the original nonantibiotic medium.
    d. demonstrated the mutagenic effects of penicillin.

12. Which of the following correctly describes the status among modern biologists of the theory of evolution by natural selection?
    a. Most accept it, though there is little concrete evidence either for or against it.
    b. It is a useful theory, even though it is not consistent with the fossil record.
    c. Many different fields of research are providing support for it.
    d. The theory may be invalid—several of its basic assumptions have been found to be incorrect.

13. Which of the following questions is still attracting the attention of evolutionists today?
    a. By what mechanism does evolution proceed in a given population?
    b. Why are there so many species?
    c. How old is the earth?
    d. Why are there no fossils of species other than those that are alive today?
    e. Both a and b are pertinent questions today.

14. The most accurate estimate of the age of the earth has been obtained by:
    a. the calculations of Lord Kelvin.
    b. radioactive dating of geologic strata.
    c. analysis of the fossil record.
    d. piecing together genealogical information.

15. A sample of liquid tritium (^3H) has a half-life of 12.25 years. This means that half of the _____ after 12.25 years.
    a. liquid will become gas
    b. valence electrons will become excited
    c. radioactive atoms will be stable
    d. volume of the sample will be gone

16. Match each scientist with his or her contribution.

    Charles Darwin _____
    Alfred Wallace _____
    Thomas Huxley _____
    H.B.D. Kettlewell _____
    Joshua and Esther Lederberg _____
    Lord Kelvin _____

    a. made the first scientific attempt to calculate the earth's age
    b. published an evolutionary theory in *The Origin of Species*
    c. used the fossil record of the horse to support Darwin
    d. demonstrated how bacteria become resistant to penicillin
    e. showed the consequences of industrialization for populations of the peppered moth
    f. came to the same conclusions as Darwin after similar travels

17. Match each species or group of animals with its evolutionary significance.

    marsupials _____
    Galapagos finches _____
    Eohippus _____
    Archaeopteryx _____
    whales _____

    a. ancestor of the modern horse
    b. the only indigenous mammals of remote islands
    c. possessed birdlike and reptilian features

Biston betularia
_____

bats _____

d. the major mammalian forms of Australia
e. provide a striking example of graded diversity within a small area
f. retain vestigial structures of terrestrial vertebrates
g. a now undisputed example of natural selection by industrial melanism

18. Outline the basic framework used by Darwin and Wallace to support their concept of the process of evolution.
19. What lines of evidence for evolution were available to Darwin at the time he formed his theory?
20. Explain how the work of either Kettlewell or the Lederbergs demonstrates that natural selection can bring about evolutionary change in a population.

PERFORMANCE ANALYSIS

1. Darwin would not have agreed with choices a, b, or d. Choice **c**, however, embodies Darwin's central idea about how the species present on the earth at his time had come to exist.
2. By the middle of the nineteenth century, taxonomists (**a**) had identified an enormous number of species. This raised several basic questions: If each species was created for a particular way of life, why are geographically similar areas inhabited by different species? Why are some species restricted to certain parts of the world? Why does the fauna of islands differ from that of nearby continents? In the face of these questions, it became more and more difficult for creationists to explain why a divine being would have created so many functionally similar species when one or a few would have been sufficient.
3. Creationists used choices a and b to deal with the species problem presented by the naturalists. They attacked the idea of adaptation on the ground that any intermediate stages in the development of an exquisite structure like the vertebrate eye would have been useless. Therefore, choice **d** is correct.
4. Choice **b** is correct.
5. Choice b is consistent with the doctrine of special creation. Choices a and c are observations that Darwin made and considered to be inconsistent with special creation. Therefore, **d** is correct.
6. In developing the concept of natural selection, Darwin was unaware of mutation as the source of inheritable variations. He was on the right track, however, in attributing such variation to chance (**b**).
7. Darwin was aware of choices c, d, and e and used them to support his theories. Had Darwin been aware that abrupt changes in genetic material (mutations) occur, his theory of natural selection would have been more cohesive. However, Kelvin's conclusion that the earth was only about 100 million years old (**b**) posed difficulties for Darwin, for he knew that this was an insufficient time for the whole of biological evolution to have taken place.
8. As phenotypes of individuals are selected by nature, the genetic constitution of a population changes, and the population (**d**) becomes better adapted (in an evolutionary sense) to its environment.
9. Darwin concluded that adaptations result from gradual improvements (**b**) and are only as good as they have to be.
10. Kettlewell assumed that the black form of the peppered moth was caused by a rare mutation and showed that this form had become more abundant as the English countryside became covered with soot from factories. Predatory birds were unable to detect the black forms and, therefore, ate only lighter-colored moths. Choice **c** is correct.
11. Bacterial colonies plated onto the medium containing penicillin were clones of colonies that had never been exposed to penicillin. Some of the clones were able to survive in the antibiotic medium. This indicates that the parent cells had possessed the capacity for penicillin resistance (**b**).
12. Many different fields of research, including biochemistry, cytology, and paleontology (the study

of fossils), have provided support for evolutionary theory. Choice **c** is correct.

13. Choices a and b both represent current fields of inquiry by evolutionists. The age of the earth is now generally accepted to be about 4.6 billion years old. Choice d is of course nonsensical. Choice **e** is correct.

14. Choice **b** is correct. See the essay, The Record in the Rocks, pages 886–887.

15. Rates of decay of radioactive substances are measured in terms of half-lives. This is the amount of time (for tritium, 12.25 years) necessary for half of the atoms in a sample of an element to lose their radioactivity and become stable. Choice **c** is correct.

16. The correct answers in order are **b, f, c, e, d**, and **a**.

17. The correct answers in order are **d, e, a, c, f, g**, and **b**.

18. Darwin and Wallace based their ideas concerning the mechanism of evolution on four premises: (1) Organisms beget like organisms; (2) in any population, there are chance variations among individuals that are inheritable; (3) in most species, only a few of the individuals produced actually survive and reproduce; and (4) natural selection is a major factor determining which individuals survive.

19. Five lines of evidence were available to Darwin for the development of his theory: (1) the large number and diversity of species; (2) the distribution of many species, which were apparently confined within particular areas by natural barriers; (3) the fossil record, which showed that organisms had a history and that they had changed in the course of time; (4) homologous structures, which implied a common ancestry; and (5) adaptations, which indicated that changes occurred in populations in response to selective forces in the environment.

20. The evidence for evolution presented by Kettlewell is pointed out in the analysis for Question 10; the research done by the Lederbergs is interpreted in the analysis for Question 11.

CHAPTER 47

The Genetic Basis of Evolution

CHAPTER ORGANIZATION

I. Introduction

II. The concept of the gene pool
  A. Introduction
  B. Alleles and genotypes: The Hardy-Weinberg principle
    1. General information
    2. Multiple alleles
    3. Essay: Survival of the fittest

III. The agents of change
  A. Introduction
  B. Mutations
  C. Migration: Gene flow
  D. Sampling errors: Genetic drift
    1. Introduction
    2. The founder effect
    3. The bottleneck
  E. Nonrandom mating

IV. Summary

MAJOR CONCEPTS

For a population geneticist, evolution may be defined as a change in the composition of a gene pool.

The Hardy-Weinberg principle states that in a given population under certain conditions, the proportions of genotypes and the frequencies of alleles will remain constant from one generation to the next; that is, reproduction alone (which involves the segregation and recombination of alleles) does not change the composition of the gene pool. The principle applies under the following conditions: (1) The population is large; (2) random mating occurs; (3) there is no emigration or immigration; (4) mutations do not occur, or mutation rates are very low; and (5) natural selection does not occur.

Natural selection is the principal cause of change in the Hardy-Weinberg equilibrium. Mutations provide the raw materials (variations) upon which natural selection acts.

Gene flow may introduce new alleles or alter the proportions of alleles already existing in the population. Gene flow tends to counteract natural selection.

Genetic drift occurs when a very small population breaks off from a larger population and chance causes changes in the allelic frequencies of the smaller group.

Nonrandom mating causes changes in the proportions of genotypes but does not necessarily cause changes in allelic frequencies.

HOW TO STUDY THE CHAPTER

Read the chapter, focusing on the major concepts.

Reread the chapter, using the questions that follow to help you focus on details as they relate to major concepts. Answer the questions on a separate sheet of paper. Your answers will provide a valuable study aid in preparing for examinations.

FOCUS ON CHAPTER DETAILS

I. Introduction (pages 892–893)

1. What consequences did the work of Mendel have for Darwinian theory?
2. What were the attitudes of the early geneticists toward Darwin's theory? What did they believe to be responsible for changes that might occur within a species?
3. What is neo-Darwinian theory? Why is it also known as the synthetic theory?
4. Define the following in the context of population genetics: population; gene pool; allele frequencies; fitness; evolution.
5. Why do allele frequencies change?

II. The concept of the gene pool (pages 893–894)

A. Introduction

6. Allelic frequencies are numerically equivalent to what?
7. What factors might cause these frequencies to change between the first sampling of a population and a later sampling?

B. Alleles and genotypes: The Hardy-Weinberg principle

General information

8. What question about populations does the Hardy-Weinberg principle attempt to answer?
9. Write the Hardy-Weinberg equation that represents all of the possible combinations of two alleles in a population.
10. What do p and q represent? What information does each term of the equation provide about the population?

Multiple alleles

11. Cite an example of a trait in fruit flies that is determined by multiple alleles.
12. How can the Hardy-Weinberg formula be expanded to include the case of three alleles?

Essay: Survival of the fittest (page 894)

13. How was the Darwinian idea of "survival of the fittest" exploited by the social Darwinists?
14. Why is this phrase an inaccurate description of the means by which evolution proceeds in nature?
15. How do population geneticists measure fitness?

III. The agents of change (pages 895–897)

A. Introduction

16. Under what five conditions can the Hardy-Weinberg principle be used to predict the proportions of genotypes in a population?
17. Of what usefulness is the Hardy-Weinberg principle if the five conditions cannot be met?
18. What agents are responsible for the changes in allele frequencies in a population? According to Darwinian theory, which of these is the most important?

B. Mutations

19. How does a population geneticist define mutation?
20. Do environmental factors, such as radiation, influence the type or the rate of mutations? Explain your answer.
21. What is the range of the average rate of mutation detectable per generation?
22. Why is such a low incidence of mutation of importance in bringing about variation within a population?
23. What is the role of mutation in evolution?

C. Migration: Gene flow

24. What is gene flow?
25. What are the effects of gene flow and natural selection on the differences between populations?

D. Sampling errors: Genetic drift

Introduction

26. Why does the Hardy-Weinberg principle apply only to large populations?
27. Explain how allele frequencies may change drastically as a result of genetic drift.

The founder effect

28. Why are small populations that splinter from a large population usually genetically different from the large population?

29. Why are such populations likely to have a high degree of homozygosity?

30. How do the Amish of Pennsylvania provide an example of the founder effect?

The bottleneck

31. Define population bottleneck.

32. How can drastic reductions in population size lead to the loss of genetic variability?

E. Nonrandom mating

33. Cite some examples of nonrandom mating in plants and animals.

34. What effect does nonrandom mating have on the Hardy-Weinberg equilibrium?

35. How can nonrandom mating result in changes in the genotype frequencies without changing the allele frequencies?

IV. **Summary** (pages 897–898): Read the Summary. If you are familiar with the essential features of the material presented there, you are ready to take the following diagnostic examination.

TESTING YOUR UNDERSTANDING

After you have completed the following examination, compare your answers with the annotated key in the Analysis section.

1. The Hardy-Weinberg principle applies only to populations that:
   a. are small.
   b. have equal frequencies of alleles.
   c. are not acted on by agents of change.
   d. have moderate rates of mutation.

2. A gene has two alleles that are equal in frequency in a population that is in Hardy-Weinberg equilibrium. The proportion of individuals with a heterozygous genotype is _____ percent.
   a. 100
   b. 75
   c. 50
   d. 25
   e. 0

3. The most certain consequence of a drastic reduction in a population's size is:
   a. a higher incidence of disease.
   b. extinction.
   c. an increase in homozygosity.
   d. an increase in mutation rates.
   e. nonrandom mating.

4. Immigration of individuals into a population may cause changes in allele frequencies. This situation is referred to as:
   a. genetic drift.
   b. the founder effect.
   c. gene flow.
   d. nonrandom mating.
   e. speciation.

5. A gene pool is an abstraction that applies to a(n):
   a. organism.
   b. genotype.
   c. breeding group.
   d. mating pair.
   e. species.

6. The synthetic theory of evolution is a merging of the ideas of:
   a. Darwin and Lamarck.
   b. Huxley and Wallace.
   c. Darwin and Mendel.
   d. Hardy and Weinberg.
   e. Mendel and Morgan.

7. What is the evolutionary role of mutations?
   a. They are the source of variability in a population.
   b. They cause striking changes in the frequencies of alleles.
   c. They usually lead to reductions in a population's fitness.
   d. Mutations occur in such very low frequencies that they have no important evolutionary role.
   e. Both a and b are correct.

8. Which of the following is true concerning the occurrence of mutations within a population?
   a. Different genes have different rates of mutation.
   b. The rate of mutation may be influenced by environmental factors.

c. The incidence of mutation in any given gene is low.
   d. The number of new mutations in a generation is very high.
   e. All of the above are correct.

9. In the context of population genetics, an organism is considered to be fit if it:
   a. is in good physical health.
   b. is well adapted to its environment.
   c. lives to a ripe old age.
   d. makes a contribution to the gene pool of the next generation.
   e. is heterozygous at a large number of loci.

10. The Old Order Amish provide an example of:
    a. a population bottleneck.
    b. the founder effect.
    c. a population with a very high rate of mutation.
    d. evolution by natural selection.
    e. gene flow.

11. Polydactylism is more prevalent among the Amish of Pennsylvania than among their European ancestors because:
    a. the Amish of Pennsylvania have been subject to higher mutation rates.
    b. the founder population was not genetically representative of the European population.
    c. polydactylism has been selected for in the Pennsylvania Amish.
    d. the Pennsylvanian Amish population has been heavily affected by gene flow.
    e. Both b and d are correct.

12. According to neo-Darwinian theory, the chief agent of evolutionary change is:
    a. mutation.
    b. gene flow.
    c. natural selection.
    d. sampling error.
    e. nonrandom mating.

13. If the frequency of homozygous recessives (aa) in a population is 0.16, then the frequency of allele a is:
    a. 0.16.
    b. 0.0016.
    c. 0.256.
    d. 0.4.
    e. 0.04.

14. If a gene has 4 alleles, then the number of genotypes possible is:
    a. 4.
    b. 6.
    c. 8.
    d. 10.
    e. 16.

15. Without a knowledge of _____, Darwin was unable to explain why variability does not disappear within a few generations.
    a. mutations
    b. natural selection
    c. Mendelian inheritance
    d. the Hardy-Weinberg principle

16. Discuss the synthesis of concepts that gave rise to the field of population genetics.

17. Describe the ways in which the frequency of a given phenotype within a population may change.

18. How was Darwin's theory misapplied by those who defended the inequalities of industrial society?

PERFORMANCE ANALYSIS

1. The Hardy-Weinberg principle states that allele frequencies and the proportions of genotypes in a population will remain constant provided the population is large enough (so that sampling error is avoided), mating is random, and agents of change (mutation, migration, and natural selection) are not in effect. Choice **c** is correct.

2. If the two alleles are equal in frequency, the frequency of each must be 0.5. To find the proportion of the population that is heterozygous for this gene, use the middle term of the Hardy-Weinberg equation: $2pq = 2 \times 0.5 \times 0.5 = 0.50$, or 50 percent. Choice **c** is correct.

3. When a population is drastically reduced in size, alleles relatively low in frequency may be completely lost, so that the genetic variability of the population is greatly reduced. This means that the population will become more homozygous (**c**).

4. Immigration of individuals into a population may result in changes in the allele frequencies of that population if the immigrants are drawn from a

population whose allele frequencies are different from those of the population they enter. This type of evolutionary change is referred to as gene flow (**c**).

5. A gene pool is defined as the sum total of all the alleles of all the genes of all the individuals in a population. In practice, however, it is only those individuals that are members of the breeding group (**c**) who actually make contributions to the gene pool.

6. The synthetic theory combines Darwin's theory of evolution based on natural selection with the principles of Mendelian inheritance. From the synthetic theory arose the field of population genetics. Choice **c** is correct.

7. Mutations are the source of the variability within a population (**a**) upon which natural selection acts.

8. Environmental factors may influence mutation rates. Also, different genes do have different rates of mutation. Normally, mutations occur at very low rates, but since the number of genes an organism carries may be large, the number of new mutations in each generation of the entire population may be very high. Choice **e** is therefore correct.

9. According to Darwin, fitness meant an organism's ability to produce offspring that could reproduce. This has been amended slightly by population geneticists, who define a fit organism as one that makes a contribution (passes on its alleles) to the gene pool of the next generation (**d**).

10. The Old Order Amish of Pennsylvania provide an example of a small population that branched from a larger one and was not genetically representative of that population. This is known as the founder effect (**b**). Among the founding Amish, for example, recessive alleles for dwarfism and polydactylism were much higher in frequency than among their European ancestors. As a result, the Pennsylvanian Amish have a higher incidence of dwarfs with more than five fingers.

11. Choice **b** is correct. See the analysis for Question 10.

12. Neo-Darwinian theory explains evolution in terms of variation and natural selection (**c**). The theory considers natural selection to be the chief agent of change within a population, as did Darwin himself. As we shall see in a later chapter, this idea is being seriously questioned.

13. Choice **d** is correct. From the Hardy-Weinberg equation, the frequency of homozygous recessives is equal to q^2. Therefore, if $q^2 = 0.16$, then $q = \sqrt{0.16} = 0.4$.

14. The number of possible genotypes may be found by modifying the Hardy-Weinberg equation to include 4 alleles. Thus, $(p + q + r + s)^2$. The term expands as follows: $p^2 + q^2 + 2pq + 2qr + 2qs + 2rs + 2ps + 2pr + r^2 + s^2$. There are 10 (**d**) terms in the expansion, each of which corresponds to a different genotype.

15. At the time Darwin proposed the theory of natural selection, biologists explained inheritance in terms of a blending theory. Darwin saw that by blending the genetic material of a population, its variability would soon be reduced to zero. Mendel (**c**) provided the solution to this problem by showing how alleles are hidden in one generation but reappear in the next.

16. The synthetic, or neo-Darwinian, theory of evolution is a merger of Darwin's theory of variation and selection with the principles of Mendelian inheritance, which provide a genetic explanation for population change. Biologists began to think in terms of the frequency of alleles within a population's gene pool, rather than in terms of individual organisms and their genotypes. This new perspective gave rise to the field of population genetics.

17. Changes in the proportions of phenotypes may result from changes in allele frequency caused by mutation, gene flow, natural selection, or genetic drift. Or they may be due to nonrandom mating, which does not require that allele frequencies change.

18. Social Darwinists misapplied the phrase "survival of the fittest." By "fitness," Darwin meant the ability to produce many offspring that could, in turn, reproduce. The social Darwinists measured "fitness" as an innate superiority that allowed certain people (the "fit") to compete successfully in the social and commercial worlds at the expense of others (the "unfit").

CHAPTER 48

Variability: Its Extent, Preservation, and Promotion

CHAPTER ORGANIZATION

I. Introduction
II. **The extent and origins of variations**
   A. Introductory remarks
   B. Artificial selection
      1. General information
      2. Bristle number in *Drosophila*
   C. Quantifying variability
   D. Explaining the extent of variation
III. **Preservation and promotion of variability**
   A. Sexual reproduction
   B. Mechanisms that promote outbreeding
   C. Diploidy
IV. **Natural selection and variability**
   A. Introductory remarks
   B. Balanced polymorphism
      1. Introduction
      2. Shell color in snails
      3. Human blood groups
   C. Heterozygote superiority
      1. Introduction
      2. Sickle cell anemia
      3. Heterosis or hybrid vigor
   D. Geographic variations: Clines and ecotypes
   E. Frequency-dependent selection
V. **Variation and the eukaryotic chromosome**
VI. **Summary**

MAJOR CONCEPTS

The extent of the genetic variability present in a population is a major factor in determining that population's capacity to undergo evolutionary changes.

Genetic variability in populations is preserved by several mechanisms, including sexual reproduction, outbreeding, and diploidy.

The means by which natural selection promotes genetic variability include balanced polymorphism, heterozygote superiority, heterosis, clines, and selection by frequency.

HOW TO STUDY THE CHAPTER

Read the chapter, focusing on the major concepts.

Reread the chapter, using the questions that follow to help you focus on details as they relate to major concepts. Answer the questions on a separate sheet of paper. Your answers will provide a valuable study aid in preparing for examinations.

FOCUS ON CHAPTER DETAILS

I. **Introduction** (page 900)
   1. What role do variations play in the history of a species?
   2. Modern population geneticists are especially interested in what aspects of variability?

II. **The extent and origins of variations** (pages 901-905)
   A. Introductory remarks
      3. How did creationists and Darwin differ in their interpretations of variability?

367

B. Artificial selection

General information

4. How did the practice of artificial selection influence Darwin's thinking about evolution?

Bristle number in Drosophila

5. How were high and low bristle numbers artificially produced in fruit flies?
6. What eventually happened to the low-bristle-number line?
7. What happened to bristle number when a high-bristle-number population was allowed to interbreed on its own?
8. What is the probable explanation for the attainment of a high bristle number without the side effect of sterility in a second artificial breeding trial?
9. What principle of selection cannot be ignored by those attempting to select a desired trait artificially?

C. Quantifying variability

10. How is electrophoresis of proteins employed to assess the extent of variability in a population?
11. What is an isozyme?
12. How is the number of isozymes related to the degree of heterozygosity in a population?
13. What is the estimated degree of heterozygosity in populations of humans? Fruit flies?
14. What is the technique of DNA sequencing expected to reveal about variability? Why?

D. Explaining the extent of variation

15. Why did most geneticists think that there should be little genetic variability left within most populations? What did the Hubby and Lewontin experiments reveal?
16. How do selectionists and neutralists differ in their viewpoints?

III. **Preservation and promotion of variability** (pages 905–907)

A. Sexual reproduction

17. What are the three ways in which sexual reproduction promotes variation?
18. Why is genetic variability limited in a population that reproduces asexually?
19. What are some disadvantages of sexual reproduction that are thought to be outweighed by its single advantage—the promotion of variation?

B. Mechanisms that promote outbreeding

20. Describe those mechanisms that encourage cross-pollination in plants.
21. What behavioral strategies promote outbreeding among animals?
22. How are taboos against incest variously interpreted by psychologists and biologists?

C. Diploidy

23. How does being diploid (as opposed to being haploid) promote the genetic variability of a population?
24. How successful would a controlled breeding or sterilization program be in completely ridding a population of "undesirable" recessive alleles? What is the reason for this?

IV. **Natural selection and variability** (pages 907–911)

A. Introductory remarks

25. Why did anti-Darwinists think that natural selection would have an anti-evolutionary effect? What is the view of modern population geneticists?

B. Balanced polymorphism

Introduction

26. What is polymorphism?
27. What two phenomena can account for it?

Shell color in snails

28. What two distinct forms are found among snails of the genus *Cepaea*?
29. What factor is responsible for balanced polymorphism in this case?

Human blood groups

30. Why is it thought by some scientists that

the polymorphism in human blood groups might be maintained by natural selection? What argument has been presented against this proposal?

31. The irregular distribution of blood group alleles may result from what?

C. Heterozygote superiority

Introduction

32. Define the concept of heterozygote superiority.

Sickle cell anemia

33. Why are mutations unlikely to account for the stability of the sickling allele in African tribes?

34. Why is the heterozygote in these tribes at a selective advantage over other genotypes?

Heterosis or hybrid vigor

35. Why is heterosis believed to be an important mechanism for maintaining polymorphism in a population?

36. Why are hybrids likely to be more vigorous than either of the parent strains from which they are derived?

D. Geographic variations: Clines and ecotypes

37. What are clines? What factors might account for them?

38. Cite some examples of north-south clines, emphasizing how the adaptive value of a phenotype might change with latitude.

39. What is an ecotype?

40. How did researchers at the Carnegie Institution show that the differences among ecotypes of *Potentilla* were genetic differences? Why was this outcome expected?

E. Frequency-dependent selection

41. Cite an example from the study of animal behavior in which the more common allele is selected against.

42. Explain how self-sterility alleles are an example of frequency-dependent selection.

V. **Variation and the eukaryotic chromosome** (page 912)

43. How could the complex genomes of eukaryotes have evolved from a very few protogenes? How would such a process have been important in evolutionary change?

VI. **Summary** (page 912): Read the Summary. If you are familiar with the essential features of the material presented there, you are ready to take the following diagnostic examination.

TESTING YOUR UNDERSTANDING

After you have completed the following examination, compare your answers with the annotated key in the Analysis section.

1. Darwin was aware of the extent of variability in animal populations because he had observed the effects of:
   a. selection on bristle number in *Drosophila*.
   b. artificial breeding on pigeons.
   c. environmental factors on mutation rates.
   d. asexual reproduction.

2. When selection for bristle number in *Drosophila* was carried out artificially:
   a. bristles would often disappear entirely.
   b. a high number of bristles was always accompanied by sterility.
   c. bristle number could be increased but not decreased.
   d. bristle number stabilized above the original norm when random breeding was allowed.

3. Experimentation has shown that, on the average, human populations are heterozygous for about _____ percent of their genes.
   a. 50
   b. 25
   c. 15
   d. 7
   e. 1

4. A major problem associated with the inbreeding of organisms for desirable traits is:
   a. heterosis.
   b. sterility.
   c. variability.
   d. genetic drift.

5. The greatest extent of inheritable variation in an organism would be reflected by its:
   a. amino acid sequences.
   b. DNA nucleotide sequences.
   c. relative numbers of isozymes.
   d. complement of tRNA molecules.
   e. Both a and c are correct.

6. A population contains more than two isozymes for the same enzyme. This indicates the presence of:
   a. multiple alleles.
   b. nonfunctional enzymes.
   c. deleterious mutations.
   d. occasional asexual reproduction.
   e. Both a and b.

7. Selectionists and neutralists represent opposing schools that try to explain how _____ in a natural population.
   a. recessive alleles are conserved
   b. variation is maintained
   c. natural selection acts
   d. fitness may be altered

8. Which of the following are mechanisms that favor outbreeding?
   a. self-sterility alleles
   b. dispersal of offspring
   c. incest taboos
   d. Only b and c are correct.
   e. All three answers (a, b, and c) are correct.

9. Variability is preserved in a diploid species because _____ are not always exposed to natural selection.
   a. heterozygous organisms
   b. recessive alleles
   c. mutant phenotypes
   d. all organisms in a population

10. Which one of the following characteristics provides proof that a heterogeneous habitat may be responsible for balanced polymorphism?
    a. wing color in peppered moths
    b. shell color in snails
    c. bristle number in fruit flies
    d. human blood groups
    e. body size in house sparrows

11. Which of the following is correct regarding polymorphism in human blood groups?
    a. Those with type O blood are at a selective advantage.
    b. The distribution of blood types in certain parts of the world is correlated with the incidence of particular diseases.
    c. It is not certain how polymorphism is maintained.
    d. There are clear-cut disadvantages associated with particular blood types.

12. The sickle cell allele has not disappeared in the North American black population chiefly because:
    a. of heterozygote superiority.
    b. it is difficult to eliminate recessive alleles completely.
    c. of the interference of genetic counselors.
    d. those having anemia are still able to reproduce.

13. Hybrid plants are very often hardier than either of the two strains from which they are bred. This phenomenon is referred to as:
    a. unbalanced polymorphism.
    b. frequency-dependent selection.
    c. heterosis.
    d. diploidy.
    e. Both a and c are correct.

14. Which of the following is an example of a cline?
    a. distributions of human blood groups
    b. shell color in snails of the genus *Cepaea*
    c. body size in male North American house sparrows
    d. seasonal changes in feather color of snow geese

15. Ecotypes may be:
    a. found living in identical habitats.
    b. phenotypically similar.
    c. genetically different.
    d. incapable of interbreeding.

16. In a situation where polymorphism is maintained by frequency-dependent selection, the most common _____ is often at a selective disadvantage.
    a. allele
    b. genotype
    c. heterozygote
    d. phenotype

17. It is now believed that the chief reason for the expansion in the complexity of genomes throughout evolutionary history is due to:
    a. DNA replication.

b. sexual reproduction.
c. gene duplication.
d. transcription of DNA segments.

18. Discuss the role of each of the following in maintaining genetic variability in a population: sexual reproduction, outbreeding, and diploidy.

19. What factors introduce new variations into a population?

PERFORMANCE ANALYSIS

1. Darwin was an expert on pigeon breeding (**b**) and was aware, from the artificial selection of desirable traits, that animal populations must have a great deal of variability hidden in their gene pools.

2. See Figure 48-5 and its caption. Note that selection for high bristle number was at first accompanied by sterility. When the number of bristles reached 56, sterility became such a problem that the high-bristle line was allowed to interbreed randomly. After several generations of random interbreeding, bristle number stabilized at a level higher than the original norm. Choice **d** is correct.

3. At least one-quarter of the genes in any human population are represented by two or more alleles. On the average, an individual is heterozygous at about 7 percent of its loci. Choice **d** is correct.

4. Artificial selection for desirable traits is often complicated by sterility (**b**), as was seen in the *Drosophila* bristle-number experiments (see the analysis for Question 2).

5. DNA sequencing (**b**) is expected to demonstrate the extent of variation in a population more accurately than amino acid sequencing can, since a change in the nucleotide sequence of DNA does not always lead to a change in an amino acid sequence. Also, electrophoresis is not able to detect all changes in amino acid sequences.

6. Isozymes are enzymes that are functionally identical but structurally distinct. Since the structural differences are genetically determined, the number of different isozymes of a given enzyme is proportional to the number of alleles at the locus responsible for synthesis of the enzyme. For example, three isozymes isolated from a population would indicate that three alleles are present; therefore, **a** is correct.

7. Selectionists claim that variation is maintained (**b**) in a population by the forces of natural selection, whereas neutralists attribute variation to the accumulation of random mutations that do not affect fitness and have no value in terms of selection.

8. Choice **e** is correct. Self-sterility alleles prevent self-pollination in plants. Dispersal of offspring from a family group or herd results in their mating with members of other groups. Incest taboos in human societies discourage mating with close relatives.

9. In an organism that is diploid, the effects of a recessive allele (**b**) are often masked in the heterozygote. Thus, even if the homozygous recessive individuals are selected against, the allele can be maintained in the population (although it will gradually decrease in frequency).

10. Two shell forms, banded and unbanded, are maintained in constant proportions in populations of snails (**b**) of the genus *Cepaea*, since each form has a selective advantage in a different part of the habitat. Birds that eat snails are less likely to see unbanded shells in habitats with uniform backgrounds and less likely to see banded forms in habitats with mottled backgrounds.

11. There does not appear to be a clear-cut advantage to having a particular blood type, and correlations between blood type and resistance to disease have not been firmly established. Since distributions of blood types throughout the world are irregular, some selective force may be at work, or gene flow or genetic drift may be involved. As yet, however, the way in which this polymorphism is maintained is not understood (**c**).

12. Choice **b** is correct; see the analysis for Question 9. (See also the table on page 907.) It has been shown that the sickle cell allele is maintained in black populations of Africa because of heterozygote superiority.

13. Because hybrid plants are heterozygous at more

loci than the parent strains from which they are bred, they are protected more often from the occurrence of deleterious alleles in the recessive state. The overall superiority of the hybrid over either parent is called heterosis (**d**).

14. A cline is a geographically graded variation in a trait. A good example is body size of male house sparrows in North America (**c**).

15. Ecotypes are phenotypically different populations of a species living in different habitats. Although these differences may be related to environmental factors, some ecotypes, such as *Potentilla glandulosa* (see Figure 48–11), have been shown to be genetically different (**c**) as a result of different selective pressures in the different habitats.

16. Polymorphism may be maintained by selection against the more common alleles, resulting in the favoring of those alleles originally low in frequency (frequency-dependent selection). But natural selection acts on the phenotype (**d**) that expresses the most common genotype rather than directly on the alleles.

17. It is now believed that the duplication (**c**) and modification of a few protogenes was responsible for the construction of complex genomes.

18. Sexual reproduction provides recombination of alleles by independent assortment, crossing over, and contributions from two different parents. Outbreeding (contrasted with self-fertilization) provides alleles from two different parents or groups. Mechanisms among plants include self-sterility alleles, maturation of pollen at a time when the plant's own stigma is unreceptive, anatomical arrangements that inhibit self-fertilization, and single-sex individuals. Among animals, particularly mammals, outbreeding (as contrasted with inbreeding of a small group) maintains more variation than the possible combinations provided by two parents. Various behavioral strategies promote outbreeding in mammals. The effect of diploidy is discussed in the analysis for Question 9.

19. Factors that introduce new variations into the gene pool of a population are the recombination of alleles associated with events of sexual reproduction, mutation, and gene flow.

CHAPTER 49

Natural Selection

CHAPTER ORGANIZATION

I. Introduction
II. Types of selection
III. What is selected?
IV. Evolution and the idea of progress
    A. Introductory remarks
    B. Fecundity and longevity
    C. Developmental and structural constraints
    D. Eyeless arthropods and other degenerates
    E. The Red Queen effect
    F. Sexual selection
V. Patterns of evolution
    A. Introductory remarks
    B. Divergent evolution
    C. Convergent evolution
    D. Parallel evolution
VI. Coevolution
    A. Introductory remarks
    B. Milkweed, monarchs, and mimics
VII. Summary

MAJOR CONCEPTS

Natural selection is the process whereby different genotypes are produced as a result of interactions between individual organisms and their environment. Natural selection leads to evolution and adaptation.

The three major types of selection are stabilizing (selection against extremes), disruptive (selection for two extremes at the expense of the intermediates), and directional (selection for one extreme at the expense of the other).

Natural selection acts only on characteristics that are expressed in the phenotype. Therefore, the unit of natural selection is the individual organism.

Evolution does not necessarily produce organisms that are perfectly adapted to their environment. The evolutionary potential in a population is limited by the extent of variability in the gene pool and by the structural and developmental capabilities of individual organisms.

Three observable patterns of evolution are divergent evolution (related species become more dissimilar), convergent evolution (unrelated species become more similar), and parallel evolution (separate phylogenetic lines evolve in a similar way).

Coevolution is the result of interactions between two or more groups of organisms such that each group is a selective force on the other.

HOW TO STUDY THE CHAPTER

Read the chapter, focusing on the major concepts.

Reread the chapter, using the questions that follow to help you focus on details as they relate to major concepts. Answer the questions on a separate sheet of paper. Your answers will provide a valuable study aid in preparing for examinations.

FOCUS ON CHAPTER DETAILS

I. **Introduction** (page 913)

1. How did Malthus's essay on population growth influence Darwin's thinking about evolution?
2. Define natural selection as it is applied to population genetics. What effect does it have on a population's gene pool?

II. **Types of selection** (pages 914-916)

3. Describe how the three types of selection operate within a population. Give examples of each type.
4. Which of these three types can lead to the formation of two new species?

III. **What is selected?** (pages 916-917)

5. Natural selection acts on the phenotype. What is the population geneticist's concept of phenotype?
6. How does the IQ score reflect the complex relationship between genotype and phenotype?
7. What do identical twins reveal about the relationship between genotype and phenotype?
8. In terms of the evolutionary history of a species, why are individual genes of more importance than either genotypes or phenotypes?
9. Use the example of industrial melanism in the peppered moth to illustrate that a winning phenotype is rarely the result of a single allele, even when a conspicuously important trait is determined by a single gene.
10. What is a coadaptive gene complex?
11. What are supergenes? How might they be protected from recombination?

IV. **Evolution and the idea of progress** (pages 917-920)

A. Introductory remarks

12. Does evolution always result in increased complexity and a greater degree of adaptation to a specific environment? Explain?

B. Fecundity and longevity

13. Explain how natural selection acts on fecundity.
14. Why is longevity generally *not* an adaptive trait? Under what conditions are post-reproductive individuals of value to a society?

C. Developmental and structural constraints

15. Cite some examples to support the following statement: Evolution is conservative in that it can build only on past history.

D. Eyeless arthropods and other degenerates

16. Cite some examples of organisms that are highly evolved (highly specialized) as a result of having lost certain characteristics.

E. The Red Queen effect

17. How is the comment by the Red Queen in Lewis Carroll's *Through the Looking Glass* (Figure 49-9) applicable to the process of natural selection?
18. What quality must a population possess in order to remain adapted in a changing environment?

F. Sexual selection

19. How did Darwin define sexual selection?
20. Why is it usually the male that must compete for a mate and not the female? What forms does this competition take?
21. What is sexual dimorphism? What usually accounts for it?
22. Cite some extreme examples of sexual dimorphism.
23. Under what conditions will the sexes be nearly alike? Under what conditions may the sexes exhibit pronounced dimorphism?
24. In what sense can conspicuous dimorphic characteristics be considered maladaptive?

V. **Patterns of evolution** (pages 921-923)

A. Introductory remarks

25. What general patterns of evolution may result from natural selection?

B. Divergent evolution
  26. Define divergent evolution. What is responsible for its occurrence?
  27. Explain how the speciation of the polar bear from brown bears is an example of divergent evolution.
C. Convergent evolution
  28. What is convergent evolution? Under what conditions can it occur?
  29. Cite some examples of convergence in the plant and animal kingdoms.
  30. What evidence from the field of biogeography indicates that convergent evolution is a widespread phenomenon?
D. Parallel evolution
  31. When is the evolution of two species considered to be parallel?
  32. What is the most striking example of parallel evolution?

VI. **Coevolution** (pages 924–926)
A. Introductory remarks
  33. Define coevolution.
B. Milkweed, monarchs, and mimics
  34. What is the relationship between the milkweed plant, the monarch caterpillar, and insectivorous birds?
  35. What defensive strategy has evolved as a means of deterring predators? Cite some common examples.
  36. What is mimicry? Outline the evolution of the two types of mimics of the monarch butterfly.
  37. How was Jane Brower able to demonstrate the selective value of Batesian mimicry?
  38. Why is it disadvantageous for a species to be a model for a mimic?
  39. Under what conditions will a mimic species be at its greatest advantage?

VII. **Summary** (pages 926–927): Read the Summary. If you are familiar with the essential features of the material presented there, you are ready to take the following diagnostic examination.

TESTING YOUR UNDERSTANDING

After you have completed the following examination, compare your answers with the annotated key in the Analysis section.

1. A type of natural selection that eliminates any extreme phenotype is called _____ selection.
   a. directional
   b. disruptive
   c. stabilizing
   d. sexual
   e. polymorphic

2. Which of the following is an example of disruptive selection?
   a. clutch size in starlings
   b. birth weight in humans
   c. degeneration of eyes in cave-dwelling arthropods
   d. industrial melanism in peppered moths
   e. None of the above are correct.

3. Natural selection operates directly on:
   a. alleles.
   b. genotypes.
   c. phenotypes.
   d. populations.
   e. gene pools.

4. Individuals that are genetically identical must be:
   a. phenotypically identical.
   b. homozygous at all gene loci.
   c. produced from the same fertilized ovum.
   d. ecotypes.
   e. similarly adapted.

5. Which of the following characteristics of a population is *not* apt to be determined by natural selection?
   a. phenotype
   b. fecundity
   c. extreme longevity
   d. sexual dimorphism
   e. male/female ratios

6. The panda's thumb is believed to demonstrate that:
   a. exquisite adaptations are abundant in nature.
   b. evolution can build only on past history.
   c. natural selection results in perfection.
   d. traits are selected as coadaptive complexes.
   e. Both a and c are correct.

7. One consequence of a species becoming very well adjusted to its environment over a long period of time is:
   a. structural complexity.
   b. loss of genetic flexibility.
   c. a high reproductive rate.
   d. a long life span.
   e. Both c and d are correct.

8. Which of the following statements about the evolution of sexual dimorphism is *not* correct?
   a. It is most apparent when only a small fraction of the males mate.
   b. Selection for secondary sex characteristics is always strongest among males.
   c. Conspicuous appearance may be maladaptive in some respects.
   d. It is now considered to be under the influence of natural selection.

9. Batesian mimics realize a great advantage when they:
   a. are more abundant than the model.
   b. are as unpalatable as the model.
   c. emerge after the model.
   d. resemble the model only to a slight degree.
   e. Both a and b are correct.

10. A population geneticist would define natural selection as:
    a. any change in the frequency of an allele within the gene pool.
    b. differential net reproduction of genotypes.
    c. directional selection.
    d. survival of the fittest.
    e. the adjustment of a species to its environment.

11. Coevolution may be considered an example of:
    a. convergent evolution.
    b. parallel evolution.
    c. directional selection.
    d. stabilizing selection.
    e. disruptive selection.

12. If gene flow between two populations becomes restricted, divergent evolution may occur provided that:
    a. selection pressures are different for each population.
    b. there is no mutation.
    c. the gene pools differ initially.
    d. some crossbreeding is still possible.

13. In Jane Brower's laboratory confirmation of the value of deceptive mimicry, green-banded mealworms dipped in quinine served as _____, and green-banded mealworms dipped in distilled water served as _____.
    a. Müllerian mimics; models
    b. Batesian mimics; models
    c. models; Müllerian mimics
    d. models; Batesian mimics

14. A population may never become finely adjusted to its environment because:
    a. of the Red Queen effect.
    b. the gene pool may not be sufficiently variable.
    c. of the problems posed by sexual dimorphism.
    d. the environment may be constantly changing.
    e. of all of the above reasons.

15. Malthus had a profound impact on Darwin by pointing out the disparity between _____ and _____.
    a. artificial selection; natural selection
    b. rate of increase; availability of resources
    c. genotype; phenotype
    d. organism; environment
    e. survival; fitness

16. Genes involved in a chromosomal inversion are thought to be:
    a. duplications or modifications of a single gene.
    b. responsible for the expression of the same trait.
    c. nonessential to the existence of the organism.
    d. supergenes.
    e. maladaptive.

17. Match each type of evolution with the species that exemplify it.

    divergent evolution _____
    convergent evolution _____
    Batesian mimicry _____
    Müllerian mimicry _____
    parallel evolution _____

    a. porpoises and sharks
    b. marsupials and placentals
    c. bees, hornets, and wasps
    d. monarch and viceroy butterflies
    e. polar bears and brown bears

18. Distinguish between parallel and convergent evolution. Use an example if you wish.
19. Why is it incorrect to assume that evolution always involves the improvement of a species?
20. The diagram below shows the effects of the three different types of natural selection on a population. Identify each type of selection, and describe how it operates within the population.

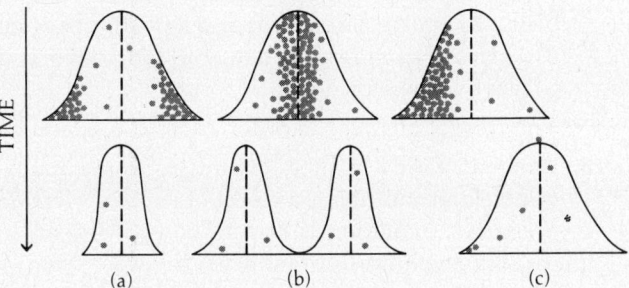

PERFORMANCE ANALYSIS

1. Choice **c** is correct. In stabilizing selection, extreme phenotypes are eliminated and the most frequently occurring phenotypes are maintained.
2. Choices a and b are examples of stabilizing selection; choices c and d are examples of directional selection. Therefore, **e** is correct.
3. Natural selection operates directly on the phenotype (**c**) of an organism. Remember that the phenotype is a composite of all the organism's observable traits.
4. Only identical twins are genetically identical because they are produced from the same fertilized ovum (**c**). Even identical twins may be phenotypically different as a result of environmental influences such as nutrition.
5. Natural selection determines the distribution of only those traits that have adaptive value. In most species, extreme longevity (**c**) has little adaptive value, since individuals that live after their last offspring reach maturity do not usually contribute to the gene pool and can deprive the reproductive individuals in the population of essential resources. Exceptions to this generalization are those cases in which older members of a population contribute to the parental care and education of the offspring of younger members.
6. Stephen Jay Gould points to the panda's thumb as an example of a structure that is "imperfectly" adapted for its functions. These examples of "imperfectly" adapted structures are strong indications that evolution can only build on past history (**b**). They thus provide, as Gould points out, more convincing evidence that evolution has occurred than do the more commonly cited examples of exquisite adaptation.
7. If natural selection acts continuously on a species in an unchanging environment, its genetic variability and thus its flexibility (ability to remain "in tune" and to adjust as the environment changes) will disappear (**b**).
8. Sexual selection, a form of natural selection, is most intense when only a small fraction of those individuals seeking a mate will be able to find one. Sexual dimorphism will be most pronounced in the sex that must do the courting, which, as the case of the phalarope illustrates, is not always the male (**b** is the correct answer). Secondary sexual characteristics may be maladaptive to the individual, since traits such as bright colors and lengthy or noisy displays can attract predators. However, if an individual produces greater numbers of surviving offspring, the secondary characteristics are considered to be "fit."
9. A Batesian mimic is one that closely resembles its model but is palatable to predators. It is best protected if the predator encounters the model first. The probability that this will happen can be maximized if the mimic emerges after the model (**c**) or is generally less abundant than the model.
10. Choice a is incorrect because agents other than natural selection may cause allele frequencies to change. Directional selection (**c**) is the most common form of selection, causing a species to adjust gradually to changes in its environment (adaptation). A strict Darwinian evolutionist might see natural selection merely as survival of the fittest, but the population geneticist regards it from the standpoint of differential net reproduction (**b**). Those individuals with a selective advantage leave more offspring.

11. Common examples of coevolution include flowers and their pollinators, and predators and their prey. In these situations each member of the relationship evolves in a directional fashion (**c**) toward one extreme type. Flowers become more specialized to ensure pollination, predators become more effective hunters, and prey animals evolve more effective means of escaping or deterring predators.

12. If gene flow between two populations becomes reduced, as in the case of populations separated by a geographic barrier, the two groups may begin to show a divergence in one or more characteristics that are under different selective pressures (**a**) in the two habitats.

13. Mealworms dipped in quinine, a substance that tasted bitter to the starlings (as it would to us), served as the unpalatable model. Mealworms of a similar size and color but treated only with distilled water were Batesian mimics. Therefore, **d** is correct. Müllerian mimics are groups of unrelated species that have come to resemble one another in their warning characteristics.

14. If natural selection acts on a population as a steady, unidirectional force, the population may begin to adjust but will always be somewhat behind the changes (the Red Queen effect). A population can keep up with environmental changes only if it has sufficient genetic variability (i.e., genetic flexibility). Selection for secondary sexual characteristics may interfere with the fine tuning of other traits. Constant environmental change, as already implied, will prevent even an adapting population from being absolutely in tune. Therefore, **e** is correct.

15. Malthus pointed out in his essay on the growth of human populations that if animal populations increased geometrically while their resources increased only linearly, the populations would soon outstrip available resources (**b** is correct). Darwin used this idea in formulating his theory of natural selection. Since many species of animals reproduce prolifically, many of the offspring die before reaching reproductive age. Those that survive perpetuate their own type in the next generation because they possess advantageous characteristics.

16. Some of the chromosomal inversions observed in the giant salivary chromosomes of *Drosophila* have been shown to contain supergenes (**d**), genes on the same chromosome that work together as a coadaptive gene complex.

17. The correct answers in order are **e**, **a**, **d**, **c**, and **b**.

18. When distantly related organisms come to resemble one another as a result of similar selection pressures, evolution is said to be convergent. In parallel evolution, organisms diverge from a common ancestor and continue on a similar evolutionary course as a result of similar environmental pressures.

19. Species adapted to their environment are often burdened by developmental or structural constraints (such as in the case of the panda's thumb), limited by sexual dimorphism, and kept from being perfectly in tune with the environment because of the Red Queen effect. Therefore, evolution does not always improve a species toward some perfect form.

20. a. Stabilizing selection: Extreme individuals are continually eliminated.
    b. Disruptive selection: The intermediate forms in the population are eliminated, producing two divergent populations.
    c. Directional selection: One phenotype is gradually eliminated in favor of another, producing adaptive change.

CHAPTER 50

On the Origin of Species

CHAPTER ORGANIZATION

I. **Introduction**

II. **Modes of speciation**
   A. Introduction
   B. Allopatric speciation: Geographic races
   C. Essay: The breakup of Pangaea
   D. Sympatric speciation
      1. Hybridization
      2. Disruptive selection
   E. Maintaining genetic isolation
      1. Introduction
      2. Premating isolating mechanisms
      3. Postmating isolating mechanisms
      4. Essay: Creating sexual chaos
   F. An example: Darwin's finches

III. **The evidence of the fossil record**
   A. Introductory remarks
   B. Phyletic change
   C. Cladogenesis
   D. Adaptive radiation
   E. Extinction
   F. *Equus*: A case study

IV. **On the imperfection of the geological record**

V. **Punctuated equilibria**

VI. **Summary**

MAJOR CONCEPTS

Speciation occurs either allopatrically, in which the populations are geographically isolated, or sympatrically. Two sympatric methods are hybridization followed by polyploidy, which has occurred frequently in plants, and disruptive selection, which is difficult to document.

The essential element in the formation of a new species is its genetic isolation from all other species. Once speciation has occurred, closely related groups can inhabit the same area and still maintain genetic isolation from each other through elaborate premating and postmating mechanisms.

Four major patterns of evolutionary change are phyletic change, or anagenesis (change within a single evolutionary line); cladogenesis (the branching off of one population from another); adaptive radiation (the rapid formation of several species from a single ancestral group, usually associated with the opening of new biological environments); and extinction (the disappearance of a species from the earth).

According to the concept of punctuated equilibrium, the fossil record is an accurate reflection of the evolutionary process. Proponents of the concept argue that speciation occurs relatively rapidly, without transitional forms. Currently, the concept also proposes that major changes in evolution can result from natural selection among species as well as within species.

HOW TO STUDY THE CHAPTER

Read the chapter, focusing on the major concepts.

Reread the chapter, using the questions that follow to help you focus on the details as they relate to major concepts. Answer the questions on a separate sheet of paper. Your answers will provide a valuable study aid in preparing for examinations.

FOCUS ON CHAPTER DETAILS

I. Introduction (page 928)
1. What central question links population genetics with paleobiology?
2. Write working definitions of macroevolution and microevolution.

II. Modes of speciation (pages 928–938)

A. Introduction
3. Define species as first presented in Chapter 19 and as restated in this chapter.
4. How does the population geneticist define a species?
5. What two modes of speciation have been proposed?

B. Allopatric speciation: Geographic races
6. What is a race? A subspecies?
7. Under what conditions might geographic races become distinct species?
8. Cite some examples of geographic barriers.
9. Why is an isolated population likely to differ genetically from the parent population? All isolated populations do *not* become new species. Why not?

C. Essay: The breakup of Pangaea (pages 930–931)
10. What is the theory of continental drift?
11. Explain how current landforms might have arisen from a supercontinent called Pangaea.
12. How does continental drift explain the fossil distribution of *Mesosaurus* and of marsupials in the Southern Hemisphere?

D. Sympatric speciation

Hybridization
13. What is a hybrid? Name some examples of hybrids from the plant and animal kingdoms.
14. Why are certain hybrids considered species in one sense but not in another?
15. Why are hybrids often sterile?
16. Explain how an infertile hybrid plant may, in nature, become fertile. How can fertility be induced artificially?
17. What has been the impact of polyploidy within the plant kingdom?

Disruptive selection
18. Why has speciation due to disruptive selection been difficult to document?
19. Describe one example that shows how sympatric speciation might actually occur.

E. Maintaining genetic isolation

Introduction
20. What two mechanisms maintain genetic isolation between closely related species that live together? Which of the two appears to be the more significant?

Premating isolating mechanisms
21. What types of animal behaviors often serve to maintain sexual isolation between groups of closely related species?
22. Why is it that artificial breeding experiments are not considered valid means for determining if two individuals are of the same species?
23. How do pheromones maintain sexual isolation?
24. How do temporal differences help to keep species separate?

Postmating isolating mechanisms
25. What anatomical or physiological incompatibilities serve to keep species genetically isolated?
26. What is the role of postmating mechanisms?
27. Why is there strong selection pressure for premating isolating mechanisms?

Essay: Creating sexual chaos (page 935)
28. Why have agriculturists been seeking an alternative to insecticides for the control of bollworms and budworms?
29. How were synthetic pheromones expected to limit budworm populations?
30. How did the synthetic pheromones lead to a

breakdown in the isolating mechanisms of the two species of *Heliothis*? What potential for pest control might this have?

F. An example: Darwin's finches
31. What is believed to be the origin of the Galapagos finches?
32. What type of environment did the finches encounter in the Galapagos archipelago?
33. How was the distance between the islands critical to the process of speciation that Darwin thought had taken place?
34. How do the thirteen species of Galapagos finches differ from one another?
35. What mechanism provides for the sexual isolation of the different species?

III. **The evidence of the fossil record** (pages 938–942)

A. Introductory remarks
36. How must macroevolutionary theory be verified?

B. Phyletic change
37. What is phyletic change (anagenesis)? What type of selection produces this change?

C. Cladogenesis
38. What is cladogenesis?
39. How does Ernst Mayr view the role of cladogenesis in evolution?
40. How might cladogenesis account for the sudden appearance of new species in the fossil record?
41. Are anagenesis and cladogenesis mutually exclusive? Explain.

D. Adaptive radiation
42. What is adaptive radiation? Under what conditions is it believed to occur?
43. What kinds of evolutionary change are at work during adaptive radiation?
44. What are some of the more prominent examples of adaptive radiation?

E. Extinction
45. How frequent is the phenomenon of extinction?
46. What are some examples of mass extinctions? What factors might have been responsible?
47. What is the role of extinction in evolution?

F. *Equus*: A case study
48. Trace the evolution of the modern horse *Equus* from the primitive herbivore *Hyracotherium*.
49. Describe how the changes in the evolving horse can be correlated with changes in the environment.
50. Why is it believed that both anagenesis and cladogenesis contributed to the evolution of the horse?

IV. **On the imperfection of the geological record** (pages 942–943)
51. What kind of evidence did Darwin fail to find in the fossil record?
52. What reason did Darwin advance to explain why the fossil record did not give strong support to his theories?
53. What is the current state of the fossil record? Until very recently, how has this state been interpreted?

V. **Punctuated equilibria** (pages 943–944)
54. Describe the fossil records of the species studied by Eldredge and Gould. What explanation did they give for the "flaws" in the record?
55. How can the mechanism of allopatric speciation account for rapid evolutionary changes?
56. What is the chief difference between the theory of punctuation, as it was first proposed in 1972, and neo-Darwinism?
57. How was the punctuated equilibrium model modified by the radical ideas of Steven Stanley?

VI. **Summary** (page 945): Read the Summary. If you are familiar with the essential features of the material presented there, you are ready to take the following diagnostic examination.

TESTING YOUR UNDERSTANDING

After you have completed the following examination, compare your answers with the annotated key in the Analysis section.

1. Which of these best demonstrates the effects of microevolution?
   a. ecotypes of *Potentilla*
   b. formation of separate marsupial and placental lines
   c. a single species radiating into a number of unoccupied habitats
   d. development of three genera through anagenesis and cladogenesis

2. Which of these geographic settings is potentially capable of maintaining the genetic isolation of two populations?
   a. an oceanic island
   b. a mountaintop
   c. an isthmus
   d. a forest
   e. Only a and c are correct.
   f. The first four answers (a, b, c, and d) are correct.

3. The result of a cross between plants of different species may develop into a fertile hybrid as a result of:
   a. postmating isolating mechanisms.
   b. polyploidy.
   c. meiosis.
   d. further crossbreeding with one of its parents.

4. Which of these is considered to be the most common method by which speciation occurs?
   a. allopatric speciation
   b. disruptive selection
   c. hybridization
   d. directional selection

5. Which of the following is true regarding reproductive isolating mechanisms?
   a. Premating mechanisms reinforce postmating mechanisms.
   b. Postmating mechanisms are often of a temporal nature.
   c. Postmating mechanisms are rarely tested in nature.
   d. Postmating mechanisms generally involve a visual display or some elaborate behavior.
   e. All of the above are correct.

6. Which of the following may serve as a premating isolating mechanism?
   a. sterility
   b. failure to reach sexual maturity
   c. anatomical incompatibilities of the genitalia
   d. pheromones
   e. Both b and c are correct.

7. Which of the following is an example of adaptive radiation?
   a. presence of *Mesosaurus* on distant continents
   b. diversification of mammals following extinction of dinosaurs
   c. splitting of the polar bear from the brown bear
   d. evolution of the horse
   e. Both a and b are correct.

8. Darwin showed that the thirteen species of finches present on the Galapagos islands differ in:
   a. their ability to fly long distances.
   b. the size and shape of their beaks.
   c. their ancestry.
   d. the type of food they eat.
   e. Both b and d are correct.

Questions 9 and 10 refer to the diagram below.

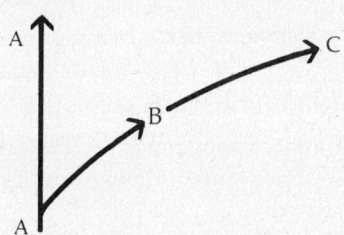

9. The formation of species C from species B is certainly an example of:
   a. phyletic evolution.
   b. cladogenesis.
   c. microevolution.
   d. allopatric speciation.
   e. sympatric speciation.

10. The formation of species B from species A is certainly an example of:
    a. cladogenesis.

b. microevolution.
   c. anagenesis.
   d. sympatric speciation.
   e. allopatric speciation.
11. The opening of biological frontiers plays a major role in:
    a. anagenesis.
    b. microevolution.
    c. extinction.
    d. adaptive radiation.
    e. the founder effect.
12. Which of the following represents adaptations of modern horses to the present grassland habitat?
    a. grinding teeth
    b. eyes high in the head
    c. a spring-footed gait
    d. reduction in number of toes
    e. All of the above are correct.
13. All models of speciation propose that speciation cannot occur unless there is _____ isolation of the two populations.
    a. geographic
    b. genetic
    c. sympatric
    d. Both a and b are correct.
14. Which of these people referred to the fossil record as "a history of the world imperfectly kept"?
    a. George G. Simpson
    b. Ernst Mayr
    c. Stephen Gould
    d. Steven Stanley
    e. Charles Darwin
15. Which of these subjects is *not* specifically discussed in *The Origin of Species*?
    a. the imperfection of the fossil record
    b. the relation of artificial selection to natural selection
    c. the process by which speciation has occurred
    d. the processes by which microevolution has occurred
16. The theory of punctuated equilibrium assumes that:
    a. there have been long periods during which relatively little change has occurred.
    b. evolution is accounted for by periods of rapid changes.
    c. the fossil record may be an accurate reflection of the changes that have occurred.
    d. speciation could occur over a period of just 10,000 years.
    e. All of the above are correct.
17. Steven M. Stanley has radicalized the punctuated equilibrium model of evolution by suggesting that:
    a. little of what Darwin originally said is still valid.
    b. speciation is the principal mode of evolution.
    c. natural selection acts on species as well as on individuals.
    d. adaptive radiation has probably never occurred except on a small scale.
18. According to the theory of plate tectonics, plate boundaries:
    a. separate the continents from the oceans.
    b. are regions of intense geologic activity.
    c. are continually moving away from one another to create more ocean area.
    d. are the primary areas where fossils are found.
19. Synthetic sex pheromones introduced to control budworm populations unexpectedly:
    a. broke down premating isolating mechanisms between the budworm and bollworm.
    b. produced more matings than would normally have occurred.
    c. isolated the gene pool of the budworm from that of the bollworm.
    d. accounted for the introduction of new and damaging pests.
    e. Both b and d are correct.
20. Match each term with a description or example.

    cladogenesis _____
    premating isolating mechanisms _____
    punctuated equilibrium _____
    sympatric speciation _____
    anagenesis _____
    allopatric speciation _____

    a. change within a lineage through directional selection
    b. hybrid polyploidy
    c. pheromones
    d. sudden appearance of new species
    e. lineage branches
    f. rapid diversification associated with vacating of environments by other species

adaptive radiation   g. geographic barriers

21. What evidence does biogeography offer to support the theory of continental drift?
22. Explain how the course of evolution might be consistent with the fossil record.
23. Explain how both anagenesis and cladogenesis could have been responsible for the evolution of the thirteen species of Galapagos finches from a mainland ancestor.

PERFORMANCE ANALYSIS

1. Microevolution refers to genetic changes within populations, that is, to changes below the species level. The ecotypes of *Potentilla* have different genotypes, but they are still able to interbreed; therefore, **a** is correct. Macroevolution concerns itself with evolution above the species level, that is, with speciation. Choices **b** and **d** therefore fall within the realm of macroevolution. Choice **c** is a process that might eventually result in either microevolution or macroevolution.

2. Choices a through d are all correct (**f**). Oceanic islands can isolate terrestrial populations as can a submerged isthmus. An emerged isthmus can separate marine populations. Forest-dwelling organisms may be unable to migrate (or disperse their pollen) across a large plain that separates wooded areas. For a plant that is specialized for living at high altitudes, a mountaintop may be an island.

3. Normally, hybrids are infertile because the chromosomes from the two parent strains do not pair in meiosis. If, however, the chromosome number should double, as occurs in polyploidy (**b**), meiosis could occur normally, and the hybrid could give rise to a fertile line (see Figure 50-4).

4. Most cases of documented speciation have involved the geographic isolation of two populations (allopatric speciation, **a**) followed by genetic divergence.

5. In nature, premating isolating mechanisms are generally sufficient to maintain sexual isolation between closely related species, so that postmating mechanisms are rarely tested (**c**). Individuals of two species that happen to interbreed are subject to these postmating mechanisms, which act to reinforce premating mechanisms, ensuring that such individuals do not have the opportunity to pass on the tendency to mate outside the species.

6. Choices a, b, and c represent postmating mechanisms. Pheromones (**d**) are an example of a premating isolating mechanism because these chemicals are used in attracting a mate of the same species.

7. Choice **b** is correct. During the Permian period, the dinosaurs vacated a considerable number of ecological niches. (An ecological niche is a way of life; it includes types of food, place and manner of obtaining food, and living space.) Mammals quickly diversified into groups that filled the vacant niches. Choice a, which refers to fossil evidence that South America was once joined to Africa, is an example of a geographic barrier splitting a population. The speciation of the polar bear (c) is an example of cladogenesis. The evolution of the horse (d) involved both anagenesis and cladogenesis, but apparently without the rapid and almost simultaneous formation of many new species that is characteristic of adaptive radiation.

8. Darwin supposed that the thirteen species of finches on the Galapagos islands evolved from a single mainland ancestor, perhaps even from a single pregnant female, that had made its way to the Galapagos archipelago. (It is unlikely that many birds would have made the trip because South American finches are not able fliers.) Darwin found distinct differences among the finch species in the size and shape of the beak (b) and associated these differences with the types of food (d) eaten by the various species. Choice **e** is correct.

9. Species C is in the same phylogenetic lineage as species B. Such direct descent is the result of the gradual process known as phyletic evolution (**a**). The diagram supplies no information regarding choices d and e.

10. Species B has split from the phylogenetic lineage of species A. This type of evolution is called cladogenesis (**a**). Again, there is no information regarding d and e.

11. See the analysis for Question 7. Adaptive radiation (**d**) is the relatively rapid evolution of a single species into a variety of groups with different habitats and life styles. Two prerequisites for adaptive radiation are (1) the opening of biological frontiers (ecological niches) and (2) the genetic capacity to evolve into species that can fill the niches.

12. All of these characteristics (**e**) are adaptations of modern horses to life on the grasslands. Grinding teeth are useful in chewing the coarse leaves typical of monocots. The position of the eyes high in the head is useful for detecting predators while grazing. A spring-footed gait, facilitated by a reduction in the number of toes (horses have only one), is an adaptation for transporting a large body over relatively hard ground.

13. Speciation cannot take place unless the gene pools of populations become distinct—in other words, unless the populations become genetically (**b**) isolated.

14. Darwin (**e**) referred to the fossil record as an imperfect history of the world because of the scarcity of intermediate forms, which made it difficult to document gradual evolutionary change. Mayr, Gould, and Stevens suggest that the fossil record *is* an accurate reflection of the process of evolution because changes have been abrupt and not gradual. Simpson emphasizes that adaptive radiation is the major pattern of macroevolution.

15. In *The Origin of Species*, Darwin speculated on how changes might have occurred within populations but ignored processes of speciation (**c**).

16. Niles Eldredge and Stephen Jay Gould have proposed that the fossil record is an accurate reflection of the course of evolution. They argue that there are long periods during which little evolutionary change occurs and that these long stretches are punctuated by short periods of rapid change; this accounts for the abrupt appearances and disappearances of fossil species. Followers of this theory of punctuated equilibrium also claim that speciation can occur in as short a time as 10,000 years. Therefore, **e** is correct.

17. In order to explain the existence of a number of closely related species at the same point in the fossil record, Steven Stanley suggests (as Mayr did before him) that speciation is the principal mode of evolution and, more radically, that natural selection acts on species in the same way that it acts on individuals. Choice **c** is correct.

18. According to the theory of plate tectonics, the earth's crust is divided into plates that move about, pull apart, slide against one another, and move under and over one another. At the boundaries of adjacent plates, earthquakes and volcanic activity occur. Choice **b** is correct.

19. The pheromones caused individuals of the two species to enter into a copulatory embrace from which they could not disengage. Choice **a** is correct.

20. The correct answers in order are **e**, **c**, **d**, **b**, **a**, **g**, and **f**.

21. Biological evidence for continental drift can be seen in the distribution of marsupials on the Australian and Antarctic continents and the presence of *Mesosaurus* in South Africa and Brazil. (See The Breakup of Pangaea, pages 930–931.)

22. Your discussion should follow the lines of the analysis for Question 16.

23. Adaptive radiation, such as that which produced Galapagos finches, generally involves both cladogenesis and anagenesis. Birds on a single island could have evolved many times to assume their present form. Different species living on different islands attest to cladogenesis (assuming there was but a single ancestral species).

SECTION 8 Ecology

CHAPTER 51

Population Dynamics and Life-History Strategies

CHAPTER ORGANIZATION

I. Introduction
II. Properties of populations
  A. Introduction
  B. Patterns of growth
  C. Carrying capacity
  D. Mortality patterns
  E. Age structure
  F. The asexual advantage
    1. Vegetative reproduction
    2. Parthenogenesis
III. Life-history strategies
  A. Introduction
  B. The alternatives
    1. Introduction
    2. Early or late
    3. Small or large
  C. Some consequences
IV. Summary

MAJOR CONCEPTS

Ecology is the study of the interactions of organisms with each other and with their environment. In ecology, a population may be defined as a group of organisms sharing a location and belonging to the same gene pool.

The carrying capacity of a population is the number of individuals of that particular species that can be supported by local resources.

There are two basic patterns of population growth. The first is exponential growth, in which the population increases at a constant rate apparently without limits. This pattern is seen among bacteria cultured in the laboratory and, for relatively short periods, among opportunistic species colonizing a newly opened resource. The second pattern of growth is represented by a sigmoid (S-shaped) curve, in which the population increases until it reaches the carrying capacity, then stabilizes.

The mortality pattern of a population indicates the relationship between age and risk of death. The age structure of a population is the proportion of individuals of different ages.

Asexual reproduction can increase the total production of young at the cost of loss of genetic variability among the offspring.

Life-history strategies are genetically determined and subject to natural selection. At one extreme are r-selected (prodigal or opportunistic) strategies—many young; small young; rapid maturation; high early mortality rate; little or no parental care; and reproduction once in a lifetime. At the other extreme are K-selected (prudent or equilibrium) strategies—few young; large young; slow maturation; intensive parental care; and reproduction many times during a lifetime.

HOW TO STUDY THE CHAPTER

Read the chapter, focusing on the major concepts.

Reread the chapter, using the questions that follow to help you focus on the details as they relate to major concepts. Answer the questions on a separate sheet of paper. Your answers will provide a valuable study aid in preparing for examinations.

FOCUS ON CHAPTER DETAILS

I. **Introduction** (page 951)
   1. Define ecology.
   2. What information do ecologists seek?
   3. Why is ecology considered to be both a very old and a very young discipline?
   4. Name and define the levels of biological organization that are studied by ecologists.

II. **Properties of populations** (pages 951–956)
   A. Introduction
   5. What properties of life emerge at the population level that do not apply to individual organisms?
   B. Patterns of growth
   6. How is the rate of increase of a population expressed?
   7. What is a population's intrinsic rate of growth?
   8. Define exponential growth, and write a simple formula that expresses it.
   9. Look at the form of the exponential growth curve of Figure 51-2. If r is a constant, why does the slope of this growth curve increase as the population expands?
   10. Cite some examples of species that might grow exponentially under certain conditions.
   11. What factors prevent populations from continuing to increase exponentially?
   12. What is an opportunistic species? Cite some examples of opportunists.
   13. Why do certain pest populations often "crash"? How can such populations be controlled?
   C. Carrying capacity
   14. What is the carrying capacity of a population? What usually determines the carrying capacity for animal populations? For plant populations?
   15. Cite several types of seasonal variations that can limit a population's size.
   16. Write an equation that expresses limited growth of a population. Explain each term of the equation. Why is such a growth pattern represented by an S-shaped (sigmoid) curve? (See Figure 51-4.)
   17. According to the sigmoid growth curve, what is the most practical way of controlling vermin? Why? What does the curve suggest for harvesting?
   D. Mortality patterns
   18. Compare the mortality patterns of sparrow, oyster, and *Hydra* populations.
   E. Age structure
   19. What is the age structure of a population? How is it useful in making predictions?
   20. Why has the U.S. population continued to increase despite the current decrease in the birth rate?
   F. The asexual advantage
   Vegetative reproduction
   21. Why are the chances of survival better for a plantlet produced by vegetative growth than for one produced from a seed?
   Parthenogenesis
   22. What is parthenogenesis? What are its advantages over vegetative methods of asexual reproduction? Its disadvantages?
   23. Name some organisms that reproduce parthenogenetically.
   24. Under what conditions do some organisms alternate asexual phases with sexual ones?

III. **Life-history strategies** (pages 956–960)
   A. Introduction
   25. How does the biological meaning of the word "strategy" differ from its everyday meaning?
   26. What is a life-history strategy?
   27. Why do life-history strategies vary within populations as well as from one population to another?

B. The alternatives

Introduction

28. What terms have been used to classify alternative life-history strategies?
29. Compare the two extreme types of life-history strategies.
30. What factors determine which strategy along a continuum of possibilities will be exploited by a population?

Early or late

31. Using the examples in the text, explain what factors might determine the reproductive strategies used by organisms.
32. Describe precocious development in the fungus-eating gall midge. What happens when the parthenogenic cycle breaks?
33. Why is precocity risky for the larger mammals?

Small or large

34. Make a general comparison of seed size in goldenrod growing in an old field and goldenrod growing on a prairie. Which of the two populations is nearer carrying capacity?
35. What is the relationship between the size and the number of seeds produced by a given plant? Are goldenrod seeds more *r*-selected in an old field or in a prairie?
36. How would you categorize the two genotypes present in Solbrig's survey of Ann Arbor dandelions?

C. Some consequences

37. What are the long-term risks encountered by individuals of *r*-selected species? Of *K*-selected species? What risks are encountered by entire *r*-strategy and *K*-strategy populations?

IV. **Summary** (page 960): Read the Summary. If you are familiar with the essential features of the material presented there, you are ready to take the following diagnostic examination.

TESTING YOUR UNDERSTANDING

After you have completed the following examination, compare your answers with the annotated key in the Analysis section.

1. A population that grows exponentially has a constant:
   a. size once it reaches K.
   b. equilibrium size.
   c. rate of increase.
   d. slope to its growth curve.

2. The formula for exponential growth can be modified to represent resource-limited growth by including the term:
   a. $(K-N)/K$.
   b. $(K-N)/N$.
   c. DN.
   d. rN.

3. As the growth curve of a resource limited population begins to level off, _____ approaches _____.
   a. $(K-N)/K$; 1
   b. N; K
   c. r; K
   d. dN/dt; N
   e. None of the above are correct.

4. The birth rate minus the death rate determines a population's:
   a. age structure.
   b. fecundity.
   c. longevity.
   d. intrinsic rate of increase.
   e. carrying capacity.

5. A population that exhibits constant mortality for all age groups has a survivorship curve in the form of a:
   a. convex curve.
   b. concave curve.
   c. straight line of positive slope.
   d. straight line of negative slope.
   e. horizontal line.

6. A population will eventually reach a stable age structure if:
   a. $r = 0$.
   b. $r = 1$.

   c. *r* remains constant.
   d. its male/female ratio is 1:1.
   e. mortality approaches zero.
7. Organisms that reproduce by vegetative methods have _____ than those that reproduce parthenogenetically.
   a. higher juvenile survival rates
   b. a need for more gametes
   c. less genetic similarity
   d. a better chance for dispersal
   e. higher reproductive rates
8. G. E. Hutchinson classified life-history strategies as either prodigal or prudent. These strategies are also called _____ and _____.
   a. *r*-selected; *K*-selected
   b. opportunistic; equilibrium
   c. asexual; sexual
   d. precocious; mature
   e. Both a and b are correct.
9. Which of these is typically a characteristic of a *K*-selected species?
   a. a single reproductive effort
   b. large young
   c. precocious development
   d. mechanisms for wide dispersal
   e. Both b and d are correct.
10. A weed that takes root in a field after harvest usually will produce:
   a. many large seeds.
   b. many small seeds.
   c. a few large seeds.
   d. a few small seeds.
   e. The life-history strategy of such a species is far too variable to support such a general statement.
11. A group of coadapted traits affecting reproductive survival is known as a:
   a. reproductive adaptation.
   b. genetic isolating mechanism.
   c. life-history strategy.
   d. coadaptive gene complex.
12. A *K*-selected population is often at greater risk than an *r*-selected one because a *K*-selected population:
   a. is genetically less variable.
   b. may have difficulty recovering from catastrophic mortality.
   c. has high age-specific mortality rates.
   d. is more likely to outstrip its resources.
13. Which of the following organisms would leave *no* offspring if it were unable to complete metamorphosis?
   a. a gall midge
   b. a fly
   c. a gypsy moth
   d. Only b and c are correct.
   e. A, b, and c are correct.
14. In a species in which the female is homogametic (XX), parthenogenesis would result in _____ offspring.
   a. all female
   b. all male
   c. both male and female
   d. hermaphroditic
15. A species that overshoots its carrying capacity will certainly:
   a. become extinct.
   b. turn to alternative resources.
   c. "crash."
   d. It is not possible for a species to overshoot its carrying capacity.
16. On three separate graphs, draw (a) the growth curve of an *E. coli* population living under optimum conditions in a laboratory culture; (b) the age pyramid of a typical underdeveloped country; and (c) the survivorship curve of a population of oysters. Label the axes of the graphs appropriately, and interpret the data depicted.

17. What methods of asexual reproduction provide a population with a distinct advantage over a similar population that reproduces sexually? When might an asexually reproducing population begin to reproduce sexually?

The chart below shows the current population of India, grouped by age and sex. The total number of people is approximately 600 million. Let us assume that the goals of those fighting disease, famine, and other evils of poverty and overpopulation in India are suddenly realized: the young no longer die—death becomes, to all intents and purposes, a function primarily of age. Under such circumstances, the chart, over the course of time, would reflect the continuous elimination of those in the oldest age groups. Assuming this basic change in death rates, use the chart below to answer Questions 18–20.

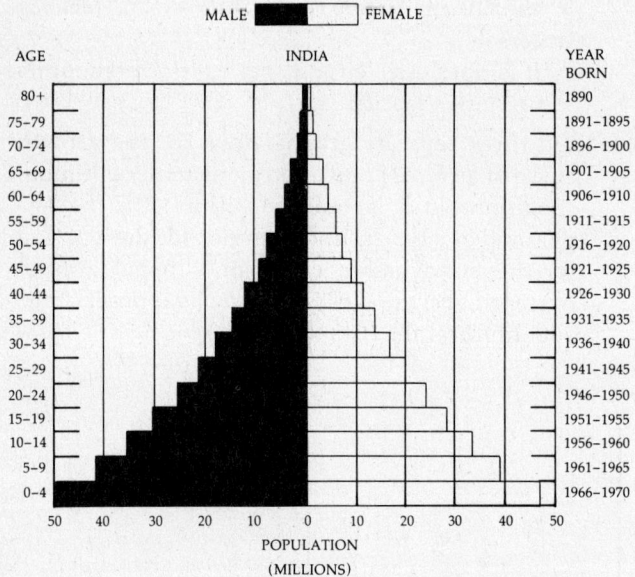

18. Suppose that there were no more births from the moment this chart was drawn. How much smaller would the population of India be 20 years from now?

19. Suppose that for the next 20 years, each male-female pair in the 20–24 age group had exactly two children (just enough to replace themselves).

   a. What would the bottom four bars of the graph of the population look like 20 years from now?
   b. Would the population in 20 years be smaller or larger than it is now?
   c. What would the bottom four bars of the graph look like 40 years from now?

20. Suppose each male-female pair in the 20–24 age group had only one child. Would the population 20 years from now be smaller or larger?

PERFORMANCE ANALYSIS

1. Exponential growth is described by the equation $dN/dt = rN$ and by the graph in Figure 51-2. In this equation, r, the rate of increase, is a constant, and the growth rate of the population, represented by dN/dt, depends on N, the size of the population. The slope of an exponential growth curve is ever-increasing. Choice **c** is correct.

2. The carrying capacity of a population is represented by K. As N approaches K, there are fewer resources per individual in the population, and its growth rate slows down. Incorporating the term $(K-N)/K$ into the exponential equation reflects this slowdown in growth rate; therefore **a** is correct.

3. It can be seen from the equation for growth under resource-limited conditions, $dN/dt = rN(K-N)/K$, that as N approaches K, the term $(K-N)/K$ approaches zero, as does dN/dt. This is indicated by the flattening of the curve in the graph of Figure 51-4. Choice **b** is correct.

4. In the absence of net emigration or immigration, the intrinsic rate of increase (**d**) of a population is determined by the difference between the birth rate and the death rate. Rate of increase can be defined as the increase in the number of individuals in a given unit of time per individual present.

5. Choice **d** is correct. See the survivorship curve of *Hydra* in Figure 51-5.

6. A stable age structure will be reached when a population is no longer increasing in number, that is, if r, the rate of increase, equals zero. Choice **a** is correct.

7. Individuals produced from a parent by vegetative reproduction are genetically identical. Partheno-

genesis, because it requires meiosis for gamete production, can result in offspring that are variable. The juvenile survival rate is higher (**a**) among organisms produced by vegetative reproduction because the parent provides nutrition for the developing offspring. Parthenogenetic organisms trade off individual survival for greater reproductive potential (larger numbers of offspring) and wider dispersal of the young.

8. A prodigal life-history strategy has also been termed r-selected and opportunistic and applies to populations that are good colonizers and have a high rate of increase. A prudent strategy can also be called an equilibrium and a K-selected strategy and applies to populations that reach equilibrium size at or near carrying capacity. Choice **e** is correct.

9. Choice **b** is correct. Table 51-2 lists the characteristics of r-selected and K-selected species.

10. Plants that colonize open habitats usually follow r-strategies, which allow them to exploit new opportunities quickly. A large number of small seeds is characteristic of r-selected plants. Choice **b** is correct.

11. Choice **c** is correct.

12. If a K-selected population is reduced to a low population density, it may have difficulty returning to its equilibrium size because of its low rate of increase; therefore, **b** is correct. Choices a, c, and d apply to r-selected populations.

13. Many insect species must undergo metamorphosis from larva to adult before they reproduce. Such is the case for both flies and moths (**d** is correct). The gall midge is unusual in that it can skip both fertilization and metamorphosis, thereby producing a new generation very rapidly.

14. If the female of a species is homogametic (XX), she can produce only gametes having an X chromosome. Eggs that develop parthenogenetically have only X chromosomes and therefore develop to be females (**a**).

15. Species that reproduce at an exponential rate may overshoot their carrying capacities. When resources are exhausted, large numbers of individuals die, and the population is said to experience a "crash" (**c**). The population may become extinct if there are no survivors, if recolonization from an adjacent area does not take place, or if the remaining individuals cannot reproduce quickly enough to bring the population to a stable level. However, extinction is not a certainty.

16. Your graphs should resemble (a) Figure 51-2; (b) Figure 51-6; and (c) Figure 51-5. Refer to the captions of the figures for the interpretations.

17. Vegetative reproduction usually involves nutritional support from the parent and maximizes the probability of juvenile survival. Parthenogenesis results in large numbers of offspring and usually provides for wide dispersal of the young. Asexual reproduction tends to increase the number of individuals and is therefore advantageous when a new resource becomes available. However, asexual methods do not allow for genetic flexibility, since individuals are genetically similar. When environmental conditions change, an asexual population with the ability to reproduce sexually will begin to do so.

18. The population would be reduced by only 15 to 20 million people.

19. a. They would be a repeat of the bottom four bars in today's graph.
    b. The population would be very much larger.
    c. They would be the same as the bottom four bars in today's graph.

20. It would still be larger. You can see from the chart that even half the number of people of childbearing age is still larger than the number of people in the older categories.

CHAPTER 52

Interactions in Communities: Competition, Predation, and Symbioses

CHAPTER ORGANIZATION

I. **Introduction**

II. **Competition**
   A. Introduction
   B. The principle of competitive exclusion
      1. General information
      2. Testing Gause's principle
   C. The ecological niche
      1. General information
      2. Fundamental and realized niche
      3. Niche shift
   D. Family rivalries
   E. The African ungulates
   F. Winner takes all

III. **Predation**
   A. Introductory remarks
   B. Predation and numbers
   C. Escape from predation
   D. Predation and species diversity
   E. The arms race
      1. Introduction: Chemical defenses
      2. Bats, moths, and biosonar

IV. **Symbiosis**
   A. Introductory remarks
   B. Parasitism
   C. Mutualism
      1. Background
      2. Ants and acacias

V. **The number of species**

VI. **Summary**

MAJOR CONCEPTS

An ecological community can be defined as a group of locally interacting populations. Three primary types of interaction between the species in a community are competition, predation, and symbiosis.

According to Gause's principle of competitive exclusion, if two species in a community are competing for an identical set of resources, one species will eventually exclude the other species from the community.

Each species in a community occupies a particular ecological niche, which is the way of life of the species, including the resources it uses and the ways in which it uses them. A fundamental niche represents the resources that a species is theoretically capable of utilizing; a realized niche represents the resources it actually utilizes. Niche overlap refers to the situation in which two or more species utilize the same limited resource and may cause the realized niche of a species to be smaller than its fundamental niche. Character displacement may result from the selection pressure against niche overlap.

Interactions between predator and prey populations may increase species diversity, may result in coevolution of the two populations, and/or may affect the growth of both populations.

The three forms of symbiosis are parasitism (one partner benefits and the other is harmed), mutualism (both partners benefit), and commensalism (one partner benefits and the other is not affected).

The number of species in a community apparently reaches an equilibrium level at which the species immigration and extinction rates are equal.

HOW TO STUDY THE CHAPTER

Read the chapter, focusing on the major concepts.

Reread the chapter, using the questions that follow to help you focus on the details as they relate to major concepts. Answer the questions on a separate sheet of paper. Your answers will provide a valuable study aid in preparing for examinations.

FOCUS ON CHAPTER DETAILS

I. Introduction (page 962)

1. Define an ecological community.
2. What characteristics of populations are influenced by interactions at the community level?
3. What image did G. E. Hutchinson use to describe the relationship between ecology and evolution?

II. Competition (pages 962–968)

A. Introduction

4. How does an ecologist define competition?
5. What resources may be limiting for plants? For animals?
6. Distinguish between interspecific and intraspecific competition.
7. What factors affect the intensity of competition between two populations?
8. What fundamental question does the science of ecology attempt to answer?

B. The principle of competitive exclusion

General information

9. State the principle of competitive exclusion as formulated by Gause.
10. How did Gause go about testing this principle? What were the results?
11. What concept is demonstrated by the outcome of competition between *Lemna gibba* and *L. polyrrhiza*?
12. How would it be possible to alter the outcomes of the *Lemna* and *Paramecium* experiments? In what way is the outcome predictable regardless of alterations?

Testing Gause's principle

13. What mechanism of coexistence did Robert MacArthur discover in closely related species of warblers living in the same forest habitat?
14. How do similar species of mosses of the genus *Sphagnum* coexist within their bog environment?

C. The ecological niche

General information

15. Provide a simple definition and a more complex working definition of a niche. Distinguish between an organism's niche and its habitat.

Fundamental and realized niche

16. What advantages do barnacles offer to scientists studying the effects of competition on resource utilization?
17. How are the two dominant species of intertidal barnacles distributed within the intertidal habitat? What factors did Connell show to be important in determining this distribution?
18. Describe the fundamental and realized niches of both barnacle species.
19. What is niche overlap?

Niche shift

20. Explain the phenomenon of character displacement.
21. How do Darwin's finches exemplify character displacement?

D. Family rivalries

22. What sets Henry Horn's study of the competition of trees for light apart from other studies of competitive interactions?

23. What did Horn conclude about the relative competitive advantages of multilayered and monolayered trees?

E. The African ungulates

24. Explain how a number of diverse browsing species are able to coexist in an East African ecosystem.
25. How do the grazing species partition the available resources?

F. Winner takes all

26. Why are examples of competitive exclusion in nature more difficult to document than those of coexistence?
27. Describe two well-known examples of competition that resulted in the elimination of one or more species. What was the limiting resource in each example?

III. Predation (pages 968–974)

A. Introductory remarks

28. What is a food web? (Figure 52–10 gives you an idea of what a relatively simple food web looks like. See page 981 for a formal definition.)
29. What are the influences of predation on various aspects of the community?

B. Predation and numbers

30. What influence may predation have on a prey population?
31. What individuals in a prey population are most susceptible to predation?
32. Use two examples to demonstrate that the size of the prey population has a major impact on the carrying capacity of the predator population.
33. What three interpretations have been offered to explain the 10-year cycles in snowshoe hare and lynx population densities?
34. Prey species may be brought under control by the introduction of alien predators. Describe a prominent example.
35. Why are predator-prey relationships often more complex than those described so far?

C. Escape from predation

36. What are some adaptations of prey organisms that help them elude their predators? (Pay special attention to the two temporal strategies described in the text.)

D. Predation and species diversity

37. Cite three examples of the influence of predation on species diversity.

E. The arms race

Introduction: Chemical defenses

38. What are the coevolutionary adaptations that have resulted from the predator-prey relationship between the *Eleodes* beetle and the grasshopper mouse? Between legumes and bruchid weevils?

Bats, moths, and biosonar

39. Describe the experimental work that led to our knowledge of how bats navigate in the darkness (echolocation).
40. What adaptations do moths exhibit for avoiding predation by insectivorous bats?

IV. Symbiosis (pages 974–977)

A. Introductory remarks

41. What is symbiosis? What three types of symbiotic relationships are recognized?

B. Parasitism

42. How may parasitism be likened to predation?
43. What influences are infectious diseases apt to have on a population?
44. Why is it disadvantageous to both host and parasite for the parasite to be too virulent?
45. Use the example of the myxoma virus and Australian rabbits to explain how less virulent strains of parasites are selected for.

C. Mutualism

Background

46. Cite some examples of important mutualistic relationships.

Ants and acacias

47. What structural adaptations are found

among acacia trees that grow where there are large herbivores?

48. What benefit do ants derive from their association with acacias?
49. How do acacias benefit from their association with ants?
50. What happens to acacia trees if their ant symbionts are experimentally removed?
51. Describe the case of mimicry that has coevolved with the ants and acacias.

V. **The number of species** (pages 977–978)

52. Describe the model proposed by MacArthur and Wilson to determine the number of species that can live in a given community.
53. How did Wilson and Simberloff go about testing the MacArthur-Wilson model?
54. What were the results of the mangrove experiment? Why do these results support the model?
55. How is the equilibrium concept supported by the change in the distribution of American mammals after the emergence of the Isthmus of Panama?

VI. **Summary** (pages 978–979): Read the Summary. If you are familiar with the essential features of the material presented there, you are ready to take the following diagnostic examination.

TESTING YOUR UNDERSTANDING

After you have completed the following examination, compare your answers with the annotated key in the Analysis section.

1. According to Gause's principle, if two species are competing for the same set of limited resources:
   a. neither will be able to survive.
   b. one must become extinct.
   c. one will be excluded from part or all of its niche.
   d. character displacement will be observed.
   e. only the fastest growing species will persist.
2. Robert MacArthur showed that five species of New England warblers were able to coexist primarily because:
   a. they searched for food in different kinds of trees.
   b. they occupied different feeding zones within a single tree.
   c. they ate different kinds of food.
   d. they were active at different times of day.
   e. Both a and c are correct.
3. An organism's niche is most simply defined by:
   a. its behavior.
   b. what it eats.
   c. its role in the ecosystem.
   d. its interactions with other organisms.
4. Two species of Galapagos finches differ more in beak size when they inhabit the same island than when they inhabit different islands. This phenomenon is referred to as:
   a. competitive exclusion.
   b. character displacement.
   c. species diversity.
   d. mutualism.
   e. adaptive radiation.
5. In Connell's study of intertidal barnacles, the _____ niche of _____ was determined by interactions with a competing species.
   a. realized; *Chthamalus*
   b. realized; *Balanus*
   c. fundamental; *Chthamalus*
   d. fundamental; *Balanus*
6. Which of the following pairs of species coexist as a result of the partitioning of resources along the vertical dimension of the habitat?
   a. bluebirds and starlings
   b. rhinos and gerenuks
   c. *Paramecium aurelium* and *P. caudatum*
   d. bog species of *Sphagnum* moss
   e. Both b and d are correct.
7. Which of the following statements about the influence of predation on a prey population is *not* correct?
   a. Prey organisms in poor physical condition are taken by predators in disproportionately high numbers.
   b. Predation is a major factor in determining species diversity within a community.
   c. Most prey populations are limited below their carrying capacities by their predators.
   d. Age-specific predation may influence life-history strategies of the prey.

8. Which of the following represents a temporal strategy of prey organisms for reducing the impact of predation?
   a. synchronization of births
   b. noxious secretions
   c. camouflage
   d. warning coloration
   e. evasive behavior

9. Moths avoid predation by bats by:
   a. detecting bat cries.
   b. jamming the bat's sonar.
   c. flying in the opposite direction.
   d. diving to the ground.
   e. All of the above are correct.

10. The association between the components of a lichen is an example of:
    a. parasitism.
    b. mutualism.
    c. commensalism.
    d. resource partitioning.
    e. competitive exclusion.

11. The introduction of rabbits infected with myxoma virus onto the Australian continent:
    a. led to the complete elimination of Australian rabbits.
    b. had little effect on the rabbit problem.
    c. stabilized the rabbit population at a lower number of individuals.
    d. caused a virulent myxoma strain to infect sheep.

12. Ants and the acacia trees they inhabit are an example of mutualism. The ants provide _____ for the acacia, and the trees supply the ants with _____.
    a. nutrients; a refuge from predators
    b. protection from herbivores; sugar
    c. relief from competition; nesting sites
    d. mycorrhizae; water
    e. Both b and c are correct.

13. After fumigating mangrove islands, Wilson and Simberloff allowed the islands to be recolonized. Which of these results did they see?
    a. The islands returned to their former species composition.
    b. The islands regained the same equilibrium number of species.
    c. The island fauna was much more diverse than before.
    d. The island fauna was much less diverse than before.

14. Following the emergence of the Isthmus of Panama 2 million years ago, the number of mammalian families in North America:
    a. remained stable.
    b. initially increased, then returned to its original level.
    c. increased and stabilized, although the sizes of the individual populations were reduced.
    d. showed a steady decline due to extinction.

15. MacArthur and Wilson have tried to answer why a given community holds a certain number of animal species by using _____ as model ecosystems.
    a. lakes and ponds
    b. oceanic islands
    c. land bridges
    d. artificial habitats
    e. forests

16. When a starfish predator was experimentally removed from a rocky intertidal community:
    a. certain prey species were eliminated.
    b. species diversity greatly increased.
    c. species composition remained the same, though all prey populations increased in size.
    d. competition among prey species was greatly reduced.

17. Based on Henry Horn's data on the photosynthetic rates of leaves of multilayered and monolayered trees, monolayered trees would be better competitors in:
    a. an open woodland.
    b. an understory of a forest.
    c. a fence row.
    d. Both a and c are correct.

18. Competition is expected to be most intense between two species that occupy the same:
    a. ecosystem.
    b. food web.
    c. trophic level.
    d. habitat.
    e. community.

19. Match each group of organisms with the phenomenon the group exemplifies.

snowshoe hares _____

intertidal barnacles _____

legumes that produce chemical inhibitors _____

mycorrhizae _____

Paramecium aurelia and *P. caudatum* _____

Galapagos finches _____

the starfish *Pisaster* _____

human hunters _____

a. competitive exclusion leading to disappearance of one species
b. character displacement
c. realized niche determined by a competitor
d. fluctuating population size correlated with changes in predator population size
e. reducing population fitness by unwise artificial selection
f. adaptation of a prey population in response to predation
g. predation maintains species diversity
h. mutualism

(b)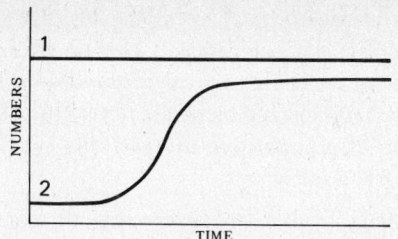

(c)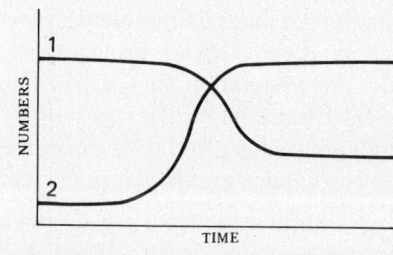

(d)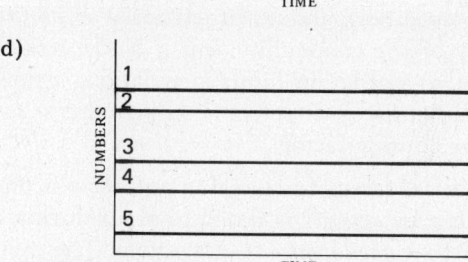

20. Describe in detail two examples of the coexistence of similar species in the same area.

21. Cite two examples in which predators help to maintain species diversity. Explain what happens to prey diversity if predators are experimentally removed.

22. Compare the relationship of a parasite and its host with that of a predator and its prey.

23. What do each of the following growth curves tell you about the species involved? How many niches are depicted in each situation?

(a)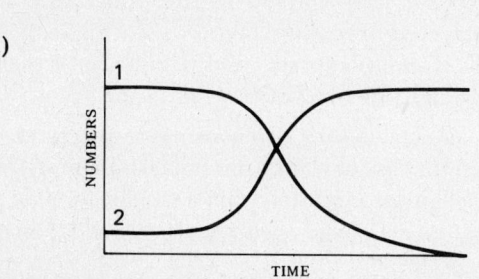

PERFORMANCE ANALYSIS

1. According to the competitive exclusion principle formulated by Gause, if two species are in competition for the same set of resources, one will eliminate, or exclude, the other; therefore, **c** is correct. Under experimental conditions, the losing species may become extinct, whereas in nature its realized niche may be narrowed.

2. Choice **b** is correct. The results of MacArthur's investigation are shown in Figure 52-4.

3. Choices a, b, and d are dimensions of a niche. A simple definition of a niche is the role that an organism plays in the ecosystem (**c**).

4. When ecologically similar species inhabit the same geographic area, there will be a selection pressure against those characteristics that overlap, such as beak size in Galapagos finches. As a result, there will be a divergence in these characteristics. This type of niche shift is referred to as character displacement (**b**).

CHAPTER **52** *Interactions in Communities: Competition, Predation, and Symbioses* **399**

5. The fundamental niche of a species is determined by its physical tolerance. Connell showed that *Chthamalus* was able to grow in the lower intertidal zone, but its realized niche did not include this zone because of competitive interactions with *Balanus*; therefore, **a** is correct.

6. Choices a and c are examples of competitive exclusion. Mosses of the genus *Sphagnum* partition resources along a moisture gradient that is related to elevation. East African browsing ungulates subdivide the vegetation in a vertical manner. Giraffes eat from tall branches, gerenuks feed on the middle branches, and rhinos browse near ground level. Choice **e** is therefore correct.

7. Predation is known to affect a prey population's size, its evolution, and its fitness, as well as the diversity of the community, but it has not been shown that predation limits populations below their carrying capacity (**c**). Food resources are often the limiting factor.

8. A temporal strategy for avoiding predation is one in which prey activity is timed to occur during a period when predators are not active. The synchronization of births (**a**) increases the likelihood of individual survival: Because there are more prey available to predators at the same time, the chances that a particular prey individual will survive are maximized. Also, the predators become less active once they are satiated.

9. Choice **e** is correct. The ears of moths are most sensitive to the frequencies of sound emitted by bats. Moths are able to produce their own high frequency sounds, which jam the bat's echolocation system. Behavioral maneuvers include evasive flying or, when the bat is closing in, diving.

10. A lichen is a symbiotic association between an alga and a fungus. Since both members benefit, the association is termed mutualistic (**b**).

11. The infected rabbits transferred the myxoma virus to the Australian rabbits, and initially the Australian rabbit population was greatly reduced. But, as is the case with parasites, a less virulent strain of the virus was selected for; the rabbit population has now reached an equilibrium level, one that is apparently tolerable to humans. Choice **c** is correct.

12. Ants nest in the thorns of acacia trees and obtain nourishment from sugars in the nectaries and from proteins and oils in the Beltian bodies. Ants are aggressive toward herbivorous insects that feed on the trees, and the ants also maul vegetation that sprouts nearby; therefore, **e** is correct.

13. After allowing the mangrove islands to be recolonized, Wilson and Simberloff observed that the former equilibrium number of species (**b**) was again attained, although species different from the original ones were responsible.

14. Emergence of the Isthmus of Panama 2 million years ago created a land bridge between North and South America that allowed mammals to move between the two continents. As a result, the number of mammalian families increased (**b**), only to return to its former level following extinctions. This "natural" experiment is thought of as a test of the equilibrium concept of diversity proposed by MacArthur and Wilson.

15. MacArthur and Wilson's equilibrium model (see Figure 52–19) was formulated for ocean islands (**b**) of a given size. Islands are highly suitable models for terrestrial ecosystems because of their well-defined boundaries.

16. When R. T. Paine removed the principal predator (the starfish *Pisaster*) from a rocky intertidal community, the number of prey species declined from 15 to 8 (**a** is correct). Removing predators might be expected to reduce species diversity because prey populations that are held in check by predation will increase in size. The ensuing competition results in the exclusion of species.

17. Multilayered trees outcompete monolayered trees in habitats where the light intensity is great, such as open woodlands and along fence rows. Multilayered trees may be dominant in the forest canopy. But below them in the understory (**b**), monolayered trees are favored, since the inner leaves of multilayered trees would not receive enough light to support their growth.

18. Since similar species often compete for food (or light, in the case of plants), interspecific competition is usually most intense within a trophic level (**c**).

19. The correct answers in order are **d, c, f, h, a, b, g,** and **e**.

20. Some of the examples you could choose are (a) the partitioning of food resources along a vertical dimension by East African grazing and browsing ungulates; (b) character displacement in beak size and shape in Galapagos finches; (c) partitioning of a bog habitat of species of *Sphagnum* moss; and (d) occupation of different feeding zones in spruce trees by New England warblers.

21. The way in which predators maintain species diversity has been demonstrated by Paine in intertidal communities where the principal predator is the starfish *Pisaster*. When *Pisaster* was removed, the number of prey species dropped from 15 to 8, and the sizes of the mussel and barnacle populations (two of the prey species) increased considerably. A second example is drawn from chalky soils in England where grasses grow more vigorously than flowering annuals do. Severe reduction of the predator (rabbit) population was followed by extirpation of the flowering annuals. A third example is provided by the experiments done by Lubchenco in a community of algae grazed by snails.

22. Predation and parasitism, with regard to their effects on prey species, are analogous in a number of ways. Both are responsible for the removal of the weak or diseased individuals in a population. In both, prey and predator populations reach equilibrium numbers. Predator-prey and host-parasite populations are constantly coevolving. And of course, both predators and parasites benefit nutritionally.

23. a. Species 2 was introduced into a community where species 1 was established and in balance with its environment. Species 2 immediately began to increase in population size and had the competitive advantage, eventually displacing species 1 from its niche. One niche.
    b. Species 2 was introduced into the community where species 1 was already established. Species 2 did not compete with species 1 but was equally successful in the environment. Two niches.
    c. Species 2 was introduced into the community and proved more successful than species 1 in its particular portion of the environment. Species 1 was displaced to a less favorable portion of the environment. Two niches.
    d. The five species involved are all established and in balance with their environments and are occupying different niches. Five niches.

CHAPTER 53

Ecosystems

CHAPTER ORGANIZATION

I. Introduction
II. The flow of energy
   A. Productivity
   B. Trophic levels
      1. Background information
      2. Producers
      3. Essay: A chemosynthetic ecosystem
      4. Consumers
      5. Detritivores
   C. Efficiency of energy transfer
   D. Essay: Energy costs of food gathering
III. Biogeochemical cycles
   A. Introduction
   B. The nitrogen cycle
   C. Recycling in a forest ecosystem
IV. Concentration of elements
V. Ecological succession
   A. Introduction
   B. Results of succession
VI. Summary

MAJOR CONCEPTS

An ecosystem is composed of all the abiotic and biotic components of a particular environment and is characterized by interactions between them. As a result of these interactions, there is a flow of energy through the ecosystem and a cycling of materials within it.

The organisms in an ecosystem are assigned to various trophic (feeding) levels. Primary producers are always autotrophs (usually photosynthetic), primary consumers are herbivores that consume the primary producers, and secondary consumers are carnivores that feed on the primary consumers. Of the total energy transferred from one trophic level to the next, only about 10 percent is actually stored as tissue.

Productivity is the total amount of energy converted to organic compounds in a given period of time; net productivity is productivity minus energy expended in metabolic activities. Biomass is the total dry weight of the organisms.

Inorganic constituents of the ecosystem, such as water, carbon, nitrogen, and minerals, move between the abiotic and biotic components of the ecosystem in biogeochemical cycles.

Ecological succession refers to the sequential replacement over time of one community by another. As ecosystems mature, the number of species and the total amount of biomass tend to increase, and net productivity tends to decrease.

HOW TO STUDY THE CHAPTER

Read the chapter, focusing on the major concepts.

Reread the chapter, using the questions that follow to help you focus on the details as they relate to major concepts. Answer the questions on a separate sheet of paper. Your answers will provide a valuable study aid in preparing for examinations.

FOCUS ON CHAPTER DETAILS

I. Introduction (page 980)

1. What are the roles of autotrophs, heterotrophs, and decomposers in cycling energy and materials?
2. Define an ecosystem. Cite some examples of ecosystems that are suitable for various studies.

II. The flow of energy (pages 980–986)

A. Productivity

3. How does energy enter ecosystems?
4. How much of the sun's energy is actually converted into useful chemical energy?
5. Define productivity, net productivity, and biomass. What is the relation of productivity to biomass?

B. Trophic levels

Background information

6. Define food chain, trophic level, and food web.
7. How many species can be involved in a food web? How many trophic levels are generally found in a food chain? What is the maximum number of levels that has been found?

Producers

8. What activities occur at the first trophic level of a food chain?
9. In terms of biomass, how important are primary producers?
10. What is primary productivity? Net primary productivity?

Essay: A chemosynthetic ecosystem (page 982)

11. How is the Galapagos Rift exceptional as an ecosystem?
12. What conditions in the rift favor chemosynthesis by bacteria?
13. What biochemical reaction provides energy for carbon fixation by chemosynthetic bacteria?
14. Describe a typical food chain in this ecosystem.
15. What is the trophic role of tube worms in the rift ecosystem?

Consumers

16. What is the role of primary consumers (herbivores)? List a variety of herbivores.
17. Why is only a fraction of the material ingested by an herbivore converted to animal biomass?
18. How is increase in animal biomass defined?
19. What are secondary consumers? Name a few.
20. How much of the biomass of a primary consumer is incorporated into the biomass of the secondary consumer?

Detritivores

21. What energy sources are exploited by detritivores? Name several types of detritivores.
22. In what sense may detritivores be considered to be consumers?
23. Which detritivores are often called decomposers? What specializations have many decomposers developed?
24. Do detritivores or herbivores play the larger role in the forest community?
25. Trace the flow of energy once it has entered the detritivore level.
26. Why does energy tend to remain stored in the detritus of a peat bog?

C. Efficiency of energy transfer

27. What is the average efficiency of energy transfer from one trophic level to another? Account for the energy that is "lost" at each level.
28. Compare the Cayuga Lake food web data (page 984) with a typical energy pyramid (Figure 53-5). Note that the 10 percent rule of thumb is subject to variation.
29. Compare the numbers pyramid for a grassland to that for a forest. (See Figure 53-6.) What is the reason for the difference?
30. When might a pyramid of biomass be

CHAPTER 53 *Ecosystems* 403

inverted? Cite a specific example of an inverted pyramid.

31. In what way do pyramids of energy differ from pyramids of numbers and pyramids of biomass?

D. Essay: Energy costs of food gathering (page 985)

32. What is the typical ratio of calories gained to calories spent in getting food for most animals and for nontechnological human societies?
33. How do technological societies differ in this respect? What costs account for our poor return on our agriculture expenditures?
34. How have humans temporarily freed themselves from a dependence on the sun for energy? What are the possible consequences?

III. Biogeochemical cycles (pages 986–990)

A. Introduction

35. How does the flow of materials through an ecosystem differ from the flow of energy?
36. Name some substances that are cycled through an ecosystem.
37. What is a biogeochemical cycle? What are the geological components of such a cycle?
38. Describe the role of the detritivores in biogeochemical cycles.
39. Outline the terrestrial and the aquatic phosphorus cycles.

B. The nitrogen cycle

40. Why is nitrogen generally the major limiting factor in plant growth, even though it is abundant in the atmosphere?
41. Describe ammonification, nitrification, and assimilation. Include relevant formulas and equations.
42. What is denitrification? Why does it occur to a greater extent in poorly aerated soils? What are some other causes of loss of nitrogen from the soil?
43. What is nitrogen fixation?
44. Outline the nitrogen cycle. Include the three basic stages, all geological components, denitrification, nitrogen fixation, and other major ways in which the soil gains and loses nitrogen.

C. Recycling in a forest ecosystem

45. How were researchers at Hubbard Brook able to establish a mineral budget for an experimental forest?
46. Generally speaking, what change occurred in the mineral budget when all trees and shrubs were cut down? How much nitrate did the soil gain or lose before clear-cutting? After clear-cutting? Explain why the change in the nitrate budget was so dramatic.
47. What did the researchers demonstrate about the role of the vegetation in conserving minerals?

IV. Concentration of elements (pages 990–991)

48. Give two reasons why elements become concentrated in living tissues. What are two dramatic examples of such concentration?
49. Explain why DDT becomes more concentrated as it moves up each step of the food chain. (See Figure 53-11.)
50. Explain how strontium-90 has been transported from atmospheric fallout into the bones of children in the Northern Hemisphere.
51. Why have Eskimos accumulated even more strontium-90 in their bones than inhabitants of the temperate regions?
52. What is the potential danger of strontium-90 accumulation in bone tissues?

V. Ecological succession (pages 992–993)

A. Introduction

53. Describe the process by which bare land is gradually repopulated.
54. How is the traditional view of succession being challenged?

B. Results of succession

55. What four trends are apparent during any successional process?

56. Give some examples of fragility in both young and mature ecosystems. What question do these examples raise?
57. What is the position of opponents of the island biogeography model on species diversity as an ecosystem matures? How is their argument supported by ants that mimic those in the mutualistic ant-acacia relationship?

VI. **Summary** (pages 993–994): Read the Summary. If you are familiar with the essential features of the material presented there, you are ready to take the following diagnostic examination.

TESTING YOUR UNDERSTANDING

After you have completed the following examination, compare your answers with the annotated key in the Analysis section.

1. In any ecosystem, all organisms have a direct role in:
   a. converting outside energy sources to chemical energy.
   b. transferring energy and cycling materials.
   c. processes of decomposition.
   d. nitrification.
2. Primary productivity is the total amount of energy:
   a. available to consumers in a food web.
   b. fixed by autotrophs into organic compounds.
   c. used at all trophic levels in carrying out metabolic activities.
   d. that falls on a given stand of vegetation.
3. Ecological pyramids are arranged according to:
   a. species.
   b. trophic levels.
   c. community type.
   d. importance of the organisms.
   e. net productivity.
4. Which of the following ecological pyramids is *never* inverted?
   a. energy
   b. numbers
   c. biomass
   d. Neither a nor b can be inverted.
   e. None of the above can be inverted.

Questions 5, 6, and 7 apply to the following hypothetical food chain:
red clover ⟶ grasshopper ⟶ toad ⟶ black snake ⟶ hawk.

5. The toads are:
   a. primary consumers.
   b. secondary consumers.
   c. top predators.
   d. secondary producers.
   e. detritivores.
6. If the energy content of red clover is approximately 5,000 calories per square meter of land surface, then the energy content of black snakes should be approximately _____ calories per square meter.
   a. 500 to 1000
   b. 100 to 500
   c. 10 to 100
   d. 1 to 10
   e. much less than 1
7. Which of the following statements about the hypothetical food chain is *false*?
   a. All of the energy entering the chain is eventually passed to detritivores.
   b. The chain has three secondary consumer levels.
   c. There are probably many more grasshoppers in this community than toads.
   d. Efficiency of energy transfer is about equal at all levels.
8. The energy that passes through a herd of herbivores in a year may be broken down as follows: (1) total calories ingested; (2) calories not digested; (3) calories converted to organic compounds; (4) calories expended in metabolic processes; (5) calories incorporated into new tissue. Which of the following statements is *false*?
   a. Net productivity = 5.
   b. Productivity = 1.
   c. Net productivity = 3 − 4.
   d. Decomposers allow 2 to continue flowing through the energy cycle.
9. Which of the following represents a reaction that provides the form of nitrogen most commonly used by plants?
   a. $2NH_3 + 3O_2 \longrightarrow 2NO_2^- + 2H^+ +$ water
   b. $2H_2 + 6H_2O \longrightarrow 4NH_3 + 3O_2$
   c. urea $\longrightarrow CO_2 + NH_3$
   d. $2NO_2^- + O_2 \longrightarrow 2NO_3^-$

10. Which of the following represents a significant loss of nitrogen from the soil due to biological activity?
    a. leaching
    b. fixation
    c. denitrification
    d. decomposition
    e. ammonification

11. What soil conditions favor the process described in Question 10?
    a. high soil pH
    b. high oxygen content of soil
    c. poorly drained soil
    d. soil to which commercial fertilizers have just been applied

12. The channeling effect observable as one proceeds upward along a food chain means that:
    a. biomass will be greatest at the uppermost trophic level.
    b. elements will become concentrated in carnivores.
    c. net productivity increases with succession.
    d. there is little loss of energy as exchanges occur from one level to another.

13. The accumulation of strontium-90 in the world's ecosystems is hazardous because strontium-90:
    a. behaves, elementally, much like a necessary nutrient.
    b. is concentrated at each successive trophic level.
    c. has a relatively long half-life.
    d. is a radioactive element.
    e. All of the above are correct.

14. Clear-cutting (cutting down all trees and shrubs) in the Hubbard Brook Experimental Forest adversely affected:
    a. nitrogen concentrations only.
    b. concentrations of negatively charged ions only.
    c. nitrate concentrations to a marked degree.
    d. concentrations of positively charged ions only.

15. Detritivores are most important in biogeochemical cycles because they:
    a. establish a soil reservoir of nutrients.
    b. release elemental nutrients from organic compounds.
    c. increase the oxygen content of the soil.
    d. supply nutrients directly by symbiotic associations with plants.

16. The current view ecologists take toward the process of ecological succession is that it is:
    a. regular and predictable in nature.
    b. analogous to the developmental processes of an organism.
    c. the outcome of a series of contests.
    d. rarely seen under natural conditions.
    e. Both a and c are correct.

17. The number of calories that a present-day technological society receives from its agricultural system is about _____ the number of calories spent.
    a. 20 times
    b. 5 to 10 times
    c. twice
    d. equal to
    e. one-tenth

18. What four trends are normally seen during the process of ecological succession?

PERFORMANCE ANALYSIS

1. The two basic consequences of the interactions of organisms within a community are the flow of energy through the system and the cycling of minerals within it; therefore, **b** is correct.

2. Primary productivity is the total amount of the sun's energy that plants are able to convert into the chemical energy of organic compounds (**b** is correct). Discounting losses due to respiration, this energy becomes available to the consumers of a food web. This is referred to as net primary productivity.

3. Choice **b** is correct. See Figures 53–5, 53–6, and 53–7.

4. Because energy is lost at each transfer from one trophic level to another, an energy pyramid (**a**) always has an upright shape (see Figure 53–5).

5. Secondary consumers (**b**), such as the toad in this food chain, make up the first level of carnivores in the chain—they feed directly on the primary consumers, the herbivores.

6. Since energy transfers between trophic levels are only about 10 percent efficient, the amount of energy present at the fourth level of the chain (the black snakes in our hypothetical chain) is 1/1,000 of

the energy of the producers, the red clover. The closest estimate of this figure is choice **d**.

7. Choice **a** is the false statement. Much of the energy that enters a food chain is used in metabolic processes.

8. Choice **b** is the false statement. Of the total calories ingested by a herd of herbivores in a year, some will be converted to organic compounds. This is referred to as productivity. Net productivity represents the calories actually incorporated into new tissue, and is obtained by subtracting the total number of calories expended in metabolic processes from productivity. Calories that are not digested are excreted as waste substances and recycled by decomposers.

9. The oxidation of nitrites (NO_2^-) by aerobic bacteria in the soil releases nitrates (NO_3^-), the form in which nitrogen is made available to plants. This is represented by the chemical reaction in choice **d**.

10. Denitrification (**c**) is the process whereby nitrates in the soil are reduced by bacteria requiring an oxygen source. This process, the reverse of nitrification, releases nitrogen gas back to the atmosphere, decreasing the amount of nitrogen immediately available to plants.

11. Denitrification is favored in poorly drained soils because of their low oxygen content (**b**).

12. Substances that are not readily excretable tend to become concentrated (**b**) as they are taken up at each successive trophic level. As a result, top-level carnivores tend to have the greatest amounts of such substances in their body tissues. An example of this is the Arctic food chain—lichens, caribou, Eskimos—in which the radioactive pollutant strontium-90 became highly concentrated in the bones of Eskimos eating caribou meat.

13. Strontium-90, because it has the same valence number as calcium, an essential nutrient, is often incorporated into compounds in place of calcium. As pointed out in the analysis for Question 12, the concentrations of strontium-90 increase as one moves upward through a food chain. Strontium-90 is radioactive and because of its long half-life may pose a threat to those organisms in which it accumulates. Choice **e** is correct.

14. In the Hubbard Brook Experimental Forest, concentrations of all minerals were reduced after clear-cutting, but the effect on nitrate concentrations was most marked (**c**). This was attributed to the absence of plants that would otherwise have taken up the negatively charged ion, which was not held in the soil.

15. Detritivores break down the waste products of both the autotrophic and heterotrophic parts of the food chain. Their metabolic activities (collectively known as decomposition) result in the release of nutrients from organic compounds in a form that is absorbable by plants. Choice **b** is correct.

16. Succession is considered to be both regular and predictable in nature. But rather than being considered a developmental process in which each successional stage makes way for the next, it is now viewed by ecologists as the outcome of a series of contests between organisms competing for resources (**e** is correct).

17. You might be surprised to find that choice **e** is correct if you have not read the essay Energy Costs of Food Gathering, page 985.

18. As succession proceeds, there is usually an increase in total biomass, a decrease in net productivity per unit of biomass, an accumulation of nutrients, and an increase in species diversity.

CHAPTER 54

Biosphere

CHAPTER ORGANIZATION

I. Introduction
II. The sun
III. Climate, wind, and weather
  A. Essay: The atmosphere
  B. General information
IV. The earth's surface
V. Biomes
  A. Introduction
  B. Temperate deciduous forest
  C. Coniferous forests
    1. The taiga
    2. The Pacific Northwest
  D. The tundra
  E. Temperate grasslands
  F. Tropical grasslands: Savannas
  G. Chaparral
  H. The desert
  I. Tropical rain forest
    1. General information
    2. Other tropical forests
VI. Effects of altitude
VII. Summary

MAJOR CONCEPTS

The biosphere is the portion of the earth, including the atmosphere, that supports living organisms.

Solar radiation powers the biosphere. The ozone-oxygen layer of the outer atmosphere prevents most ultraviolet radiation from reaching the surface of the earth; carbon dioxide and water reduce the amount of infrared (heat) radiation that is reflected back into space. Uneven heating of the earth, including seasonal variation due to the tilt of the planet's axis, is responsible for the earth's climates.

A biome is a distinctive community of organisms that extends over a large area. A biome is identified by its dominant vegetation and is the result of a specific climate. The major biomes of the earth include the temperate deciduous forest, the taiga, the Pacific Northwest coniferous forest, the tundra, the temperate grasslands, the savanna, the chaparral, the desert, and the tropical rain forest.

High altitudes and high latitudes often support similar types of organisms.

HOW TO STUDY THE CHAPTER

Read the chapter, focusing on the major concepts.

Reread the chapter, using the questions that follow to help you focus on the details as they relate to major concepts. Answer the questions on a separate sheet of paper. Your answers will provide a valuable study aid in preparing for examinations.

FOCUS ON CHAPTER DETAILS

I. **Introduction** (page 995)
   1. What are the limits of the biosphere?

II. **The sun** (pages 995–997)
   2. What happens to the portion of the incoming solar radiation that does not reach the surface of the earth?
   3. Why did organisms probably live below the surface of the earth until photosynthesis produced an oxygen-ozone layer in the outer atmosphere?
   4. What portion of solar radiation actually reaches the surface of the earth? Describe how this portion powers the water cycle.
   5. What two phenomena cause ocean currents?
   6. Describe the greenhouse effect, and explain what causes it.
   7. What are some changes that would disrupt the global equilibrium between heat loss and heat gain?

III. **Climate, wind, and weather** (pages 997–999)
   A. Essay: The atmosphere (page 996)
      8. What are the limits of the troposphere? What percentage of the atmosphere's molecules does it contain? What is the origin of its name, and why is the name appropriate?
      9. Explain the changes in temperature as one moves from the surface of the earth to the outer boundary of the thermosphere.
   B. General information
      10. What factors account for the daily and seasonal variations in energy received by different parts of the earth's surface?
      11. Why does hot air rise, why does it expand as it rises, and what eventually causes it to cool? What happens to the moisture in rising air?
      12. How does the uneven distribution of energy at different points of the earth's surface account for weather patterns at the equator, at 30° latitudes, at 60° latitudes, and at the poles?
      13. What causes the major wind patterns of the world?
      14. Explain how local modifications of regional weather patterns account for the "rain shadow" effect.
      15. What local effects do bodies of water have on regional weather patterns?

IV. **The earth's surface** (page 1000)
   16. Describe the composition of the earth.
   17. What processes are responsible for changes at the earth's surface?
   18. What two geological features have an important consequence for living organisms?

V. **Biomes** (pages 1000–1014)
   A. Introduction
      19. What is a biome?
      20. What physical factors determine the distribution of biomes on our planet?
      21. How are the communities that make up a biome similar? How are they different?
      22. What are the limitations of the biome concept?
      23. How can the biome concept help to account for the phenomenon of convergent evolution?
   B. Temperate deciduous forest
      24. Under what climatic conditions does a deciduous forest grow?
      25. Outline the costs and benefits associated with leaf-shedding.
      26. Describe the four plant layers that make up the temperate deciduous forest.
      27. What are the dominant tree species found in the various temperate deciduous forests of North America?
      28. Name a few of the representative herbivores and carnivores that inhabit deciduous forests. On what food does each depend?
      29. Describe the soil beneath deciduous forests. What activities occur there that contribute to the soil's fertility?

CHAPTER 54 *The Biosphere* 409

30. Why does land where deciduous forests have stood make good farmland?

C. Coniferous forests

The taiga

31. To what climatic conditions are conifers adapted? What are the adaptations involved?
32. Define taiga.
33. Describe the climate and the plant life of the taiga.
34. What are the principal herbivores and carnivores found in the coniferous forest? On what food does each depend?
35. Why are invertebrates less abundant in the soil of coniferous forests than in deciduous forest soil?

The Pacific Northwest

36. What is the general climate of our Pacific Northwest?
37. Why are evergreens better suited than deciduous trees to grow in this region?
38. Why are these forests much older than other North American forests?

D. The tundra

39. Where is the tundra found?
40. What is the tundra's most characteristic feature? What seasonal changes take place in the soil beneath the tundra?
41. What factors account for the stunted appearance of tundra vegetation?
42. Describe the vegetation one would expect to find in this biome.
43. What are the principal herbivores and carnivores of the tundra?
44. What animals populate the tundra during its short summer growing season?

E. Temperate grasslands

45. Under what climatic conditions do grasslands flourish? Where are the major grasslands of the earth?
46. Describe typical grassland vegetation.
47. How do the North American grasslands change as one moves from east to west?
48. What are the important herbivores of the grasslands? What animals prey upon the herbivores?
49. What factors deter the encroachment of trees onto grasslands?

F. Tropical grasslands: Savannas

50. Why may a savanna be thought of as a transitional biome?
51. What factors account for the savanna?
52. How are grasses well adapted to life on the savanna?
53. What factors can cause the destruction of the savanna?
54. What is unique about the fauna supported by the vegetation of this biome? Name some of the chief herbivores.

G. Chaparral

55. What climatic conditions give rise to the chaparral?
56. What are the characteristics of chaparral vegetation?
57. Identify those regions of the world where chaparral is found.
58. In what ways do plants of the chaparral exemplify convergent evolution?
59. What animals inhabit the chaparral?

H. The desert

60. What meteorological phenomena account for the dryness of the deserts? For the daily fluctuations in desert temperatures?
61. How have humans contributed to the spread of the desert?
62. What are the special adaptations exhibited by desert plants?
63. What adaptations are found among animals that inhabit the desert?

I. Tropical rain forest

General information

64. Under what climatic conditions do tropical forests thrive?

65. How does the number of tree species in the tropical rain forest differ from that in other forests?
66. For what resource is competition among tropical trees most intense? Describe the adaptations of trees that form the forest canopy.
67. What other plants are associated with trees of the canopy? Describe their particular adaptations.
68. What animals inhabit the tree tops?
69. What are conditions like for plants living on the forest floor?
70. Characterize the nutrient cycles in the tropical rain forest. What are the observable consequences?
71. What are the characteristics of tropical rain forest soils?
72. Why is the tropical rain forest poorly suited to agriculture?
73. Where are most of the nutrients located?
74. Why are tropical habitats currently endangered?
75. What hypotheses have ecologists formulated to account for the diversity of life in this biome?

Other tropical forests

76. What adaptations are exhibited by trees restricted to seasonally dry areas of the tropics?

VI. **Effects of altitude** (page 1014)

77. What is the relationship between changes in latitude and changes in altitude? What are the consequences?
78. What are the important differences between high-altitude habitats and high-latitude habitats?

VII. **Summary** (page 1015): Read the Summary. If you are familiar with the essential features of the material presented there, you are ready to take the following diagnostic examination.

TESTING YOUR UNDERSTANDING

After you have completed the following examination, compare your answers with the annotated key in the Analysis section.

1. The greenhouse effect is due to the absorption of _____ energy by _____ in the earth's atmosphere.
   a. ultraviolet; ozone
   b. ultraviolet; water vapor
   c. infrared; carbon dioxide
   d. infrared; water vapor
   e. Both c and d are correct.

2. Equatorial regions receive the maximum daily quantity of sunlight at:
   a. summer solstice.
   b. vernal equinox.
   c. autumnal equinox.
   d. all times of the year.
   e. Both b and c are correct.

3. Warm air along the equator rises and cools as it _____. This cooled air then _____.
   a. expands; gives up its moisture over tropical areas
   b. contracts; absorbs moisture over desert areas
   c. contracts; gives up its moisture over tropical areas
   d. Both b and c are correct.

4. In temperate regions of the Northern Hemisphere, arid regions are frequently found just to the east of high coastal mountain ranges. The phenomenon responsible for this is called the:
   a. greenhouse effect.
   b. rain shadow.
   c. doldrums.
   d. summer solstice.

5. Biomes are categories based on:
   a. geographical factors.
   b. vegetation type.
   c. regional climate.
   d. local weather patterns.

6. Taiga commonly occurs in the Northern Hemisphere at latitudes of 50° and 60°. It might also be found:
   a. 15° to 20° more northerly.
   b. in a mountainous area of the temperate zone.

CHAPTER 54 *The Biosphere* 411

   c. near the equator, if seasonality is pronounced.
   d. where the soil is rich and continually moist.
   e. Both a and c are correct.

7. Which of the following biomes would be best suited for cropland if cleared?
   a. Canadian taiga
   b. Brazilian equatorial forest
   c. temperate deciduous forest
   d. Californian chaparral
   e. low-latitude tundra

8. In a tropical rain forest, the bulk of mineral nutrients is present in:
   a. the soil.
   b. the subsoil.
   c. decaying leaves.
   d. plant biomass.
   e. animal feces.

9. If extraterrestrials decided to land their exploration craft at 30° north of the equator, the animal life they would encounter would probably be:
   a. grazing on pastures.
   b. relatively small and nocturnal.
   c. living in the treetops.
   d. covered with dark, heavy fur.
   e. primarily soil invertebrates.

10. In which of these biomes does the amount of seasonal rainfall strongly affect the balance between trees and grasses?
   a. the North American grasslands
   b. an African savanna
   c. an equatorial rain forest
   d. the Arctic tundra
   e. the Canadian taiga

11. Which of the following is an adaptation of a typical desert plant?
   a. C_4 metabolism
   b. buttressing
   c. extraordinarily deep taproots
   d. large leaves with a thick, waxy cuticle
   e. fire-resistant seeds

12. The most abundant and diverse assemblage of large herbivores is found in (on):
   a. the tropical rain forest.
   b. the temperate deciduous forest.
   c. the Arctic tundra.
   d. the African savanna.
   e. the temperate grasslands.

13. The _____ biome best exemplifies the intense competition among plants for light. The same is true for the _____ with respect to water.
   a. tropical rain forest; temperate grasslands
   b. tropical rain forest; savanna
   c. tundra; deciduous forest
   d. tundra; coniferous forest

14. In which of these biomes is the diversity of tree species the greatest?
   a. temperate deciduous forest
   b. taiga
   c. tropical rain forest
   d. tropical savanna
   e. chaparral

15. Overgrazing favors the turning of _____ into _____.
   a. savanna; woodland
   b. temperate grasslands; desert
   c. tundra; coniferous forest
   d. grasslands; tundra
   e. Both a and b are correct.

16. Match each biome with the appropriate type of vegetation or plant adaptation.
   desert _____
   chaparral _____
   Pacific Northwest forests _____
   tropical rain forest _____
   temperate deciduous forest _____
   tundra _____
   a. epiphytes
   b. mosses and lichens
   c. extensive root systems near the surface
   d. spiny evergreen shrubs
   e. large conifers
   f. four distinct plant layers

17. Match each biome with its characteristic physical feature.
   desert _____
   tundra _____
   savanna _____
   tropical rain forest _____
   temperate deciduous forest _____
   a. laterite soils
   b. gray-brown soils
   c. permafrost
   d. pronounced wet and dry seasons
   e. pronounced daily fluctuations in temperature

18. Explain how each of the following changes would affect conditions at the earth's surface: an increase in the earth's reflectivity; an increase in cloud cover; a decrease in atmospheric carbon dioxide; and a decrease in the ozone layer.
19. What three factors are responsible for the distribution of the earth's biomes?
20. Use the biome concept to explain why structurally similar but genetically unrelated organisms might be found in two widely distant parts of the globe.

PERFORMANCE ANALYSIS

1. Some of the ultraviolet and visible light absorbed by the earth is reradiated from the earth's surface at infrared wavelengths. This heat energy is absorbed by atmospheric gases, particularly carbon dioxide and water vapor, thus warming the earth. The phenomenon is known as the greenhouse effect, since the glass of a greenhouse allows light to pass through but prevents heat from escaping. Choice **e** is correct.
2. Because the earth is slightly tilted on its axis, the amount of radiation received during the day at any point on the earth's surface varies (see Figure 54-3b). On June 21-22 (summer solstice in the Northern Hemisphere), the tropical regions of the Northern Hemisphere are directly under the sun's rays. On December 21-22 (winter solstice in the Northern Hemisphere), the tropical regions of the Southern Hemisphere lie directly under the sun. At vernal and autumnal equinox, the sun is directly over the equator (**e** is correct).
3. As air over the equator is heated by the sun, it expands and cools. Since cool air holds less moisture than warm air, this rising air mass gives up its moisture over tropical areas in the form of rain. Choice **a** is correct.
4. Air that moves in from the ocean and is carried aloft over high mountain ranges gives up its moisture as it is cooled. The air that descends the eastern slopes of the ranges is warmed and absorbs moisture. Such a "rain shadow" (**b**) is responsible for arid conditions in these regions.
5. Biomes, such as the tropical forest, are categories based on type of vegetation (**b**). Choices a, c, and d are the factors that influence what type of vegetation is able to grow in a given region.
6. Because of an altitude effect, the taiga is sometimes found in mountainous regions at latitudes within the temperate zone. Choice **b** is correct.
7. The soils of temperate deciduous forests (**c**) are rich in organic material and in soil organisms and are therefore able to hold water and nutrients. The taiga and the tundra would both have too short a growing season. Tropical forests when cleared suffer from erosion and solidification of the soil, with a consequent inavailability of nutrients for plants. Californian chaparral is too dry during the summer for most crop plants.
8. Because warm temperatures and high humidity favor rapid decomposition, tropical soils do not hold a large reserve of nutrients. The bulk of the nutrients in the tropical rain forest ecosystem is tied up in plant biomass (**d**).
9. At 30° north of the equator, our extraterrestrials could expect to find desert. The typical animal forms of the desert are small and nocturnal (**b**); these features are adaptations for conserving water.
10. Grasses are well adapted to alternating periods of rain and drought. When rainfall decreases, trees die; when it increases, the trees thrive and shade out the grasses. An African savanna (**b**) is grassland dotted with patches of trees and brush.
11. Desert plants make use of the C_4 photosynthetic pathway (**a**) because it allows them to store the carbon dioxide they take in when the air is relatively cool and humid.
12. Because of the diversity of food resources and the opportunities for partitioning these resources, the African savanna (**d**) holds the greatest assemblage of large herbivores on the earth.
13. Because of the dense vegetational growth in a tropical rain forest, the plants face intense competition for light. Because of sandy soils and pronounced dry seasons, the critical competition in the savannas is for water. Choice **b** is correct.
14. Perhaps because of its age or because of environmental instability, the tropical rain forest (**c**) exhibits the greatest diversity of both trees and animals on the earth.

15. The encroachment of trees onto the savanna (a) is favored by factors that increase soil moisture, such as periods of heavy rain or overgrazing. Overgrazing in temperate grasslands encourages either desert (where there are short-grass prairies) (b) or woodland (where there are tall-grass prairies). Therefore, **e** is correct.

16. The correct answers in order are **c**, **d**, **e**, **a**, **f**, and **b**.

17. The correct answers in order are **e**, **c**, **d**, **a**, and **b**.

18. An increase in the earth's reflectivity would reduce the temperature of both the soil and the atmosphere. An increase in cloud cover would decrease the amount of solar radiation reaching the surface of the earth, thereby decreasing the temperature and the amount of energy available to plants. A decrease in carbon dioxide would lessen the greenhouse effect, lowering atmospheric temperatures. A decrease in the ozone layer would mean that more ultraviolet light would reach the surface; ultraviolet light damages organic molecules.

19. The three factors are (1) distribution of solar radiation, (2) global patterns of air circulation, and (3) geologic formations such as mountains and large bodies of water.

20. Organisms that are genetically unrelated are often similarly adapted because of the similar features of the environments in which they live. This phenomenon is called convergent evolution; Australian marsupials and North American placentals provide a vivid example.

CHAPTER 55

The Evolution of Social Behavior

CHAPTER ORGANIZATION
I. **Introduction**
II. **Insect societies**
   A. Introductory remarks
   B. Stages of socialization
   C. Honey bees
      1. Introduction
      2. The queen
      3. The annual cycle
III. **Vertebrate societies**
   A. Introductory remarks
   B. Pecking orders
   C. Territories and territoriality
      1. Definitions
      2. Territoriality in birds
      3. Territorial defense
IV. **Kin selection**
V. **The selfish gene**
VI. **Conflicts of interest**
   A. Introduction
   B. Essay: The kinship of lions
   C. Male vs. female
   D. Essay: Arts and crafts of bowerbirds
VII. **Dominance hierarchies and the selfish gene**
   A. General information
   B. Essay: Reproductive strategies in male baboons
VIII. **Evolutionarily stable strategies**
IX. **Reciprocal altruism**
X. **The biology of human behavior**
XI. **Summary**

MAJOR CONCEPTS

Ethology, the study of the evolution of behavior, emphasizes the ecological and genetic foundations of behavior.

The complex social organizations among eusocial insects are exemplified by the honey-bee society, in which behaviors are genetically programmed and the hive is controlled by pheromones from the queen.

Dominance hierarchies and territoriality are common among vertebrates. In dominance hierarchies, the higher-ranking individuals have first access to the best resources. A territory may be defined as any defended area and may be used for any or all of the following activities: courtship and mating, feeding, shelter, and rearing the young.

Two hypotheses that have been proposed to explain the existence of altruistic behavior in some animal societies are (1) the concept of group selection (rejected because alleles for restraint-of-breeding would not be passed on), and (2) the concept of kin selection, which contends that a nonreproducing individual can make a genetic contribution to the next generation by assisting in the rearing of young relatives.

According to the concept of "selfish genes," individuals are genetically programmed to turn out the maximum number of gene copies regardless of the cost to the organism. The kin-selection hypothesis is based on the selfish-gene concept. The selfish-gene concept has also been used to explain why dominance hierarchies are an evolutionarily stable strategy. An evolutionarily stable strategy is one that resists change when an alternative strategy is being utilized by an invader.

In reciprocal altruism, an individual performs an altruistic act with the expectation that it will be returned.

HOW TO STUDY THE CHAPTER

Read the chapter, focusing on the major concepts.

Reread the chapter, using the questions that follow to help you focus on the details as they relate to major concepts. Answer the questions on a separate sheet of paper. Your answers will provide a valuable study aid in preparing for examinations.

FOCUS ON CHAPTER DETAILS

I. **Introduction** (page 1017)
1. Why can behavioral characteristics be used to construct phylogenies?
2. What is ethology? What is social behavior? Why have ethologists chosen to study elaborate social behavior patterns?
3. How is the evolution of behavioral characteristics related to that of any other trait? What factors govern this type of evolution?
4. What is altruistic behavior? Cite some examples of altruism. Why are these behaviors disturbing to proponents of evolutionary theory?

II. **Insect societies** (pages 1019–1022)

A. Introductory remarks
5. Which insect groups are social?

B. Stages of socialization
6. Outline the reproductive patterns that characterize the solitary, subsocial, and eusocial stages in the socialization of bees and wasps.

C. Honey bees

Introduction

7. Trace the events in the life cycle of a worker bee.
8. What occupations does the adult worker assume?

The queen

9. What nutritional factor determines which single worker larva will become the queen?
10. What effect on the hive do the queen's pheromones have?
11. What happens if the queen of the hive is lost?

The annual cycle

12. How does the hive maintain its temperature during the cold of winter?
13. What happens to the hive when nectar is abundant during the spring?
14. What reproductive events are triggered by the departure of the old queen from the hive?
15. What is the reproductive significance of the nuptial flight of the young queen?
16. What is the contribution of drones to the hive? What is their fate?

III. **Vertebrate societies** (pages 1022–1025)

A. Introductory remarks
17. How do interactions between members of vertebrate societies influence comparative reproductive success?

B. Pecking orders
18. Using chickens as an example, describe the interactions that determine a pecking order.
19. How do the individuals of different ranks of a pecking order differ in their privileges? How do they differ in appearance?
20. When does fighting occur among members of the pecking order? How is the status of an individual recognized?
21. Which individuals in the pecking order are allowed to breed?

C. Territories and territoriality

Definitions

22. Define a territory. What is territoriality?

Territoriality in birds

23. How do birds establish their territories?
24. What activities may be carried out in a territory?
25. What benefits does a mating pair derive from territorial behavior? What is the fate of males that are unable to establish a territory?

Territorial defense

26. What behaviors do birds use in defending their territories?
27. Cite an example of how the number of animals in a communal territory is regulated.

IV. **Kin selection** (pages 1025–1028)
28. What was Wynne-Edwards' hypothesis of group selection? Why was it significant, even though it was later rejected?
29. Why was the idea of group selection rejected?
30. How did Darwin explain the evolution of sterile castes within insect societies?
31. In what way did W. D. Hamilton alter the concept of fitness to accommodate Darwin's ideas?
32. Define kin selection.
33. Calculate your degree of relatedness to a parent and to a sibling.
34. Describe the haplodiploidy found in some hymenopterans.
35. Calculate the degree of relatedness (r) between: a female worker bee and her mother; two worker bees; a female worker and her haploid brother; a haploid male and his daughter.
36. Are the r values you have calculated consistent with the behavior of workers and haploid males in the hive? Explain.
37. What criticism has the application of Hamilton's ideas to insect societies evoked?
38. Compare Darwin's criterion for fitness with that of inclusive fitness (Hamilton's concept).
39. Explain how Moehlman's study of social behavior among silver-backed jackals supports Hamilton's hypothesis of kin selection.
40. How is the idea of kin selection used to predict territorial behavior among Florida scrub jays?

V. **The selfish gene** (pages 1028–1029)
41. What does Richard Dawkins perceive to be the relationship between an organism and its genes?
42. Although an oversimplification, how has the concept of the selfish gene aided researchers?

VI. **Conflicts of interest** (pages 1029–1032)
A. Introduction
43. Cite two intergenerational conflicts that may arise due to the "selfishness" of genes.
B. Essay: The kinship of lions (page 1029)
44. What social interactions are seen among the females of a lion pride?
45. What types of behavior are found among adult males attempting to reproduce?
C. Male vs. female
46. How do the reproductive investments of males and females differ?
47. What is the explanation for male courtship displays and/or dominance over other males in some species?
48. Cite examples of other strategies used by males in attracting a mate.
49. What conditions must exist in species in which the males contribute to the care of the young?
50. Using birds as an example, explain what benefits the male and female of a mating pair derive from a lengthy courtship.
51. What evidence is there that males may be warranted in suspecting females of "cheating"?

D. Essay: Arts and crafts of bowerbirds (page 1031)
- 52. Describe the elaborate structures used by male bowerbirds to lure a mate.
- 53. How is the splendor of the bower related to the secondary sexual characteristics of its builder?
- 54. What mating behaviors does the male display? What is the evolutionary origin of these behaviors? Of the bower itself?
- 55. Of what adaptive value is it for females to choose the most artistic male?

VII. **Dominance hierarchies and the selfish gene** (page 1032)
- A. General information
  - 56. How is the "waiting" behavior of individuals that are low on a dominance hierarchy or are unable to secure territories consistent with the concept of the selfish gene?
  - 57. How has territoriality in Florida scrub jays been used to test the "waiting" hypothesis?
- B. Essay: Reproductive strategies in male baboons (page 1033)
  - 58. What type of social organization is found among all species of baboons except the hamadryas? How are hamadryas baboons organizationally different?
  - 59. What happens if a male hamadryas baboon is caged with a female whom he has seen with a mate? What activities on the part of young males does this strategy preclude?
  - 60. How do young males get around their genetic inhibition to steal females that belong to adult males?

VIII. **Evolutionarily stable strategies** (pages 1034–1035)
- 61. What is "game theory"?
- 62. Give an operational definition of an evolutionarily stable strategy.
- 63. Explain how John Maynard Smith has applied game theory to polymorphic behaviors (see also Table 55-4). What was the outcome of encounters between hawks and doves in terms of an evolutionarily stable strategy?
- 64. What are the characteristics of a bourgeois strategy? How is the outcome between hawks and doves affected by the introduction of the bourgeois strategy?
- 65. Cite an example to demonstrate that bourgeois may be a real life strategy for territorial animals.

IX. **Reciprocal altruism** (pages 1036–1037)
- 66. What is reciprocal altruism?
- 67. Under what conditions does game theory predict that cheating will be the evolutionarily stable strategy? Under what conditions is cooperation evolutionarily stable?
- 68. How may the origin of cooperative behavior in a society be explained?
- 69. What societies are most likely to exhibit cooperation?

X. **The biology of human behavior** (page 1037)
- 70. Describe the opposing viewpoints biologists have taken concerning the genetic basis of human behavior.

XI. **Summary** (pages 1037–1038): Read the Summary. If you are familiar with the essential features of the material presented there, you are ready to take the following diagnostic examination.

TESTING YOUR UNDERSTANDING

After you have completed the following examination, compare your answers with the annotated key in the Analysis section.

1. The comparative study of behavior patterns and their evolutionary importance is called:
   a. morphology.
   b. ethology.
   c. ecology.
   d. social evolution.
   e. behavioral biology.

2. In some insect species, there is cooperation among individuals in the care of the young. This type of organization is found in species that are:

a. subsocial.
   b. presocial.
   c. eusocial.
   d. altruistic.
   e. haplodiploid.
3. In honey bees, _____ develop from fertilized eggs, and _____ develop from unfertilized eggs.
   a. sterile workers; fertile females
   b. diploid drones; haploid workers
   c. fertile males; haploid females
   d. female workers; haploid drones
4. Which of the following prevents sexual development in worker honey bees?
   a. genotype
   b. nutrition
   c. pheromones
   d. location in the hive
   e. All of the above are correct.
5. In honey bees, altruistic behavior is most apparent among:
   a. workers.
   b. fertile males.
   c. drones.
   d. potential queens.
   e. larvae.
6. Once a pecking order has been established:
   a. fighting or other acts of aggression are necessary to maintain it.
   b. an individual's chances for reproduction have been determined.
   c. individuals of low rank leave the society.
   d. outside individuals are not allowed to compete for rank.
7. A territory is an area that is defended against:
   a. predators.
   b. competitors.
   c. siblings.
   d. all invaders.
   e. Both a and b are correct.
8. Which of the following is true regarding an animal's defense of its territory?
   a. Territories are marked or clearly delineated in some way.
   b. The owner of a territory if virtually undefeatable on its home ground.
   c. The boundaries of a territory are recognizable to its owner.
   d. Only b and c are correct.
   e. All three answers (a, b, and c) are correct.
9. Biologists readily rejected Wynne-Edwards's hypothesis of group selection because:
   a. altruistic behavior has no adaptive value.
   b. an allele for not breeding cannot be maintained in a gene pool.
   c. Hamilton's theory of kin selection fits the facts much better.
   d. members of a given group are generally not closely related genetically.
10. Hamilton replaced the idea of group selection with the proposition that altruistic behavior might be selected for in:
    a. monogamous males.
    b. cooperative societies.
    c. close relatives.
    d. territorial animals.
    e. situations in which danger threatens the survival of an entire population.
11. In the sense of inclusive fitness, an organism is fit if:
    a. its offspring survive to reproduce.
    b. its alleles are passed on to the next generation.
    c. it is able to survive through the next generation.
    d. its reproductive strategy is evolutionarily stable.
12. The degree of relatedness between siblings of a diploid mother and a haploid father is:
    a. 0.25.
    b. 0.5.
    c. 0.75.
    d. 0.375.
    e. 0.625.
13. The degree of relatedness between you and a sibling is:
    a. less than that between you and one of your parents.
    b. greater than that between you and your children.
    c. the same as that between you and one of your parents.
    d. the same as that between you and your children.
    e. Both c and d are correct.

14. The notion that individual organisms are merely machines by which genes replicate themselves is the central idea of:
    a. the theory of group selection.
    b. the kin-selection hypothesis.
    c. the selfish-gene concept.
    d. reciprocal altruism.
15. According to the selfish-gene concept, a low-ranking individual in a dominance hierarchy should:
    a. fight to gain a higher rank.
    b. sacrifice itself for the good of the group.
    c. attempt to mate regardless of its submissive position.
    d. simply wait its turn.
16. Which of the following is true with regard to the relative reproductive investments made by males and females?
    a. Females have a higher reproductive potential than males.
    b. Gamete production is more expensive for males.
    c. Females always select the highest-ranking male.
    d. Males tend to be monogamous only if it is necessary for the survival of the young.
    e. Both a and c are correct.
17. An animal shares its food with another with the expectation that it will receive the same treatment. This type of behavior is referred to as:
    a. bourgeois.
    b. reciprocal altruism.
    c. the unselfish-gene concept.
    d. subsocial.
18. An evolutionarily stable strategy is a:
    a. single behavioral strategy with the highest selective value.
    b. strategy that is resistant to upset by alternative strategies used by invaders.
    c. strategy not subject to the forces of natural selection.
    d. hypothetical strategy that is never used under natural conditions.
19. Game theory predicts that _____ will be the evolutionarily stable strategy when individuals are able to recognize each other on a second encounter.
    a. cheating
    b. cooperation
    c. indifference
    d. a combination of cheating and cooperation
20. Those who believe that human behavior is genetically determined and not subject to moral judgment have attributed little importance to:
    a. the selfish-gene concept.
    b. the effects of culture.
    c. group selection.
    d. altruistic behavior.
    e. behavioral polymorphism.
21. How does a kin-selection hypothesis explain the social organization of honey bees?
22. What are the sources of conflicts of interest between the members of a mating pair?
23. Explain the adaptive value of territoriality.
24. Contrast simple altruistic behavior with reciprocal altruism. Provide one example of each.

PERFORMANCE ANALYSIS

1. Choice **b** is correct.
2. Eusocial (**c**), or "truly social," insects exhibit a division of labor and cooperate in caring for their young.
3. Choice **d** is correct. The female workers are diploid, and the male drones are haploid.
4. Pheromones released by the adult queen inhibit sexual maturity in workers, thereby preventing rivalry. Choice **c** is correct.
5. Adult workers (**a**) "sacrifice" the opportunity to reproduce (altruism) and attend instead to the queen, drones, and larvae.
6. An individual's rank in a dominance hierarchy determines its chances for mating (**b** is correct). Low-ranking individuals do not fight to gain access to reproductive rights; they merely wait their turn.
7. A mating pair defends its territory against other individuals that compete for resources. Competitors are usually individuals of the same or of a closely related species. Territories may also be defended against predators; therefore, **e** is correct.
8. Even though territorial boundaries are not always physically apparent, territory owners are able to

recognize such boundaries. Against a rival, the territory owner is virtually undefeatable on its own territory; therefore, **d** is correct.

9. A group-selection hypothesis would require that altruistic individuals pass on their genetic tendency for altruistic behavior. Clearly, alleles rendering an individual a nonbreeder cannot be passed on (**b** is correct).

10. Since close relatives (**c**) share many of the same genes, altruistic behavior within families could conceivably be selected for. The altruistic individual sacrifices its chances for reproduction in order to ensure that the genes it shares will be passed on.

11. Choice **a** expresses Darwin's concept of fitness. The criterion of inclusive fitness is the relative number of an individual's alleles that are passed on to the next generation (**b**), either through its own efforts or the efforts of a close relative.

12. The probability that one of the siblings will receive an allele from its diploid mother is 0.5, so the probability that the two siblings will share that allele is 0.25. The probability that siblings will share an allele from their haploid father is 0.5. The degree of relatedness, then, is 0.75 (0.25 + 0.5). Choice **c** is correct.

13. Choice **e** is correct. See Table 55–1.

14. According to Richard Dawkins, since genes, not organisms, persist from one generation to the next, organisms may be viewed merely as machines that turn out copies of genes. This is the selfish-gene concept (**c**).

15. An individual that fights to achieve access to breeding may completely forfeit its chance to reproduce. Self-sacrifice, except for the benefit of close relatives, does not represent a genetic investment. If a low-ranking individual merely waits its turn (**d**), it still has at least some chance of contributing to the gene pool.

16. Choices **a** and **b** are stated for the incorrect sex. Choice **c** is false; when parental care of the young is important, the best strategy for a female may be to choose a mate that will provide this care. Such mating is often preceded by a lengthy courtship, which serves to assure the female that the male will be monogamous (**d** is correct).

17. Game theory has shown that cooperation is selected for in situations in which organisms can recognize one another. In such societies, reciprocal altruism (**b**) is selected for.

18. A strategy is defined as evolutionarily stable if it resists upset by an invading strategy (**b**).

19. When individuals recognize each other on a second encounter, Tit for Tat (doing what the other player did on the last move) ensures that cooperation is rewarded (**b**) and cheating is punished.

20. A new breed of social Darwinists has seized on the selfish-gene concept to make apologies for social exploitation. But humans are more than products of genes; their behavior is also influenced by culture (**b** is correct).

21. In a haplodiploid species, such as the honey bee, females share half of their genes with their mother, but sisters share three-fourths of their genes with each other. So workers, in tending to the queen (a sister), pass on more copies of their genes than they would if they produced offspring of their own. Because female workers share only one-fourth of their genes with their brothers, they invest much less energy in caring for their brothers than in caring for the queen. A male bee that mates, since he is haploid, passes on all of his genes to his daughters and provides little service for the hive.

22. Conflicts of interest between mates arise because females make a heavier gametic investment than males and, in most instances, must care for the young. The male is interested in passing on his own genes by mating with as many females as possible.

23. Territoriality assures a mating pair of food and a place to reproduce and rear young. Individuals that do not or cannot secure territories usually do not reproduce.

24. Altruistic behavior is adaptive in situations in which an organism is able to pass on its genes by caring for another organism (a close relative) that shares the same genes. An example is provided by the worker honey bee. The reciprocal altruist cooperates only if it expects to receive something in return. An example of reciprocal altruism is food sharing among chimpanzees.

CHAPTER 56

Human Evolution and Ecology

CHAPTER ORGANIZATION

I. **Introduction**
II. **Trends in primate evolution**
   A. Introductory remarks
   B. The primate hand and arm
   C. Visual acuity
   D. Care of the young
   E. Uprightness
III. **Major lines of primate evolution**
   A. Prosimians
   B. Monkeys
      1. General information
      2. Essay: Cultural evolution among macaques
   C. Apes
   D. *Ramapithecus*
   E. Essay: Anthropoid apes
   F. The first hominids
   G. How did it happen?
IV. **The emergence of** *Homo sapiens*
   A. Introductory remarks
   B. *Homo erectus*
      1. Introduction
      2. Hunting
      3. Tools
      4. Fire
      5. Dwelling places
      6. Increase in intelligence
   C. *Homo sapiens*
      1. Introductory remarks
      2. The Neanderthals
      3. The Cro-Magnons
   D. The agricultural revolution
   E. Essay: The Ice Ages

V. **The population explosion**
   A. Introduction
   B. Life vs. death
   C. The rich and the poor
   D. Birth rates, death rates, and social security
VI. **Our hungry planet**
VII. **Science and human values**
VIII. **Summary**

MAJOR CONCEPTS

The major trends in primate evolution are apparently the result of adaptations to arboreal life. These trends include hands with five digits (usually), one of which is a divergent thumb; increased visual acuity with a decreased dependence on olfaction; increased care of the young; and an upright posture. The two major groups of primates are the prosimians (exemplified by tarsiers and lemurs) and the anthropoids (monkeys, apes, and humans).

Some scientists consider members of the genus *Ramapithecus* to be ancestors of humans. *Ramapithecus* had a smaller, broader dental arch than apes have, and the shape and condition of its teeth suggest that it matured more slowly than did the other hominoids of its time. Also, it was bipedal or well on its way to being bipedal. However, evidence from molecular clocks seems to indicate that hominids diverged from other hominoids more recently.

Hominids of the genus *Australopithecus* were bipedal, and

some evidently used simple tools. At least two species are recognized: the lighter *A. africanus* and the heavier *A. robustus*. Some scientists recognize five species, including *Homo habilis*, which was the most lightly built and had a markedly greater cranial capacity than other species. The line that led to modern humans is thought to have derived from the gracile australopithecines.

Homo erectus was a species of hominids that had a skeleton and stride much like our own and whose cranial capacity overlapped that of modern humans. *H. erectus* apparently hunted large animals. The hand ax is associated with this species.

Two early varieties of the species *Homo sapiens* were the Neanderthals, who lived in caves, used stone tools and fire, buried their dead, and probably wore clothing, and the Cro-Magnons, who used a variety of blade tools and had a complex culture, as evidenced by their art.

The beginning of agriculture more than 10,000 years ago was the single most important event in the cultural evolution of humans.

The human population has increased from about 5 million around 8000 B.C., when agriculture began, to 90 million by 4000 B.C., 500 million by 1650 A.D., and 4.5 billion by 1980. It continues to grow.

HOW TO STUDY THE CHAPTER

Read the chapter, focusing on the major concepts.

Reread the chapter, using the questions that follow to help you focus on the details as they relate to major concepts. Answer the questions on a separate sheet of paper. Your answers will provide a valuable study aid in preparing for examinations.

FOCUS ON CHAPTER DETAILS

I. **Introduction** (pages 1039–1040)
 1. When did the first mammals evolve?
 2. What clues do fossils provide about these early mammals?
 3. What factors allowed for the explosive radiation of the mammals? When did it occur?
 4. Describe the three major groups into which mammals are classified.
 5. Which familiar mammalian groups are included among the placentals?
 6. How do humans fit into the classification of mammals?

II. **Trends in primate evolution** (pages 1040–1042)
 A. Introductory remarks
  7. What event marked the beginning of primate evolution?
  8. To what life style are the trends in primate evolution adapted?
 B. The primate hand and arm
  9. Describe the primate hand. What abilities does it have? Why is it considered to be relatively unspecialized?
  10. What special abilities do the forearms and shoulder joints of primates have?
  11. For what activities are nails adapted?
 C. Visual acuity
  12. How has the primate dependence on vision (rather than on smell) affected the anatomy of the head?
  13. Describe the retina of a primate.
 D. Care of the young
  14. What are the consequences of increased maternal care of the young?
 E. Uprightness
  15. Describe the posture of arboreal primates. What is the consequence of this posture for the orientation of the head?

III. **Major lines of primate evolution** (pages 1042–1054)
 A. Prosimians
  16. Into what two major groups are **primates** divided? Name several representatives of each group. In what habitats are prosimians found?
  17. Describe a typical modern prosimian.
 B. Monkeys
  General information
  18. Compare monkeys with prosimians.

19. Describe the social organization of monkeys.
20. What is the evolutionary origin of the monkeys?
21. What are the two principal groups of monkeys? Name several representatives of each group, and describe the features of each group. Why are these groups considered an example of parallel evolution?

Essay: Cultural evolution among macaques (page 1044)

22. What are "closed" and "open" behavior programs? What are the consequences of these programs for life span and juvenile dependency?
23. In what way did feeding behavior among macaques change as a result of interference by researchers?
24. Describe how the changes in behavior spread among the members of the group.

C. Apes
25. Which primates are classified as hominoids?
26. What does the fossil record tell us about hominoid evolution?
27. Into what four genera are the great apes divided?
28. Describe the anatomical features of the apes that distinguish them from other primates.
29. Describe the social structure and behavior found among the ape genera.

D. *Ramapithecus*
30. What fossil evidence points to *Ramapithecus* as the first hominid?
31. What evidence indicates that *Ramapithecus* may have been a bipedal inhabitant of the African savanna?
32. What evidence do molecular clocks provide that *Ramapithecus* is not a direct ancestor of humans?
33. What other explanation for the hominidlike aspects of *Ramapithecus* has been proposed?

E. Essay: Anthropoid apes (pages 1048–1049)
34. Describe the characteristic structural features, habitats and food preferences, and social organization and behaviors of each of the four genera of apes.

F. The first hominids
35. What feature of the *Australopithecus* skull found in South Africa distinguished *Australopithecus* from apes?
36. How was the fossil discovery initially received by the scientific community?
37. How did the australopithecines move about? Were they hominids?
38. What does *Australopithecus* indicate about the evolutionary relationship between upright posture and brain size within the hominid line?
39. Contrast *A. africanus* (and/or *A. afarensis*) with *A. robustus* (and/or *A. boisei*). To which group does Lucy belong? To which group has *Zinjanthropus* (the Leakeys' "nutcracker man") now been generally assigned?
40. On what basis did Leakey place *Homo habilis* in the same genus as modern humans?
41. What other evolutionary positions have been proposed for *Homo habilis*?
42. Which general type of early hominid was an evolutionary dead end? Which general type was probably the ancestor of modern humans?

G. How did it happen?
43. What environmental changes during the Miocene epoch are associated with the origin of bipedalism?
44. What are the adaptive advantages of an upright stance for life on the savanna? What activities of early humans might have been associated with bipedalism's freeing of the hands?
45. What were the disadvantages of bipedalism for our ground-dwelling ancestors?

IV. **The emergence of** *Homo sapiens* (pages 1054–1060)

A. Introductory remarks
46. What two fossil species are indisputably members of the genus *Homo*?

B. *Homo erectus*

Introduction

47. Compare the skeletal features of *H. erectus* with those of modern humans.
48. Where and during what period of time did *H. erectus* flourish?

Hunting

49. What evidence does the fossil record provide about the type of game hunted by *H. erectus* and the methods used to pursue it?

Tools

50. What is the difference between the pebble tools used as early as 2.5 million years ago (see Figure 56-14) and the hand ax used as early as 1 million years ago and associated with *H. erectus*?
51. Why is the hand ax used by *H. erectus* considered significant in terms of human cultural evolution?

Fire

52. What experience might early hominids have had with fire?
53. What is the earliest evidence of the use of fire?
54. What pronounced effects might the use of fire have had on the life style of hominids?

Dwelling places

55. Describe the nomadic pattern (including campsites) of early hominids.
56. What was the probable prerequisite for cave dwelling?

Increase in intelligence

57. What does the ratio of brain to body size indicate about selection pressures imposed on early hominids?
58. When did the greatest increase in relative size of the hominid cranium take place?

C. *Homo sapiens*

Introductory remarks

59. How old are the earliest *H. sapiens* fossils? How do they compare with *H. erectus* fossils?

The Neanderthals

60. Where and during what period of time did the Neanderthals flourish?
61. How do the anatomical features of Neanderthals compare with those of modern humans?
62. For what purposes did Neanderthals employ tools?
63. Why is it believed that the Neanderthals had some understanding of abstract concepts?

The Cro-Magnons

64. What reasons have been suggested to explain why Neanderthals did not survive?
65. What refinements in tools were made by Cro-Magnons? How did these refinements affect their life style?
66. Describe the cave drawings of Cro-Magnons. What cultural significance is attached to these drawings?

D. The agricultural revolution

67. How might geologic events have contributed to the development of an agricultural way of life?
68. Describe the first evidence of an agricultural society.
69. Trace the spread of agriculture from its origin in the Near East.

E. Essay: The Ice Ages (page 1061)

70. What is an Ice Age? What climatic changes bring about an Ice Age?
71. Trace Ice Age history from the beginning of the Paleozoic era up to the last period of glaciation. Describe the effects of each glaciation on the distributions of plants and animals.
72. What is an interglacial? How do populations become redistributed during these periods?
73. What factors may be responsible for the temperature changes that usher in an Ice Age? Where would you place the present decade on the graph of Ice Ages?

V. **The population explosion** (pages 1060–1063)
   A. Introduction
      74. Using the data given in this section, roughly graph the growth of the human population from 25,000 years ago to today. What events in human cultural evolution are associated with periods of rapid growth?
      75. What is the prediction for population growth in the last two decades of this century?
   B. Life vs. death
      76. What factor has been largely responsible for the world population explosion we are witnessing?
      77. How is the growth rate of a population influenced by its age structure?
   C. The rich and the poor
      78. How is the population problem complicated by the fact that the highest rates of increase occur in those populations least prepared to cope with such increases?
      79. According to Jean Mayer, why is it more important to control the number of rich than the number of poor?
   D. Birth rates, death rates, and social security
      80. Why do experts believe that the population explosion could be checked most effectively by lowering the death rate in underdeveloped countries? How do current conditions in India serve as an instructive example?
      81. What general effect has the implementation of birth-control measures had on population growth?

VI. **Our hungry planet** (pages 1063–1064)
   82. What problems have resulted directly from the population explosion?
   83. How widespread a phenomenon is malnutrition?
   84. What is the Green Revolution? What are some of its most notable successes?
   85. What are some of the serious problems associated with the Green Revolution?
   86. What problems is the Green Revolution unable to solve?

VII. **Science and human values** (pages 1064–1065)
   87. Why is science considered to be materialistic?
   88. What place do value judgments have in science? Explain.
   89. Why are many people frustrated at the apparent inability of science to solve human problems?
   90. Why is it that science cannot provide answers to such problems?
   91. How has the attitude of scientists changed with regard to political activism? What are some notable examples of this shift?
   92. What hope do we have for solving the technological and moral problems that beset us in this "age of science and materialism"?

VIII. **Summary** (pages 1065–1066): Read the Summary. If you are familiar with the essential features of the material presented there, you are ready to take the following diagnostic examination.

TESTING YOUR UNDERSTANDING

After you have completed the following examination, compare your answers with the annotated key in the Analysis section.

1. The evolution of primates began in _____ habitats.
   a. forest
   b. grassland
   c. coastal
   d. scrubland
   e. several different kinds of

2. Evolutionarily speaking, the primate hand is:
   a. highly specialized.
   b. an elaboration on a primitive reptilian pattern.
   c. poorly suited for grasping.
   d. a modification of the forepaws of four-legged animals.

3. Which of these characteristics distinguishes a New World monkey from an Old World monkey?

a. Old World monkeys are arboreal.
   b. New World monkeys have prehensile tails.
   c. Old World monkeys have platyrrhine noses.
   d. New World monkeys walk on all fours.

4. The radius and ulna are bones whose arrangement allows a primate to:
   a. rotate the hand through a full semicircle.
   b. rotate the upper arm.
   c. achieve an erect posture.
   d. raise and lower its head.

5. A small arboreal primate that eats mostly fruits or insects would be classified as a:
   a. hominid.
   b. hominoid.
   c. anthropoid.
   d. prosimian.

6. Mammals that live in trees generally have:
   a. stereoscopic vision.
   b. sharp olfactory senses.
   c. poor color vision.
   d. claws that have tactile surfaces.
   e. All of the above are correct.

7. _____ are the largest forms among the hominoids, and _____ are the primates most closely related to humans.
   a. Gibbons; monkeys
   b. Orangutans; gorillas
   c. Gorillas; chimpanzees
   d. Apes; orangutans
   e. Gorillas; prosimians

8. Which of the following pieces of fossil evidence indicates that *Ramapithecus* was probably not an ape?
   a. the shape of the dental arch
   b. the curve of the spine
   c. the size of the cranium
   d. the length of the bones in the forearm

9. Evidence from fossils of the genus *Australopithecus* indicates that _____ was probably an ancestor of modern humans.
   a. *A. robustus*
   b. *A. afarensis*
   c. the nutcracker man
   d. *A. boisei*
   e. Cro-Magnon man

10. A characteristic that differentiates hominids from other hominoids is:
    a. cooperation in capturing prey.
    b. frontally directed vision.
    c. bipedalism.
    d. manual dexterity.
    e. a narrow dental arch.

11. The skull of *H. erectus* reveals that in comparison to modern humans, early hominids had a:
    a. protruding chin.
    b. broader pelvis.
    c. smaller cranial capacity.
    d. low forehead.
    e. Both a and c are correct.

12. Which of the following fossil discoveries is considered to mark the beginnings of a clear cultural tradition among early hominids?
    a. the use of fire
    b. stone hunting tools
    c. the hand ax
    d. cave paintings
    e. an alphabet

13. Which of these hominids was (were) the first to use fire?
    a. *Australopithecus afarensis*
    b. *Ramapithecus*
    c. the Neanderthals
    d. *Homo erectus*
    e. the Cro-Magnons

14. The great increase in brain size seen among hominids is believed to:
    a. have developed simultaneously with bipedalism.
    b. have taken place very rapidly.
    c. have been associated with the development of sophisticated weapons.
    d. correlate more closely with increasing body size than with intelligence.

15. The beginning of agriculture is most closely correlated in the historical record with the:
    a. invention of tools.
    b. appearance of campsites along riverbanks.
    c. extensive use of fire.
    d. end of Paleolithic cave art.

16. Which of the following is true regarding the evolutionary position of the Neanderthals among the hominids?

a. Neanderthals gave rise to Cro-Magnons by a process of phyletic evolution.
   b. Neanderthals, apparently an evolutionary dead end, died out abruptly.
   c. The evolution of Neanderthals paralleled that of Cro-Magnons but took place in a colder climate.
   d. Neanderthals are believed to be the direct ancestors of modern humans.

17. There is evidence from ancient Neanderthal settlements that Neanderthals:
    a. had a written language.
    b. believed in an afterlife.
    c. forged tools from metals.
    d. were vegetarians.
    e. There is evidence for all four characteristics.

18. The first traces of agriculture date back about _____ years to _____.
    a. 10,000; the Near East
    b. 30,000; southern Europe
    c. 7,000; the western Mediterranean
    d. 18,000; South Africa

19. The recent population explosion is primarily a result of:
    a. a decline in death rates among the young.
    b. an increase in birth rates among the poor.
    c. an increase in the postreproductive life span.
    d. the failure of people in developed countries to use birth-control measures.

20. Which of the following is true regarding the success of the Green Revolution?
    a. It has been a failure, largely due to irrigation problems.
    b. It has eliminated much starvation but may be unable to continue to do so.
    c. It has solved not only the hunger problem but many other problems as well.
    d. It has redistributed both food and wealth.

21. Match each life form with its evolutionary description or role.

    Dryopithecus _____
    Ramapithecus _____
    Homo habilis _____
    Australopithecus _____
    Homo erectus _____
    Neanderthals _____
    Cro-Magnons _____

    a. lighter skeleton and larger brain than *A. africanus* but genus disputed
    b. a cave-dwelling hunter with artistic talents
    c. earliest known *Homo sapiens*
    d. a forest-dwelling ape of the Miocene
    e. a bipedal apelike fossil
    f. the most ancient hominid
    g. inventor and distributor of the hand ax

22. How does the unequal distribution of wealth complicate the problems created by the population explosion?

23. Explain why Barry Commoner believes that the growth of the human population in underdeveloped countries can be curbed by further reducing the death rate.

24. What is the role of science in solving social problems?

PERFORMANCE ANALYSIS

1. As the major trends in primate evolution reveal, primates evolved from small shrewlike mammals that took to the trees of the forest (**a**).

2. The primate hand is well suited for grasping and manipulating objects. It is an elaboration of the five-digit foot of the reptiles (**b**) and early mammals.

3. All species of New World monkeys dwell in trees and use their prehensile tails (**b**) for hanging from limbs. Old World monkeys include both arboreal and terrestrial species and use their tails not as prehensile organs but for balance.

4. The radius and ulna are bones of the forearm that allow the wrist to turn the hand through a full semicircle (**a**).

5. The first primates, the prosimians (**d**), were small nocturnal creatures that lived primarily on a diet of plants and insects.

6. Mammals that live in trees usually have stereoscopic vision (**a**), which allows them to judge distances. This feature is obviously essential for moving safely from branch to branch.

7. There are four genera of hominoids (apes): gibbons, orangutans, gorillas, and chimpanzees. Gorillas are the largest apes; chimpanzees are the closest relatives of humans. Therefore, **c** is correct.

8. Because of their smaller and broader dental arch (**a**), ramapithecines are not considered to be apes.

9. Modern humans are believed to have evolved from one of the gracile forms of *Australopithecus*, either *A. afarensis* or *A. africanus*. The nutcracker man is generally classified with the more heavily built forms, *A. boisei* and *A. robustus*. The robust types seem to represent an evolutionary dead end. Choice **b**, then, is correct.

10. Bipedalism (**c**) is one of the distinguishing features of hominids. Cooperation in capturing prey has been observed in chimpanzees and other animal groups, such as wolves. Evolutionary trends toward finer manipulative ability (due to the divergent and opposable thumb) and frontally directed eyes have been observed among all the primates. A narrow dental arch characterizes the large contemporary primates while a smaller and broader dental arch has been observed in *Ramapithecus*, generally believed to be the first known hominid.

11. The skull of *Homo erectus* reveals that early hominids were chinless and had a fairly large brain and low forehead (**d**). Because they were bipedal, their pelvis was narrow like that of modern humans.

12. The hand ax (**c**) has been found at an extensive number of *Homo erectus* fossil sites. Because of its widespread distribution and the uniformity of its design, it is generally considered to be the accomplishment that marks the beginnings of human culture.

13. The earliest use of fire is attributed to forms of *Homo erectus* (**d**) living in the Rift Valley of East Africa.

14. The great increase in cranial capacity seen in the fossil skulls of hominids is correlated with intelligence. This increase apparently took place fairly rapidly (**b**), well after the evolution of bipedalism and well before the development of sophisticated weapons.

15. About 10,000 years ago, Paleolithic cave art (**d**) came to an end. The end of this era of art is generally correlated with the development of an agricultural way of life.

16. Neanderthals disappeared from the fossil record about 35,000 years ago, at about the same time that evidence of another form of *Homo sapiens*, the Cro-Magnons, appeared. The Neanderthals died out, perhaps because of disease or competition with the Cro-Magnons, and today are considered to have been an evolutionary dead end (**b**).

17. Because anthropologists have uncovered food, weapons, and even flowers in the graves of Neanderthals, they hypothesize that Neanderthals must have believed in an afterlife (**b**).

18. The earliest traces of agriculture (evidence of stored grains and domesticated animals) have been found in the Near East and have been dated at about 10,000 years ago (**a**). By 7,000 years ago, an agricultural way of life had spread to the western Mediterranean.

19. Demographers consider the decline in death rates among the young of developing countries (**a**) to be the most significant factor contributing to the population explosion.

20. The Green Revolution has succeeded in providing food for starving people, largely through the development of new methods of fertilization and irrigation. However, there are several major problems that the Green Revolution may not be able to solve: mass agricultural efforts can only be handled by the wealthy landowners, and, in the near future, population growth is expected to outstrip agricultural resources despite the best of human efforts. Choice **b** is correct.

21. The correct answers in order are **d**, **e**, **a**, **f**, **g**, **c**, and **b**.

22. The affluent residents of developed countries consume more than their share of resources, and their affluent life style removes them psychologically from the problems of the poor.

23. Ecologist Barry Commoner suggests that steps taken to lower death rates will result in a lowering of birth rates as well. This has been true in industrialized nations throughout history and can be seen today in some underdeveloped nations where standards of living are just beginning to rise.

24. Science is materialistic; it deals in observable phenomena and not in value judgments. Science spawns technological achievements, but since it does not and cannot dictate how these achievements are to be applied, it cannot actually solve the moral and social problems that are often the inevitable corollaries of technological "progress."